THE EXCITED STATE
IN CHEMICAL PHYSICS

PART 2

ADVANCES IN CHEMICAL PHYSICS

VOLUME XLV

ADVANCES IN CHEMICAL PHYSICS

EDITORS

I. Prigogine

University of Brussels, Brussels, Belgium

S. Rice

Department of Chemistry and The James Franck Institute,
University of Chicago, Chicago, Illinois

THE EXCITED STATE
IN CHEMICAL PHYSICS

PART 2

Edited by J. Wm. McGOWAN

*Department of Physics and the Centre for
Interdisciplinary Studies in
Chemical Physics
The University of Ontario
London, Ontario, Canada*

Volume XLV

AN INTERSCIENCE® PUBLICATION
JOHN WILEY AND SONS
New York · London · Sydney · Toronto

AN INTERSCIENCE® PUBLICATION

Copyright © 1981 by John Wiley & Sons, Inc.

Library of Congress Catalog Card Number: 58-9935
ISBN 0-471 05119-5

Printed in the United States of America

10 9 8 7 6 5 4 3 2 1

CONTRIBUTORS TO VOLUME XLV

C. E. BRION, Department of Chemistry, University of British Columbia, Vancouver, Canada

H. HABERLAND, Fakultät für Physik der Universitat Freiburg, Freiburg im Breisgau, West Germany

A. HAMNETT, Department of Chemistry, University of British Columbia, Vancouver, Canada

I. V. HERTEL, Fachbereich Physik der Freien Universitat Berlin, D1000 Berlin 33 Germany

Y. T. LEE, Department of Chemistry and Lawrence Berkeley Laboratory, University of California, Berkeley, California

CHAVA LIFSHITZ, Department of Physical Chemistry, The Hebrew University of Jerusalem, Israel

H. HARVEY MICHELS, United Technologies Research Center, East Hartford, Connecticut

A. NIEHAUS, Department of Atomic Physics, Fysisch Laboratorium, Princetonplein, Utrecht, The Netherlands

P. E. SISKA, Department of Chemistry, University of Pittsburgh, Pittsburgh, Pennsylvania

THOMAS O. TIERNAN, Department of Chemistry and Brehm Laboratory, Wright State University, Dayton, Ohio

INTRODUCTION

This volume of the Advances in Chemical Physics is the second to be devoted entirely to studies of the excited states of molecules. Since the publication of the first volume, there has been continued expansion of the subject. The contributions in this volume, which cover a variety of topics, supplement the earlier articles and report the results and interpretations based upon later technology. Just as for the earlier volume it is hoped that this and succeeding volumes will supplement the rather broadly scattered literature and provide an introduction both for the interested student and the working scientist.

S. RICE

PREFACE

Following the direction established by its predecessor, this second volume of *The Excited State in Chemical Physics* further summarizes theoretical and experimental information available from a variety of sources. It deals with the production of excited atoms, ions, and molecules; the elastic and inelastic scattering of these species; and the production of excited products following collision.

In the five years since the first volume was published, there has been increased interest in the chemistry within gas lasers and the chemistry induced by laser radiation, the kinetics and photochemistry within fusion and industrial plasmas, as well as in the normal and perturbed lower and upper atmosphere. And, since the Three Mile Island accident there has been renewed interest in radiation damage to living and nonliving things. This state of affairs has not only precipitated a variety of spectroscopic studies, but has also brought more attention to the nonspectroscopic aspects of excited state production and the interaction of excited species. The latter topic was stressed in the earlier volume and the emphasis is retained here.

Each chapter was prepared by one or more authorities in excited state chemistry and physics, who summarize much of the latest work and new technology and review research in their areas of expertise. The choice of material and approach is as timeless as it is timely, since the experimental and theoretical techniques reviewed can be applied much more broadly than just within the immediate context.

The combination of theory with experiments dealing mainly with the excited state makes this volume invaluable for the research student as well as for the seasoned scientist, especially in such areas as laser development and laser chemistry, the chemical physics and kinetics of the atmosphere, studies of flames, and related topics.

This project has continued to receive the support of many groups and has been completed largely because of the assistance granted by the Office of Standard Reference Data, National Bureau of Standards. To this office and to many others we owe much.

As editor of this volume I must express my most sincere appreciation to those who have worked hard on the various chapters, who have reviewed the material with me, and who have been patient as this volume has slowly come together.

J. William McGowan

London, Ontario
February 1981

CONTENTS

CONTINUUM OPTICAL OSCILLATOR-STRENGTH MEASUREMENTS BY ELECTRON
SPECTROSCOPY IN THE GAS PHASE
By C. E. Brion and A. Hamnett   2

ROLE OF EXCITED STATES IN ION–NEUTRAL COLLISIONS
By T. O. Tiernan and C. Lifshitz   82

ELECTRONIC EXCITED STATES OF SELECTED ATMOSPHERIC SYSTEMS
By H. H. Michels   225

COLLISIONAL ENERGY-TRANSFER SPECTROSCOPY WITH LASER-EXCITED
ATOMS IN CROSSED ATOM BEAMS: A NEW METHOD FOR INVESTIGATING THE
QUENCHING OF ELECTRONICALLY EXCITED ATOMS BY MOLECULES
By I. V. Hertel   341

SPONTANEOUS IONIZATION IN SLOW COLLISIONS
By A. Niehaus   399

SCATTERING OF NOBLE-GAS METASTABLE ATOMS IN MOLECULAR BEAMS
By H. Haberland, Y. T. Lee, and P. E. Siska   487

Contents

I INTRODUCTION   3
   A Oscillator Strengths and Electron Spectroscopy   3
   B Energy Transfer in Electron and Photon Experiments   5
   C Scope of This Review   8

II THEORETICAL BACKGROUND   9
   A Introduction   9
   B Molecular Processes   12
   C Born Approximation   13
   D Derivation of $f(0)$ from Electron-Impact Measurements   17
   E Continuum Effects and (e, 2e) Coincidence Experiments   20

III CALCULATION OF OSCILLATOR STRENGTHS   21

IV EXPERIMENTAL CONSIDERATIONS   24
   A Electron Analyzers and Transmission Efficiency   24
   B Electron Detectors and Signal Processing   33
   C Coincidence Methods   34
   D Absolute Oscillator Strengths   38

V EXPERIMENTAL MEASUREMENTS   41
   A Introduction   41
   B The Noble Gases   41
   C Hydrogen   50
   D Nitrogen and Carbon Monoxide   54
   E Methane   69
   F Other Molecules   72

VI CONCLUSIONS   73

CHAPTER ONE

CONTINUUM OPTICAL OSCILLATOR-STRENGTH MEASUREMENTS BY ELECTRON SPECTROSCOPY IN THE GAS PHASE

C. E. Brion and A. Hamnett

Department of Chemistry, University of British Columbia, Vancouver, B. C., Canada, V6T 1W5

I. INTRODUCTION

A. Oscillator Strengths and Electron Spectroscopy

The absolute intensity of dipole-allowed transitions is conveniently expressed in terms of $f(0)$, the optical (dipole) oscillator strength. The quantity $f(0)$ is simply related to the absorption coefficient and also to the cross section for absorption or emission of radiation (see Section II). The concept of oscillator strength developed from the classical picture of the atom in which the electrons were envisaged to be in free oscillation at given frequencies about an equilibrium position with respect to the massive nucleus. The oscillator strength $f(0)$ was defined as the number of electrons in free oscillation at a given frequency and was thus considered to be related to the intensity of absorption at a given frequency. The total oscillator strength was thus equal to the total number of electrons in the atom. Subsequently, the quantum theory of atomic structure emerged, giving a description of the atom involving both bound (discrete) and continuum (ionized) states. A very large number of transitions can occur between these states—far in excess of the number of atomic electrons. Nevertheless, the historical oscillator strength terminology has been retained in assigning an absolute scale for the probability of transition between two energy states. Thus for all possible processes n, the total sum of oscillator strengths is such that $\sum_n f_n(0)$ equals the total number of electrons in the atom. The sum is over all discrete and continuum states and is discussed further in Section II.

Optical oscillator strengths have traditionally been measured by optical absorption and lifetime methods.[1] Such measurements have of necessity been restricted to those regions of the electromagnetic spectrum where suitable exciting photon sources are available. Consequently, relatively few measurements have been made in the vacuum ultraviolet (UV) and soft-X-ray regions since sufficiently intense tunable (continuum) light sources have not generally been available, at least above 20 eV. It is now apparent that electron storage rings will, in principle, supply sufficiently intense continuum radiation to allow a wide range of such measurements to be made. However, even with the increasing availability of synchrotron radiation to date, such information is scarce and many experiments, particularly those involving photoionization phenomena, are extremely difficult to carry out. Quantitative spectroscopy at ultraviolet UV and X-ray energies is of fundamental chemical interest since it involves many of the higher electronic states of valence electrons and all inner-shell excitations as well as all ionization processes. This type of quantitative information is necessary for a more complete understanding of processes such as those that occur in radiation-induced decomposition, aeronomy, high-temperature chemistry, and discharge phenomena. Furthermore, such experimental data are urgently needed for the formulation and evaluation of quantum-mechanical procedures.

In the last decade electron spectroscopy in its various forms has been proven a valuable spectroscopic tool for determining the bound and ionized energy levels of atoms and molecules. The principal achievements of photoelectron, electron impact, and Auger and Penning ionization electron spectroscopy are given in a number of recent publications.[2] Of these methods, gas-phase photoelectron spectroscopy (PES) has provided by far the largest amount of data. However, the attention of photoelectron spectoscopists has been primarily focused on the energies of the ejected electron, and little quantitative work on absolute intensities has been reported, in part because of the paucity of calibrated tunable light sources in the far UV, as discussed earlier. Another problem has been that, for ease of construction, most PES spectrometers have been designed to sample ejected electron spectra at 90° to the photon beam. The intensity is thus modulated by the variation of the (usually unknown) asymmetry parameter, β, for each state, with energy. Although this effect can be avoided by sampling at the so-called magic angle[3] (54.7° for unpolarized radiation), few spectrometers have been constructed in this configuration. However, the most serious problem in PES is that little attempt has been made to correct for the electron-transmission efficiency of the analyzer. As discussed in Section IV, such corrections may be very significant, even over a few electron volts, but it has been only recently that such corrections have been considered. The few quantitative PES experiments that have been reported are discussed in Section V.

By contrast, a growing number of quantitative fast-electron-impact experiments have been used in recent years as an alternative method of obtaining optical oscillator strengths. One of the major aims of this chapter is to draw attention to these methods, to review their principal achievements and potentialities, and, where possible, to make comparisons with directly determined optical data. Using the Born approximation, Bethe[4] laid the theoretical groundwork in 1930, showing that a quantitative relationship exists between photon absorption and the scattering of fast charged particles (see Section II). More recently, these ideas have been discussed in depth by Inokuti[5] and Kim.[6] The early work of Franck and Hertz[7] had shown in many ways the qualitative similarity between electron impact and photon absorption. For example, the processes of excitation, ionization, and dissociation could all be induced by electron bombardment of molecules. The first quantitative evidence demonstrating this relationship came from the total cross-section measurements by Miller and Platzmann.[8] However, it was the pioneering differential electron-scattering studies by Lassettre and his co-workers 9-11 that gave the main impetus leading to the development of modern quantitative electron-impact spectroscopy. Lassettre and his group at the Mellon Institute have subsequently made the dominant contribution to the measurement of optical oscillator strengths for discrete transitions by electron-energy-loss spectroscopy. This work has involved the extrapolation of generalized oscillator strengths to zero momentum transfer to obtain the optical oscillator strength (see Section II).

More recently these ideas have been extended very effectively by van der Wiel and his co-workers, 12-17 giving rise to the development of a variety of "photon-simulation" experiments using high-energy, small-angle electron scattering. These studies are significant in that the chosen experimental conditions approach the optical limit (see Section II) sufficiently closely so that dipole oscillator strengths are measured directly. These experiments have provided the major portion of the continuum oscillator strength data that are available in the literature to date. Much of this work is discussed in Sections IV and V.

B. Energy Transfer in Electron and Photon Experiments

The ability of fast electrons to excite dipole-allowed (optical) transitions can be qualitatively understood in terms of what has been called "pseudophotons" or the "virtual photonfield." Figure 1 illustrates the principal effects occurring when a fast electron interacts with a target molecule via a distant collision (large impact parameter and thus small scattering angle). As the electron passes by, the target experiences a sharply pulsed electric field of which the perpendicular component is significant. Ideally, in the limit the E field will approach a δ function that, if Fourier transformed into the frequency domain, would afford the perfect spectroscopic "light" source consisting of a continuum composed of all frequencies at equal intensities. In practice, the pulse will have a narrow but finite width, and there will be a falloff in intensity of "pseudophotons" at high frequencies. (It should be stressed that this method does not simulate photons; rather, under the appropriate conditions, the electron-impact differential cross section is related to the optical cross section by kinematic factors alone.[5,14]) Nevertheless, a sufficiently wide spectral range can readily be achieved in the laboratory, and the effective high-energy transfer limit in electron-scattering experiments is usually determined by other factors (see Section II).

For a molecule AB, we may compare the processes of photoabsorption and electron-impact excitation as follows:

$$h\nu(E) + AB \rightarrow AB^* \qquad \textit{Photoabsorption} \qquad (I.1)$$
$$e(E_0) + AB \rightarrow AB^* + e(E_0 - E) \qquad \textit{Electron-impact excitation} \qquad (I.2)$$

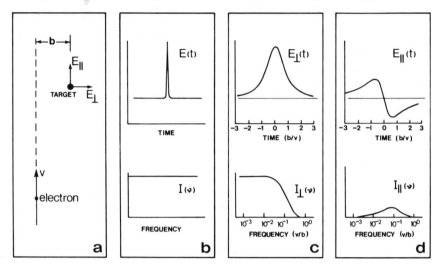

Figure 1. Electric field, $E(t)$, and corresponding frequency spectrum, $I(\nu)$, associated with distant collision of fast electron and molecular target: (*a*) collision parameters—v, electron velocity and b, impact parameters; (*b*) idealized case for very fast electron; (*c, d*) realistic picture.

where E is the energy of quantum state AB* and E_0 is the impact energy of the electron. From these expressions it is apparent that the resonant photon energy E in photoabsorption is analogous to the energy loss of the incident electron when scattered by the target in the electron-impact process. In other words, we may "equate" the photon energy with energy loss.

It should be noted that the electron-impact process is nonresonant; the important consequences of this are discussed in the text that follows. In effect, a fast electron (E_0) offers the target a "white-light" continuum of "pseudophotons" that are absorbed with a frequency-dependent probability that can be quantitatively related to the optical oscillator strength via the Bethe–Born relation (see Section II). The *net* result is that under the appropriate experimental conditions we may perform *quantitative* measurements equivalent to photoabsorption and so on using techniques of fast-electron-impact and electron-energy-loss spectroscopy (see Section III). As discussed earlier,[14] it is of particular advantage to exploit this relationship in the UV and soft-X-ray regions of the spectrum where continuum light-source availability is very restricted, with the exception of synchrotron radiation. However, the latter source is only available at a few locations and is of enormous expense. Even where synchrotron radiation is available, use of the photons is still subject to the well-known difficulties of optical spectroscopy at short wavelengths,[14] which include low reflectivity, order overlapping, and efficient monochromation. Dispersion of synchrotron radiation leads to changes in polarization and also modifications of photon flux that necessitate calibration of the effective photon intensity for ionization oscillator-strength measurements.[18] It has been pointed out by Inokuti[5] that use of electron impact instead of photons can have some additional advantage because of the nonresonant nature of electron-impact excitation. This property eliminates line-saturation effects that occur when the natural line width is narrower than the experimental photon bandwidth.

Since ionization is only a special case of excitation it is also possible to simulate the photoionization process using fast electrons. We may compare the processes

$$hr(E) + AB \rightarrow [AB^+ + e_{ej}] \qquad \text{Photoionization} \qquad (I.3)$$
$$e(E_0) + AB \rightarrow [AB^+ + e_{ej}] + e_{sc}(E_0 - E) \qquad \text{Electron-Impact}$$
$$\text{ionization} \qquad (I.4)$$

where e_{ej} is the electron ejected from AB on ionization and e_{sc} is the fast-scattered electron.

It is apparent that in both cases energy E is deposited in $[AB^+ + e_{ej}]$ and that, as in the case of excitation, the photon energy is analogous to the electron energy loss. However, since there are now two electrons sharing the excess energy in electron-impact ionization, it is necessary to use time correlation (coincidence techniques) for the simulation of photoionization

Table I.
Photon-Simulation Experiments[a]

PHOTON EXPERIMENT	ELECTRON-IMPACT EQUIVALENT	TYPICAL REFERENCES
Photoabsorption	Electron-energy-loss spectroscopy	25
Photoionization mass spectrometry (fragmentation)	Electron–ion coincidence (e–ion)	149,166
Total photoionization	Electron energy loss—total ejected electron, coincidence (e, 2e)	24
Photoelectron spectroscopy	Electron energy loss—selected ejected electron, coincidence (e, 2e)	23,194
Photofluorescence (of ionic states)	Electron-energy-loss—electron–ion–photon (triple) coincidence	191

[a]Various aspects of the instrumentation and techniques used in these experiments as well as conventional PES are discussed in Section IV.

experiments. The possible photoexcitation and ionization experiments that have been simulated are summarized in Table I.

C. Scope of This Review

This article is essentially restricted to a consideration of the measurement of optical oscillator strengths by methods employing various forms of electron spectroscopy. Targets are restricted to atoms and molecules in the gas phase, and only absolute or relative *quantitative* measurements are generally included (i.e., only those experiments in which electron transmission and ion kinetic energy are accounted for, at least on a relative basis). Generalized oscillator strengths are discussed only insofar as they relate to the derivation of optical oscillator strengths. Detailed accounts of generalized oscillator strength measurements can be found in the work of Lassettre et al.[9-11] and Bonham and Fink,[19] who have studied various aspects of the Bethe surface. Oscillator-strength measurements for discrete transitions using electron-impact spectroscopy have been the subject of two recent detailed reviews by Lassettre.[10,11] Some notable studies have also been made by Geiger.[20] Therefore, this chapter emphasizes optical oscillator-strength measurements for continuum processes involving energy transfers in excess of ~ 10 eV. Attention is focused on: (1) photoabsorption $f(0)$ measurements by electron-impact spectroscopy, (2) partial ionization (electronic state) $f(0)$ measurements using photoelectron spectroscopy and electron–electron coincidence, (3) total ionization $f(0)$ measurements using total (e, 2e) and (e–ion) coincidence with mass analysis, and (4) partial ionization (fragmentation) $f(0)$ measurements using (e–ion) coincidence with mass analysis.

Attention is given to results for the noble gases (including multiple ionization processes) as well as energy transfer and fragmentation in diatomic and small polyatomic molecules. In Section II the theoretical background of the methods is discussed and a sufficient framework developed to enable the laboratory worker to design experiments as well as to understand and interpret the data. The current status of oscillator-strength calculations is discussed in Section III. Section IV discusses some aspects of the experimental methods used in oscillator-strength measurements by electron spectroscopy. Finally, a discussion is given in Section V illustrating some of the more significant oscillator-strength measurements that have been made to date using electron-spectroscopic techniques. No attempt has been made to give tables of oscillator-strength data in this review. In many cases this most useful form of data is available in the original published articles. We would like to take this opportunity to exhort authors of forthcoming publications to provide such tables of data in addition to diagrams since this greatly facilitates the use and comparison of oscillator strengths.

II. THEORETICAL BACKGROUND

A. Introduction

In optical experiments the absorption of radiation is governed by the Beer–Lambert law. If $I_0(E)$ is the measured intensity of a beam of electromagnetic radiation of energy E and $I(E)$ that following absorption by a gas of thickness L containing n molecules per unit volume, then:[21]

$$I(E) = I_0(E)\exp\left[-\sigma_t(E)nL\right] \qquad (\text{II.1})$$

where $\sigma_t(E)$ is the optical cross section for absorption that has units of area.

An analogous expression holds for electron scattering, except that in such an experiment we usually desire to measure that proportion of the incident beam (of impact energy E_0) that has lost energy E. In general, the intensity of such an inelastically scattered beam will depend on the polar angles θ_s, ϕ_s (with respect to the main beam) at which it is measured. If the incident electron beam has an intensity I_0, the inelastically scattered beam will have an intensity

$$I_{sc}(\theta_s, \phi_s, E, E_0) = \int_{\Delta\Omega_d} I_0 nL \frac{d^2\sigma_{el}}{d\Omega}(\theta_s, \Phi_s, E, E_0)\, d\Omega \qquad (\text{II.2})$$

where $\Delta\Omega_d$ is the solid angle subtended by the detector at the scattering region and σ_{el} is the electron-impact cross section. Both expressions (II.1) and (II.2) are idealized in that corrections are usually needed in practice for pressure-dependent effects. These corrections are especially significant in optical work in the near UV since $\sigma_t \ll \sigma_{el}$.

The cross section σ is a fundamental property of the molecule and as such is related to the molecular wave functions for the two states between which a transition is induced. Hence it is desirable to separate the contributions to σ that arise from purely kinematic quantities such as the impact energy of the electron beam from those that depend solely on the properties of the molecule. To this end, a dimensionless quantity, the oscillator strength, is introduced in optical absorption spectroscopy, defined by the relation[22]

$$f_{0m}(0) = 2E \sum_\alpha A_\beta \left| \hat{\mathbf{e}} \cdot \left\langle \psi_{m\alpha} \left| \sum_{i=1}^{N} \mathbf{r}_i \right| \Psi_{0\beta} \right\rangle \right|^2 \qquad (\text{II.3})$$

in which $\Sigma_\alpha A_\beta$ denotes a summation over the degenerate excited states $\psi_{m\alpha}$ and an average over the degenerate initial states $\Psi_{0\beta}$, $\hat{e}$ is a unit vector in the direction of the electric field, and the equation is written, for convenience, in atomic units. The summation within the bracket is over the N electrons in the molecule.

An analogous quantity, the generalized oscillator strength, is found to be useful in electron-scattering theory. It is a function of the momentum $\mathbf{K}$ transferred from the incident electron to the molecule and has the form[5]

$$f_{0m}(\mathbf{K}) = \frac{2E}{|\mathbf{K}|^2} \cdot \sum_\alpha A_\beta \left| \left\langle \Psi_{m\alpha} \left| \sum_{i=1}^{N} e^{i\mathbf{K}\cdot\mathbf{r}_i} \right| \Psi_{0\beta} \right\rangle \right|^2 \tag{II.4}$$

The cross sections $\sigma_t(E)$ and $\sigma_{el}(\theta_s, \phi_s, E, E_0)$ are related[5, 21] to those two oscillator strengths through the equations

$$\sigma_t(E) = \frac{\pi}{c^2} \cdot f(0) \tag{II.5}$$

and

$$\sigma_{el}(\theta_s, \phi_s, E, E_0) = \frac{2}{E} \cdot \left| \frac{\mathbf{k}_f}{\mathbf{k}_i} \right| \cdot \frac{|f(\mathbf{K})|}{|\mathbf{K}|^2} \tag{II.6}$$

where $\mathbf{k}_i$ and $\mathbf{k}_f$ are the incident and scattered electron momenta, respectively. It is clear from equation (II.4) that

$$\underset{|\mathbf{K}|\to 0}{Lt} f_{0m}(\mathbf{K}) = 2E \sum_\alpha A_\beta \left| \hat{\mathbf{K}} \cdot \left\langle \Psi_{m\alpha} \left| \sum_{i=1}^{N} \mathbf{r}_i \right| \Psi_{0\beta} \right\rangle \right|^2 \tag{II.7}$$

where $\hat{\mathbf{K}}$ is unit vector in the direction $\mathbf{K}$. Comparison of equations (II.7) and (II.3) shows that they are numerically identical, and thus in low-momentum-transfer scattering experiments we may replace the unit electric vector by the unit vector $\mathbf{K}$. It is this close analogy between electron scattering and optical absorption spectra that we wish to exploit in the simulation experiments (see Table I).

If the energy transfer E is sufficiently large, ionization occurs and the oscillator strength and cross section become continuous functions of E. To preserve the simplicity of equations (II.1) and (II.2), relationships (II.5) and (II.6) must be modified such that[5, 21]

$$\sigma_t(E) = \frac{\pi}{c^2} \cdot \frac{df(0)}{dE} \tag{II.8}$$

$$\sigma(\theta_s, \phi_s, E, E_0) = \frac{2}{E} \cdot \left| \frac{\mathbf{k}_f}{\mathbf{k}_i} \right| \cdot \frac{1}{|\mathbf{K}|^2} \cdot \frac{df(\mathbf{K})}{dE} \tag{II.9}$$

Although the oscillator strengths are dimensionless, their derivatives with respect to E are not, and units become a significant problem in continuum absorption. In cgs (electrostatic) units, equation (II.8) reads[21]

$$\frac{df(0)}{dE} = \frac{mc}{e^2\pi h}\sigma_t(E)$$ (II.10)

where $\sigma_t(E)$ is expressed in square centimeters and E in ergs. More normally, E is expressed as a frequency $\bar{\nu}$ in wave numbers, when we have

$$\frac{df(0)}{d\bar{\nu}} = \frac{mc^2}{e^2\pi}\sigma_t(\bar{\nu})$$ (II.11)

The cross section $\sigma_t(E)$ may also be expressed in units of megabarns (1 Mbarn $= 10^{-18}$ cm^2). Under these circumstances, inverting (II.10) we have

$$\sigma(\text{Mbarns}) = 1.76 \times 10^{-10}\frac{df}{dE(\text{ergs})}$$ (II.12)

If E is expressed in electron volts (eV), we find

$$\sigma(\text{Mbarns}) = 109.75\frac{df}{dE(\text{eV})}$$ (II.13)

It should be noted that these expressions hold only for excitation to the continuum and that comparison of electron scattering and optical absorption for discrete transitions is only possible if the resolution of both types of experiment is known.

The problem with equations (II.8) and (II.9) is that the definitions of the two cross sections appear to change abruptly with passage through the ionization threshold. A full discussion of this has been given by Fano and Cooper,[29] and it suffices here to point out that we may define differential cross sections

$$\frac{d\sigma_t(E)}{dE} = \frac{\pi}{c^2}\frac{df(0)}{dE}$$ (II.14)

$$\frac{d\sigma_{el}(\theta_s, \phi_s, E, E_0)}{dE} = \frac{2}{E}\cdot\left|\frac{k_f}{k_i}\right|\frac{1}{|K|^2}\cdot\frac{df(K)}{dE}$$ (II.15)

The cross section defined in (II.15) is related to the observed intensity of the scattered beam by the expression

$$\int_{E}^{E+\Delta E} \frac{dI_{sc}(\theta_s,\phi_s,E,E_0)dE}{dE} = \int_{E}^{E+\Delta E}\int_{\Delta\Omega d} I_0 nL \frac{d^3\sigma_{el}}{dE\,d\Omega}(\theta_s,\phi_s,E,E_0)dE\,d\Omega$$

$$(\text{II}.16)$$

A similar expression may be written for optical absorption when the exponent in (II.1) is sufficiently small. Substituting equation (II.15) in (II.16), we may obtain the generalized oscillator strength over some region ΔE by simple integration.

B. Molecular Processes

The energy transferred by the photon or electron to the molecule may be lost by the latter in a number of different ways. If this energy is lower than the ionization threshold, the usual pathways are (1) reradiation or (2) dissociation into neutral fragments or charged fragments (from ion-pair processes); in other words:

$$A-B \xrightarrow{E} (AB)^* \xrightarrow{h\nu} AB$$
$$\longrightarrow A^{\cdot}+B^{\cdot}, A^{+}+B^{-}, A^{-}+B^{+}, \cdots$$

If $A^{\cdot}$ or $B^{\cdot}$ are not in their ground state, further fragmentation or radiation may occur. Radiation from AB^* may not give the ground state, and further fluorescence may take place. In the target region energy may also be lost by collision with other molecules or with the walls of the target chamber.

If the energy is higher than the ionization threshold, the preceding processes may still occur, but we also have the possibility of two types of ionization process:

$$AB \xrightarrow{E} \begin{matrix} AB^{+}+e_{ej} & \text{Molecular ionization} \\ A^{+}+B+e_{ej} & \text{Dissociative ionization} \end{matrix}$$

In energy transfers above the ionization threshold, it is usual for several different ionization processes to occur with probabilities depending on E. The measured optical oscillator strength for absorption is thus a sum corresponding to a variety of different processes. Denoting this total optical oscillator (ionization potential): strength by $df(0)/dE$, we have, for

$E > IP$

$$\frac{df(0)}{dE} = \sum_i \frac{df^i(0)}{dE} + \sum_n \frac{df^n(0)}{dE} \qquad (II.17)$$

where we have partitioned the oscillator strength into two sets of processes, those involving ionization (i) and those involving neutral processes (n). The ionization efficiency η is then given by

$$\eta = \left(\sum_i \frac{df^i(0)}{dE} \right) \left(\frac{df(0)}{dE} \right)^{-1} \qquad (II.18)$$

The individual partial oscillator strengths $df^i(0)/dE$ for each ionization process are individual functions of E. It is usual to define the branching ratio b_i as

$$b_i = \left[df^i(0)/dE \right] \left[\sum_i df^i(0)/dE \right]^{-1} \quad \text{and} \quad \sum b_i = 1 \qquad (II.19)$$

Characterization of the neutral processes is far more difficult, and little information is available at present. However, studies on some simple molecules (see Section V) have indicated that the ionization efficiency approaches unity quite rapidly as the energy loss increases above threshold, suggesting that, except where transitions to Rydberg states just below a new ionization threshold are significant, the dominant mode of energy loss in the far UV is by ionization often accompanied by molecular fragmentation.

C. Born Approximation

Consider a beam of electrons incident along the $+z$ direction. The wave function is of the form $\exp(i|\mathbf{k}_i|z)$, and the current is given by[26]

$$\mathbf{j}_i(\mathbf{r}) = Re\left(\psi^* \frac{\nabla}{i} \psi \right) = |\mathbf{k}_i|\hat{z} \qquad (II.20)$$

where $\hat{z}$ is a unit vector in the $+z$ direction. This corresponds to a current of one electron per second with momentum $\mathbf{k}_i$. If the electron is scattered inelastically, the outgoing wave function will be a spherical wave, emanating from the scattering center. Such a wave function may be represented by a function that behaves asymptotically at large values of r as

$$\psi_f = \frac{\exp(i\mathbf{k}_f\mathbf{r})}{r} \cdot f(\theta,\phi) \qquad (II.21)$$

where $f(\theta,\phi)$ is some angular function. The scattered current may be calculated using equation (II.20) and is

$$\mathbf{j}_{sc}(r)dA = |\mathbf{k}_f|\hat{\mathbf{r}}| \left|\frac{f(\theta,\phi)}{r^2}\right|^2 dA \tag{II.22}$$

where dA is the area of the detector.

The cross section $\sigma_{el}(\theta,\phi,E,E_0)$ is measured using the detectors, and since we expect it to be proportional to both the solid angle subtended by the detector and to the incident current, we have

$$\mathbf{j}_{sc}(\mathbf{r})dA = \mathbf{j}_i(\mathbf{r})\sigma_{el}(\theta,\phi,E,E_0)\cdot\frac{dA}{r^2} \tag{II.23}$$

where $\sigma_{el}(\theta,\phi,E,E_0)$ is introduced as a proportionality constant. We have, comparing (II.22) and (II.23) and using (II.20),

$$\sigma_{el}(\theta,\phi,E,E_0) = \frac{|\mathbf{k}_f|}{|\mathbf{k}_i|}|f(\theta,\phi)|^2 \tag{II.24}$$

Thus to calculate the cross section, we need only calculate the asymptotic part of the scattered electron wave function. This is straightforward, at least for first-order perturbation theory. Provided that the time scale during which the perturbation of the molecule by the impact electron occurs is small compared to the time scale for electronic motion, we find[26]

$$f(\theta,\phi) = \frac{-1}{2\pi}\int \exp(i\mathbf{K}\cdot\mathbf{r})\langle\Psi_m|V|\Psi_0\rangle d\mathbf{r} \tag{II.25}$$

where the perturbing potential V is given by

$$V = \frac{-\sum_{p=1}^{M}Z_p}{|\mathbf{r}-\mathbf{R}_p|} + \sum_{i=1}^{N}\left(\frac{1}{(\mathbf{r}-\mathbf{r}_i)}\right) \tag{II.26}$$

in atomic units, $\mathbf{R}_p$ are the nuclear coordinates, and $\mathbf{r}_i$ are those of the electrons.

In addition to the assumptions underlying the use of first-order perturbation theory, a number of other assumptions underlie equation (II.25).

1. In effect, the equation has been derived under the assumption of infinite nuclear mass. This is accurate enough for electron scattering, but for proton and atom scattering a coordinate transformation is needed, the details of which are given in Mott and Massey.[26]

2. All relativistic effects have been ignored. Corrections for impact energies of less than 10 keV are very small, but above this level a careful distinction must be drawn between the velocity and energy of the electron. Corrections to first order have been given by Inokuti.[5]
3. All exchange effects have been neglected. For fast incident electrons inducing discrete transitions, this is a very accurate assumption, but if ionization occurs, and especially if the energies of the two outgoing electrons are comparable, such effects are likely to be of great importance. For the most part we are concerned here with ionization to energy levels only a few tens of volts into the continuum using electron energies of very high incidence and, under these circumstances, the effects of exchange are practically negligible.[26-28]

To derive an expression for the oscillator strength, we must consider equation (II.25) more closely. Assuming that the electronic part of the wavefunctions Ψ_m and Ψ_0 are orthogonal, integration over the nuclear attraction part of V(II.26) will vanish. The electron-repulsion part may be simplified by interchanging the order of integration in (II.25) and considering first the integration over $\mathbf{r}$. We find

$$\sum_i \int \exp(i\mathbf{K}\cdot\mathbf{r}) \int \cdots \int \Psi_m^* \left(\frac{1}{|\mathbf{r}-\mathbf{r}_i|}\right)\Psi_0 \, dr_1 \cdots dR_M \, dr$$

$$= \sum_i \int \cdots \int \Psi_m^* \left(\int \frac{\exp(i\mathbf{K}\cdot\mathbf{r})}{|r-r_i|} \, dr\right)\Psi_0 \, dr_1 \cdots dR_M dr \quad \text{(II.27)}$$

The integral in brackets is well known to have the value $(4\pi/|\mathbf{K}|^2)\exp(i\mathbf{K}\cdot r_i)$, so[26]

$$f(\theta,\phi) = (-2/|\mathbf{K}|^2) \sum_i \langle \Psi_m | \exp(i\mathbf{K}\cdot r_i) | \Psi_0 \rangle \quad \text{(II.28)}$$

Recalling the definition of the generalized oscillator strength (II.4) and using equation (II.24), we find

$$\sigma_{el}(\theta,\phi,E,E_0) = \frac{2}{E}\cdot\frac{|\mathbf{k}_f|}{|\mathbf{k}_i|}\cdot\frac{1}{|\mathbf{K}|^2}\cdot f_{0m}(\mathbf{K}) \quad \text{(II.29)}$$

provided we sum and average equation (II.28) in the usual way.

The generalized oscillator strength defined by (II.4) has a number of important properties that have been listed by Inokuti.[5] Of great importance practically are the sum rules

$$S(\mu,K) = \left(\sum_m + \int_m\right) E_m^\mu f_{0m}(K) \quad \text{(II.30)}$$

where $S(\mu, K)$ is convergent for $-1 \leqslant \mu \leqslant +2$ and E_m is the energy loss associated with the transition $0 \rightarrow m$. For our purpose the most significant is

$$S(0, K) = \left(\sum_m + \int_m \right) f_{0m}(K) = N \tag{II.31}$$

where N is the total number of electrons in the molecular system. This is an analogue of the well-known optical sum rule

$$S(0, 0) = \left(\sum_m + \int_m \right) f_{0m}(0) = N \tag{II.32}$$

Its importance lies in the fact that it may be used, at least for simple molecules, to place the measured photoabsorption spectrum on an absolute scale. However, great care should be exercised in the use of (II.31) and (II.32) since the summation includes transitions from valence orbitals to inner orbitals already occupied. Such transitions cannot be seen, of course, and thus calibration of subshell photoabsorption spectra by this method will give results that are only approximate.[123]

Clearly, if $|K|$ is small, the exponential term in (II.4) may be expanded and we obtain a series of the form[9,30]

$$f_{0m}(\mathbf{K}) = 2E\left[\varepsilon_1^2 + \left(\varepsilon_2^2 - 2\varepsilon_1\varepsilon_3 \right)|\mathbf{K}|^2 + \left(\varepsilon_3^2 - 2\varepsilon_{24} + 2\varepsilon_{15} \right)|\mathbf{K}|^4 + \cdots + \right]$$

$$\tag{II.33}$$

$$= f_{0m}(0) + |\mathbf{K}|^2 f_{0m}^{(1)}(0) + |\mathbf{K}|^4 f_{0m}^{(2)}(0) + \cdots + \tag{II.34}$$

where

$$\varepsilon_n = \frac{1}{n!} \sum_i \langle \Psi_m | (\hat{\mathbf{K}} \cdot \mathbf{r}_i)^n | \Psi_0 \rangle$$

and the unit vector $\hat{\mathbf{K}}$ is defined as previously. Two things are important here. The first is that, as foreshadowed earlier, the limiting value of $f_{0m}(\mathbf{K})$ as $\mathbf{K} \rightarrow 0$ is indeed the optical value. The second is that, as a consequence of sum rules (II.31) and (II.32), we have, at once

$$\left(\int_m + \sum_m \right) f_{0m}^{(1)}(0) = 0 \tag{II.35}$$

The sum rule has been checked by Backx et al., who found, in addition that $K^2 \cdot f_{0m}^{(1)}(0)$ was much smaller than $f_{0m}(0)$ for the molecules studied.[31]

D. Derivation of f(0) from Electron-impact Measurements

The quantity $\mathbf{K}$, the momentum transferred to the molecule, is, from the preceding paragraphs, a fundamental quantity in the theory of electron scattering. It may be obtained from purely kinematic considerations. Consider Fig. 2a (i.e., Fig. 1a of Hamnett et al.[23]). If $\mathbf{k}_i$ and $\mathbf{k}_f$ are the initial and scattered momenta of the incident electron, we have, from conservation of momentum,

$$\mathbf{K} = \mathbf{k}_i - \mathbf{k}_f \qquad (\text{II.36})$$

$$|\mathbf{K}|^2 = |\mathbf{k}_i|^2 + |\mathbf{k}_f|^2 - 2|\mathbf{k}_i||\mathbf{k}_f| \cos\theta \qquad (\text{II.37})$$

where θ is the angle shown. Further, conservation of energy tells us that

$$|\mathbf{k}_i|^2 - |\mathbf{k}_f|^2 = 2E \qquad (\text{II.38})$$

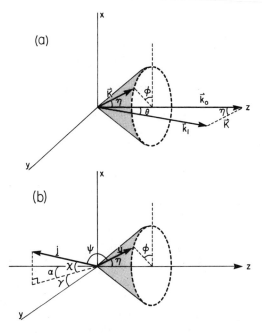

Figure 2. (a) Illustration of forward scattering geometry: $\mathbf{k}_0$ is momentum of the electrons in main beam defining z axis; $\mathbf{k}_1$ is momentum of forward scattered electron, with polar angles $\theta, \phi + \pi$; $\mathbf{K}$ is momentum transfer vector with polar angles η, ϕ; (b) geometry of ejected electrons, collected in narrow slit along z axis. Corresponding acceptance angles are χ and $\gamma(\chi \ll \gamma)$; j and u are unit vectors in directions of ejected electron and momentum transfer $\mathbf{K}$, respectively.

where E is the energy loss. If E_0 is the incident energy

$$|\mathbf{k}_i|^2 = 2E_0$$

$$|\mathbf{K}|^2 = 2E_0 + (2E_0 - 2E) - 2\sqrt{2E_0} \cdot \sqrt{2E_0 - 2E} \cdot \cos\theta \qquad (\text{II}.40)$$

Since E, E_0, and θ are experimentally measurable, $\mathbf{K}$ may be obtained.

Now, if we desire to make $|\mathbf{K}|^2$ as small as possible, we desire E to be very small compared to E_0 and θ also to be very small. Under these circumstances, (II.40) may be written to second order in θ and the dimensionless parameter $x = (E/2E_0)$

$$|\mathbf{K}|^2 = 2E_0(x^2 + \theta^2) \qquad (\text{II}.41)$$

If $|\mathbf{K}|$ is so small that we may replace $f_{0m}(\mathbf{K})$ by $f_{0m}(0)$ as in equation (II.34), the cross section σ may be written from (II.29) as

$$\frac{d^2\sigma_{el}}{d\Omega_{sc}}(\theta,\phi,E,E_0) = \frac{2}{E} \cdot \left|\frac{\mathbf{k}_f}{\mathbf{k}_i}\right| \cdot \frac{1}{2E_0(x^2 + \theta^2)} f_{0m}(0) \qquad (\text{II}.42)$$

Integrating over ϕ is straightforward, and we find

$$\frac{d\sigma_{el}}{d\theta} = \frac{4\pi}{E} \left|\frac{\mathbf{k}_f}{\mathbf{k}_i}\right| \cdot \frac{\sin\theta}{2E_0(x^2 + \theta^2)} \cdot f_{0m}(0) \qquad (\text{II}.43)$$

If the half angle of acceptance of our detector is θ_0 and we are concerned with θ_0 very small, that is, with forward scattering, integration with respect to θ may be carried out analytically. We find[23]

$$\sigma_{el}(E,E_0) = \left[\frac{2\pi}{EE_0} \cdot \left|\frac{\mathbf{k}_f}{\mathbf{k}_i}\right| \log_e \left(\frac{x^2 + \theta_0^2}{x^2}\right)^{1/2}\right] f_{0m}(0) \qquad (\text{II}.44)$$

Thus the forward scattered cross section may be seen as a product of the oscillator strength and a purely kinematic factor. Thus from a knowledge of the inelastic scattering cross section σ in the forward direction, we may, provided $\theta_0 \ll 1$ and $x \ll 1$, obtain *optical* oscillator strengths. Comparing (II.44) and (II.5), the cross section σ_{el} can be seen to be much greater than the optical cross section σ_t for small values of E, but that σ_{el} falls away much more rapidly than does σ_t with increasing values of E. This may be brought out clearly by considering the behavior of (II.44) if θ_0 is very small with respect to x. We then have

$$\sigma_{el}(E, E_0) \sim \frac{\pi}{E_0 E} \cdot \left| \frac{\mathbf{k}_f}{\mathbf{k}_i} \right| \cdot \frac{\theta_0^2}{x^2} \cdot f_{0m}(0) = \frac{4\pi E_0 |\mathbf{k}_f| \theta_0^2}{|\mathbf{k}_i|} \cdot \frac{f_{0m}(0)}{E^3} \qquad (II.45)$$

and comparison with (I.5) shows that $(\sigma_{el}/\sigma_t) \sim (1/E^3)$. If $\theta_0 \approx x$, the decrease in σ_{el} with increasing E is less dramatic. In practice, x and θ_0 may easily be made sufficiently small for equation (II.44) to constitute a very good approximation. Impact energies of a few kiloelectron volts upward have been routinely used, and θ_0 values of less than 10^{-2} radians are sufficient for equation (II.44) to be valid for up to ~ 100-eV energy loss.[13,14] First-order corrections to (II.44) using equation (II.34) have been derived experimentally by Backx et al.[31] and found to be insignificant for the gases studied (H_2, He, and CH_4). The oscillator strengths derived may be normalized using the sum rule of equation (II.32) or by comparison with absolute optical data at a single point.

In situations where the impact energy has been insufficiently high to allow the use of equation (II.44), equation (II.29) must be used. The generalized oscillator strength $f(\mathbf{K})$ may then be obtained as a function of $\mathbf{K}$. Clearly, a sensitive test of the Born approximation will be the fact that $f(\mathbf{K})$ must be independent of the incident energy E_0. Lassettre and Skerbele[10] have recently reviewed work in this field and have shown that, although this independence holds at quite low impact energies for many transitions in the discrete part of the absorption spectrum, it is by no means universally true. It is clear, however, that we may invert equation (II.29) and use it as phenomenological definition of the generalized oscillator strength

$$f_{0m}(\mathbf{K}) = \left| \frac{\mathbf{k}_i}{\mathbf{k}_f} \right| \cdot \frac{|\mathbf{K}|^2 E}{2} \cdot \sigma_{el}(\theta, \phi, E, E_0) \qquad (II.46)$$

Lassettre et al.[33] were able to show that even if the Born approximation does not hold, we still find

$$\underset{|\mathbf{K}| \to 0}{Lt} f_{0m}(\mathbf{K}) = f_{0m}(0) \qquad (II.47)$$

Unfortunately, in practice the lowest value of $|\mathbf{K}|$ that may be conveniently measured at impact energies less than 500 eV is still too large for the limit described by (II.47) to be obtained accurately. However, the approach has been proven sufficiently useful for a number of absolute measurements to be made, and comparison with optical data has been very fruitful.[10]

E. Continuum Effects and (e, 2e) Coincidence Experiments

The equations used to define $f_{0m}(0)$ involve averaging over all initial degenerate states of the molecule and summing over all final states as in (II.4). If the energy loss is sufficiently large that ionization occurs (and assuming $\eta = 1$), the effect of summing over all final states is, in effect, to sum over all possible angles of ejection of the ionized electron.[5] Clearly, however, the very act of measuring the intensity of the ejected electrons at a given angle selects out of all possible directions of ejection certain specific ones. Thus the preceding formulas must be modified if we wish to account for the results of photoelectron or dipole-electron coincidence experiments in which the intensities of forward scattered and ejected electrons are measured simultaneously. We need, in fact, an expression for the oscillator strength differential in the polar angle of ejection.

The kinematics of the situation for the case of "optical limit" type (e, 2e) experiments are illustrated in Fig. 2b (Fig. 1b of Hamnett et al.[23]), which shows the direction of the ejected electron $\mathbf{j}$ as a function of the two polar angles χ and γ. The angle between $\mathbf{j}$ and the vector $\mathbf{K}$ is denoted by ψ. Provided the forward scattering kinematics are such that $|\mathbf{K}|$ is small and we may approximate $f(\mathbf{K})$ by $f_{0m}(0)$, then,* as is well known, regardless of the detailed form of the continuum wave function, provided that Ψ_m is orthogonal to Ψ_0[3,35]

$$\frac{d^3 f^i(0)}{dE\,d\Omega_j} = \frac{1}{4\pi}\frac{df^i(0)}{dE}\left(1 + \frac{\beta_i}{2}(3\cos\psi - 1)\right) \qquad \text{(II.48)}$$

Where β_i is the usual asymmetry parameter and is a function, in general, of E. An analogous expression holds in PES except that ψ is then the angle between the direction of ejection and the electric vector of the incident radiation.

For the (e, 2e) experiment we may, by analogy with equation (II.29), define a cross section

$$\frac{d^5\sigma_{el}}{dE\,d\Omega_s\,d\Omega_j} = \frac{2}{4\pi E}\cdot\left|\frac{\mathbf{k}_f}{\mathbf{k}_i}\right|\cdot\frac{1}{|\mathbf{K}|^2}\left(1 + \frac{\beta_i}{2}(3\cos^2\psi - 1)\frac{df^i(0)}{dE}\right) \qquad \text{(II.49)}$$

This fivefold differential cross section† thus measures the probability of

---

*For clarity of notation in discussion, we now denote the differential oscillator strength $df_{0m}(0)/dE$ for the ionization process $(0 \rightarrow m)$ by $df^i(0)/dE$ as given in 2.18). The notation of the type $df_{0m}(0)$ is used in Sections II.C and II.D since this is more commonly employed in scattering theory.

†The term σ_{el} is a function of $E, E_0, \theta_s, \phi_s, \chi$ and γ, but the variables have been omitted for simplicity.

simultaneously detecting an electron scattered into the direction $\theta, \phi(\theta \ll 1)$ and an electron ejected in the direction (χ, γ) at a given E, E_0. Integration of (II.49) over the acceptance angles for the forward and ejected electron detectors leads to a rather complicated expression for the resultant coincidence intensity, which is usually written as[23]

$$I_{\text{coinc}}(\alpha, E, i) \sim \left[1 + C_\alpha(E)\beta_i \right] \frac{df^i(0)}{dE} \tag{II.50}$$

where $C_\alpha(E)$ is a purely kinematic function defined elsewhere,[23] and $df^i(0)/dE$ is the oscillator strength for that ionization process defined by the energy of the ejected electron. The function $C_\alpha(E)$ has the interesting property that for angle α (defined in Fig. 2b) equal to 54.7°

$$C_\alpha(E) = 0 \tag{II.51}$$

to first order. Thus[23]

$$I_{\text{coinc}}(54.7°, E, i) \sim \frac{df^i(0)}{dE} \tag{II.52}$$

and for a given energy loss, we may measure the coincidence intensity for all ionization processes i and derive the branching ratios b_i defined in equation (II.19). Knowledge of the total oscillator strength also enables calculation of the ionization efficiency defined by (II.18). It can then be seen that the coincidence technique provides a flexible alternative to photoelectron spectroscopy in that it is experimentally far easier to vary the energy loss E than the incident photon frequency at a known (relative) ionizing flux.

III. CALCULATION OF OSCILLATOR STRENGTHS

We recall that the definition of the generalized oscillator strength may be written [see (II.4)]

$$f_{0m}(\mathbf{K}) = \frac{2E}{|\mathbf{K}|^2} \cdot \sum_\alpha A_\beta \left| \left\langle \Psi_{m\alpha} \left| \sum_{i=1}^N \exp(i\mathbf{K} \cdot \mathbf{r}_i) \right| \Psi_{0\beta} \right\rangle \right|^2$$

We have used the Born–Oppenheimer approximation to factor $\Psi_{0\beta}, \Psi_{m\alpha}$ into electronic and nuclear parts and have further assumed that the former are orthogonal to enable us to reduce V. Both wave functions may be approximated by products of electronic, nuclear rotation and vibrational wave functions. The last of these may be factored out at once, and integration over the $3N-6$ normal coordinates will lead to the usual Franck–Condon factors. Rotation is more difficult. If, as is usual, the rotational states are not resolved, summing and averaging over the rotational wave functions is, at normal temperatures, equivalent to the situation where molecules would be classically rotating. Thus we must consider all possible orientations of $\mathbf{r}_i$ with respect to $\mathbf{K}$, which is equivalent to considering $\mathbf{K}$ to have all possible orientations with respect to the molecule. The effect of this has been discussed by a large number of authors, and the resultant equations are well known in the literature.[3, 35-37]

These expressions may be considerably simplified if we assume that Ψ_0 and Ψ_m may be expanded as antisymmetrized products of spin orbitals. Making the further assumption that no relaxation occurs, that is, that Ψ_m and Ψ_0 may be described by the *same set* of spin orbitals, we have

$$\left\langle \Psi_{m\alpha} \left| \sum_{i=1}^{N} \exp(i\mathbf{K}\cdot\mathbf{r}_i) \right| \psi_{0\beta} \right\rangle = \int \phi_{m\alpha}(\mathbf{r})\left[\exp(i\mathbf{K}\cdot\mathbf{r})\right]\phi_{0\beta}\,d\mathbf{r} \qquad \text{(III.1)}$$

corresponding to a transition from an orbital $\phi_{0\beta}$ occupied in the ground state to $\phi_{m\alpha}$, which is unoccupied in the ground state. This is usually termed the *single-electron approximation* and has been extensively used as a first-order theory.[36-38] The optical analogue is obtained by allowing $|\mathbf{K}|\to 0$, and clearly

$$f_{0m}(0) = 2E\sum_\alpha A_\beta|\hat{\mathbf{K}}\cdot\langle\phi_{m\alpha}|\mathbf{r}|\phi_{0\beta}\rangle|^2 \qquad \text{(III.2)}$$

For continuum transitions an analogous expression must be used. In both discrete and continuum transitions several problems may arise:

1. The excited state is, in general, an open-shell system, and $\phi_{m\alpha}$ must be coupled correctly to the remaining orbitals. This coupling is straightforward for transition to non-Rydberg-like orbitals, but for highly excited Rydberg levels coupling the angular momentum correctly is a source of some difficulties, as Veldre et al. have pointed out.[39] In the case of ionization, especially from open-shell ground states, additional selection rules, based on fractional parentage coefficients, must be included. These have been discussed recently by Cox et al.[40]
2. A fundamental requirement of the derivation of (III.1) and (III.2) is that $\phi_{m\alpha}$ must be orthogonal to $\phi_{0\beta}$. If they are not initially orthogonal, they must be Schmidt orthogonalized in the normal way. However, such orthoganalization may lead to serious errors even in simple systems, as Bell and Kingston found for the helium $1^1S\to 3^1S$ transition.[41]

3. The most serious problem in continuum transitions is the form of the radial part of $\phi_{m\alpha}$. An immense amount of work has been reported on possible forms for $\phi_{m\alpha}$, and the conclusion seems to be that for ionized electrons with kinetic energies above a few hundred electron volts, approximating $\phi_{m\alpha}$ by a plane wave $\exp(i\mathbf{k}_{ej} \cdot \mathbf{r})$ is fairly satisfactory numerically. Unfortunately, the use of simple plane waves is attended by a fundamental disadvantage, namely that the angular dependence of the photoelectron signal is predicted to be of the form[42]

$$I \sim I_0 \cos^2 \psi \qquad \text{(III.3)}$$

which comparison with (II.48) shows to be generally incorrect (i.e., except where β is always equal to 2). It has been pointed out that this defect may be remedied by orthogonalizing the plane wave to the orbitals as discussed in paragraph 2 of this list.[36] Unfortunately, there are no measurements of β at sufficiently high energies for this procedure to be checked experimentally. It is clear from the preceding discussion that the normal plane-wave method will not, in general, yield good results near the ionization threshold. Two avenues of improvement have been suggested within the single-particle approximation framework.

a. For diatomic molecules exact one-electron functions are available for H_2^+ in ellipsoidal coordinates. These are used as the basis for a much more accurate calculation of the continuum wave function. This approach has been used by Tuckwell for N_2 and O_2 [43] and by Itikawa for H_2.[44] Similar calculations for H_2 have also been reported by Shaw and Berry.[45] Very recently Hirota[46] has succeeded in extending these methods to CO and has reported distorted Coulomb wave calculations that are in encouragingly good agreement with experiment.[18,171,172] These calculations are limited to the energy region below 20 eV.

b. Single-center expansion methods have been explored by Burke and co-workers.[47-50] The molecular wave function is expanded about the center of mass of the molecule, and the problem is treated as a pseudoatomic one. Although the scattering equations are substantially simplified by this approach, the computation becomes formidable for more complex molecules.

A related but somewhat different approach has been suggested by the recently introduced MS–Xα method. Calculation of the continuum functions within this method is relatively straightforward,[51,52] and the method has been applied to the K-shell X-ray absorption spectrum[53] of N_2 and very recently to photoionization of N_2 and CO near threshold.[54] This last calculation is very encouraging in that close agreement at an absolute level with many of the features seen in the photoelectron spectra at different incident photon energies may be obtained.

It has become clear from recent theoretical work that many of the effects seen in the variation of oscillator strength with energy loss cannot be explained within the single-particle model. All the inert gases show complicated resonances extending over many electron volts, which result from correlation effects involving the remaining "passive" electrons. A major theoretical advance was the development and use of the random-phase approximation with exchange (RPAE) or time-dependent Hartree–Fock perturbation theory by Amusia and Cherepkov.[55] Within this theory we must return to equation (II.4) and include all electrons in the calculation. Very impressive agreement has been obtained by the Russian group,[55] not only for optical oscillator strengths, but also for generalized oscillator strengths for the inert gases. Recently this method has been extended[56, 57] to open-shell atoms and calculations on the photoionization cross section of chlorine have also appeared.[58] Unfortunately, extension to molecules has not yet proved possible. Related to these methods are those of Kelly and co-workers, who have calculated cross sections for a number of atoms and, using a one-center expansion technique, for CH.[59, 60] However, the computational complexities of this Hartree–Fock perturbation method seems to preclude their general application to molecules at the moment.

Although several computational advances have been made in recent years, it appears that in the foreseeable future, calculations of the oscillator strengths for molecules will remain firmly grounded within the one-electron framework. The very encouraging success of Davenport's calculations[54] on N_2 and CO using the MS–Xα technique may signpost the best route until the computational problems of extending the RPAE method to molecules have been overcome. It should also be noted that the recent moment-theory calculations of oscillator strengths for photoabsorption and partial photoionization reported by Langhoff, McCoy, and co-workers hold great promises in that the results are significantly more accurate than the MS–Xα calculations.[53, 54] Impressive results have been obtained for valence shells of N_2[216] and CO[217] and also for the K-shell of N_2.[218]

IV. EXPERIMENTAL CONSIDERATIONS

A. Electron Analyzers and Transmission Efficiency

Many detailed descriptions of devices and techniques used in electron spectroscopy are to be found in the literature. Accordingly, in this section we give only a general discussion highlighting items of special concern in making quantitative measurements. Scattered or ejected electrons may be energy analyzed by means of electric and/or magnetic fields. The properties and performance of many commonly used types of electron analyzer are to be found elsewhere.[61] For quantitative electron spectroscopy, electrostatic selectors are generally preferred over magnetic analyzers because of the problems of fringing fields. Fringe magnetic fields can interfere with the performance of associated electron-optical elements. The most generally useful types of electrostatic analyzer are the 180° hemispherical,[62] the 127° cylindrical,[63] and the cylindrical mirror analyzer.[64] However, Wien filters (crossed electric and magnetic fields) have also been used effectively for quantitative work at low resolution by Van der Wiel.[65] Very high resolution (0.004 eV) has also been achieved by Geiger[20] using a Wien filter and higher (10 to 50 keV) impact energies.

Operation of electron analyzers at high resolving power necessitates considerable attenuation (to less than a few milligauss) of the earth's magnetic field together with any other AC or DC (alternating or direct current) magnetic fields that may be present. This may be achieved with Helmholtz coils and/or mumetal shielding (mumetal must be hydrogen annealed after all fabrication procedures have been completed). Careful attention must also be given to minimizing AC pickup by the spectrometer and its associated electronic controls. In particular, pickup must be suppressed by eliminating ground loops and using coaxial and triaxial shielding where necessary as well as the employment of suitable filtering. To reduce ripple, care must be taken to ensure that all floated power supplies have a low impedance path to ground.

For quantitative analysis of ejected and scattered electrons, it is usually necessary that the resulting spectra be corrected for the transmission efficiency of electrons, which will often vary significantly with energy. This "chromatic aberration" is the result of the electron optical lens effects (change of focal length) that occur whenever electrons are accelerated or retarded into an analyzer operating with a constant pass energy. These effects can be considerable, as is shown in Fig. 3, which shows the transmission curve obtained[66] for a 127° analyzer with a simple double-aperture retarding system. The transmission factor varies by at least an order of magnitude over a few volts. It should be noted that the exact form of the curve will depend not only on the particular geometry, but also on the pass energy of the electron spectrometer.

Although transmission effects can be accounted for more readily by scanning the analyzer electric field while allowing the electrons to enter on a field free path, this mode of operation is not usually convenient since it results in a changing energy resolution ($\Delta E/E$ is constant rather than ΔE) over the spectrum. An alternative procedure to minimize differences in

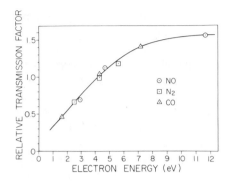

Figure 3. Relative transmission correction factor for an electron spectrometer.[66]

transmission is to construct the spectrometer inputs and/or outputs with multielement (zoom) electron lenses. An excellent compilation of design data for a wide variety of electron optical lenses suitable for electron spectrometers is to be found in a recent book by Harting and Read.[67] Other useful lens designs have been given by Heddle.[68] Detailed and useful treatments of electron-spectrometer design have been given by Lassettre[10], Kuyatt and Simpson,[69] Read et al.,[70] and Noller et al.[71]

In the case of high-resolution electron-impact spectroscopy (i.e., if the exciting bandwidth < 0.4 eV) it is necessary to monochromate the incident electron beam. A schematic representation of the typical requirements for a high-resolution electron-impact energy-loss spectrometer is shown in Fig. 4. Electrons from either an indirectly heated oxide cathode or a directly heated filament are produced in the gun with a Maxwell–Boltzmann distribution of energies, with a full width at half maximum (FWHM) of 0.4–1.0 eV depending on the cathode material, temperature, and gun design. For spectroscopy in which only modest resolution is needed, the incident unmonochromated beam may be used directly in the collision chamber. However, if information is required for more closely spaced electronic or vibrational states, an energy analyzer is needed to function as a monochromator to select a suitably narrow slice from the distribution. The electron beam is then accelerated to a final energy E_0 at the collision region, where it is scattered by a high-density gas target produced either by a jet or by using a "gas-tight" collision chamber. The latter has generally been found superior for quantitative work since a more homogeneous gas density is obtained. A rotatable "gas-tight" collision chamber allowing wide angular variation has been described by Tam and Brion.[72] Electrons inelastically scattered or ejected in the collision region are sampled by an energy analyzer. The scattered (ejected) electrons are accelerated (or retarded) to the appropriate constant pass energy of the analyzer by the application of a voltage [which effectively compensates the energy loss (E)

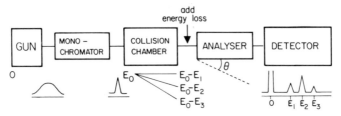

Figure 4. Schematic diagram of energy-loss electron scattering.

in the case of scattered electrons]. For fast electron impact, this energy loss is analogous to the photon energy in optical experiments (Section I). The energy loss is conveniently provided by a suitable programmable DC power supply that thus acts as a simple "analogue" of an optical monochromator. Furthermore, in electron impact a single spectrometer can cover an extremely wide spectral range as illustrated in Fig. 5, which shows the energy-loss spectrum of CF_4 from the IR through to the soft-X-ray region determined on a single electron-impact spectrometer[43] at a modest resolution (0.5 eV). Several optical spectrometers and light sources would be needed to cover the same range. Of course, at low energy (long wavelength) the attainable optical resolution would be far superior. However, as shown in Fig. 6, the constant energy resolution attainable with electron impact becomes relatively more advantageous as we go to higher energies.[14] Three cases are indicated, an FWHM of 0.5 eV (no monochromation), 0.1 eV (modest monochromation), and 0.01 eV (state of the art

2.5 keV ELECTRON ENERGY LOSS SPECTRUM OF CF_4

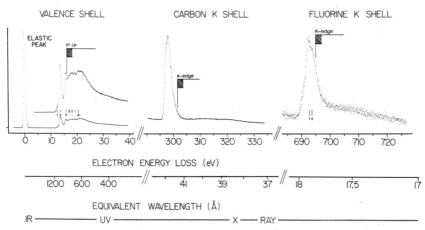

Figure 5. Schematic diagram of 2.5-keV electron energy-loss spectrum of CF_4.

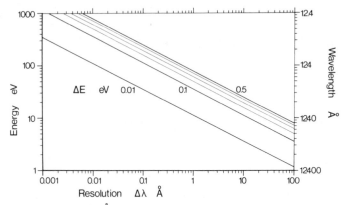

Figure 6. Resolution $\Delta\lambda$ (Å), plotted against energy for fixed values of resolution, ΔE (0.01 to 0.05 eV).

monochromation). At higher energies (i.e., corresponding to inner-shell excitations) it is apparent that the attainable electron-impact resolution may exceed that realized in optical spectroscopy. For example, the nitrogen K-shell electron energy-loss spectrum[73] obtained without monochromation (FWHM = 0.5 eV] compares favorably with the corresponding synchrotron photoabsorption spectrum[74] as shown in Fig. 7. The optical spectrum exhibits complete absorption as a result of the high oxygen partial pressure that was needed to filter out higher-order radiation.

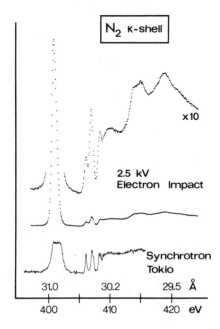

Figure 7. Comparison of the K-shell energy-loss spectra of molecular nitrogen obtained using electron impact and synchrotron radiation.

The monochromator (M)–analyzer (AN) combination in an electron-impact spectrometer results in a combined energy halfwidth of $(\Delta E_M^2 + \Delta E_{AN}^2)^{1/2} = 1.4\Delta E$, where geometrically identical analyzers are operated with the same pass energies.[10] The effective energy spread will include a contribution from the Maxwell–Boltzmann gun distribution when a monochromator is not used. In the latter case there is little point in operating the analyzer at very high resolution since this would only result in a large loss in signal with only a very small improvement in overall resolution. It should be noted that the resolution of an electron selector is in part determined by the acceptance angles as well as the particular geometry, apertures, and pass energy.[6] It is thus necessary to restrict opening angles and beam divergence not only with angular stops, but wherever possible by electron optical focusing to maintain optimum intensity.[67] As has been shown in Section II, electron-impact spectra can be obtained sufficiently close to the optical limit by using high impact energies (several kiloelectron volts) and small scattering angles ($\theta < 10^{-2}$ radians). Under these conditions electron impact spectrometers are very much simpler to operate since fast electron beams are better behaved with regard to space-charge "blowup" and focusing. Highly collimated (< 0.2-mm-diameter) electron beams can readily be produced from television-tube-type guns[23,65] at several kiloelectron volts with currents of 0.1 A. Schmoranzer et al.[75] have also described the use of a telefocus electron gun. Field-emission electron guns have been used extensively in electron microscopy, and the improved energy spread could offer advantages in electron spectroscopy. In the case of forward scattering at several kiloelectron volts the energy loss is only a very small percentage of impact energy, and as such the effective lens voltage ratio for a simple two-cylinder retarding lens changes very little over the range of a given energy-loss spectrum. A system of this type used under these conditions[23,31] is not subject to any significant chromatic aberration, and thus zoom lenses or correction factors are not needed to give quantitative spectra. Figure 8 show the oscillator-strength spectrum for helium obtained by Backx et al.[31] using electron energy-loss spectroscopy. After correction for the acceptance angle of the forward analyzer (i.e., integration over a finite θ) and using the Bethe–Born relation [see equation (II.44)] it can be seen that the electron-impact values of $f(0)$ for helium are in excellent agreement with theory[76] and with photoabsorption experiments.[77, 78]

In the case of lower impact energies in forward scattering and for ejected electron analyzers used in photoelectron as well as electron-impact spectroscopy, the larger changes in voltage ratios at the analyzer input result in transmission variations (with electron energy) that may not be readily corrected with zoom lenses (this is particularly true where slits are

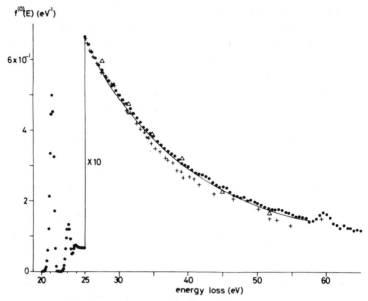

Figure 8. Optical oscillator strengths of He: ●, e–ion),[31] normalized on $1s$–$2p$ transmission; +, photoabsorption;[77] △ photoabsorption;[78] _____, dipole length approximation.[76]

used and also for cylindrical mirror analyzers that require conical geometry). In addition, geometric restrictions often preclude the use of (lengthy) multielement lenses, and a simple aperture entry system must be used. If a zoom lens is used, it should be remembered that it will have a finite "constant" transmission energy range.[67] In all cases the analyzer transmission efficiency must be checked in the laboratory and in most cases a correction curve must be used so that meaningful spectral intensities can be derived from the measurements. The transmission of electron analyzers may be calibrated in a variety of ways:

1. Ejected electron analyzers can be calibrated at lower energies (<25 eV) using UV photoelectron spectroscopy and comparison with quantitative photoelectron spectra. The intensity ratios provide a relative transmission function (T_{ej}) directly. Quantitative (relative) photoelectron spectra have been reported by Hotop and Niehaus[79] at an ejection angle of 90°, and these results have been used by Yee et al.[66] to calibrate a 127° analyzer for which the correction curve has already been shown in Fig. 3. More recently Gardner and Samson[80] reported quantitative (relative) photoelectron spectra that can be used as a standard for analyzer

transmission calibration (at 54.7°). These data can be used at other angles if the β dependence is included.

2. Gardner et al.[81] have recently described how analyzer transmission can be measured using a different method also based on photoelectron spectroscopy. In this work photoelectron ejection from a rare gas was studied using a calibrated photon source at a series of wavelengths. In combination with measurements of the total photoabsorption, the results can be used to derive a transmission function. A cylindrical mirror analyzer sampling at an ejected angle of 54.7° was used thus assuring independence from the asymmetry parameter β (which is known at very few energies for molecules and hence where possible PES should be run at the "magic" angle). The method of Gardner et al.[81] is not easily carried out in routine situations because of the difficulty of photon-flux measurement.

3. Secondary-electron yield measurements from electron-impact ionization have been used. Opal et al.[82] and Oda[83] have reported secondary-electron yields for a number of gases at variety of electron-impact energies (500–2000 eV) and over a wide range of scattering angles. These data[82, 83] have been used to determine analyzer transmission function by Branton and Brion[84] and give good argreement with the earlier (e, 2e) method.[14, 32, 94] This method, using ejected electron spectra, is more versatile than the PES methods since data for a wide range of sampling angles and ejected energies have been reported. These spectra are also continuous and hence more useful than the discrete PES spectra.

4. A novel coincidence method for analyzer transmission calibration at 54.7° has recently been described by Hamnett et al.[23] for use in the dipole (e, 2e) electron-impact simulation of PES. The theory[23] predicts (see Section II) a "magic" angle at 54.7° for the ejected electrons (i.e., at this angle the intensities are independent of the asymmetry parameter β and can thus be used to derive relative total partial cross sections). From the theory it can be seen[23] that the ratio I_{coinc}/I_{sc} as a function of energy loss E ("photon" energy), should be the same for all rare-gas atoms in a given energy range (i.e., in those energy regions where only a single ionization process occurs and where total ionization equals total absorption). Furthermore, the theory[23] shows that this ratio should also be directly proportional to T_{ej}. The predicted results have been verified,[23] as shown in Fig. 9, which shows the curve to be close agreement with the value for T_{ej} derived using the ejected electron spectra of Opal et al.[82] This method has two special features of importance for coincidence spectrometers. First, it is a true coincidence method and thus ensures that the sampling volume and angles for the ejected analyzer are the correct ones for the coincidence experiment. Second, for very

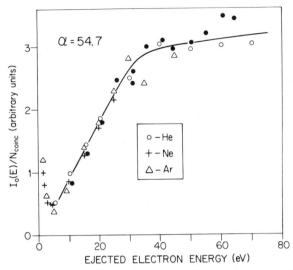

Figure 9. Schematic of $I_0(E)/N_{coinc}$ for gases helium ($\bigcirc$), neon ($+$), and argon ($\triangle$) at an ejection angle of $\alpha = 54.7°$; $\bullet$, transmission factors as obtained using only ejected-electron channel and interpolated cross-section data for He.[82]

low kinetic-energy electrons (< 5 eV) the optimum delay time (see text that follows) varies significantly with energy.[23] However, for the data in Fig. 9 the delay was maintained constant, and the transmission curve is seen to rise at low energies to compensate for this. It is much easier to operate the spectrometer[23] in this fashion and to apply the delay correction as part of T_{ej}. This method is rather similar to Gardner and Samson's PES method (paragraph 2, this list), except that the constant intensity of the virtual photon field eliminates the difficult photon-flux measurement.

5. In principle, transmission functions could also be obtained by using cross sections for elastic scattering.[86]

In obtaining a transmission function by any of these methods, great caution must be exercised at lower kinetic energies ($\leqslant 5$ eV) with regard to pressure-dependent effects. At low electron energies the cross sections for elastic electron scattering[86] and vibrational excitation[87] by electron impact are very large for many molecules, and significant losses of ejected electrons can occur. In this region intensities must be determined as a function of pressure[23] and then extrapolated to zero pressure. Streets et al.[88] have clearly shown in the photoelectron spectrum of N_2 the effect of vibrational excitation via resonance formation. Resonances are the subject of a recent review by Schulz.[89]

In photoelectron spectroscopy the same considerations discussed in the preceding section apply with regard to the collection efficiency of electrons ejected in the photoionization process. To date only a very limited amount of quantitative work has been done on oscillator strengths by PES because of the limited availability of continuum light sources and where also, at least as a minimum requirement, the *relative shape* of the continuum must be known *after* dispersion in the UV optical monochromator.[18] Some work has been done on the noble gases by Codling et al.[90,91] [128] on the Daresbury synchrotron. Samson and Gardner[92] have used a series of UV lines up to 40.8 eV to obtain a number of points on the partial oscillator-strength curves for states of CO^+. This experiment necessitated the measurement of photon flux for each line which is a difficult and tedious procedure. Siegbahn[93] has discussed the potentialities of synchrotron radiation applied to PES, whereas the recent experiments of Plummer et al.[18] have illustrated the possibility of using such a light source for UV molecular photoelectron spectroscopy.

B. Electron Detectors and Signal Processing

In all types of electron spectroscopy it is often necessary to detect electrons over a wide range of intensities. Since very small currents must usually be detected with rapid time response it is generally necessary to use electron multipliers, pulse-counting techniques, and some type of signal averaging. The most generally used type of detector is the channel electron multiplier, which has the enormous advantage of a high gain ($\leqslant 10^8$), which is not degraded by repeated exposure to air. The high gain results in ready realization of a saturated count rate desirable for accurate quantitative work. The small size of these devices renders them suitable for incorporation in complex geometric situations.

Using a normal preamplifier, amplifier, and discriminator arrangement, channeltrons can be used for count rates up to about 20,000 counts per second (cps). At higher count rates the pulse height distribution is drastically reduced and results in loss of counts. However, use of very high gain (>2000) fast amplifiers (e.g., SSR 1110) enables count rates up to at least 10^6 cps to be handled[23, 94] with channel multipliers. At such high count rates it is found that after some usage loss of gain by channeltrons can become a problem. In the authors' laboratory it has been found that loss of gain occurs more rapidly when hydrocarbon gases are used instead of, for instance, rare gases. The frequently reported fatiguing effects have not been observed in a clean vacuum environment. From these observations we have concluded that the drop in gain is caused by surface contamination resulting from absorbed gases. Under continuous electron bombardment absorbed hydrocarbons will probably become polymerized by processes similar to those induced by UV radiation. We have found that the performance of channel multipliers can be repeatedly restored (in the case of Mullard B419AL) by cleaning with kitchen scouring powder and pipe cleaners. Alternatively, high count rates can be achieved with Johnston focused mesh electron multipliers,[96] and these have the useful feature of a large active input area ($\sim$5 cm^2). More recently, microchannel plate multipliers have become available,[97] and when used in tandem these have sufficient gain to detect quantum events by pulse counting. The large active area permits total collection over a wide area or, alternatively, position-sensitive detection. Position-sensitive detection permits a very high sensitivity since "all of the spectrum is recorded all of the time" instead of the more usual practice of scanning the spectrum over a slit. Moak et al.[98] have described a position-sensitive channel plate detector using a resistive plate for direct electrical readout. This is to be preferred over the more conventional system using a fluorescent screen readout and a television camera.[99] Position-sensitive detection not only offers greatly increased sensitivity, but also provides a fast, on-line integrating detector that eliminates effects of beam and pressure fluctuations. As such, it is far superior to the use of photographic plates.

In most applications the low signal levels necessitate the use of signal-averaging techniques. A wide variety of multichannel analyzers, mini computers, and microprocessors suitable for this application are now commercially available. Frequently these devices must issue commands and/or receive information from parts of the spectrometer that are at elevated voltage levels, and this can best be achieved by means of optical coupling.

C. Coincidence Methods

Many experiments (see Section I.B) require the energy analysis and detection of two or more particles with time correlation; in other words, coincidence counting techniques must be used. Coincidence methods have long been used in nuclear physics because of the convenient fast detectors that have long been available. The more recent availability of fast, high-gain electron multipliers has created the possibility of coincidence measurements in electron spectroscopy. Various aspects of coincidence measurements have been discussed elsewhere.[100–102]

Two general types of arrangement have been used for the detection of time-correlated events in electron-spectroscopy experiments. Conventional methods using time-to-amplitude convertors (TAC) and single-channel analyzers have been described by several groups[101–104] and a typical schematic electronic arrangement[104] is shown in Fig. 10. In this arrangement, to isolate the true coincidences, two single-channel analyzers (SCA)

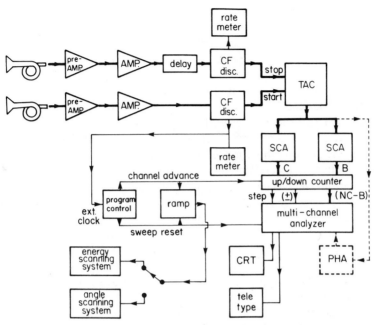

Figure 10.   Time-to-amplitude convertor coincidence electronics.

are set up to pass pulses occurring within the windows labeled C and B in the inset to Fig. 8. The width of the window B is N times (typically $N = 8$) that of C. Clearly, a large value of N enables the accidental coincidence rate to be determined with greater statistical accuracy. A quantity ($NC - B$), where C and B now refer to the number of counts passed by the respective SCAs, is accumulated for a fixed period in a specially constructed up–down counter. At the end of a counting period the "channel-advance" pulse from the program control causes the counter to read out serially ($NC - B$) pulses into the multichannel analyzer, together with a "sign" signal determining whether they are to be added to or subtracted from the memory contents. This coincidence timing and background-subtraction system is similar to that used by Weigold et al.[103] To accurately subtract out the accidental coincidences, so as to achieve a net average result of zero when there are no true coincidences present, great stability is required for the SCAs. In experiments where the accidental coincidence rate (determined by the product of the two singles count rates and the time resolution) remains effectively constant during a scan, any subtraction error caused by imprecise setting of the SCA windows will only produce a flat (positive or negative) background, which can be determined by including in a scan a region where no true coincidences are possible. In situations where the coincidence signal is strong with low singles count rates, the subtraction error is in any case negligible. Time-to-amplitude conversion has also been used by Backx and Vander Wiel[102] for recording simulated "photoionization" mass spectra as a function of energy loss ("photon energy") in an electron–ion (e–ion) coincidence experiment.

An alternative method is the double-delay scheme for coincidence counting described by Ikelaar et al.[65, 105] which has been used extensively at the FOM Institute in Amsterdam by Van der Wiel et al.[102, 149] and also in the authors' laboratory.[23, 94] A typical arrangement (Fig. 11) has been used in the (e, 2e) simulation of PES.[94] Delayed pulses from the ejected electron channel are reflected by an open-ended coaxial cable such that each event in the ejected channel is presented to the first coincidence gate (GATE I) as a double pulse consisting of the straight through pulse followed 300 nsec later by the reflected pulse. The output of GATE I consists (see insert) of n_1 (true + accidental) plus n_2 (accidental). This signal, together with the direct ejected electron signal, is presented to GATE II, for which the output will be equal to n_1 (since there are no reflected pulses). The true coincidence count rate, equal to $n_1 - n_2$ can be obtained directly using automatic subtraction by means of an appropriate add–subtract interface.[94] This method is particularly suitable when one of

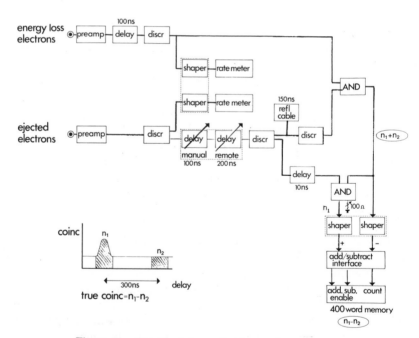

Figure 11.   Double-delay coincidence electronics.

the signal channels varies in intensity during a scan, as occurs with the ejected electron signal in the (e, 2e) experiment.[94] Furthermore, in the dipole (e, 2e) method a very high singles count rate ($\sim 10^6$ cps) is required in the forward channel, and use of the double-delay method places less stringent timing requirements on the final data-collection system because of the relatively low number of true coincidence counts. This method avoids the drifts that may occur in the level and window settings of the TAC and also eliminates TAC dead-time corrections. The double-delay coincidence system described by Ikelaar et al.[65, 105] (see following paragraph) does not determine the accidental coincidences with as good statistical accuracy as does the TAC system. It is, however, the preferred system for experiments,[23, 94] in which the accidental rate varies across the spectrum and when dead-time corrections are large.

In a coincidence experiment the accidental coincidence rate $n_{acc} = n_1 n_2 \tau$, where n_1, n_2 are the singles count rates and τ is the time resolution. Therefore, it is important to reduce the time resolution as much as possible. The contribution due to differing path lengths in electrostatic analyzers and electronics should be minimized by appropriate design and operating parameters. However, a major contribution to the time resolution can also occur from the electron multiplier. Typical time resolutions are 10–20 nsec (FWHM). Recently Dillon and Lassettre[106] reported an (e, 2e) experiment where the time resolution was considerably improved, and this should result in a considerable improvement in statistics for recording the true coincidence spectrum. Commercial channel plate detectors have a very fast time response (~ 2 nsec) and should thus prove very suitable for coincidence experiments.

Van der Wiel[12] has pointed out that there are two inherent limitations to intensity (with given statistics) in the coincidence experiment. First, it can be seen (Fig. 11) that the statistical relative error

$$\frac{\Delta n_{true}}{n_{true}} = \frac{\sqrt{(2 n_{acc} + n_{true})}}{n_{true}}$$

Since for a given measuring time $n_{true} \sim I_{el}$ and $n_{acc} \sim I_{el}^2$, where I_{el} is the electron beam current, it can be seen that when $n_{acc} \gg n_{true}$ the statistical error is optimal and independent of I_{el}. Second, the coincidence rate is restricted by the maximum allowable singles count rates in either channel ($\sim 10^6 - 10^7$ cps, as discussed earlier).

A number of optical photoion–photoelectron coincidence experiments have been reported[107] giving qualitative information on molecular breakdown. However, in these experiments little or no attempt has been made to correct for either the transmission efficiency of the photoelectron spectrometer or the ion kinetic-energy discrimination. Fragment ions usually possess considerable kinetic energies (0–20 eV), resulting in severe loss of energetic ions with most mass spectrometers. This effect is dramatically illustrated in the case of CO and N_2 when the early (e–ion) coincidence experiments [kinetic energy $(KE) \leqslant 0.5$ eV] of El-Sherbini and Van der Wiel[108] are compared with more recent work by Wight, et al.[25] In the former work[108] a conventional magnetic analyzer was used, whereas in the latter work[25] a total ion collection ($KE \leqslant 20$ eV) time-of-flight tube and extraction system were employed.[109]

D. Absolute Oscillator Strengths

The measurement of absolute cross sections in electron scattering has been discussed by Kessler.[110] It is important to distinguish between *absolute measurements* based solely on experimental parameters from those based on *relative measurements* that have been normalized either to other experimental data or to calculations. To make an absolute measurement, it is necessary to accurately determine the absolute intensity of the ionizing (exciting) beam as well as the target particle density. In the case of electrons the flux measurement is fairly straightforward, provided that reflection and secondary-electron production are suitably controlled. Calibration difficulties are more severe for UV and soft X-ray photons emerging from the dispersing monochromator.

Absolute electron-impact cross sections have been measured by Chamberlain, et al.[111] for excitation of the $2'P$ and $2'S$ states of helium at a scattering angle of 5° in the energy range 50–400 eV (these conditions are far from the optical limit). Measurements of this type are extremely difficult but with care may approach an accuracy of 5 to 10%. Such determinations are useful for normalizing relative measurements.

A number of methods have been used to derive absolute data from relative measurements for which it is only necessary to have (or correct for) constant particle flux and gas pressure. This is much easier to effect in the laboratory than an absolute measurement. If the spectral shape of an oscillator strength can be obtained, the curve may then be normalized at a single point when suitable data are available. This approach has been taken by Lassettre[10, 11] and Bromberg,[112] who have developed a method that depends on the construction of one apparatus for the measurement of *absolute* elastic cross sections[112] and a second machine for the determination of *relative* elastic and inelastic cross sections.[10] The absolute elastic cross-section measurement is less difficult to make. However, the method is only suitable for scattering angles greater than a few degrees (as a result of the main electron beam) and is thus not suitable for direct optical oscillator-strength measurements. A different instrument has been developed by Geiger and co-workers,[113–115] who have developed a method for deriving $f(0)$ values at very high impact energies.

Optical oscillator strengths have also been derived from differential angular scattering cross-section measurements (since $|\mathbf{K}|^2$ varies with θ) by Lassettre et al.[10, 11] and also by Hertel and Ross.[116] In this method the cross section is determined as a function of the momentum transfer and extrapolated to $|\mathbf{K}|^2 = 0$. However, the method is somewhat tedious, and absolute determinations require some normalization procedure for the measured generalized oscillator strengths. Furthermore, in some cases Lassettre et al.[10, 11] have shown that the behavior of $f(\mathbf{K})$ is complicated as the $|\mathbf{K}|^2 = 0$ limit is approached, and this may lead to significant errors if measurements are not made at sufficiently small $|\mathbf{K}|$. Lassettre has shown[10,117] that, regardless of whether the Born approximation applies, the apparent generalized oscillator strength will still extrapolate to the optical oscillator strength at $|\mathbf{K}|^2 = 0$. This limit theorem has been discussed in Section II. Relative optical oscillator strengths for photoabsorption derived from energy-loss spectra measured by fast-electron impact at small scattering angles (i.e., with experimental conditions sufficiently close to the optical limit) may be normalized at a single energy using absolute photoabsorption data from the literature. For example, Fig. 12 shows a photoabsorption measurement for N_2 measured by Wight, et al.[25] using fast-electron impact. The curve has been normalized at one point only to the

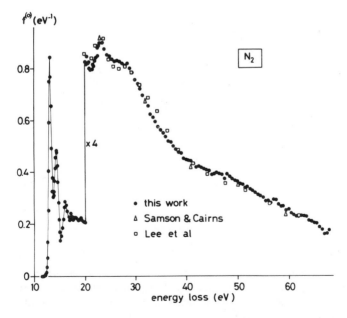

Figure 12. Absorption oscillator strength of N_2; ●, (e,ion) coincidence;[25] △, From Samson and Cairns; [118] □, From Lee et al.[119]

conventional optical work of Samson and Cairns.[118] It can be seen that over the whole energy range the data are in excellent agreement with the data due to Samson and Cairns,[118] as well as the more recent synchrotron work by Lee et al.[119] This type of method has also been used by Huebner et al.[120] in normalizing apparent oscillator strengths determined by low-energy (100 eV) electron impact at zero degrees scattering angle. Since this work covers the discrete region of the spectrum and only a few electron volts of the continuum, it is not discussed futher. Some useful sources of photoabsoption data are listed under Ref. 121.

Total ionization oscillator strengths (see Section V) may be put on an absolute scale by normalizing to the total absorption at energies where the ionization efficiency (see Section II) is unity. For most species this is the case above ~20 eV.[25, 102] Alternatively, absolute photoionization cross sections [122] may be used for normalization.

A second method that exists for normalization of relative absorption curves is dependent on the Thomas–Reiche–Kuhn (TRK) sum rule (see Section II). The "zeroth" sum rule simply states that the total oscillator strength for photoabsorption is equal to the number of electrons in the atom or molecule. In cases where the electronic shells are well separated

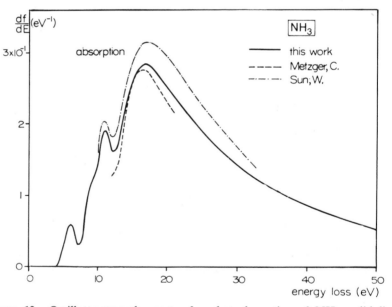

Figure 13. Oscillator-strength spectra for photoabsorption of NH_3: solid line, electron-impact[32] absorption; smooth curve through 400 data points, normalized to a total f value of 8; dashed curve, Metzger and Cook;[124]. Dashed–dotted curve, Sun and Weissler.[125].

we may apply this rule to each shell with a suitable correction for already occupied levels, as discussed by Wheeler and Bearden.[123] Using the constant-intensity virtual photon field (see Section I), it is possible to obtain essentially the whole relative absorption spectrum for a given shell (a small extrapolation error is involved when the spectrum is extended over 100–150 eV above threshold). An example of the use of sum-rule normalization is shown for the valence shell of ammonia[32] in Fig. 13. Excellent agreement is obtained with two absolute optical measurements[124, 125] that are limited in energy range. Results of similar accuracy have also been reported for the absolute oscillator strength for photoabsorption by H_2O.[84] Bonham has also considered sum-rule normalization of electron-impact spectra.[126]

V. EXPERIMENTAL MEASUREMENTS

A. Introduction

In this section we discuss the more important experimental results for continuum oscillator strengths measured by electron spectroscopy that have been reported up to mid 1978. The discussion is divided on the basis of target species rather than the type of experiment since this stresses the interrelation and complementary nature of many of the experiments. As the experimental work is far from complete in many cases, only a limited picture of the overall breakdown processes is available at present. In particular, a very limited amount of work has been reported for inner shells. More data are generally available for mass fragmentation (photoionization mass spectrometry) than for partial ionization cross sections (photoelectron spectroscopy).

B. The Noble Gases

As of mid 1978 only a few quantitative photoelectron spectroscopic measurements had been made for the noble gases. Samson et al.[127] have measured the $2p_{3/2,1/2}$ partial photoionization cross-section ratios for xenon, krypton, argon, and neon at a number of wavelengths down to 304 Å (40.8 eV), finding values somewhat less than the statistical ratio. Within resonances the ratio was found to vary dramatically. Values for the partial ionization cross sections of $Xe^+(2p_{3/2,1/2})$ were given for the limited wavelength range 460–570 Å. Very recently Wuilleumier et al.[215] have measured the energy dependence of the $2p_{3/2}/2p_{1/2}$ branching ratio of xenon up to 107 MeV.

Valence-shell ns electron ionization of argon[128–130] and neon[131] have been studied by Codling and his co-workers using radiation from the Glasgow[128] and Daresbury[129-131] synchrotrons. These studies have been prompted by theoretical investigations of the cross sections for photoejection of *ns* electrons. Hartree–Fock calculations by Kennedy and Manson for argon had shown a cross section rising smoothly from threshold.[132]. However, more recent calculations by Amusia and Cherepkov[55] using the many-body RPAE method, which takes account of electron-correlation effects, indicate that a deep minimum should occur in the cross section just above the 3s threshold. This effect is caused by interaction with the 3p channel. The most recent experimental work on argon by Houlgate et al.[130] used a 127° electron analyzer to measure the argon 3s/3p branching ratio at 90° ejection angle over the photon energy range up to 90 eV. The total photoionization cross-section data of West and Marr[133] were than used, together with measured values of $\beta(3p)$ and an assumed value of 2 for $\beta(3s)$, to derive the 3s partial cross sections from the branching ratios. The results for the argon 3s partial cross section are shown in Fig. 14 and confirm the existence of a minimum near threshold as a result of inter-channel interaction as predicted by the R matrix[134] and RPAE calculations.[55] These latter calculations are also seen to represent the cross section more accurately at higher photon energies. The argon 3s data are in good agreement with earlier conventional PES experiments at a few isolated wavelengths by Samson and Cairns[135] and Samson and Gardner.[136]

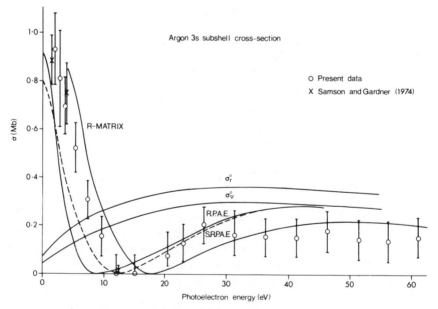

Figure 14. Partial photoionization cross section for 3s electron of argon.[130]

More recently Tan and Brion[199] have made a more direct measurement of the argon $3s/3p$ branching ratios and the partial photoionization cross section for the $3s$ electron of argon using magic-angle dipole (e, 2e) spectroscopy. The results are quite similar to those reported by Houlgate et al.,[130] but with slight differences in the cross sections and the position of the Cooper minimum. The differences may, at least in part, be explained by the assumption of $\beta(3s) = 2$ made by Houlgate et al.[130] This assumption is questionable in view of the strong intershell correlation effects that cause the $3s$ electron to "follow" the $3p$ behavior to a considerable extent. Corrections[199] for multiple photoionization were made using the (e, e + ion) oscillator-strength measurements by Wight and Van der Wiel.[144] The (e, 2e) result[199] for the $3s$ cross section agrees best with the SRPAE calculation.[200]

Samson and Gardner[136] have also obtained a few data points near the threshold and at 40.8 eV for the $4s$ and $5s$ cross section of krypton and xenon, respectively. The shapes of the krypton and xenon cross sections are in good agreement with the RPAE theory,[55] except that in the case of xenon the theoretical cross sections are approximately twice the experimental values. Gustafsson,[201] has measured the partial photoionization cross section of the $5s$ electron in xenon from 25 to 40 eV using synchrotron radiation. Some dependence on the asymmetry parameter, β, may be expected because of the type of geometry employed. Good agreement is obtained with RPAE calculations.[55]

The $2s$ partial ionization cross section for neon has been studied from threshold out to about 130 eV by Codling et al.[131] using synchrotron radiation and two types of photoelectron spectrometer (127° and CMA). The results are more closely represented by R-matrix,[128] or RPAE,[55] theory than by the Hartree–Fock calculations.[132] Wuilleumier and Krause[137] have studied Ne(2s) ionization with a few soft-X-ray lines, and these results, together with a single measurement near threshold by Samson and Gardner[136] are compatible with the synchrotron measurements.[131]

More recently, using synchrotron radiation and magic-angle PES, West et al.[138] have also studied the $4d$, $5s$, and $5p$ partial photoionization cross sections of xenon in the region 60–135 eV (i.e., above the $4d$ threshold). Typical photoelectron spectra are shown in Fig. 15a, and the partial cross sections are shown in Fig. 15b. The total absorption spectrum of xenon shows a broad maximum centered about 100 eV,[139] which has been attributed to electron–electron correlation effects as a result of many-body calculations by Amusia and Cherepkov[55] and Wendin.[140] Recent calculations[55] suggest that the partial cross sections for the $5s$ and $5p$ electrons should be enhanced in the region of the $4d$ maximum because of intershell interaction. In 1972 El-Sherbini and van der Wiel,[141] using coincidence between fast scattered electrons and ions, observed a peak in the oscillator

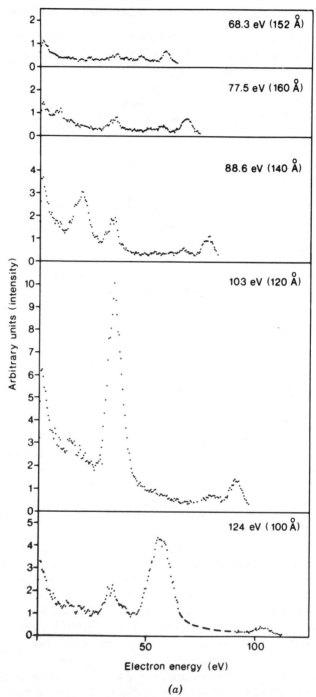

Figure 15. *a* Photoelectron spectrum of xenon at five different energies;[138] (*b*) partial cross sections[138] for $Xe(4d^{10}5s^25p^5 \rightarrow 4d^95s^25p^6, 4d^{10}5s5p^6)$ and $Xe^+(5s + 5p)$. Dashed lines are the calculations by Amusia.[55]

44

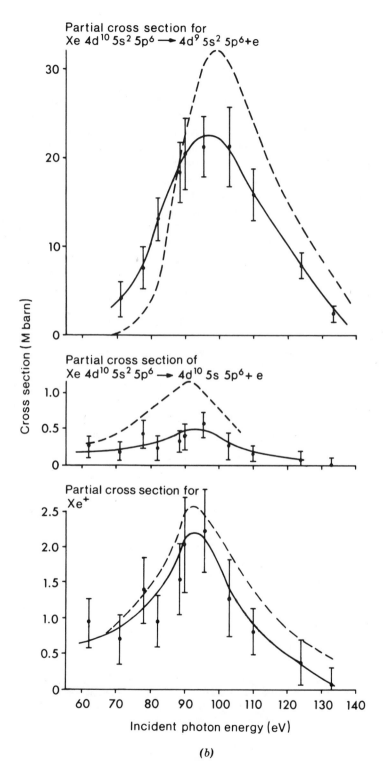

(b)

Figure 15. *continued*

45

strength for single-ion production centered at about 100 eV, which is consistent with the existence of such a phenomenon. Similar results were found for krypton.[14] As can be seen from Fig. 15b, the direct photoelectron experiment confirms the existence of $4d$ subshell correlation effects in the $5s$ and $5p$ channels. More recent measurements on krypton and xenon by Van der Wiel and Wight,[142] and on xenon by Wight and Van der Wiel,[143] using an improved electron–ion coincidence spectrometer lend further support to the existence of interchannel interaction in this system.

Double photoionization has been studied extensively in the outer shells of atoms by photon impact (for a summary, see Wight and Van der Wiel[144]). The recent electron–ion coincidence experiments by Wight and Van der Wiel,[144] for the double: single ionization ratios in helium, neon, and argon are in excellent agreement with the photoionization (synchrotron) data of Schmidt et al.[145] Earlier electron–ion coincidence work,[141] on the double:single ionization ratios appears to be in error by 25%. This discrepancy, attributed,[138] to discrimination as a result of recoil momentum transferred to the ions, has been rectified in the new experimental arrangement,[109] by more efficient (100%) ion extraction. The $2+/1+$ ratios were converted into double-ionization oscillator strengths using photoabsorption data published by Watson.[146] The result for Ne^{2+} is shown in Fig. 16.

The pioneering studies of electron–ion coincidence at small momentum transfer by Van der Wiel and his co-workers provided the first approximate data for dipole oscillator strengths for the multiple ionization of helium and neon,[8,11] argon,[149] and krypton and xenon[141] in both valence

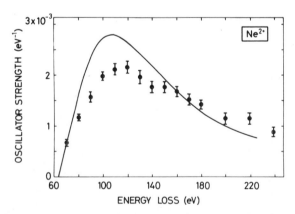

Figure 16.   Double-ionization oscillator strength of neon: ●, (e–ion) data[144] obtained using measured $(2+/1+)$ ratio and photoabsorption data due to Watson;[146] solid curve, calculation by Chang and Poe.[95]

and inner shells over a wide range of energy transfers (up $\leqslant 400$-eV "photon" energy). The data include charge states up to and including $+3$ in neon and $+4$ in argon, krypton, and xenon. Although more recent work,[144] has indicated that some of the cross sections are in error by up to 25% because of ion-recoil effects and mass discrimination, the work is still largely unique and provides much semiquantitative information. These experiments are of great significance in that they illustrated for the first time the elegance and viability of the fast-electron-impact simulation of the photoionization process using electron-impact coincidence methods. The experiments have clearly demonstrated the feasibility of working effectively at conditions sufficiently close to the optical limit to enable measurement of dipole oscillator strengths. Oscillator-strength spectra for L-shell ionization of argon to charge states $+1$ through $+4$ are shown in Fig. 17. The spectrum in Fig. 18 shows the total "photoabsorption" obtained by summing the different charge-state spectra. It can be seen that there is good agreement with optical data,[150–151] except for a small discrepancy in the L-shell region. A renewed normalization of the $2+$, $3+$, and $4+$ data on the recent $(2+/1+)$ ratio for the outer shell[144] removes this discrepancy (M. J. Van der Wiel, private communication). More recently, Wight and Van der Wiel[202] have reported partial oscillator-strength measurements for production of Xe^{1+}, Xe^{2+}, and Xe^{3+} in the photon energy range 30–80 eV using the $(e, e + ion)$ coincidence method. The role of intershell correlation between the $4d$ and $n = 5$ shells, as evidenced in the Xe^{2+} channel, has been discussed by Van der Wiel and Chang.[203]

It is appropriate to note here that in the region of inner-shell thresholds charge-state oscillator strengths may exhibit unexpected shape modulations[149] from the effect of postcollision interaction (PCI). This phenomenon has been observed in electron-impact excitation of short-lived autoionizing states near threshold.[152] The effect may result in "shake down"[152] with a resultant redistribution of oscillator strength. Such effects might be expected to occur near inner-shell edges because of the short lifetime of inner hole states with respect to Auger decay. As a result, the fast Auger electron may gain energy at the expense of the slow photoelectron. Such Auger line shifts have been observed by Ohtani et al.,[153] and recently Van der Wiel[154] has reviewed the evidence for PCI associated with inner-shell ionization. In particular, Van der Wiel et al. have reported such effects in the L-shell ionization of argon[155] and for the M-shell of xenon.[154,202]

Generalized oscillator strengths in the helium continuum have been reported by Lassettre[156] and by Silverman and Lassettre[157] using an electron-energy-loss method. The limiting oscillator strengths were obtained by extrapolating to zero momentum transfer, and, as discussed by Lassettre,[156] the values are in good agreement with both theory[158] and the results of optical experiments.[159]

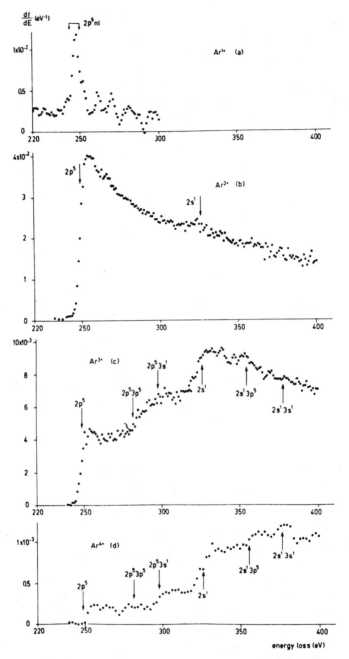

Figure 17. Oscillator-strength spectra for argon *L*-shell ionization, leading to charge states 1 to 4.[149]

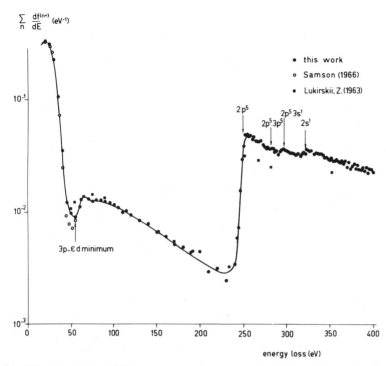

Figure 18. Photoabsorption spectrum of argon comparing (e, ion) coincidence data[149] with optical work.[150, 151]

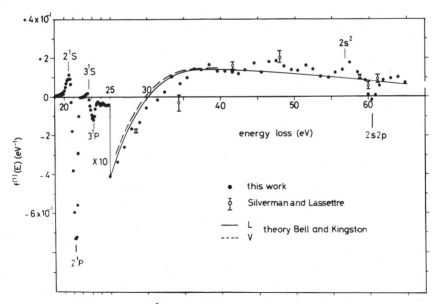

Figure 19. Derivative to K^2 (au) of generalized osicllator strength at $K=0$: ●, (e, ion) coincidence;[31] ○, from extrapolation of data by Silverman and Lassettre.[157]

Backx et al.,[31] using the improved electron–ion coincidence machine[24,31] at the FOM institute in Amsterdam have studied further important aspects of the ionization of helium by fast electrons. The forward scattering energy-loss spectrum (noncoincident) was measured at two small angles (and thus at two values of $|\mathbf{K}|^2$, thus allowing estimation of the $f^{(1)}(0)$ term in the expansion of the generalized oscillator strength [equation (II:34)]). The result which is in excellent agreement with theory,[160] is shown in Fig. 19 and is significant in that it allows assessment of the contribution of $\mathbf{K}^2 \cdot f^{(1)}(0)$ [see equation (II:34)] to the scattering intensity under the conditions of the small-angle scattering experiment. Thus for most practical purposes, any correction for $f^{(1)}(0)$ is likely to be less than the experimental uncertainty in determining the oscillator strength. The oscillator strength for total photoabsorption by helium as measured by Backx et al.[31] using the forward scattering of fast electrons was shown in Fig. 8. Excellent agreement is obtained with both theory[158] and optical absorption.[159]

C. Hydrogen

The interaction of the hydrogen molecule and its deuterated analogues with electromagnetic radiation has been studied extensively, not least because of the tractability of calculations for these molecules. Photoionization and photoabsorption studies up to 60 eV have been reported by Cook and Metzger[161] and by Samson and Cairns.[118] The region of the photoionization threshold has been the object of a detailed study by Chupka and Berkowitz.[162] Mass-spectrometric photoionization studies over a limited energy range have been reported by Comes and Lessmann,[163] Dibeler et al.,[164] and Browning and Fryar.[165]

Prior to the work of Backx et al.[166] there were no continuous dipole oscillator-strength data over a wide energy range for production of H_2^+ and H^+. Backx et al.[24,166] have described an apparatus (shown in Fig. 20) in which fragment ions analyzed by the time-of-flight (TOF) method are collected with 100% efficiency for kinetic energies of fragmentation less than 20 eV (see Section V.E for a description of the method used). This consideration is very important for molecular systems where fragment ions may be produced with high kinetic energies as a result of dissociation from repulsive electronic states. Failure to account for this will result in a loss of the intensity arising from energetic fragments. These effects are well illustrated by comparing the fragment-ion spectra from N_2 and CO in the work (discrimination for $KE \leqslant 0.5$ eV) of El-Sherbini and Van der Wiel[108] with the more recent work of Wight et al.[25] where fragments with $KE < 20$ eV are totally collected. Backx et al.[166] have reported dipole oscillator strengths for absorption, ionization, and fragmentation of H_2, HD, and D_2

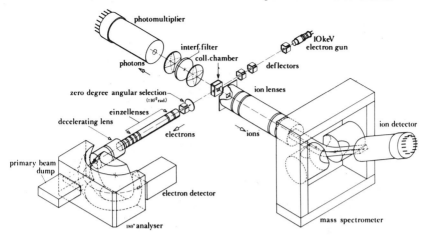

ELECTRON-ION-PHOTON COINCIDENCE EXPERIMENT

Figure 20.   Electron–ion–photon coincidence experiment.[24,166]

over the energy-transfer range 10–70 eV using the electron–ion coinci-dence method to simulate photoionization mass spectrometry. Since there are no repulsive states of H_2^+ that give rise to protons with kinetic energies in excess of 20 eV, the oscillator-strength spectrum for H^+ formation is free of ion-transmission effects. The absorption oscillator strength for H_2 (Fig. 21) was obtained from the forward-scattering (noncoincident) energy-loss spectrum corrected for the $f^{(1)}(0)$ term as discussed in the case of helium (see earlier). The spectrum was normalized using the TRK sum rule (see Section II). Excellent agreement is obtained with optical work, whereas the earlier electron-impact measurements (extrapolation to $K^2 = 0$) by Lassettre and Jones [167] are somewhat higher near threshold.

The $f(0)$ for ionization of hydrogen was obtained[166] as follows. The parent (molecular) H_2^+ ions, selected by TOF criteria, were measured in coincidence with the forward-scattered electrons as a function of energy loss (for details of different modes of operation, see Backx and Van der Wiel[102] using the double-delay method (see Section IV). The coincidence spectrum was then converted into an oscillator-strength curve (see Section II). Following this the mass spectra were recorded as a function of energy loss ("photon energy"), using a TAC for TOF measurement. The start and stop pulses came from the electron and ion channels, respectively. The TAC output was stored by a minicomputer operated as a pulse-height analyzer. Spectra were corrected for accidental coincidences and also for the TAC dead time. The fractional abundances H^+/H_2^+ were derived

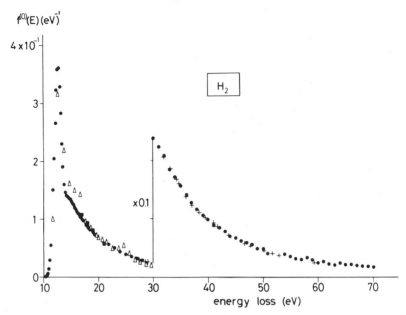

Figure 21. Absorption oscillator-strength spectrum of H_2: ● electron impact[166] normalized on an integral value of 2 (TRK sum rule) with 1% correction for energies beyond 80 eV: +, photoabsorption;[188] ■, data from Cook and Metzger;[161] △, extrapolation of Generalized Oscillator Strength obtained at 461-eV impact energy.[167]

from the TOF mass spectra and used together with the H_2^+ oscillator strength to calculate the total ionization oscillator strength. The oscillator-strength ratio of total ionization: total absorption gives the ionization efficiency η (see Section II), which is found to be close to unity after a few volts above threshold, as shown in Fig. 22. It should be noted that in a short region just above the ionization potential (~15.4 eV) the ionization efficiency is less than unity, indicating the existence of superexcited states converging on vibrationally excited states of H_2^+.[168] At about 37-eV energy loss (Fig. 21) η is also less than unity, and this may be a result of neutral dissociation of a doubly excited state. The difference between $f(0)$ values for absorption and ionization gives the oscillator-strength curve for discrete excitation shown in Fig. 23. The integrated oscillator strength for excitation in H_2 is 0.86. Slight variations occur for HD and D_2 because of differences in Franck–Condon factors.

The oscillator-strength distributions[166] for H_2^+ and H^+ are shown in Fig. 24 in comparison with the *total* photoionization measurements of Samson et al.[169] The H^+ spectrum shows considerable structure in the regions of

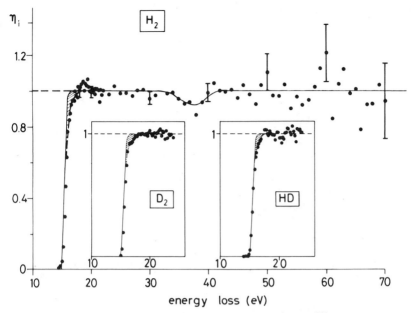

Figure 22. Ionization efficiency of hydrogen.[166]

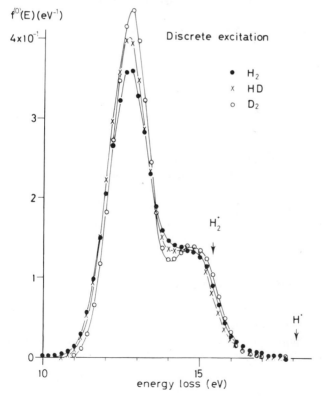

Figure 23. Difference of absorption and ionization oscillator strengths: ●, H_2; x, HD and ○; D_2, from Backx et al.[166]

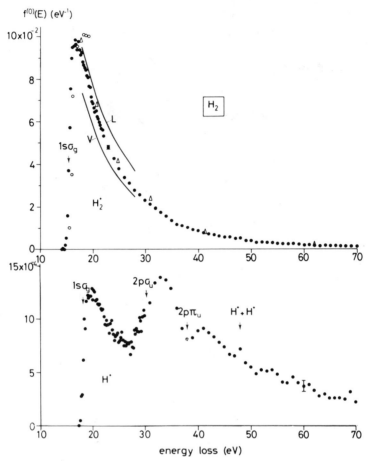

Figure 24. Oscillator-strength spectra for H_2^+ and H^+ from reference.[166] Open triangles are photoionization data due to Samson et al.[169]

the onsets to the higher repulsive states of H_2^+.[170] The hydrogen data are discussed extensively by Backx et al.,[166] who conclude that autoionization plays a significant role in the dissociative ionization of H_2 and its deuterated analogues.

D. Nitrogen and Carbon Monoxide

The isoelectronic molecules N_2 and CO are conveniently discussed together. Comprehensive investigations of N_2 and CO have been made using both the (e–ion)[25] and (e, 2e)[171] methods to probe the broad spectral features of the oscillator-strength distributions for both ion formation and production of the electronic states of the ions. These molecules have also been studied by photoelectron spectroscopy using conventional light sources[172,204] and very recently with synchrotron radiation.[18,205] These latter results have confirmed the accuracy of the earlier (e, 2e) simulation experiments.[171]

Electron–ion coincidence studies of both the valence[108] and inner-shell[173] ionization of N_2 and CO were first carried out by El-Sherbini and Van der Wiel. However, as described earlier in this review, the fragmentation spectra obtained with this instrument represent only dissociation products with kinetic energies below about 0.5 eV. Furthermore, the observed ion intensities may be modified by ion-recoil effects.[144] Nevertheless, El-Sherbini and Van der Wiel[108] produced dipole oscillator strengths for molecular ionization in close agreement with more recent coincidence studies[25] as well as direct optical measurements using synchrotron radiation.[174] Furthermore, the fragment-ion spectra provide useful information concerning the low kinetic-energy fragments. The inner-shell studies[173] provide a useful semiquantitative picture of some of the decomposition pathways resulting from excitation and ionization of the K-shell electrons. Studies were made for the nitrogen K-shell of N_2 and for the carbon K-shell of CO. The oxygen K-shell was not studied in coincidence because of the difficulties of working at higher energy losses [see equation (II.45)]. The information provided by these studies is complementary to that obtained by Auger and photoelectron spectroscopy. The results for the carbon K-shell of CO are shown in Fig. 25 and show the ion yields ($KE \leqslant 0.5$ eV) coincident with electrons that have lost energies in the range 280–315 eV. The K-shell electron energy-loss spectra (noncoincident) of these molecules have been studied in detail by Van der Wiel et al.[175] and by Wight et al.[176] The carbon K-shell energy-loss spectrum of CO reported by Wight et al.[176] (see Fig. 26) is almost identical to that for the nitrogen K-shell of N_2 (see Wight et al.[176] and Fig. 7). These spectra, obtained at 2500 eV impact energy and at an average scattering angle of 2×10^{-2} radians (and thus not strictly under dipole conditions), closely resemble the optical absorption spectra.[177] The spectra were interpolated on the basis of core analogies, valence, and Rydberg excitation. The recent multiple scattering calculations of Dehmer and Dill[53,178] have provided a different semiquantitative description in terms of shape resonances. The electron–ion coincidence spectra Fig. 25 indicate some of the decomposition pathways resulting from autoionization or Auger decay (in 10^{-14} secs) of the inner-shell hole states represented in Fig. 25. (The probability of decay of a K-shell hole in carbon or nitrogen by X-ray emission is $\leqslant 1\%$.[179]) It can be seen that autoionization of K-shell excited states leads mainly to singly charged fragment and molecular ions, whereas the K-continuum states

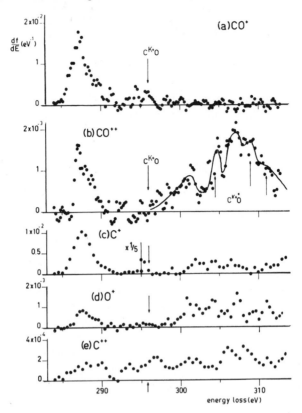

Figure 25. Carbon K-shell oscillator-strength spectra of CO. In C^+ spectrum (c) above 295 eV scale has been changed by a factor of 5.[173]

principally undergo Auger decay to CO^{++} or two fragment ions. Van der Wiel and El-Sherbini[173] summed the contributions from the large peak in the discrete part of the spectrum and obtained lower limits (because of the possibility of high-kinetic-energy fragments) for the oscillator strength of 4×10^{-2} eV^{-1} and 7×10^{-2} eV^{-1} per each K-electron for nitrogen in N_2 and carbon in CO, respectively. Krause and Wiulleumier[180] had earlier estimated a value of 7×10^{-2} eV^{-1} for the main $N(K)$ absorption.

It can be seen from Fig. 25 that there is no evidence for any significant increase in oscillator strength at the K-ionization edge in the CO^{++} spectrum, and this, in part, may well be the result of PCI (see Refs. 152–155 and also section V.B). Similar results were obtained for nitrogen,[173] but the interpretation is less certain because of the superimposition of N^+ and N_2^{++}. More definitive quantitative work on the inner-shell

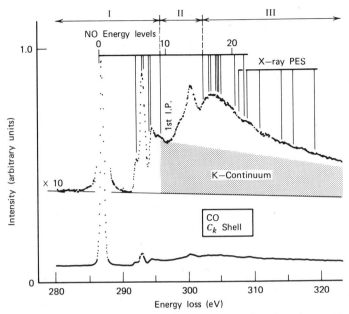

Figure 26.   Carbon K-shell energy-loss spectrum of carbon monoxide.[176]

oscillator strengths with total fragment ion collection (irrespective of kinetic energy) has recently been reported by Kay et al.[206,207] using the electron-impact simulation method. Absolute oscillator strengths for the carbon and nitrogen K-shell spectra[206] of CO and N_2 have been measured as well as those for the formation of the various charged products[207] arising from decay of the hole states. A comparison is made with the resonance model proposed by Dehmer and Dill.[178]

Recently, using fast-electron impact, Wight et al.[25] have reported electron–ion coincidence studies of dipole excitation, ionization, and fragmentation of N_2 and CO in the 10–60 eV region. This work has been done with improved instrumentation[24] in which a quantitative collection has been made of the fragment ions regardless of their kinetic energy. The various types of data were measured using the methods already discussed in the case of hydrogen. Oscillator strengths for photoabsorption were obtained from the forward scattering energy-loss spectra, and the result for N_2 has been shown in Fig. 12. The measurements were made absolute by normalizing to optical data at a single energy. Over the entire range the spectra for CO and N_2 are in excellent agreement with the most recent photoabsorption work using synchrotron radiation.[174]

The parent-ion oscillator-strength spectra[25] for N_2^+ and CO^+ are shown in Figs. 27 and 28, respectively. The absolute scales were obtained as follows. Using the relative ion abundances from the TOF mass spectra, the fractional contributions of dissociative processes and double ionization were added to the (relative) parent-ion oscillator strength to give a relative total ionization spectrum. The ratio of this relative total ionization to the total absorption (previous paragraph) is found to become constant above 20 eV. Hence this ratio, the relative ionization efficiency, is normalized to unity in that range. The parent-ion oscillator strengths (Figs. 27 and 28) are quite close to those reported earlier by El-Sherbini and Van der Wiel.[108] Onsets in both parent-ion spectra occur at the thresholds for ion formation in the X, A, and B one-electron states. Another structure in this region has been ascribed to autoionizing Rydberg states.[25]

The fragment-ion oscillator strengths may be generated from the fragment-ion ratios and parent-ion oscillator strengths. The results are shown in Figs. 29 and 30, although it should be noted that N_2^{2+} and the dissociative double ionization process $N^+ + N^+$ will also contribute to $(m/e) = 14$ at higher energy losses. In both cases the earlier data[108] on ions of low kinetic energy are included, as is the estimated contribution of N_2^{2+} (thermal energy only). The gross features of the N^+ spectrum (Fig. 29) suggest that at least three electronic states of N_2^+ are responsible for the observed fragmentation. These have been indicated in Fig. 31, in which the lowest state at 25 eV is the well-known "two-electron" state designated as

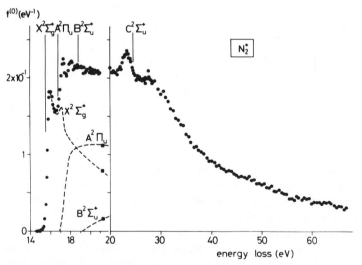

Figure 27. Oscillator strength of N_2^+ formation.[25]

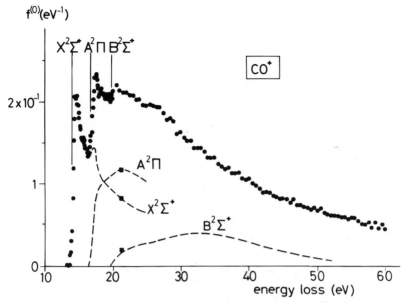

Figure 28. Oscillator strength of CO^+ formation.[25]

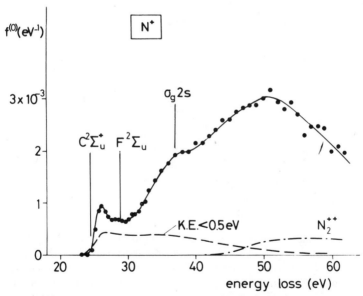

Figure 29. Oscillator strength for dissociative of N_2: ●, From Wight et al.;[25] broken and chain curves, El-Sherbini and Van der Wiel[108] using mass spectrometer with strong discrimination against ions with more than 0.5 eV of kinetic energy.

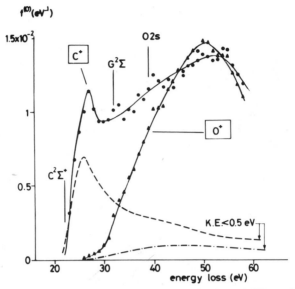

Figure 30. Oscillator strengths for dissociation ionization of CO: ●, From Wight et al.;[25] broken and chain curves, El-Sherbini and Van der Wiel[108] using mass spectrometer with strong discrimination against ions with more than 0.5 eV of kinetic energy.

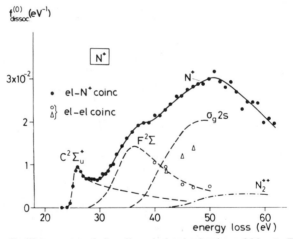

Figure 31. Oscillator strength for dissociative ionization of N_2: ●, From Wight et al.;[25] broken curves, partial oscillator strengths for three dissociative ion states, with C state dissociative only from $v=3$ up, which makes total C-state spectrum approximately 10% higher than spectrum shown here. The three broken curves combined equal total spectrum and give a best fit with other data points; dissociative double ionization ($N^+ + N^+$) is expected to set in at 48 eV; ○ and △, from branching ratios measured in an electron–electron coincidence experiment.[171] Chain curve (N_2^{2+}) from El-Sherbini and Van der Wiel.[108]

$C^2\Sigma_u^+$. The two higher onsets indicated in Fig. 31 correspond to the maxima of the only other two prominent bands recorded in that region in the (e, 2e) PES simulation experiment[171] (see following paragraph). The breakdown picture (Fig. 31) has been described[25] as follows.

The C state is known to decay mainly through predissociation for vibrational levels with $v > 3$[181]; however, since the integrated FC factors for the levels $v < 3$ amount to less than 10%,[182] the partial oscillator strength for the C state is assumed to be equal to its major, dissociative fraction. Since the FC zone extends only 3 eV above the lowest dissociation limit, no fragments of more than 1.5-eV energy are produced, and actually most ions have considerably lower kinetic energy. This is in keeping with the fact that even in the earlier[108] low-ion-energy work (broken curve of Fig. 29), a significant fraction of the intensity at the first onset at 24 eV was observed. The two higher states clearly produce N^+ ions of considerably kinetic energy, in view of the large discrimination against energetic fragments in our previous result.[108]

To assess the relative contribution of the three electronic states as a function of the energy loss, use is made of PES branching ratios at 40.8 eV. For the C state, only a single value of 1% at 90° photoelectron angle is available.[183] In general, the branching ratios of the latter authors[183] for the higher states appear to be rather too low when compared with other work.[171,184] By comparison with the C state in CO (see later), a value of 2% is adopted for the branching ratio of the C state in N_2 at 40.8 eV. This percentage, multiplied by the total ionization oscillator strength at 40.8 eV, gives a value for the C-state partial oscillator strength. On this basis the onset can reasonably be extrapolated to produce a spectrum for the C state (dashed curve in Fig. 31). Given that curve, the difference between the C state and the total for dissociative processes must be the result of $F^2\Sigma^+$ ionization up to an energy of 36 eV, where the next higher state sets in. Photoelectron spectroscopy branching values at 40.8 eV for these two states show considerable discrepancies, probably because of large and uncertain background corrections. For the 29eV ($F^2\Sigma^+$) state the values range from 3.3%[183] to 6.3%,[184] whereas for the state near $\sim$36 eV these same authors report values of 2.2% and 2%, respectively. Since Wight et al.[25] observe a total fraction of dissociative ionization at 40.8 eV equal to 19%, of which only 2% results from the C state, neither of the PES measurements is compatible with the present work (see Fig. 31). The (e 2e) measurements,[171] however, give branching ratios in the 40–50-eV region, which account reasonably well for the correct amount of dissociation. It should be noted that the (e, 2e) experiment (discussed later) was performed at 90° ejection angle of the secondary electrons. This leads to a dependence of the results on the angular distribution parameters.[94] However, if the β values for the various states are not very different, the effect on the branching ratios will not be large. In any case the (e, 2e) measurements are in excellent agreement with magic angle PES[172] as well as β-dependent studies using synchrotron radiation.[18] From the (e, 2e) branching ratios, a set of spectra for the $F^2\Sigma^+$ and $(\sigma_g 2s)^{-1}$ states can be constructed, which are consistent with the total dissociation oscillator strength, whereas particularly the $F^2\Sigma^+$ spectrum around 36 eV shows a reasonable continuation of its threshold region. Around 48 eV dissociative double ionization sets in,[185] which means that the three-state breakdown picture must be adapted here.

In the case of CO (see Fig. 32, which shows the sum of the C^+ and O^+ oscillator strengths) a rather more complete picture of the dissociative ionization breakdown can be constructed since the $C^2\Sigma$ state oscillator strength has been measured in the (e, 2e) experiment[171] in addition to other higher states. In Fig. 32 the total $C^+ + O^+$ oscillator strength measured in the (e, ion) experiment[25] is compared with the (e, 2e) partial oscillator

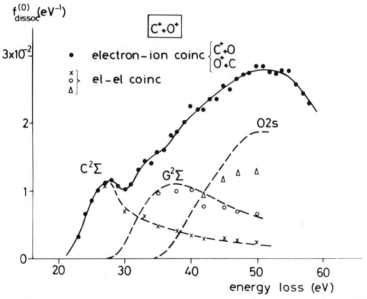

Figure 32.   Oscillator strength for dissociative ionization of CO: ●, From Wight et al.;[25] sum of $C^+ + O$ and $O^+ + C$; broken curves, partial oscillator strengths for three dissociative ion states (assuming C state to be entirely dissociative), which equal total spectrum when added up and give a best fit with other data points; dissociative double ionization ($C^+ + O^+$) is expected to set in at about 50 eV; x, O, and △ from branching ratios of electron–electron coincidence experiment.[171]

strengths for the higher electronic states.[171] Above 50 eV part of the C^+ intensity will be caused by the double dissociative process $C^+ + O^+$ (no contribution to the O^+ will be observed since the faster C^+ will stop the TAC for these events). It can be seen (Fig. 32) that there is excellent quantitative agreement.

Using the results of Fig. 32, Wight et al.[25] worked out the branching of each of the three electronic states over the two dissociation channels: $C^+ + O$ or $O^+ + C$. The C state apparently decays solely into the first channel. The remaining oscillator strength for C^+ and O^+ formed from the two higher states can be divided into fractions for each state separately by making use of the fact that the branching of any particular state over the dissociation channels becomes constant with energy loss at energies above the FC region of that state. Thus it can be shown that the data are consistent to within 5% over the whole energy range, with the following breakdown pattern: the $G\,^2\Sigma$ state produces C^+ and O^+ in the fractions 0.54 and 0.46, whereas for the $(O2s)^{-1}$ state these fractions are 0.39 and 0.61. This result is completely at variance with the conclusion of Gardner and Samson[184] that the $(O2s)^{-1}$ state produces C^+ fragments only, thus demonstrating the limitation of their method of ion-energy analysis without mass analysis.

A number of quantitative PES studies of branching ratios as a function of energy for N_2 and CO have been made by both conventional studies as well as by the (e, 2e) simulation technique. The early PES studies by Blake et al.[186] on O_2, H_2O, and N_2 and by Bahr et al.[187] on NO, N_2O, CO_2, NH_3, and CO using a conventional discharge light source and a Lozier-type tube provided branching ratios in the very limited energy range from threshold up to 20 eV. Partial cross-section measurements over a similar energy range have been reported by Schoen for O_2, CO, and N_2.[188] In 1973 Van der Wiel and Brion[94] reported the first measurements using the (e, 2e) simulation of PES. Relative populations for the X, A, and B states of CO^+ were given up to energies of 50 eV. These branching ratios did not take into account the contribution from the higher one and two electron states that have been included in the more recent (e, 2e) work by Hamnett et al.[171] Typical binding-energy spectra from the work of Hamnett et al.[171] are shown in Fig. 33. Branching ratios are shown in Fig. 34 for six states of CO^+ and five states of N_2^+ (where the C state could not be resolved). These values were derived from the (e, 2e) measurements and involved corrections for electron transmission (see Section IV) and deconvolution (in the case of N_2). It should be noted that different state designations were used by Wight et al.[25] and Hamnett et al.[171] In the case of CO^+, W^{171} is equivalent to $G\,^2\Sigma^{25}$ and for N_2^+, Z is equivalent to $F\,^2\Sigma$. Branching ratios for CO and N_2 measured by Samson and Gardner[172] using magic-angle

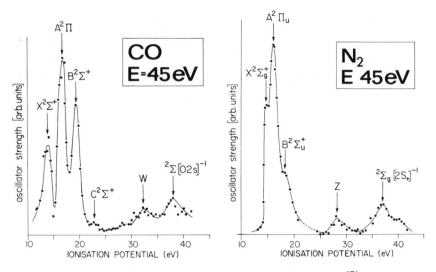

Figure 33. Binding energy spectra of CO and N_2.[171]

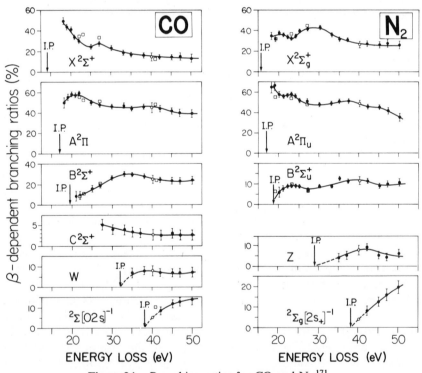

Figure 34. Branching ratios for CO and N_2.[171]

PES at a selection of wavelengths up to 27 eV and at 40.8 eV are in excellent agreement with the (e, 2e) work depicted in Fig. 34. It should be noted that the (e, 2e) work is at an ejection angle of 90°, and thus the branching ratios will include a β-dependent term. The β dependence of the (e, 2e) experiment differs from that in conventional PES, and a full discussion of this has been given by Hamnett et al.[23] The results of Samson and Gardner[172] for the branching ratios (X, A, and B states only) of CO^+ are shown in Fig. 35 in comparison with earlier (e, 2e) work.[94] These results[172] give a much more detailed picture near to the thresholds. From the agreement between the magic-angle PES[172] and the 90° (e, 2e) experiments, we may conclude that the variation of β with energy is very similar for at

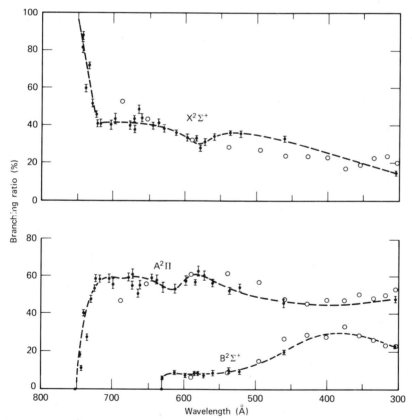

Figure 35. Branching ratios as a function of wavelength for photoionization of CO to CO^+ ($X^2\Sigma^+$), ($A^2\pi$), and ($B^2\Sigma^+$) states.[172] Open circles represent (e, 2e) data due to Van der Wiel and Brion.[94]

least the X, A, and B states of CO^+ and N_2^+. Theory[23] shows that near 21 eV the (e,2e) cross sections are independent of β, and excellent agreement is obtained at this energy with 584-Å PES.[25] Using oscillator strengths for total photoionization obtained in the (e–ion) experiment[25] or the total photoabsorption[174] in the region where the ionization efficiency is unity,[25] it is possible to generate the partial oscillator strengths for ionization from the branching ratios. Figure 36 shows the partial oscillator-strength results of Hamnett et al. obtained in this way using the (e,2e) method.[171]

A characteristic feature of the peaks with binding energy greater than 25 eV in Fig. 33 in both spectra is their extreme broadness. This suggests that the corresponding ion states are dissociating rapidly, which is known from the recent ion–electron coincidence measurements made by Wight et al.[25] Figure 37 (cf. Fig. 32) shows that up to 45 eV the total yield of fragment ions[25] from CO correlates well with the total intensity of these broad "inner" peaks measured in the (e,2e) experiment,[171] as both sets of data are on an absolute oscillator-strength scale. Given the considerable uncertainties involved in the experiment, both in the width of these peaks

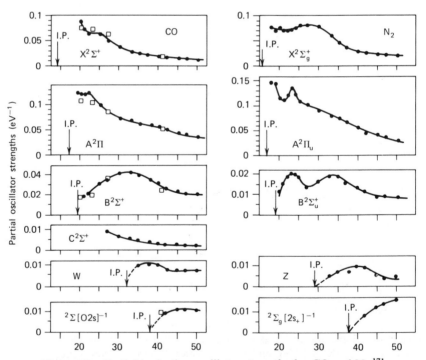

Figure 36. Partial ionization oscillator strengths for CO and N_2.[171]

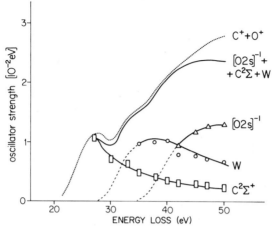

Figure 37. Partial ionization oscillator strengths for the three higher states of CO$^+$.[171] Sum (solid line) is compared with total C$^+$+O$^+$ ion yield.[25]

and the background near threshold, the agreement between the two sets of data for CO is very satisfactory and again suggests that the β dependence of the data is not very marked.

Very recently the photoelectron spectroscopy of CO and N$_2$ has been studied by Plummer et al.[18] using radiation from the Wisconsin synchrotron storage ring in the energy range 14–40 eV. In this work the photon beam was directed at 90° to the axis of a cylindrical mirror analyzer, leading to a complex dependence[18] on the values of β. This effect was neglected, and the resulting branching ratios for N$_2$ and CO are shown in Fig. 38 in comparison with (e, 2e) data of Hamnett et al.[171] The agreement is generally very good, confirming the conclusions regarding the variation of β values. Plummer et al.[18] did not attempt to record the contributions from levels above the B state, and this largely accounts for the discrepancies with the (e, 2e) data at the higher energies. The neglect of the higher states is particularly evident in the CO$^+$ branching ratios in Fig. 38. The partial cross sections derived by Plummer et al.[18] are likewise in similar agreement with the (e, 2e) work[171] and the conventional PES data of Gardner and Samson.[172] Similar results for N$_2$ have been reported by Woodruff and Marr.[205] The recent MS–Xα calculations by Davenport[54] give some degree of quantitative agreement with the measured partial cross sections for formation of CO$^+$ and N$_2^+$ in the X, A, and B states.

The $B\,^2\Sigma$ state of CO$^+$ has been the subject of several independent studies. The partial oscillator strength for CO$^+(B\,^2\Sigma)$ can be derived from the following experiments: (1) the (e, 2e) experiments of Van der Wiel and

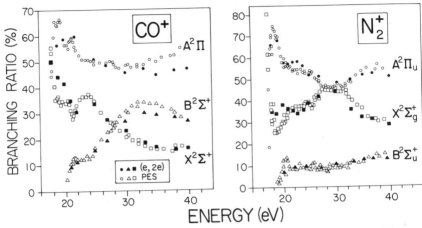

Figure 38. Comparison of synchrotron PES[18] (open points) and (e, 2e)[171] (solid points) branching ratios for CO and N_2.

Brion[94] and of Hamnett et al.[171] (2) the PES experiments of Samson and Gardner[172] and Plummer et al.[18] (3) the photofluorescence experiments of Judge and Lee[189] and the recent extended measurements by Lee et al.[190] and (4) a different type of measurement, the triple coincidence experiment by Backx et al.[191] In this work coincidences were recorded between electrons of variable energy loss, CO^+ ions, and photons from the process $CO^+(B^2\Sigma \rightarrow X^2\Sigma)$. The results of the various experiments shown in Fig. 39 indicate excellent quantitative agreement between the five different electron-spectroscopic methods. However, the photofluorescence measurements

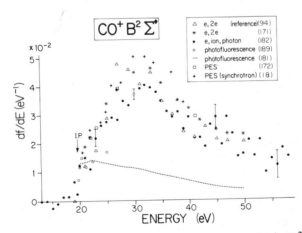

Figure 39. Oscillator strength for production of CO^+ ($B^2\Sigma^+$).

of Lee et al.[190] are very different above about 21 eV. The reason for this discrepancy is at present not clear.

E. Methane

Oscillator strengths for the methane molecule have been studied both theoretically[192,193] and also fairly extensively experimentally by electron-impact experiments. Backx and Van der Wiel[102] have reported electron-ion coincidence measurements of the oscillator strengths for fragmentation of CH_4 using methods similar to those already described for H_2, N_2, and CO. Results have been given[102] for formation of CH_4^+, CH_3^+, CH_2^+, CH^+, C^+, and H^+. Because of the simplicity of the term scheme, it was possible to derive the optical oscillator strengths for the ionization processes for the $(1t_2)^{-1}$ and $(2a_1)^{-1}$ states from the observed constant fragmentation ratios arising from each electronic state. This procedure involved three assumptions, namely;

1. The vibrational population of CH_4^+ is independent of "photon" energy for a given electronic state above the Franck–Condon region.
2. The electronic part of the transition matrix element is treated as being effectively constant for the different vibrational levels.
3. Contributions from *autoionizing* states are small.

On the basis of these assumptions, breakdown of the $(1t_2)^{-1}$ state should result in constant abundance ratios in the range 17–23 eV, as was observed.[102] These abundances were then extrapolated and subtraction of the $(1t_2)^{-1}$ contributions enabled computation of the part that arises from the $(2a_1)^{-1}$ state. These results are shown in Fig. 40, together with recent measurements by the (e, 2e) method. In this case the (e, 2e) measurements were made at the "magic" angle of 55°.[194] Figure 41 shows a typical experimental arrangement for the (e, 2e) spectrometer. Use of the "magic" angle permits measurement of branching ratios and partial oscillator strengths independent of the asymmetry parameter β. There is excellent agreement between the two sets of data (see Fig. 40), essentially justifying the assumptions made in the (e–ion) experiment. The calculations of Dewar et al.[195] using ab initio wave functions for the initial state and the plane-wave approximation for the final state are seen (Fig. 40) to give reasonable agreement for the $(1t_2)^{-1}$ state at higher energies but to be rather unsatisfactory for the $2a_1$ orbital. Typical binding-energy spectra for CH_4 obtained in the magic-angle (e, 2e) spectrometer are shown in Figure 42. Backx et al.[24] have also performed an (e, 2e) type (total) experiment for CH_4 using the electron–ion coincidence spectrometer. For these measurements the potential configuration for the ion–extraction and analysis channel was inverted to permit total collection of "photo" electrons. This

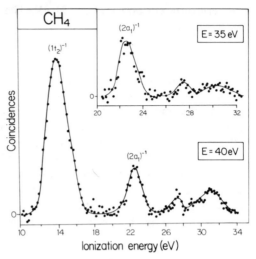

Figure 40.   Spectra of electron ejected from CH_4 at 54.7°, recorded in coincidence with scattered (3.5 keV) electrons at energy losses of 35 and 40 eV. Correction has been made for transmission of ejected electron analyzer. The 40-eV spectrum is a composite of two overlapping runs (10 to 26 eV and 20 to 34 eV).[194]

was first used in an ingenious method for calibrating the efficiency of the ion-extraction and analysis channel. Using helium, the ratio of total ionization as measured by He^+ and by total ejected electrons was determined using the two potential configurations for (e–ion) and (e–2e) coincidence measurements as a function of energy loss. For the thermal helium ions, no change of transmission could take place, whereas the energy of the ejected electrons increases with energy loss. The ratio was found to be constant up to 20 eV kinetic energy and thus indicates a constant transmission for ion-fragmentation studies in this energy range.

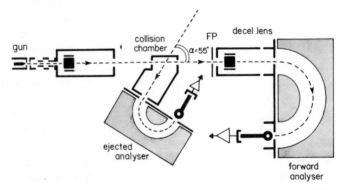

Figure 41.   Schematic of magic-angle (e, 2e) spectrometer.[209]

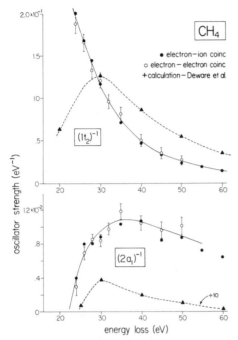

Figure 42. Partial oscillator-strength spectra for $(1t_2)^{-1}$ and $(2a_1)^{-1}$ states of CH_4^+; $\bigcirc$, e, 2e)[194]; $\bullet$ (e, ion), coincidence;[102] $\blacktriangle$, calculation.[195]

The absorption oscillator strength for CH_4 was obtained[24] from the forward scattering. The absolute ionization efficiency was then obtained by normalizing the quotient of the relative oscillator strength for ionization and absorption on a few absolute points obtained by a calibration procedure using a mixture of helium and neon. The ionization efficiency is unity from about 5 eV above threshold to the limit of the data at 80 eV.

By lowering the extraction field across the collision chamber, Backx et al.[24] were able to collect essentially only the zero energy electrons (i.e., in this mode the extraction system is used as a simple energy analyzer). The zero-energy electrons were detected in coincidence with the energy-loss electrons to provide a simulation of threshold photoelectron spectroscopy. The discrimination against non-zero-energy electrons was further enhanced by setting a narrow time window so as to pass only the slowest electrons. Although no absolute measurements were made, it is of interest to consider the relative threshold oscillator-strength spectrum since this will directly reflect the distribution of Franck–Condon factors (unlike the case in normal PES, which is a nonthreshold method). An analysis of the data gave good agreement with the threshold PES data of Stockbauer and Inghram.[196] Figure 43 shows the results of a threshold scan over an energy-loss range up to 35 eV. In addition to the $(1t_2)^{-1}$ and $(2a_1)^{-1}$

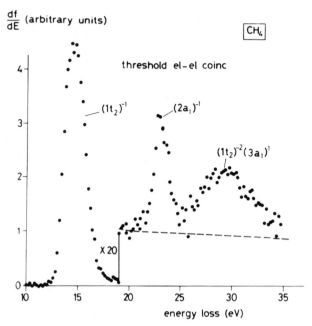

Figure 43. Dipole oscillator strengths for formation of threshold electrons from CH_4.[24]

one-electron states, a third broad band is seen centered at 29 eV. A similar band has been observed in a binary (e, 2e) study of CH_4 by Weigold et al.[197] A later, somewhat higher resolution binary (e, 2e) experiment by Hood et al.[104] shows two peaks in this band, in agreement with the earlier dipole (e, 2e) "photoelectron" spectra reported by Van der Wiel et al.[194] These two higher peaks can be attributed to two-electron processes.

F. Other Molecules*

A number of other photoelectron and (e, 2e) experiments have been reported giving limited data for branching ratios and for partial cross section for other molecules. Using the (e, 2e) method, Branton and Brion[84] have reported oscillator strengths for photoabsorption of H_2O up to 100 eV, normalized by the TRK sum rule. Good agreement was obtained with optical absorption studies. Some preliminary data were also obtained for the partial oscillator strengths for ionization of four valence orbitals of H_2O at a 90° ejection angle. However, the ionization oscillator strengths are in error because of the use of an inadequate transmission function for the ejected channel and a variable electron beam profile.[14] Furthermore,

*These include H_2O, NH_3, CO_2, N_2O, SF_6, and O_2.

there is an error in the reported [84] absorption oscillator strengths (see also Tan et al.[208]). Lassettre and White,[198] using 500-eV electrons, have measured values of continuum absorption oscillator strengths for H_2O by extrapolating generalized oscillator strengths to $K^2 = 0$. The data were normalized by using elastic scattering measurements.

Very recently Tan et al.[208] using the magic-angle (e, 2e) and (e, e + ion) spectrometers have reported absolute oscillator strengths for the photoabsorption ($\leqslant 300$ eV), partial and total photoionization ($\leqslant 60$ eV), and fragmentation ($\leqslant 60$ eV) of H_2O. The photoionization efficiency is also given. The results of the two experiments have been combined to give a detailed quantitative picture of the dipole-induced breakdown of the H_2O molecule via the four valence states of H_2O^+. Full details and tabulations of data are given by Tan et al.[208] Blake and Carver,[186] have reported partial ionization cross sections for H_2O, O_2, and N_2 at a series of energies up to 21 eV. Similar studies of NO, N_2O, CO, CO_2, and NH_3 have been reported by Bahr et al.[187] Preliminary (e, 2e) studies of NH_3[32] have given absorption oscillator strengths (see Fig. 13) and approximate data for the β-dependent partial oscillator strengths for ionization from the three valence orbitals. Subsequently, using the improved (e, 2e)[209] and (e, e + ion)[210] spectrometers, detailed studies of the photoabsorption, photoionization, and fragmentation of NH_3 have been made. A quantitative picture of the dipole-induced breakdown has been given.[209,210] Partial oscillator strengths for photoionization of CO_2 and N_2O have also recently been obtained by the (e, 2e) method[211] in the energy range up to 60 eV. These results are interesting, not least because of the large (30%) contribution from multiple electron processes at the higher energies. Gustafsson et al.[212] have made partial photoionization cross-section measurements for CO_2 up to 40 eV using synchrotron radiation.

Nishikawa and Watanabe[192] have calculated relative oscillator strengths for absorption and ionization of NH_3, H_2O, and HF. Most recently Gustafsson has reported an interesting study of the partial photoionization cross sections of SF_6 using synchrotron radiation.[212] Resonances, observed in a number of channels, are used to assign the symmetries of peaks in the photoelectron spectrum. Hitchcock et al.[213] have reported oscillator strengths for absorption and fragmentation associated with sulfur $2p$ excitation in SF_6 using fast-electron-impact simulation methods. Partial photoionization cross sections for O_2^+ have been measured by Samson et al.[214] using PES with a many-line source. Dipole (e, 2e) studies of O_2[219] and H_2S[220] have also been reported.

VI. CONCLUSIONS

Electron spectroscopy has become a useful tool for measuring optical oscillator strengths for absorption and ionization (fragment partial and total). In principle, photoelectron spectroscopy is capable of producing partial oscillator strengths for photoionization, but only a limited amount of work has been done to date. The virtual photon field provided by fast-electron impact has been shown to provide a usable continuum source of excitation energies. Use of the Bethe–Born relation has made it possible to obtain optical oscillator strengths at modest energy resolution from electron-impact measurements employing electron spectroscopy together with coincidence and mass-analysis techniques. The resulting data are of comparable quality to the optical data currently attainable.

ACKNOWLEDGMENTS

We gratefully acknowledge financial support provided by the National Research Council of Canada, The North Atlantic Treaty Organization, and the Killam Foundation (Postdoctoral fellowship to A. Hamnett) for the writing of this review and also for that part of the experimental work performed in the authors' laboratory. We are grateful to those who have given permission to quote their published work. In particular, we are deeply indebted to Dr. Marnix Van der Wiel and his co-workers at the FOM Institute, Amsterdam, who have contributed so much to this field, not only at their own institute, but also in our own laboratory in Vancouver. We would also thank Dr. Van der Wiel for stimulating our own interest in the subject and for his many helpful comments and suggestions during the preparation of this chapter.

REFERENCES

1. J. E. Hesser, *J. Chem. Phys.*, **4**, 2518 (1968).

2. D. A. Shirley, *Electron Spectroscopy*, North-Holland, Amsterdam, 1972; A. Hamnett and A. F. Orchard, in P. Day, Ed., *Electronic Structure and Magnetism of Inorganic Compounds*, Vol. 3, The Chemical Society, (London, 1974); K. Siegbahn et al., *ESCA Applied to Free Molecules*, North-Holland, Amsterdam, 1969; C. E. Brion, in A. Maccoll, Ed., *MTP International Review of Science*, Vol. 5, Butterworths, London, 1972; E. N. Lassettre and A. Skerbele, *Meth. Exp. Phys.*, **3B**, 868 (1974); T. A. Carlson, *Photoelectron and Auger Spectroscopy*, Plenum, New York, 1975.

3. J. W. Cooper and R. N. Zare, *Lect. Theor. Phys.* **11C**, 317 (1969).

4. H. Bethe, *Ann. Phy. (Leipz.)*, **5** (5), 325 (1930).

5. M. Inokuti, *Rev. Mod. Phys.*, **43**, 297 (1971).

6. Y.-K. Kim, *Phys. Rev.* **A6**, 666 (1972).

7. J. Franck and G. Hertz, *Verh. Deutsch Phys. Ges.*, **16**, 10 (1914).

8. W. F. Miller and R. L. Platzmann, *Proc. Phys. Soc. (Lond.)*, **70**, 299 (1957).

9. E. N. Lassettre, *Radiation (Res. Suppl.)*, **1**, 530 (1959).

10. E. N. Lassettre and A. Skerbele, *Meth. Exp. Phys.*, **3B**, 868 (1974).

11. E. N. Lassettre, in C. Sandorfy, P. J. Ausloos, and M. B. Robin, Eds. *Chemical*

Spectroscopy and Photochemistry in the Vacuum Ultraviolet, Reidel, Dordrecht–Boston, 1974.

12. M. J. Van der Wiel, in B. C. Cobic and M. V. Kurepa, Eds., *Invited Lectures and Progress Reports, VIII ICPEAC*, Institute of Physics, Belgrade, 1973, p. 417.

13. C. Backx and M. J. Van der Wiel, in E. E. Koch, R. Haensel, and C. Kunz, Eds., *Proceedings of the IVth International Conference on Vacuum Ultraviolet Radiation Physics*, Hamburg, 1974, p. 137. Pergamon/Vieweg.

14. C. E. Brion, *Radiat. Res.*, **64**, 37 (1975).

15. M. J. Van der Wiel, in F. J. Wuilleumier, Ed., *Photoionization and Other Probes of Many-electron Interactions*, Plenum, New York, NATO-ASI series B, 1976, Part I, 187–198, Part II, 198–208.

16. M. J. Van der Wiel, *Radiation Research Biomedical, Chemical and Physical Perspectives*, Academic, New York, 1975 p. 205.

17. M. J. Van der Wiel, in T. R. Govers and F. J. de Heer, eds., *The Physics of Electronic and Atomic Collisions*, North-Holland, Amsterdam, 1972, p. 140.

18. W. Plummer, T. Gustafsson, W. Gudat, and D. E. Eastman, *Phys. Rev.*, **A15**, 2339 (1977).

19. R. A. Bonham and M. Fink, *High Energy Electron Scattering*, Reinhold, New York, 1974.

20. J. Geiger, *Z. Phys.*, **181**, 413 (1964).

21. J. A. R. Samson, *Adv. Atom. Molec. Phys.*, **2**, 177 (1966).

22. A. S. Davydov, *Quantum Mechanics*, Pergamon, Oxford, 1965.

23. A. Hamnett, W. Stoll, G. R. Branton, C. E. Brion, and M. J. Van der Wiel, *J. Phys.*, **B9**, 945 (1976).

24. C. Backx, G. R. Wight, R. R. Tol, and M. J. Van der Wiel, *J. Phys.*, **B8**, 3007 (1975).

25. G. R. Wight, M. J. Van der Wiel, and C. E. Brion, *J. Phys.*, **B9**, 675 (1976).

26. N. F. Mott and H. S. W. Massey, *Theory of Atomic Collisions*, Clarendon, Oxford, 1965.

27. S. T. Hood, I. E. McCarthy, P. J. O. Teubner, and E. Weigold, *Phys. Rev.*, **A9**, 260 (1974).

28. S. Cvejanovic and F. H. Read, *J. Phys.*, **B7**, 1180 (1973).

29. U. Fano and J. W. Cooper, *Rev. Mod. Phys.*, **40**, 441 (1968).

30. I. C. Walker, *Chem. Soc. Rev.*, **3**, 467 (1974).

31. C. Backx, R. R. Tol, G. R. Wight, and M. J. Van der Wiel, *J. Phys.*, **B8**, 2050 (1975).

32. M. J. Van der Wiel and C. E. Brion, *J. Electron Spectrosc.*, **1**, 443 (1973).

33. E. N. Lassettre, A. Skerbele, and M. A. Dillon, *J. Chem. Phys.*, **50**, 1829 (1969).

34. P. Ikelaar, M. J. Van der Wiel, and W. Tebra, *J. Phys.*, **E4**, 102 (1971).

35. M. Peshkin, *Adv. Chem. Phys.*, **18**, 1 (1970).

36. A. D. Buckingham, B. J. Orr, and J. M. Sichel, *Phil. Transact.*, **A268**, 147 (1970).

37. J. W. Rabalais and T. P. Debies, *J. Electron Spectrosc.*, **5**, 847 (1974).

38. W. Theil and A. Schweig, *Chem. Phys. Lett.*, **16**, 409 (1972).

39. Y. Va. Veldre, A. V. Lyash, and L. L. Rabik, *Opr. Spectrosc. (USSR)*, **19**, 182 (1965).

40. P. A. Cox, S. Evans, and A. F. Orchard, *Chem. Phys. Lett.*, **13**, 386 (1972).

41. K. L. Bell and A. E. Kingston, *Adv. Atom. Molec. Phys.*, **10**, 53 (1974).

42. B. Ritchie, *J. Chem. Phys.*, **64**, 3050 (1976).

43. H. C. Tuckwell, *Proc. Phys. Soc.*, **3**, 293 (1970); *J. Quant. Spectrosc. Radiat. Transf.*, **10**, 653 (1970).

44. Y. Itikawa, *J. Electron Spectrosc.*, **2**, 125 (1973).

45. G. B. Shaw and R. S. Berry, *J. Chem. Phys.*, **56**, 5808 (1972).

46. F. Hirota, *J. Electron Spectrosc.*, **9**, 149 (1976).

47. F. H. M. Faisal, *J. Phys.*, **B3**, 636 (1970).

48. P. G. Burke and A.-L. Sinfailam, *J. Phys.*, **B3**, 641 (1970).

49. P. G. Burke and N. Chandra, *J. Phys.*, **B5**, 1696 (1972).

50. P. G. Burke, N. Chandra, and F. Gianturco, *J. Phys.*, **B5**, 2212 (1972).

51. D. Dill and J. L. Dehmer, *J. Chem. Phys.*, **61**, 692 (1974).

52. J. Siegel, D. Dill, and J. L. Dehmer, *J. Chem. Phys.*, **64**, 3204 (1976).

53. J. L. Dehmer and D. Dill, *Phys. Rev. Lett.*, **36**, 213 (1976).

54. J. W. Davenport, *Phys. Rev. Lett.*, **36**, 945 (1976).

55. M. Ya. Amusia and N. A. Cherepkov, *Case Stud. Atom. Phys.*, **5**, 47 (1975).

56. L. Armstrong, *J. Phys.*, **B7**, 2320 (1974).

57. D. J. Rowe and C. Ngo-Trong, *Rev. Mod. Phys.*, **47**, 471 (1975).

58. A. F. Starace and L. Armstrong, *Phys. Rev.*, **A13**, 1850 (1976).

59. H. P. Kelly and A. Ron, *Phys. Rev.*, **A5**, 168 (1972).

60. T. E. H. Walker and H. P. Kelly, *Chem. Phys. Lett.*, **16**, 511 (1972).

61. W. Steckelmacher, *J. Phys. E*, **6**, 1061 (1973); K. D. Sevier, *Low Energy Electron Spectrometry*, Wiley-Interscience, New York, 1972.

62. E. M. Purcell, *Phys. Rev.*, **54**, 818 (1938).

63. A. L. Hughes and V. Rojansky, *Phys. Rev.*, **34**, 284 (1929). A. L. Hughes and J. H. McMillan, *Phys. Rev.*, **34**, 291 (1029).

64. H. Z. Sar-El, *Rev. Sci. Instrum.*, **41**, 561 (1970).

65. M. J. Van der Wiel, *Physica*, **49**, 411 (1970).

66. D. S. C. Yee, W. B. Stewart, C. A. McDowell, and C. E. Brion, *J. Electron Spectrosc.*, **7**, 93 (1975).

67. E. Harting and F. H. Read, *Electrostatic Lenses*, Elsevier, Amsterdam, 1976.

68. D. W. O. Heddle, Joint Institute for Laboratory Astrophysics, University of Colorado, report No. JILA 104, 1970.

69. C. E. Kuyatt and J. A. Simpson, *Rev. Sci. Instrum.*, **38**, 103 (1967).

70. F. H. Read, J. Comer, R. E. Imhof, J. N. H. Brunt, and E. Harting, *J. Electron Spectrosc.*, **4**, 293 (1974).

71. H. G. Noller, H. D. Polaschegg, and H. Schillalies, *J. Electron Spectrosc.*, **5**, 705 (1974).

72. W. C. Tam and C. E. Brion, *J. Electron Spectros.*, **3**, 82 (1974).

73. G. R. Wight, C. E. Brion, and M. J. Van der Wiel, *J. Electron Spectrosc.*, **1**, 457 (1972/73).

74. M. Nakamura et al., *Phys. Rev.*, **178**, 80 (1969).

75. H. Schmoranzer, H. F. Wellenstein, and R. A. Bonham, *Rev. Sci. Instrum.*, **46**, 8 (1975).

76. P. G. Burke and D. D. McVicar, *Proc. Phys. Soc.*, **86**, 989 (1965).

77. J. A. R. Samson, *Advances in Atomic and Molecular Physics*, Vol. 2, (Academic, New York, 1966, p. 177.

78. J. F. Lowry, D. H. Tomboulian, and D. L. Ederer, *Phys. Rev.*, **137A**, 1054 (1965).

79. H. Hotop and A. Niehaus, *Internat. J. Mass Spectrosc. Ion, Phys.*, **5**, 415 (1970).

80. J. L. Gardner and J. A. R. Samson, *J. Electron Spectros.*, **8**, 469 (1976).

81. J. L. Gardner and J. A. R. Samson, *J. Electron Spectrosc.*, **6**, 53 (1975); **2**, 267 (1973).

82. C. P. Opal, E. C. Beaty, and W. K. Peterson, *Atom. Data*, **4**, 209 (1972).

83. N. Oda, *Radiat. Res.*, **64**, 80 (1975).

84. G. R. Branton and C. E. Brion, *J. Electron Spectrosc.*, **3**, 129 (1974).

85. M. J. Van der Wiel and C. E. Brion, *J. Electron Spectrosc.*, **1**, 439 (1972/73).

86. R. H. J. Jansen, F. J. de Heer, H. J. Luyken, B. van Wingerden, and H. J. Blaauw, *J. Phys.*, **B9**, 185 (1976); R. H. J. Jansen and F. J. de Heer, *J. Phys.*, **B9**, 213 (1976).

87. A. V. Phelps, *Rev. Mod. Phys.*, **40**, 399 (1968); G. J. Schulz, *A Review of Vibrational Excitation of Molecules by Electron Impact at Low Energies*, manuscript in preparation.

88. D. G. Streets, A. W. Potts, and W. C. Price, *Internat. J. Mass Spectrosc. Ion Phys.*, **10**, 123 (1972).

89. G. J. Schulz, *Rev. Mod. Phys.*, **45**, 378 (1973).

90. R. G. Houlgate, J. B. West, K. Codling, and G. V. Marr, *J. Phys.*, **B7**, L470 (1974).

91. J. B. West, P. R. Woodruff, K. Codling, and R. G. Houlgate, *J. Phys.*, **B9**, 407 (1976).

92. J. A. R. Samson and J. L. Gardner, *J. Electron Spectrosc.*, **8**, 35 (1976).

93. K. Siegbahn, Report, Uppsala Institute of Physics, UUIP-880, 1974.

94. M. J. Van der Wiel and C. E. Brion, *J. Electron Spectrosc.*, **1**, 309 (1972/73).

95. T. N. Chang and R. T. Poe, *Phys. Rev.*, **A12**, 1432 (1975).

96. Johnston Laboratories, Inc., Maryland (USA).

97. Galileo Electro-Optics Corporation; Varian; Mullard (UK).

98. C. D. Moak, S. Datz, F. Garcia Santibanez, and T. A. Carlson, *J. Electron Spectrosc.*, **6**, 151 (1975).

99. U. Gelius and K. Siegbahn, *Faraday. Discuss. Chem. Soc.*, **54**, 257 (1972).

100. Q. C. Kessel, in E. W. McDaniel and M. R. C. McDowell, Eds. *Case Studies in Atomic Collision Physics*, Vol. I, p. 399.

101. H. Ehrhardt, in P. G. H. Sanders, Eds. *Proceedings of the Second International Conference on Atomic Physics*, Plenum, New York, 1971, p. 141.

102. C. Backx and M. J. Van der Wiel, *J. Phys.*, **B8**, 3020 (1975).

103. E. Weigold, S. T. Hood, and I. E. McCarthy, *Phys. Rev.*, **A11**, 566 (1975).

104. S. T. Hood, A. Hamnett, and C. E. Brion, *J. Electron Spectrosc.*, **11**, 205 (1977).

105. P. Ikelaar, M. J. Van der Wiel, and W. Tebra, *J. Phys.*, E *(Sci. Instru.)*, **4**, 102 (1971).

106. M. A. Dillon and E. N. Lassettre, *J. Chem. Phys.*, **62**, 2373 (1975).

107. J. H. D. Eland, *Photoelectron Spectroscopy*, Butterworths, London, 1971; J. H. D. Eland, *Adv. Mass Spectrom.*, **6**, 917 (1974); M. E. Gellender and A. D. Baker, *Internat. J. Mass Spectrom. Ion. Phys.*, **17**, 1 (1975); R. Stockbauer and M. G. Ingrham, *J. Chem. Phys.*, **62**, 4862 (1975); B. Brehm, V. Fuchs, and P. Kebarle, *Internat. J. Mass, Spectrom. Ion Phys.*, **6**, 279 (1971).

108. Th. El-Sherbini and M. J. Van der Wiel, *Physica*, **59**, 433 (1972).

109. C. Backx, G. R. Wight, R. R. Tol, and M. J. Van der Wiel, *J. Phys.*, **B8**, 3007 (1975).

110. K. G. Kessler, "Comments," *Atom. Molec. Phys.*, **1**, 70 (1967).

111. G. E. Chamberlain, S. R. Mielczarek, and C. E. Kuyatt, *Phys. Rev.*, **A2**, 1905 (1970).

112. J. P. Bromberg, *J. Chem. Phys.*, **50**, 3906 (1969).

113. J. Geiger, *Z. Phys.*, **177**, 138 (1964).

114. J. Geiger, *Z. Phys.*, **181**, 413 (1964).

115. J. Geiger and H. Schmoranzer, *J. Molec. Spectrosc.*, **32**, 39 (1969).

116. I. V. Hertel and K. J. Ross, *J. Phys.*, **B2**, 285 (1969).

117. E. N. Lassettre, A. Skerbele, and M. A. Dillon, *J. Chem. Phys.*, **50**, 1829 (1969).

118. J. A. R. Samson and R. B. Cairns, *J. Opt. Soc. Am.*, **55**, 1035 (1965).

119. L. C. Lee, R. W. Carlson, D. L. Judge, and M. Ogawa, *J. Quant. Spectrosc. Radiat. Transf.*, **13**, 1023 (1973).

120. R. H. Huebner, R. J. Cellotta, S. R. Mielczarek, and C. E. Kuyatt, *J. Chem. Phys.*, **63**, 241 (1975); **59**, 5434 (1973); **63**, 4490 (1975).

121. J. A. R. Samson, *Adv. Atom. Molec. Phys.*, **2**, 177 (1966); "Critical Review of Ultraviolet Photoabsorption Cross Sections for Molecules of Astrophysical and Aeronomical Interest," report No. NSRDS-NBS 38, U.S. Government Printing Office, Washington, D.C., 1971; R. D. Hudson and L. J. Keiffer, *Atom. Data*, **2**, 205 (1971); L. J. Kieffer, "Bibliography of Photoabsorption Data," JILA report No. 5, 1968, Boulder, Colorado; L. C. Lee, R. W. Carlson, D. L. Judge, and M. Ogawa, *J. Quant. Spectrosc. Radiat. Transf.*, **13**, 1023 (1973); J. Lang and W. S. Watson, *J. Phys.*, **B8**, L339 (1975); W. S. Watson, *J. Phys.*, **B5**, 2292 (1972); A. Gardner, M. Lynch, D. T. Stewart, and W. S. Watson, *J. Phys.*, **B6**, L262 (1973).

122. See first four references in Ref. 121 and also: J. B. West and G. V. Marr, *Proc. Roy. Soc. (Lond.), Ser.* **A349**, 397 (1976) G. V. Marr and J. B. West, *Atomic Data and Nuclear Data Tables*, **18**, 297 (1976) C. J. Powell, *Rev. Mod. Phys.*, **48**, 33 (1976).

123. J. A. Wheeler and J. A. Bearden, *Phys. Rev.*, **46**, 755 (1934).

124. P. H. Metzger and G. R. Cook, *J. Chem. Phys.*, **41**, 642 (1964).

125. H. Sun and G. L. Weissler, *J. Chem. Phys.*, **23**, 1160 (1954).

126. R. A. Bonham, *Chem. Phys. Lett.*, **31**, 559 (1975).

127. J. A. R. Samson, J. L. Gardner, and A. F. Starace, *Phys. Rev.*, **A12**, 1459 (1975).

128. M. J. Lynch, A. B. Gardner, K. Codling, and G. V. Marr, *Phys. Lett.*, **43A**, 237 (1973).

129. R. G. Houlgate, J. B. West, K. Codling, and G. V. Marr, *J. Phys.*, **B7**, L470 (1974).

130. R. G. Houlgate, J. B. West, K. Codling, and G. V. Marr, Daresbury Laboratory, report No. DL/SRF/P21, January 1976; and *J. Electron Spectrosc.*, **9**, 205 (1976).

131. K. Codling, R. G. Houlgate, J. B. West, and P. R. Woodruff, Daresbury Laboratory, report No. DL/SRF/P19, January 1976.

132. D. J. Kennedy and S. T. Manson, *Phys. Rev.*, **A5**, 227 (1972).

133. J. B. West and G. V. Marr, *Proc. Roy. Soc.*, **A349**, 397 (1976).

134. P. G. Burke and K. T. Taylor, *J. Phys.*, **B8**, 2620 (1975).

135. J. A. R. Samson and R. B. Cairns, *Phys. Rev.*, **173**, 80 (1968).

136. J. A. R. Samson and J. L. Gardner, *Phys. Rev. Lett.*, **33**, 671 (1974).

137. F. Wuilleumier and M. O. Krause, *Phys. Rev.*, **A10**, 242 (1974).

138. J. B. West, P. R. Woodruff, K. Codling, and R. G. Houlgate, *J. Phys.*, **B9**, 407 (1976).

139. D. L. Ederer, *Phys. Rev. Lett.*, **13**, 760, 1964.

140. G. Wendin, *J. Phys.*, **B6**, 42 (1973).

141. Th. El-Sherbini and M. J. Van der Wiel, *Physica*, **62**, 119 (1972).

142. M. J. Van der Wiel and G. R. Wight, *Phys. Lett.*, **54A**, 83 (1975).

143. G. R. Wight and M. J. Van der Wiel, *Phys. Lett.*, **55A**, 335 (1976).

144. G. R. Wight and M. J. Van der Wiel, *J. Phys.*, **B9**, 1319 (1976).

145. V. Schmidt, N. Sandner, H. Kuntzemuller, P. Dhez, F. Wuilleumier, and E. Kallne, *Phys. Rev.*, **A13**, 1748 (1976).

146. W. S. Watson, *J. Phys.*, **B5**, 2292 (1972).

147. M. J. Van der Wiel, *Physica*, **49**, 411 (1970).

148. M. J. Van der Wiel and G. Wiebes, *Physica*, **54**, 411 (1971).

149. M. J. Van der Wiel and G. Wiebes, *Physica*, **53**, 255 (1971).

150. J. A. R. Samson, *Adv. Atom. Molec. Phys.*, **2**, 177 (1966).

151. A. P. Lukirskii and T. M. Zimkina, *Bull. Acad. Sci. (USSR), Phys. Ser (Engl. transl.)*, **27**, 808 (1963).

152. F. H. Read, *Radiat. Res.*, **64**, 23 (1975).

153. S. Ohtani, H. Nishimura, H. Suzuki, and K. Wakiya, *Phys. Rev. Lett.*, **36**, 863 (1976).

154. M. J. Van der Wiel, Progress Report given at Second International Conference on Inner-Shell Ionization Phenomena, Freiburg, March 1976.

155. M. J. Van der Wiel, G. R. Wight, and R. R. Tol, *J. Phys.*, **B9**, L5 (1976).

156. E. N. Lassettre, *Can. J. Chem.*, **47**, 1763 (1969).

157. S. M. Silverman and E. N. Lassettre, *J. Chem. Phys.*, **40**, 1265 (1964).

158. A. L. Stewart and T. G. Webb, *Proc. Phys. Soc. (Lond.)*, **82**, 532 (1963).

159. D. J. Baker, D. E. Bedo, and D. H. Tomboulian, *Phys. Rev.*, **24**, 1471 (1961); J. A. R. Samson, *J. Opt. Soc. Am.*, **54**, 876 (1964).

160. K. L. Bell and A. E. Kingston, Phys., **B8**, L265 (1975).

161. G. R. Cook and P. H. Metzger, *J. Opt. Soc. Am.*, **54**, 968 (1964).

162. W. A. Chupka and J. Berkowitz, *J. Chem. Phys.*, **51**, 4244 (1969).

163. F. J. Comes and W. Lessmann, *Z. Naturforsch.*, **19a**, 508 (1964).

164. V. H. Dibeler, R. M. Reese, and M. Krauss, *J. Chem. Phys.*, **42**, 2045 (1965).

165. R. Browning and J. Fryar, *J. Phys.*, **B6**, 364 (1973).

166. C. Backx, G. R. Wight, and M. J. Van der Wiel, *J. Phys.*, **B9**, 315 (1976).

167. E. N. Lassettre and A. E. Jones, *J. Chem. Phys.*, **40**, 1222 (1964).

168. W. A. Chupka and J. Berkowitz, *J. Chem. Phys.*, **51**, 4244 (1969).

169. J. A. R. Samson, G. N. Haddad, and J. L. Gardner, in E. E. Koch, R. Haensel, and C. Kunz, *Proceedings of Fourth International Conference on VUV Radiation Physics* (Hamburg), Pergamon, Viewig, Braunschweig, p. 157.

170. G. H. Dunn and L. J. Keiffer, *Phys. Rev.*, **132**, 2109 (1963).

171. A. Hamnett, W. Stoll, and C. E. Brion, *J. Electron Spectrosc.*, **8**, 367 (1976).

172. J. A. R. Samson and J. L. Gardner, *J. Electron Spectrosc.*, **8**, 35 (1976).

173. M. J. Van der Wiel and Th. El-Sherbini, *Physica*, **59**, 453 (1972).

174. L. C. Lee, R. W. Carlson, D. L. Judge, and M. Ogawa, *J. Quant. Spectrosc. Radiat. Transf.*, **13**, 1023 (1973).

175. M. J. Van der Wiel, Th. M. El-Sherbini, and C. E. Brion, Chem. Phys. Lett., **7**, 161 (1970).

176. G. R. Wight, M. J. Van der Wiel, and C. E. Brion, *J. Electron Spectrosc.*, **1**, 457 (1972/73).

177. M. Nakamura et al. *Phys. Rev.*, **178**, 80 (1969).

178. J. L. Dehmer and D. Dill, *Proceedings of Second International Conference on Inner Shell Ionization Phenomena*, Freiberg, 1976.

179. R. W. Fink, R. C. Jopson, H. Mash, and C. D. Swift, *Rev. Mod. Phys.*, **38**, 513 (1966).

180. M. O. Krause and F. Wuilleumier, Oak Ridge National Laboratories, Chemistry Division, Annual Report, May 1971.

181. C. A. Van de Runstraat, F. J. de Heer, and T. R. Govers, *Chem. Phys.*, **3**, 431 (1974).

182. L. Asbrink and C. Fridh, *Phys. Scripta*, **9**, 338 (1974).

183. A. W. Potts and T. A. Williams, *J. Electron Spectrosc.*, **3**, 3 (1974).

184. J. L. Gardner and J. A. R. Samson, *J. Chem. Phys.*, **62**, 1447 (1975).

185. P. M. Hierl and J. L. Franklin, *J. Chem. Phys.*, **47**, 3154 (1967).

186. A. J. Blake and J. H. Carver, *J. Chem. Phys.*, **47**, 1038 (1967).

187. J. L. Bahr, A. J. Blake, J. H. Carver, J. L. Gardner, and V. Kumar, *J. Quant. Spectrosc. Radiat. Transf.*, **9**, 1359 (1969); **12**, 59 (1972).

188. R. L. Schoen, *J. Chem. Phys.*, **40**, 1830 (1964).

189. D. L. Judge and LJ. C. Lee, *J. Chem. Phys.*, **57**, 455 (1972).

190. L. C. Lee, R. W. Carlson, D. L. Judge, and M. Ogawa, *J. Geophys. Res.*, **79**, 5286 (1974).

191. C. Backx, M. Klewer, and M. J. Van der Wiel, *Chem. Phys. Letters*, **20**, 100 (1973).

192. S. Nishikawa and T. Watanabe, *Chem. Phys. Letters*, **22**, 590 (1973).

193. T. Watanabe and S. Nishikawa, *Chem. Phys.*, **11**, 49 (1975).

194. M. J. Van der Wiel, W. Stoll, A. Hamnett, and C. E. Brion, *Chem. Phys. Letters*, **37**, 240 (1976).

195. M. J. S. Dewar, A. Kormornicki, W. Thiel, and A. Schweig, *Chem. Phys. Letters*, **31**, 286 (1975).

196. R. Stockbauer and M. G. Inghram, *J. Chem. Phys.*, 2242 (1971).

197. E. Weigold, S. Dey, A. J. Dixon, I. E. McCarthy, and P. J. O. Teubner, *Chem. Phys. Letters* **41**, 21 (1976).

198. E. N. Lassettre and E. R. White, *J. Chem. Phys.*, **60**, 2460 (1974).

199. K. H. Tan and C. E. Brion, *J. Electron Spectrosc.*, **13**, 77 (1978).

200. C. D. Lin, *Phys. Rev.*, **A9**, 11 (1974).

201. T. Gustafsson, *Chem. Phys. Lett.*, **51**, 383 (1977).

202. G. R. Wight and M. J. Van der Wiel, *J. Phys.*, **B10**, 601 (1977).

203. M. J. Van der Wiel and T. N. Chang, *J. Phys.*, **B11**, L125 (1978).

204. J. A. R. Samson, G. N. Haddad, and J. L. Gardner, *J. Phys.*, **B10**, 1749 (1977).

205. P. R. Woodruff and G. V. Marr, *Proc. Roy. Soc.*, **A358**, 87 (1977).

206. R. B. Kay, Ph. E. Van der Leeuw, and M. J. Van der Wiel, *J. Phys.*, **B10**, 2513 (1977).

207. R. B. Kay, Ph. E. Van der Leeuw, and M. J. Van der Wiel, *J. Phys.*, **B10**, 2521 (1977).

208. K. H. Tan, C. E. Brion, Ph. E. Van der Leeuw, and M. J. Van der Wiel, *Chem. Phys.*, **29**, 299 (1978).

209. C. E. Brion, A. Hamnett, G. R. Wight, and M. J. Van der Wiel, *J. Electron Spectrosc.* **12**, 323 (1977).

210. G. R. Wight, M. J. Van der Wiel, and C. E. Brion, *J. Phys.*, **B10**, 1863 (1977).

211. K. H. Tan and C. E. Brion, *J. Electron Spectros.* **15**, 241 (1979); *Chem. Physics.*, **34**, 141 (1978).

212. T. Gustafsson, E. W. Plummer, D. E. Eastman, and W. Gudat, *Phys. Rev.*, **A17**, 175 (1978).

213. A. Hitchcock, C. E. Brion, and M. J. Van der Wiel, *J. Phys.*, **B11**, 3245 (1978).

214. J. A. R. Samson, J. L. Gardner, and G. N. Haddad, *J. Electron Spectrosc.*, **12**, 281 (1977).

215. F. Wuilleumier, M. Y. Adam, P. Dhez, N. Sandner, V. Schmidt, and W. Mehlhorn, *Phys. Rev.*, **A16**, 646 (1977).

216. T. N. Rescigno, C. F. Bender, B. V. McCoy, and P. W. Langhoff, *J. Chem. Phys.*, **68**, 970 (1978).

217. N. Padial, G. Csanak, B. V. McCoy, and P. W. Langhoff, *J. Chem. Phys.*, **69**, 2992 (1978).

218. T. N. Rescigno and P. W. Langhoff, *Chem. Phys. Lett.*, **51**, 65 (1977).

219. C. E. Brion, K. H. Tan, M. J. Van der Wiel and Ph. E. Van der Leeuw, *J. Electron Spectrosc.*, **17**, 101 (1979).

220. C. E. Brion, J. P. D. Cook and K. H. Tan, *Chem. Phys. Letters*, **59**, 241 (1978).

CHAPTER TWO

ROLE OF EXCITED STATES IN ION–NEUTRAL COLLISIONS

Thomas O. Tiernan

Department of Chemistry and Brehm Laboratory, Wright State University, Dayton, Ohio 45431

Chava Lifshitz*

Department of Physical Chemistry, The Hebrew University of Jerusalem, Israel

---

*On sabbatical leave at Wright State University, 1976–1977.

Contents

I EXPERIMENTAL TECHNIQUES   84
  A Formation of Excited Ions and Determination of Internal-energy Distributions   84
  B Excitation of Neutral Reactants   108
  C Methods for Studying Excited Ion–Neutral Interactions   108

II INTERACTIONS OF EXCITED IONS WITH NEUTRALS   120
  A Reactive Scattering   126
  B Non-reactive Scattering—Energy Transfer   145

III REACTIONS OF IONS WITH EXCITED NEUTRALS   161

IV EXCITED PRODUCTS FROM ION–NEUTRALS COLLISIONS (ELECTRONIC, VIBRATIONAL, AND ROTATIONAL EXCITATION)   163
  A Chemiluminescent Reactions   165

V COLLISION MECHANISMS AND THEORETICAL IMPLICATIONS   196
  A General Effects of Internal Excitation   196
  B Theoretical Treatment of Energy Partitioning   199
  C Calculations of Energy States, Correlation Diagrams, and Potential Surfaces   201
  D Quasiclassical and Collinear Quantum-mechanical Trajectory Calculations   205

Several books and review chapters devoted to the field of ion–neutral reactions in the gas phase have appeared in recent years, [1a-g, j,k] some of which are concerned at least in part with the special topic of interest for the present review chapter—namely, the role of excited states in such interactions. The present review attempts to present a comprehensive survey of the latter subject, and the processes to be discussed include those in which an excited ion interacts with a ground-state neutral, interaction of an excited neutral with a ground-state ion, and on–neutral interactions that produce excited ionic products or excited neutral products. Reactions in which ions are produced by reaction of an excited neutral species with another neutral, for example, Penning ionization, are not included in the present chapter. For a recent review of this topic, the reader is referred to the article by Rundel and Stebbings.[1i] Electron–molecule interactions and photon–molecule interactions are discussed here only as they relate to the production of ions in excited states, which can then be reacted with neutral species.

Recent advances in experimental techniques, particularly photoionization methods, have made it relatively easy to prepare reactant ions in well-defined states of internal excitation (electronic, vibrational, and even rotational). This has made possible extensive studies of the effects of internal energy on the cross sections of ion–neutral interactions, which have contributed significantly to our understanding of the general areas of reaction kinetics and dynamics. Other important theoretical implications derive from investigations of the role of internally excited states in ion–neutral processes, such as the effect of electronically excited states in nonadiabatic transitions between two potential-energy surfaces for the simplest ion–molecule interaction, $H^+(H_2, H)H_2^+$,* which has been discussed by Preston and Tully.[2] This role has no counterpart in analogous neutral–neutral interactions.

From a practical standpoint, much of the interest in the role of excited states in ionic interactions stems from their importance in ionospheric chemistry.[1h] In addition, it has been realized more recently that certain ion–neutral interactions offer a comparatively easy means of populating electronically excited reaction products, which can produce chemiluminescence in the visible or UV region of the spectrum. Such systems are potential candidates for practical laser devices. Several charge-transfer processes have already been utilized in such devices, notably $He^+(I, He)I^+$ and $He_2^+(N_2, 2He)N_2^+$.[3] Interest in this field has stimulated new emphasis on fundamental studies of luminescence from ion–neutral interactions.

---

*The nomenclature used here and throughout the tables presented in this review to designate ion–neutral reactions conforms to that currently in use (see Ref. 1d, p. 104). For collisional dissociation and dissociative charge transfer, we have adopted the notations $AB^+(X; X, B)A^+$ and $A^+(BC; A, B)C^+$, respectively.

Because of the breadth of the subject matter covered in this review, much of the information on specific interactions has been presented in tabular form, and the discussion is limited to selected examples of particular interest. The choice of subjects reflects, of necessity, the taste of the authors.

I. EXPERIMENTAL TECHNIQUES

A. Formation of Excited Ions and Determination of Internal-energy Distributions

1. Electron Impact

Ionization by electron impact, employing ionizing energies of 10 to 100 eV, can produce atomic and molecular ions in a variety of excited states. Detection of the presence of electronically excited states is possible if their reactions with neutral molecules are different from those of the ground-state ions. For example, two electronic states are important in reactions of O^+ ions—the ground ^4S state and the long-lived excited ^2D state. The reaction, $O^+(N_2,O)N_2^+$, is endoergic for ground-state $O^+(^4S)$ and becomes energetically allowed for $O^+(^2D)$. As the energy of the electrons used to generate the reactant O^+ ions is increased above the lowest ionization potential, this reaction demonstrates a clear threshold, corresponding to the threshold for production of $O^+(^2D)$[4a] (Fig. 1). Correct identification of the excited state responsible for a particular ion–molecule reaction using such techniques is dependent on the ionizing electron-energy resolution that can be achieved. Since the resolution is not better than 0.5 eV under normal operating conditions, only preparation of ions in states of electronic excitation rather widely separated in energy can be controlled. In this respect, photoionization has a great advantage over electron-impact ionization (see Section I.A.2). There are, however, some ionic states that are accessible only by electron impact, because of either energy or spin multiplicity requirements. In addition, electron impact has become established as the most widely employed ionization method because of the ease of generation of electrons, the high efficiency of electron-impact ionization, and the accessibility of a wide range of energies. Two methods have been employed in most previously reported studies to control the electronic and to a limited extent the vibrational states of ions produced by electron impact: (1) varying the ionizing electron energy and (2) varying the source molecule from which the ion is produced. For example, the fraction of $O^+(^2D)$ in O^+ beams produced by dissociative ionization of O_2 can be increased from 0 to 0.3 by increasing the ionizing electron energy from 20 eV to 100 eV.[4a] Alternatively, if the ionizing electron energy is maintained constant at 60 eV, the fraction of

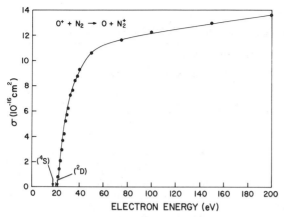

Figure 1. Dependence of cross section, σ, for N^{+2} production in O^+-N_2 collisions on energy of ionizing electrons used to produce O^+ from O_2. Results are presented for ion energy of 100 eV. Threshold electron energies required to produce the ^4S and ^2D states of O^+ are indicated.[4a]

$O^+(^2D)$ in an O^+ beam is only 0.04 if the molecule dissociatively ionized is CO_2, whereas the fraction is 0.90 if the source molecule is H_2O.[5a] The abundance of a given state in an ion beam is, of course, also dependent on the lifetime of that state and the transit time of the beam from the point of formation to the point at which the abundance is measured.

Increased ion-source pressures and/or introduction of paramagnetic molecules, such as NO, cause deactivating collisions that convert excited ions into ground-state species[13,61,63,64,66] (see Table I). The fractional abundance of excited states in an ion beam may thus be lowered without a concomitant reduction in total ion intensity, which would occur if the ionizing electron energy were simply lowered.

The fractional abundances of excited states in ion beams have been determined by making use of their different reactivity characteristics. Three methods have been employed, and these are described using O^+ ions as an example.

a. Method No. 1

If a given ion–neutral reaction is known to occur only for ground-state ions, it is quite straightforward to determine the fractional abundance of the excited state. For instance, the reaction $O^+(N_2,N)NO^+$ proceeds with $O^+(^4S)$ reactant ions but not with ions in the ^2D state.[4b,c] The cross section for this reaction has been determined as a function of the ionizing electron energy used to produce the O^+. The observed cross section at any given electron energy can be written as a sum of the contributions of the two

Table I.
Selected Electronically Excited States of Ions And Their Reactions

SOURCE GAS	ELECTRON ENERGY (eV)	APPROXIMATE FRACTIONAL ABUNDANCE	REACTIONS STUDIED	REMARKS	REFERENCES
$B^+(^3P)$; Excitation Energy[8a] = 4.6 eV; Lifetime = "Moderately Long"					
BF_3	100	Abundant	Dissociative charge transfer hydride-ion transfer, $B^+(RH,BH)R^+$	No	6
BF_3	70	0.35	$B^+(N_2O,B)N_2O^+$	Yes[a]	7a
BCl_3	70	0.13	$B^+(N_2O,N_2)BO^+$	Yes[a]	7a
BBr_3	70	0.03	$B^+(N_2O;N,B)NO^+$	Yes[a]	7a
BI_3	70	0.0	$B^+(N_2O;N_2,B)O^+$	Yes[a]	7a
BF_3	70	0.35	$B^+(D_2,D)BD^+$	Yes[a]	7b
$C^+(^4P)$; Excitation Energy = 5.3 eV; Lifetime = "Moderately Long"					
COS,CO_2,CO $CBrF_3,CBr_2F_2,$ CCl_3F	100	0.4	Dissociative charge transfer	No	6
CO	70	0.3	$C^+\left(O_2;C,O \begin{array}{c} C \\ O \\ O \end{array} \begin{array}{c} O_2^+ \\ O^+ \\ CO^+ \end{array}\right)$	Yes[a] Yes[a] Yes[a]	9a 9a 9a
			$C^+\left(N_2;C,N \begin{array}{c} C \\ N \\ N \end{array} \begin{array}{c} N_2^+ \\ N^+ \\ CN^+ \end{array}\right)$	Yes[a] Yes[a] Yes[a]	9a 9a 9a
CO	23 to 41	0.06	$C^+(H_2,H)CH^+(a^3\Pi)$	Yes[b]	9b
CO		0.16	$C^+(NO,C)NO^+$	Yes[c]	9c
$N^+(^1D)$; Excitation Energy = 1.90 eV; Lifetime = 4.13 min					
$N^+(^1S)$; Excitation Energy = 4.05 eV; Lifetime = 0.9 sec					
N_2O,NH_3		0.00 (^1D); 0.1 (^1S)	Dissociative charge transfer	No	6
N_2 (May produce N_2^{++})	50		$N^+(O_2,N)O_2^+$	Yes[a]	4b
			$N^+(O_2,O)NO^+$	Yes[a]	4b
			$N^+(NO,N)NO^+$	Yes[a]	10
			$N^+(Kr,N)Kr^+$	No	11a
N_2	>35	0.15 (^1D)	$N^+(Kr,N)Kr^+$; $N^+(CO,N)CO^+$	Yes[a]	11b
N_2	160		$N^+(H_2,H)NH^+$	Yes[b]	12a
$O^+(^2D)$; Excitation Energy = 3.32 eV; Lifetime = 5.9×10^3 $(^2D_{3/2})$; 2.1×10^4 $(^2D_{5/2})$ sec					
CO	100	0.3	Dissociative charge transfer	No	6
CO_2	100	0.3		No	6
O_2	50; 100	0.27; 0.3	$O^+(Ar,O)Ar^+$	No	4a
O_2	100	0.3	$O^+(N_2,O)N_2^+$	No	4a
O_2	40	0.2	$O^+(N_2,O)N_2^+$	Yes[a]	4c
O_2–He	40	0.38	$O^+(N_2,O)N_2^+$	Yes[a]	4c
CO_2	150	0.2	$O^+[N_2(v>0),N]NO^+$	Yes[a]	4d
O_2	100	0.44	$O^+(O_2,O)O_2^+$	Yes[a,d]	4b, 5b
			$O^+(NO,O)NO^+$	Yes[d]	5b, 10
CO_2	60	0.04	$O^+(N_2,O)N_2^+$	Yes[a]	5a
N_2O	60	0.43	$O^+(CO,O)CO^+$	Yes[a]	5a
NO_2	60	0.34	$O^+(Ar,O)Ar^+$	Yes[a]	5a
H_2O	60	0.90	$O^+(Kr,O)Kr^+$	Yes[a]	5a
CO_2	100	0.20	$O^+\left(N_2O; \begin{array}{c} O \\ O,N \end{array} \begin{array}{c} N_2O^+ \\ NO^+ \end{array}\right)$	Yes[d]	5b
			$O^+\left(CO_2, \begin{array}{c} O \\ CO \end{array} \begin{array}{c} CO_2^+ \\ O_2^+ \end{array}\right)$	Yes[a,d]	5b

SOURCE GAS	ELECTRON ENERGY (eV)	APPROXIMATE FRACTIONAL ABUNDANCE	REACTIONS STUDIED	REMARKS	REFERENCES
			$O^+\left(H_2, \begin{matrix} N \\ OH \end{matrix}\right)\begin{matrix} OH^+ \\ H^+ \end{matrix}$	Yes[c]	12b
	\multicolumn{5}{l}{$F^+(^1D)$; Excitation Energy = 2.59 eV; Lifetime = 0.26 sec}		
$SF_6, CBrF_3$	100	0.30	Dissociative charge transfer	—	6
CF_4	100	0.45	$F^+(Ne, F)Ne^+$	Yes[a]	13
CF_4–NO	100	0.0	$F^+(H_2, H)FH^+$	Yes[a]	13
			$F^+(D_2, D)FD^+$	Yes[a]	13
	\multicolumn{5}{l}{$Ar^+(^2P_{1/2})$; Excitation Energy = 0.18 eV; Lifetime = 19 sec}		
Ar	15.95[e]	0.33	$Ar^+(H_2, H)ArH^+$	Yes[e]	14
Ar	15.95	0.33	$Ar^+(N_2, Ar)N_2^+$	Yes[f]	15a
Ar	—	—	$Ar^+(Ar, Ar)Ar^+$	Yes[a]	15b, c
	\multicolumn{5}{l}{$Kr^+(^2P_{1/2})$; Excitation Energy = 0.68 eV}		
Kr	100	0.33	Dissociative charge transfer	No	6
Kr	—	—	$Kr^+(Kr, Kr)Kr^+$	Yes[a]	15c, 16
Kr	—	—	$Kr^+\left(CH_4; \begin{matrix} Kr \\ Kr, H \end{matrix}\right)\begin{matrix} CH_4^+ \\ CH_3^+ \end{matrix}$	No / No	17, 18 / 18, 19
Kr	—	—	$Kr^+(H_2, H)KrH^+$;	Yes[d,e]	20, 21
Kr	20 to 50	0.3–0.4	$Kr^+(N_2O, Kr)N_2O^+$; $Kr^+(CO_2, Kr)CO_2^+$	Yes[d]	23c, e
	\multicolumn{5}{l}{$Xe^+(^2P_{1/2})$; Excitation Energy = 1.3 eV}		
Xe	100	0.33	Dissociative charge transfer	No	6
Xe	—	—	$Xe^+(CH_4, Xe)CH_4^+$	No	19, 22, 23a
Xe	25	0.2	$(Xe, Xe)Xe^+$ $Xe^+(O_2, Xe)O_2^+$	Yes[a]	16b, 23b
Xe	—	—	$Xe^+(O_2, Xe)O_2^+$	Yes[e]	23d
	\multicolumn{5}{l}{$N_2^+(A^2\Pi_u)$; Excitation Energy[8b] = 1.12 eV; Lifetime ≃ 6×10^{-6} sec}		
	\multicolumn{5}{l}{$N_2^+(^4\Sigma_u^+)$; Excitation Energy ~6 eV, See Text; Lifetime > 10^{-2} sec}		
N_2	20 to 22; 30 to 80	(Onset); ~10^{-2}	$N_2^+(N_2, N)N_3^+$	Yes[g]	24–37
N_2	—	—	$N_2^+(N_2; N, N_2)N^+$; $N_2^+(N_2, N_3)N^+$	No	32, 35, 36
N_2	16 to 90	—	$N_2^+(N_2; N, N_2)N^+$; $N_2^+(N_2, N_2)N_2^+$	Yes[a]	38
N_2	19, 26, 59	< 0.574 ($^2\Pi_u$)	$N_2^+(N_2; N, N_2)N^+$	Yes[a]	38c
N_2	> 22.5	< 0.05 ($^4\Sigma_u^+$?)	$N_2^+(N_2; N, N_2)N^+$	Yes[a]	38c
N_2	—	—	$N_2^+(O_2, N_2)O_2^+$	Yes[a]	4b
N_2	15 to 200	—	$N_2^+(NO, N_2)NO^+$	Yes[a]	10
N_2	—	—	$N_2^+(Ar, N)ArN^+$	No	24, 27
N_2	—	—	$N_2^+(Kr, N)KrN^+$	No	24, 27
N_2	—	—	$N_2^+(Xe, N)XeN^+$	No	24, 27
	\multicolumn{5}{l}{$O_2^+(a^4\Pi_u)$; Excitation Energy = 4.04 eV; Lifetime > 10^{-3} sec}		
O_2	100	0.3–0.4	Dissociative charge transfer	No	6
O_2	—	—	$O_2^+(O_2, O)O_3^+$	Yes[g]	30–32, 39–43
O_2	16.5 to 19.0[e]	—	$O_2^+(O_2, O)O_3^+$	Yes[e]	44
O_2	16.5 to 18.7[e]	—	$O_2^+(O_2, O)O_3^+$	Yes[e]	45
O_2–He	21[h]	0.5	$O_2^+(O_2, O_2)O_2^+ (X^2\Pi_g)$	Yes[h]	46a
O_2	14 to 90	0.2	$O_2^+(O_2; O, O_2)O^+$	Yes[i]	38a, b
O_2	15 to 200	—	$O_2^+(NO, O_2)NO^+$	Yes[a]	10
O_2–He	21[h]	0.5	$O_2^+(NO, O_2)NO^+$	Yes[h]	46b

87

Table I. *continued*

SOURCE GAS	ELECTRON ENERGY (eV)	APPROXIMATE FRACTIONAL ABUNDANCE	REACTIONS STUDIED	REMARKS	REFERENCE
O_2	25; 50; 100	0.22; 0.30; 0.33	$O_2^+(H_2, ?)?$	—	4a
O_2	10; 100	< 0.03; ~0.4	$O_2^+(H_2, H)HO_2^+$	Yes[b]	47, 48
O_2	15.3 to 19.0[e]	—	$O_2^+(H_2, H)HO_2^+$	Yes[e]	49
O_2	—	—	$O_2^+(H_2, H)HO_2^+$	Yes[a,e]	21
O_2–He	21[h]	—	$O_2^+(H_2, H)HO_2^+$; $O_2^+(H_2, H_2)O_2^+(X^2\Pi_g)$	Yes[h]	46a
O_2	—	—	$O_2^+(H_2, OH)OH^+$	Yes[a,e]	21
O_2	13 to 50	—	$O_2^+(Ar; O, Ar)O^+$	Yes[a]	50
O_2	—	—	$O_2^+(Ne; O, Ne)O^+$	Yes[a]	50
O_2	90	0.51	$O_2^+(Ar, O_2)Ar^+$	Yes[a]	51
O_2–He	21[h]	—	$O_2^+(Ar, O_2)Ar^+$; $O_2^+(Ar, Ar)O_2^+(X^2\Pi_g)$	Yes[h]	46a
O_2	100	0.33	$O_2^+(Ar, O_2)Ar^+$	No	4a
O_2–He	21[h]	—	$O_2^+(CO, O_2)CO^+$; $O_2^+(CO, CO)O_2^+(X^2\Pi_g)$	Yes[h]	46a
O_2–He	21[h]	—	$O_2^+(CO, O)CO_2^+$	Yes[h]	46a
O_2	13.2 to 19.0[e]	—	$O_2^+(CO, O)CO_2^+$	Yes[e]	52a
O_2	100	0.33	$O_2^+(N_2, O_2)N_2^+$	Yes[a,d]	4a, 5b, c
O_2–He	21[h]	—	$O_2^+(N_2, O_2)N_2^+$; $O_2^+(N_2, N_2)O_2^+(X^2\Pi_g)$	Yes[h]	46a
O_2	16.5 to 18.8[e]	—	$O_2^+(N_2, O)N_2O^+$	Yes[e]	52b
O_2–He	21[h]	—	$O_2^+(CO_2, O_2)CO_2^+$; $O_2^+(CO_2, CO_2)O_2^+(X^2\Pi_g)$	Yes[h]	46a
O_2	16 to 40	—	$O_2^+(Mg, O_2)Mg^+$	Yes[a]	53
O_2	16 to 40	—	$O_2^+(Na, O_2)Na^+$	Yes[a]	54
O_2	16 to 40	—	$O_2^+(Ca, O_2)Ca^+$	Yes[a]	55
O_2	16 to 40	—	$O_2^+(Fe, O_2)Fe^+$	Yes[a]	56a
O_2	15.9 to 17.7[e]	—	$O_2^+(Xe, O_2)Xe^+$	Yes[e]	23d
O_2	40	—	$O_2^+(H_2O, O_2)H_2O^+$	Yes[a]	56b
		O_2^+ (Higher States); Lifetimes~0.67–1.2 μsec			
O_2	14 to 40	—	$O_2^+(O_2, —)O^+, O^-, O_2^+$	—	38a
		$CO^+(A^2\Pi)$; Excitation Energy = 2.53 eV; Lifetime = 2.6×10^{-6} sec;			
		$CO^+(B^2\Sigma^+)$; Excitation Energy = 5.66 eV; Lifetime = 10^{-7} sec			
		CO^+ (4) (See Text); Lifetime > 10^{-2} sec			
CO	16 to 90	—	$CO^+(CO; O, CO)C^+$	Yes[j]	38b, 57
CO	—	—	$CO^+(CO, O)C_2O^+$	Yes[g]	27, 31, 32
CO	20 to 22; 30	(Onset); > 3.8×10^{-3}	$CO^+(CO, O)C_2O^+$	Yes[g]	58
CO	—	—	Dissociative charge transfer	No	6
CO	—	—	$CO^+\left(CO; O, CO \begin{array}{c} C \\ \end{array}\right)\begin{array}{c}CO_2^+ \\ C^+\end{array}$	No	34b, 35b
CO	—	—	$CO^+\left(CO; O, CO \begin{array}{c} O \\ \end{array}\right) C_2O^+$	Yes[k]	79a
		$NO^+(^3\Delta)$; Excitation Energy = 7.4 eV; Lifetime > 10^{-3} sec and Possibly Lower Excited States—$a^3\Sigma^+$ and $b^3\Pi$			
NO, N_2O	100	—	Dissociative charge transfer	No	6
NO	50	0.42	$NO^+(H_2O, NO)H_2O^+$	Yes[a,f]	50, 56b,
NO	17 to 20	—	$NO^+\left(Ar; O, Ar \begin{array}{c}NO \\ N, Ar\end{array}\right)\begin{array}{c}Ar^+ \\ N^+ \\ O^+\end{array}$	Yes[f]	50
NO	13 to 20	—		Yes[a]	50
NO	13 to 20	—		Yes[a]	50
NO, NO_2	10 to 80	—	$NO^+(NO; O, NO)N^+$	Yes[a,j]	38b, 50,

SOURCE GAS	ELECTRON ENERGY (eV)	APPROXIMATE FRACTIONAL ABUNDANCE	REACTIONS STUDIED	REMARKS	REFERENCES
	20	—	$NO^+(Ne; O, Ne)N^+$	No	50
	20	—	$^{15}NO^+(^{14}NO; O, NO)^{14}N^+, ^{15}N^+$	No	50
NO_2	10 to 80	—	$NO^+(NO; N, NO)O^+$	Yes[a,j]	38b, 50, 60
	50	—	$NO^+(NO; NO)NO^+(X^1\Sigma^+)$	No	61
	5 to 10	<0.03	$NO^+(He; O, He)N^+$; $NO^+(He; N, He)O^+$	Yes[b]	62
			$H_2O^+(^2A_1)$; Excitation Energy = 1.1 eV; Lifetime = 10^{-5} sec		
	12 to 70	<0.6	$H_2O^+(N_2O, H_2O)N_2O^+$	No	63
			$D_3O^+(^3?)$		
	70	0.5	$D_3O^+(D_2, D_2O)D_3^+$	Yes[a]	64
-NO	70	0.0	$D_3O^+(D_2, D_2O)D_3^+$	Yes[a]	64
			$CS_2^+(^2\Pi_{3/2u})$		
	—	—	$CS_2^+(CS_2, CS)CS_3^+$	No	26, 42, 65
			$C_2H_2^+(^2\Sigma_g^+)$; Excitation Energy = 4.96 eV;		
2	—	—	$C_2H_2^+ \left(C_2H_2, \begin{array}{c} H_2 \\ H \end{array} \right) \begin{array}{c} C_4H_2^+ \\ C_4H_3^+ \end{array}$	Yes[k]	71, 72
2	>16.5	—		No	72–78
2	>16.5	—	$C_2H_2^+(C_2H_2, C_2H)C_2H_3^+$	No	74–78
2	—	—	$C_2H_2^+ \left(C_2H_2, \begin{array}{c} C_2H \\ H_2 \end{array} \right) \begin{array}{c} C_2H_3^+ \\ C_4H_2^+ \end{array}$	Yes[k]	79b
			$C_2H_4^+(^2B_3)$; Excitation Energy = 1.7 eV; Lifetime ≥ 10^{-5} sec		
4	12.2 to 15[e]	0.70	$C_2H_4^+(C_2H_4, CH_3)C_3H_5^+$	Yes[e]	80, 81
4	12.8	0.70	$C_2H_4^+(C_2H_4, C_2H_4)C_2H_4^+(X)$	No	80
			$C_3H_5^+$; Excitation Energy = ?		
4	—	—	$C_3H_5^+(C_2H_4, C_2H_2)C_3H_7^+$	Yes[l]	82
			$C_6H_6^+$; Excitation Energy = 3.5 ± 0.5 eV		
6 (Benzene)	~12.7 to 80	—	$C_6H_6^+(C_6H_6, —)C_{12}H_{12}^+$	Yes[m]	83
			Fe^+(Large Number of Unidentified States), Lifetimes ≥ 3 μsec		
vapor	10 to 70	Unknown	$Fe^+(H_2O, Fe)H_2O^+$	No	84a
vapor	10 to 70	Unknown	$Fe^+(O_2, Fe)O_2^+$	No	
vapor	10 to 70	Unknown	$Fe^+(Aniline, Fe)Aniline^+$	No	
			$C^-(^2D)$; Excitation Energy = 1.21 eV; Lifetime > 10^5 sec		
, CO_2, CO	300[n]	1.0 (CH$_4$) 0.4 (CO)	$C^-(O_2, C)O_2^-$	Yes[n]	84b

action cross section studied as a function of relative translational energy of the reactants
citation function") and compared with that of the ground state.

locity vector distributions (contour maps) were determined for various relative transla-
al energies of the reactants and the dynamics for ground and excited state ions compared.

ission in the UV from the reaction products: $C(^1P^0)$, $OH^+(A^3\Pi)$ and $OH(A^2\Sigma^+)$,
ectively, was monitored in these experiments (see also Section IV.A).

te coefficients for ground and excited ions compared.

ns produced by photon impact; analysis of photoionization efficiency curves and compari-
with ground-state cross section.

alysis of electron-impact-ionization efficiency curves.

e discussion in text.

Table I. *continued*

[h]Reactant ions produced by Penning ionization of O_2 by $He(2^3S)$; rate constant measured in flow-drift tube for the relative kinetic energies in the range 0.04 to 2 eV.

[i]The Aston band technique was employed to determine the abundances of excited states O_2^+.

[j]Product-ion velocity distributions were determined for collision-induced dissociations 2000-eV ions.[38b]

[k]The ion $C_2H_2^{+*}$ was obtained by charge transfer with ions having recombination energies the range $RE = 15$ to 20 eV, and CO^{+*} was obtained by charge transfer with ions having recombination energies in the range $RE = 16.5$ to 17.5 $(A^2\Pi)$ and 18 to 20 $(B^2\Sigma^+)$ eV.

[l]Photoexcitation of ions produced by electron-impact ionization.

[m]The reactant ion is either an electronically excited benzene ion or an open-chain C_6H isomer.

[n]Primary electron energy; dissociative electron capture results from low-energy secondary electrons released from a metal surface.

states accessible for the energy range employed. That is,

$$\sigma = \sigma(^4S)f(^4S) + \sigma(^2D)f(^2D) \qquad (I.1)$$

where $\sigma(^4S)$ and $\sigma(^2D)$ are the cross sections for the NO^+ producing reactions of the ^4S and ^2D states, respectively, and $f(^4S)$ and $f(^2D)$ are the fractional abundances of these states. At relatively low energy (20 eV) where only $O^+(^4S)$ is present, $\sigma = \sigma(^4S)$. At higher energies, where $f(^2D)$ is finite, the effect of $O^+(^2D)$ is to reduce σ by a percentage equal to its fractional abundance since $\sigma(^2D) \ll \sigma(^4S)$.

b. Method No. 2

The attenuation method [4a] is more widely applicable. It is based on attenuation of an ion beam on passage through a chamber containing a gas, the pressure of which can be varied. In this experiment the disappearance of the reactant ion is followed, rather than the appearance of a particular product. When O^+ ions are allowed to pass through argon, [4a] the beam is attenuated as a result of reactive $O^+(Ar,O)Ar^+$ collisions as well as nonreactive scattering.[5a] The O^+ ion current, I, reaching the detector after attenuation of the beam is obviously smaller than I_0, the current measured when a scattering gas is not present in the collision chamber. A plot of $\log(I/I_0)$ as a function of argon pressure in the collision chamber yields a straight line when the electron energy producing O^+ is 20 eV. At higher electron energies the corresponding plots are curved [4a] (Fig. 2). This is explained by the fact that at 20 eV the only state present in the beam is $O^+(^4S)$ and a single disappearance cross section $\sigma(^4S)$ is applicable because of all the processes by which $O^+(^4S)$ ions are

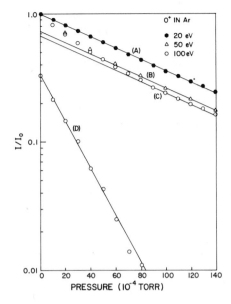

Figure 2. Attenuation curve measured for O^+ in argon. Curves A, B, and C are for ionizing electron energies of 20 eV, 50 eV, and 100 eV, respectively. Curve D represents decay of only excited O^+ ions that are present in ion beam for 100-eV electron-energy case. Curve D is obtained by subtracting from low-pressure data points of C, the linear extrapolation of high-pressure portion of C.[4a]

lost upon collision with argon. At higher electron energies the O^+ ion current detected after scattering on argon is given by the expression

$$I = f(^4S)I_0 \exp\left[-n\sigma(^4S)l \right] + f(^2D)I_0 \exp\left[-n\sigma(^2D)l \right] \qquad (I.2)$$

where n is the number density of argon, l is the length of the collision chamber, and $\sigma(^2D)$ and $\sigma(^4S)$ are the disappearance cross sections for $O^+(^2D)$ and $O^+(^4S)$, respectively, in argon. Since $\sigma(^2D) > \sigma(^4S)$, the limiting slope of the plot of $\log(I/I_0)$ at low pressures is larger at higher electron energies than it is at 20 eV. Since such a plot yields a straight line at high pressures, the second term in equation (I.2) is negligible in this region, and so I/I_0 can be extrapolated to zero pressure to give $f(^4S)$ directly.

c. Method No. 3

Assume that the fractional abundance of an excited ionic state produced by electron impact on a given molecule under specified reaction conditions is known and that an ion–neutral reaction is known to occur preferentially with that ionic state. Then, by determining the cross section σ for that reaction, with the reactant ion produced from a series of different molecules, and comparing these with σ for the reaction using ions generated from the source molecule that yields the known fractional abundance of the excited state, it is possible to determine the fractional abundances of the excited ionic state in beams formed from the other source molecules.[5a] This method was applied to determine the abundance of $O^+(^2D)$ in O^+ ion beams from various source molecules by measuring the cross section for the reaction, $O^+(N_2,O)N_2^+$. The fractional abundance $O^+(^2D)$ from dissociative ionization of O_2 had been previously determined.[4a] σ was measured for the O^+ reaction indicated with the reactant generated from CO_2, NO_2, N_2O, and H_2O, and the fractional abundance of $O^+(^2D)$ from these molecules was thus determined.

d. Other Methods

Up to this point we have discussed methods for detecting excited ionic states and determining their fractional abundances, based solely on their specific chemical reactivity in ion–neutral interactions. Several other methods[38a–c, 67–70] have been applied to detect excited ionic states in ion beams produced by electron-impact ionization. Some of these are described in the paragraphs that follow.

Aston Banding[38a–c]

Ions undergoing dissociative collisions, after they have been fully accelerated but prior to their analysis in the magnetic field of a mass spectrometer, are focused at an apparent mass, m^*. The intensity of these nominal mass ions is monitored by varying the ionizing electron energy. The presence of excited states in the ion beam is detected by an increased ionization efficiency at the ionizing electron energy corresponding to the threshold of the excited state.

Auger Electron-emission Yields from Metal Surfaces[67]

A clean Mo molybdenum surface is bombarded by ions having kinetic energies of 30–400 eV. Auger electrons are emitted, and their yield is considerably higher for the excited state of an ion than for its ground state. As the ionizing electron energy is gradually increased while the ion energy is maintained constant, a sharp increase in the Auger electron yield indicates the threshold of formation of an excited state in the ion beam.

Velocity-distribution Measurements for Product Ions of Collision-induced Dissociations[38b]

Theoretical distribution curves are synthesized from known potential curves (of the ground state, bound metastable, and repulsive excited states) of the molecular ion and from Franck–Condon transition probabilities. The necessity of including contributions from excited metastable states in the ion beam is indicated when a fit is obtained between the calculated curve and those experimental distributions that are obtained at high ($\sim$80 eV) ionizing electron energies.

Translational-energy-distribution Measurements on Product Ions from Double Charge Transfer[68a]

High-translational-energy single-charged positive ions interact with argon atoms via double charge transfer in a single collision [X^+(Ar, X^-)Ar^+] or in multiple collisions [X^+(Ar, X)Ar^+; X(Ar, X^-)Ar^+]. The translational-energy spectrum of singly charged negative ions formed is recorded. The energy-loss spectrum demonstrates peaks at positions calculated for ground, as well as for metastable excited states, provided these are present in the original beam.

Translational-energy Distribution Measurements on Product Ions from Collisional Ionization[68b]

This technique is very similar to the one described in the preceding paragraph. Kinetic-energy-loss spectra are measured for doubly charged positive ions produced in collisional ionization of high-kinetic-energy singly charged ions.

Laser Photodissociation Spectroscopy of Ions[70a–c]

Photodissociation cross sections of ions are measured as a function of photon energy. Information is obtained on the photon-absorbing state(s) as well as on the predissociating or repulsive states formed on photon absorption. The method has not yet been employed for detection of metastable states in ion beams.

Laser Photofragment Spectroscopy[69]

A crossed-ion-beam–laser-beam experiment is performed in either of two ways—by scanning either the photofragment kinetic energy at fixed wavelength or the laser wavelength at fixed photofragment energy. From observed structure in photofragment kinetic-energy spectra, the electronic transition induced by photon absorption is clearly identified. Transitions from metastable states to higher excited states have been studied. A wealth of information on the states involved and their energies, potential curves, dissociation times, and other details is obtained from these experiments.

Laser-induced Fluorescence[70d]

Photons from a tunable dye laser are absorbed by a metastable state of the ion. Spontaneous emission at several wavelengths follows. The ionizing electron energy is varied over the region of interest. At each setting the laser is frequency scanned across the Doppler profile of the ion-beam

Table II.
Reactions of Vibrationally Excited Ions

ION	VIBRATIONAL LEVELS v'	METHOD	REACTIONS STUDIED	COMMENTS	REFERENCE
$H_2^+(X^2\Sigma_g^+)$	0 to 12	Photoionization–mass spectrometry	$H_2^+(H_2,H_2)H_2^+$	σ (Cross section for charge transfer) maximizes at $v'\simeq1,2$	85a
$H_2^+(X^2\Sigma_g^+)$	Franck–Condon distribution	Electron-impact ionization; high kinetic-energy ion beam	$H_2^+(H_2,H_2)H_2^+$ $H_2^+(D_2,H_2)D_2^+$	σ (Charge exchange) decreases with increasing ionizing electron energy	96
$H_2^+(X^2\Sigma_g^+)$	Franck–Condon distribution	Electron-impact ionization; high kinetic-energy ion beam	$H_2^+(H_2;H_2,H)H^+$	σ (Collision-induced dissociation) increases with increasing electron energy	117
$H_2^+(X^2\Sigma_g^+)$	0 to 3	Electron-impact ionization; electron velocity selector	$H_2^+(H_2,H)H_3^+$	Exothermic reaction; σ decreases with increasing v'	119
$H_2^+(X^2\Sigma_g^+)$	0 to 5	Photoionization–mass spectrometry	$H_2^+(H_2,H)H_3^+$	Exothermic reaction; σ decreases with increasing v'	85a,91
$H_2^+(X^2\Sigma_g^+)$	Franck–Condon distribution	Electron impact–mass spectrometry	$H_2^+(He,H)HeH^+$	It was demonstrated that this reaction possesses a vibrational threshold	99a
$H_2^+(X^2\Sigma_g^+)$	0 to 7	Photoionization–mass spectrometry	$H_2^+(He,H)HeH^+$	Endothermic for $v'<3$; σ increases with increasing v'; excitation functions measured for different v' values	85a,92
$H_2^+(X^2\Sigma_g^+)$	Franck–Condon distribution	Crossed ion–neutral beams; electron-impact ionization	$H_2^+(He,H)HeH^+$	Excitation functions measured for different ionizing electron energies; importance of vibrational energy decreases with increasing ion translational energy	93
$H_2^+(X^2\Sigma_g^+)$	Franck–Condon distribution	Merging ion–neutral beams; electron-impact ionization	$H_2^+(He,H)HeH^+$	Measurements of product-ion translational-energy distributions	94

Ion	Internal-energy distribution	Method	Reaction	Comments	Ref.
$H_2^+(X^2\Sigma_g^+)$	Franck–Condon distribution	Ion-cyclotron resonance (ICR); electron-impact ionization	$H_2^+(He,H)HeH^+$	Vibrational deactivation of ions observed on collisions with helium	95a
$H_2^+(X^2\Sigma_g^+)$	0 to 5	Photoionization–mass spectrometry	$H_2^+(He;H,He)H^+$	Collision-induced dissociation; excitation functions measured for different ν' values	85a, 9
$H_2^+(X^2\Sigma_g^+)$	Franck–Condon distribution	Electron impact–mass spectrometry	$H_2^+(Ne,H)NeH^+$	It was demonstrated that this reaction possesses a vibrational threshold	99a
$H_2^+(X^2\Sigma_g^+)$	0 to 7	Photoionization–mass spectrometry	$H_2^+(Ne,H)NeH^+$	Endothermic for $\nu' < 2$; σ increases with increasing ν'	85a, 9
$H_2^+(X^2\Sigma_g^+)$	Franck–Condon distribution	Electron-impact ionization; high-kinetic-energy ion beam	$H_2^+(Ar,H_2)Ar^+$	σ (Charge exchange) decreases with increasing ionizing electron energy	96
$H_2^+(X^2\Sigma_g^+)$	Franck–Condon distribution	Photoionization–mass spectrometry	$H_2^+(Ar,H)ArH^+$	Exothermic reaction; σ insensitive to ν'	85a, 9
$D_2^+(X^2\Sigma_g^+)$	Franck–Condon distribution	Electron-impact ionization; high-kinetic-energy ion beam	$D_2^+(H_2,D_2)H_2^+$	σ Decreases with increasing ionizing electron energy	96
$He_2^+(X^2\Sigma_u^+)$	Up to 1 eV of internal energy	Ion beam–collision cell; Hornbeck–Molnar process produces He_2^+ via $He^* + He \to He_2^+ + e$	$He_2^+(He;He,He)He^+$	Collision-induced dissociation; vibrational deactivation of ions observed on collision	97
$N_2^+(X^2\Sigma_g^+)$	Franck–Condon distribution	Electron-impact ionization; high-kinetic-energy ion beam	$N_2^+(N_2,N_2)N_2^+$	σ (Charge exchange) decreases with increasing ionizing electron energy (i.e., with decreasing population of $N_2^+ (\nu'=0)$	38a, 98b, 99b
$N_2^+(X^2\Sigma_g^+)$	Franck–Condon distribution	Electron-impact ionization; high-kinetic-energy ion beam	$N_2^+(Ar,N_2)Ar^+$	$N_2^+ (\nu'=1,2)$ Levels exhibit near-resonant behavior in charge transfer with argon	100
$N_2^+(X^2\Sigma_g^+)$	Franck–Condon distribution	Electron-impact ionization; high-kinetic-energy ion beam	$N_2^+(H_2,N_2)H_2^+$	Cross sections decrease slightly with increasing electron energy	101
$O_2^+(X^2\Pi_g)$	> 6	Photoionization–mass spectrometry	$N_2^+(D_2,N_2)D_2^+$ $O_2^+(CO,O)CO_2^+$	Endothermic reaction for $\nu' < 6$	52a

Table II. *continued*

ION	VIBRATIONAL LEVELS v'	METHOD	REACTIONS STUDIED	COMMENTS	REFERENCE
$O_2^+(X^2\Pi_g)$	>9	—	$O_2^+(NO,O)NO_2^+$	At least one of the products is in an excited state	102
$O_2^+(X^2\Pi_g)$	>0	Ion beam; electron impact	$O_2^+(O_2,O_2)O_2^+$	σ (Charge transfer) increases with increasing population of high v'	103
$O_2^+(X^2\Pi_g)$	>1	Ion-cyclotron resonance; electron-impact ionization	$O_2^+(Xe,O_2)Xe^+$	Reaction slightly endothermic for $v'=0$	23c
$O_2^+(a^4\Pi_u)$	5 to 10	Photoionization–mass spectrometry	$O_2^+(O_2,O)O_3^+$	Endothermic reaction for $v'<5$; σ rises with increasing v'	44,45
$O_2^+(a^4\Pi_u)$	0 to 6	Photoionization–mass spectrometry	$O_2^+(H_2,H)HO_2^+$	Exothermic reaction	49
$O_2^+(a^4\Pi_u)$	>0	Photoionization–mass spectrometry	$O_2^+(CO,O)CO_2^+$	Exothermic reaction	52a
$O_2^+(a^4\Pi_u)$	>4	Photoionization–mass spectrometry	$O_2^+(N_2,O)N_2O^+$	Endothermic reaction for $v'<3$	52b
$CO^+(X^2\Sigma^+)$	Franck–Condon distribution	Ion beam; electron impact	$CO^+(CO,CO)CO^+$	σ (Charge exchange) decreases with increasing ionizing electron energy	99b
$NO^+(X^1\Sigma^+)$	>0	Ion beam; electron impact	$NO^+(NO,NO)NO^+$	σ (Charge exchange) increases with increasing ionizing electron energy (i.e., with increasing populations of high v')	103
$NO^+(X^1\Sigma^+)$	0 to 5	Photoion–photoelectron coincidence and TOF ion analysis	$NO^+(NO,NO)NO^+$	σ (Charge transfer) maximizes at $v'\simeq1,2$	86c
$NO^+(X^1\Sigma^+)$	0 to 4	Photoionization–mass spectrometry; photoelectron spectroscopy	$NO^+(i\text{-}C_4H_{10},HNO)C_4H_9^+$	Relative abundances of vibrational states were determined by PES	86d
H_3^+ or/and D_3^+	Up to 3 eV of excess internal energy	Ion beam–collision chamber (tandem mass spectrometer)	$D_3^+(X; D_2, X)D^+$ $X\equiv$ He, Ne	Compound $H_2^+ + H_2 \rightarrow H_3^+ + H$ produces highly excited reactant ions; these are deexcited on collision with H_2 at increased source pressures	105

Species	Energy range	Method	Reactions	Comments	Ref.
H_3^+ or/and D_3^+	Up to 3 eV of excess internal energy	Ion beam–collision chamber (tandem mass spectrometer)	$H_3^+(RH;2H_2)R^+$ $RH=CH_4,C_2H_4$ $H_3^+(RH;H_2,H)RH^+$ $RH=C_2H_4,C_2H_2$ $H_3^+(RH;H_2)RH_2^+$ $RH=CH_4,C_2H_2$	H_3^+ Ions having different internal-energy contents react differently with the same neutral; internal-energy content may thus be probed by measurements of branching ratios of endothermic to exothermic reaction channels	105
H_3^+ or/and D_3^+	Up to 3 eV of excess internal energy	Ion-cyclotron resonance	$H_3^+(CH_3X;H_2)CH_3XH^+$ $H_3^+(CH_3X;H_2,H)CH_3X^+$	Charge transfer and hydride-ion abstraction dominate at high H_3^+ internal energies; proton transfer dominates at low H_3^+ internal energies	106
H_3^+, D_3^+	0 to 13 (D_3^+)	Ion-cyclotron resonance and/or tandem Dempster–ICR mass spectrometer	$H_3^+(CH_3X,2H_2)CH_2X^+$ $H_3^+(CH_2X;HX,H_2)CH_3^+$ $X=CH_3,NH_2,OH,SH,F,C$ Reactions with $CH_4,C_2H_4,NH_3,D_2,H_2O,$ $H_2S,C_2H_2,C_2H_6,CH_3OH,Ar$	Random distribution of energy in reaction producing $H_3^+(D_3^+)$; at least 41% of D_3^+ reactant ions are produced with 0.9 eV or more of excitation energy	107
H_3^+, D_3^+	0 to 13 (D_3^+)	Crossed ion–neutral beams	$H_3^+(D_2,H_2)D_2H^+$;	Vibrational populations of H_3^+ determined from measurements of Q, the translational exoergicity	108a
D_3^+	0 to 13 (D_3^+)	Tandem Dempster–ICR	$H_3^+(Ar,H_2)ArH^+$ $D_3^+(H_2,HD)HD_2^+$; $D_3^+(H_2,D_2)H_2D^+$	Ratio of H_2D^+/HD_2^+ decreases with decreasing vibrational energy of D_3^+	108b,c
YH^+, YD^+ ; $Y=Ar,D_2,N_2,$ $CO_2,N_2O,CO,$ CH_4,Kr,O_2 Ne,He,Xe	0 to 15 (ArD^+)	Tandem Dempster–ICR	Assorted proton transfer, hydride-ion abstraction, and charge transfer	Reactant ions produced by ion–molecule reactions (e.g., $D_2^+ + Ar \rightarrow ArD^+ + D$)	109
NH_3^+	0 to 11	Photoionization–mass spectrometry	$NH_3^+(NH_3,NH_2)NH_4^+$	Exothermic reaction; σ decreases with increasing ν'	85a,8
NH_3^+	0 to 11	Photoionization–mass spectrometry	$NH_3^+(H_2O,OH)NH_4^+$	Exothermic reaction; σ is insensitive to ν'	85a,8
NH_3^+	0 to 11	Photoionization–mass spectrometry	$NH_3^+(H_2O,NH_3)H_3O^+$	Endothermic for $\nu'=0$, demonstrates a vibrational threshold	85a,8

Table II. *continued*

ION	VIBRATIONAL LEVELS v'	REACTIONS STUDIED	METHOD	COMMENTS	REFERENCE
$CH_3NH_2^+$	Several (unspecified) >0	$CH_3NH_2^+(CH_3NH_2, CH_4N)CH_3NH_3^+$	Photoionization–mass spectrometry	Rate coefficient drops with increasing photon energy	114b
$(CH_3)_2NH^+$	Several (unspecified) >0	$(CH_3)_2NH^+[(CH_3)_2NH,C_2H_6N](CH_3)_2NH_2^+$	Photoionization–mass spectrometry	Rate coefficient drops with increasing photon energy	114b
$(CH_3)_3N^+$	Several (unspecified) >0	$(CH_3)_3N^+[(CH_3)_3N,C_3H_8N](CH_3)_3NH^+$	Photoionization–mass spectrometry	Rate coefficient drops with increasing photon energy	114b
$CH^+(X^1\Sigma^+)$	Unspecified	$CH^+\left(CH_4; \begin{array}{ll} H_2 & C_2H_3^+ \\ H_2, H & C_2H_2^+ \\ H & C_2H_4^+ \end{array}\right)$	Ion-cyclotron resonance; reactant ion produced by He^+ charge transfer to methane	Ions are deexcited on collision with helium	110
$CH_2^+(X^2\Pi)$	Unspecified	$CH_2^+\left\{CH_4; \begin{array}{ll} H & C_2H_5^+ \\ H_3 & \\ H_2, H & C_2H_4^+ \\ & H_3^+ \\ 2H_2 & C_2H_2^+ \end{array}\right\}$	Ion-cyclotron resonance; reactant ion produced by He^+ charge transfer to methane	Ions are deexcited on collision with helium	110
CH_4^+		$CH_4^+(CH_4, CH_3)CH_5^+$	Photoionization–mass spectrometry	Exothermic reaction; σ insensitive to v'	104
$CH_2Br_2^+$	Distribution 0.0 eV, 0.20 eV, and 0.6 eV of internal energy	$CH_2Br_2^+(CH_2Br_2; Br, CH_2Br_2)CH_2Br^+$	Photoion–photoelectron coincidence spectroscopy	Translational and internal energy are equally effective in overcoming reaction endoergicity of collisional dissociation	86b
$C_2H_2^+(X^2\Pi_u)$	Distribution 0 to 2 in ν_2 mode (C–C stretching)	$C_2H_2^+(H_2, H)C_2H_3^+$	Photoionization–mass spectrometry	Reaction slightly endothermic for $v'=0$; σ increases with increasing v'	111

$C_2H_2^+ (X^2\Pi_u)$	Distribution 0 to 2 in ν_2 mode (C–C stretching)	Photoionization–mass spectrometry	$C_2H_2^+ \left(\begin{array}{c} C_2H_2,H_2 \\ C_2H_2,H \end{array} \right) \begin{array}{c} C_4H_2^+ \\ C_4H_3 \end{array}$	Ratio of $C_4H_2^+/C_4H_3^+$ decreases with increasing ν'	112
$C_2H_4^+$	Unspecified	Ion-cyclotron resonance and electron impact	$C_2H_4^+ (C_2H_4,CH_3)C_3H_5^+$	Rate coefficient for the reaction, which proceeds via an intermediate long-lived complex, decreases with increasing electron energy	113
$C_2H_4^+$	Unspecified	Photoionization–mass spectrometry	$C_2H_4^+ \left\{ \begin{array}{ll} H & C_4H_7^+ \\ CH_3 & C_2H_5^+ \\ C_2H_4; & \\ ? & C_2H_4^+ \\ ? & C_3H_6^+ \end{array} \right.$	Rate coefficient for disappearance of $C_2H_4^+$ decreases with increasing photon energy	114a
$C_2H_4^+$	A combination of overtones; 0 to 5	Photoionization–mass spectrometry	$C_2H_4^+ (C_2H_4,CH_3)C_3H_5^+$	σ Decreases with increasing ν'	80
$C_2H_5F^+$	A combination of overtones; 0.0 to 0.53 eV of internal energy	Photoionization–mass spectrometry	$C_2H_3F^+ \left(\begin{array}{c} CHF_2 \\ C_2H_3F, CH_2F \\ CH_2 \end{array} \right) \begin{array}{c} C_3H_5^+ \\ C_3H_4F^+ \\ C_3H_3F_2^+ \end{array}$	σ Values decrease with increasing internal energy; effect on $C_3H_3F_2^+$ is especially pronounced	115a
CH_2CO^+	A combination of overtones; 0.0 to 0.41 eV of internal energy	Photoionization–mass spectrometry	$CH_2CO^+ \left(\begin{array}{c} CO \\ CH_2CO, \\ 2CO \end{array} \right) \begin{array}{c} C_3H_4O^+ \\ \\ C_2H_4^+ \end{array}$	σ Decreases with increasing internal energy σ Increases with increasing energy	115b
$C_2H_4O^+$ (Ethylene oxide)	0,–	Photoionization–mass spectrometry	$C_2H_4O^+(C_2H_4O,C_2H_3)C_5H_5O^+$ $C_2H_4O^+(2C_2H_4O; CH_2O,H,$ $C_2H_4O)C_3H_5O^+$ (two-step process)	σ Decreases with increasing internal energy σ Increases with increasing internal energy	115c 115c
$C_3H_6^+, C_4H_8^+$	Unspecified	Ion-cyclotron resonance with electron impact	$C_3H_6^+(C_3H_6,CH_3)C_5H_9^+$ $C_4H_8^+(C_3H_8,CH_3)C_7H_{13}^+$	Rate coefficient decreases with increasing electron energy	113
$C_3H_6^+, C\text{-}C_3H_6^+$ $C_3D_6^+, C\text{-}C_3D_6^+$ Unspecified	Unspecified mass spectrometry	Photoionization–other than $C_3H_6^+$ or C_{36}^+	$C_3H_6^+(C_3H_6)$ Products increasing photon energy	Rate coefficient decreases with	114a

Table II. *continued*

ION	VIBRATIONAL LEVELS v'	METHOD	REACTIONS STUDIED	COMMENTS	REFERENCE
$C_6D_6^+$ (perdeuterated benzene)	> 0	Photoionization–mass spectrometry	$C_6D_6^+(C_6D_6,-)C_{12}D_{12}^+$	The product demonstrates a third-order dependence on pressure and is formed almost exclusively from the ground vibrational state of reactant ion	85
$C_6H_6^+$ (Benzene)	> 0	Photoionization–mass spectrometry	$C_6H_6^+(C_6H_6;-)C_{12}H_{12}^+$		
SF_6^-	Unspecified	Single-source electron-impact ionization–mass spectrometry	$SF_6^-(XF_6, SF_6)XF_6^-$; X = Se, Te, U $SF_6^-(TeF_6; SF_6,F)TeF_5^-$	The ion SF_6^- produced by thermal electron capture of SF_6 in a vibrational degree of excitation equal to electron affinity of the neutral	116a
SF_6^-	Unspecified	Single-source electron-impact ionization–mass spectrometry	$SF_6^-(X, SF_6)X^-$; X = O_2, NO_2, CS_2	The ion SF_6^- produced by thermal electron capture of SF_6 in a vibrational degree of excitation equal to electron affinity of the neutral	116b

absorption while the spontaneous emission, which is proportional to the population of the metastable state, is being recorded.

Discussion

Additional details on some of these methods are described in other sections of this review. Attempts have also been made to determine excited-state populations in single-source mass-spectrometric experiments from an analysis of ionization efficiency curves.[38a, d] There are several difficulties in applying such methods. For instance, it is now known from photoionization studies that ionization processes may be dominated by autoionization. Therefore, the onset of a new excited state is not necessarily characterized by an increased slope in the electron-impact ionization-efficiency curve, which is proportional to the probability of producing that state, as had been assumed earlier. Another problem arises because of the different radiative lifetimes that are characteristic of various excited ionic states (see Section I.A.4).

Electronically excited states of ions produced by electron-impact ionization, for which various ion–neutral reactions have been studied, are listed in Table I. Additional information included in Table I, such as ion lifetimes, are discussed in subsequent sections.

By increasing the ionizing electron energy, it is possible, to shift the vibrational-state population of the ground electronic state of the ion to higher levels. Results of such experiments are included in Table II, which summarizes ion–neutral reactions of vibrationally excited ions that have been reported in the literature. Vibrational populations of ground-state ions, formed as a result of direct ionization processes, can be computed, at least for diatomic ions such as N_2^+, from the squares of the overlap integrals of the relevant vibrational wave functions of the neutral and ionic states (Franck–Condon factors).[99] This procedure is again complicated by the occurrence of autoionization and also by radiative decay from excited electronic states of the ion.

2. Photoionization

The technique of photoionization mass spectrometry is more difficult to apply experimentally than electron-impact ionization. It has, however, several advantages. First, when two reactants are introduced in a mixture into a single ion source, it is rather easy to ionize one without ionizing the other, by choosing the appropriate wavelength (i.e., photon energy) of the incident radiation. Second, the energy resolution achievable with photoionization is very high (typically 0.01 eV, but resolution of 0.0003 eV has been achieved) and the internal energy of the ionic species can thus be readily controlled. The pioneering work on internal-energy effects on ion–neutral interactions using photoionization techniques was carried out by Chupka, who has written several reviews on this topic.[85]

When a molecule M is photoionized,

$$M + h\nu \rightarrow M^+ + e \qquad (I.3)$$

the electron e can carry off varying amounts of kinetic energy, KE_e, with the only requirement that the relation

$$KE_e = h\nu - IP(M) \qquad (I.4)$$

where $IP(M)$ is the lowest adiabatic ionization potential of the molecule, must be satisfied. In photoionization mass spectrometry the ion current of M^+ is measured as a function of photon energy $h\nu$. Thus although $h\nu$ is very well defined since KE_e is unknown, the specific internal energy state of M^+ is not well defined. As demonstrated by Chupka, [85] however, it is possible to determine the distributions of internal states of reactant ions by analysis of photoionization efficiency curves. Alternatively, it is possible to state select the *ground* vibrational state of the electronic ground state of the ion.

There are two alternative methods for determining internal-state distributions in single-source photoionization experiments, depending on the prevailing type of ionization process.

a. Method No. 1

Direct ionization produces a staircaselike structure in the plot of ion current as a function of photon energy, where the height of each step is proportional to the probability of production of a certain vibronic state of the ion. Such favorable cases of staircaselike structure have been observed for ammonia[87] and acetylene.[88] The structure in ammonia is attributable to excitation of successive vibrational levels of the out-of-plane bending mode of the ion and in acetylene, to excitation of the C–C stretching mode. As a result, these molecules are favorable candidates for studying the effects of vibrational excitation on the cross sections for ion–molecule reactions.

b. Method No. 2

In many other cases, with the hydrogen molecule as the classical example, ionization is dominated by autoionization.[89] In this case the ionizing process actually occurs in two steps:

$$M + h\nu \rightarrow M^* \qquad (I.5a)$$

$$M^* \rightarrow M^+ + e \qquad (I.5b)$$

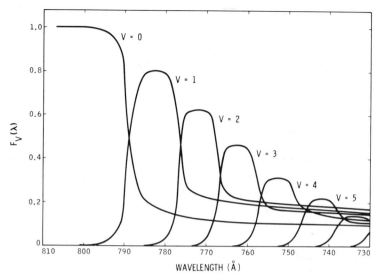

Figure 3. Photoionization of H_2. Approximate distribution of H_2^+ ions produced in vibrational state, v, as function of photon wavelength;[95b] calculated from low-resolution data.[91,92]

where M* is the excited neutral state of M. The photoionization efficiency curve in this instance consists of a large number of sharply spiked peaks ("resonances") attributable to the various M* states that autoionize. The spacings between adjacent vibrational levels of M^+ are sometimes independently known (as for H_2^+), but a staircaselike structure as a result of direct ionization producing these levels is still not observed. Interpretation of the photoionization efficiency curve then requires knowledge (from theory) of the so-called propensity rule that governs the vibrational level in which M^+ will be formed in reaction (I.5b). The propensity rules predict for this step $\Delta v = -1$ so that M^+ is preferentially formed in the highest vibrational state energetically possible.[90] Application of this rule then permits calculation of the distribution of vibrational states of photoions as a function of photon energy. This method has been applied to reactant H_2^+ ions[91,92] (Fig. 3).[95b]

c. Other Methods

In a further refinement of the photoionization method the kinetic energy of the photoelectrons produced is measured in coincidence with the photoion current; this is the so-called photoion photoelectron coincidence (PIPECO) method. By this method, reactions of state selected ions can be directly monitored. This technique has been successfully applied to study unimolecular decompositions of energy-selected ions[86a] as well as to bimolecular interactions.[86b, c] In one version of the method, ions are measured in coincidence with zero kinetic-energy electrons, so that the internal energy of the reactant ion is equal to the photon energy minus the adiabatic ionization potential of the molecule. With a diatomic molecule such as NO^+, it is possible to select the vibrational level of the reactant by varying the photon energy.[86c] This is obviously a very sophisticated method and promises to yield further interesting and valuable information. However, most of the work to date has concentrated on single-source photoionization experiments, in which a distribution of internal states of the reactant ion was employed.

With more conventional photoelectron spectroscopy (PES), the photon energy is fixed and the distribution of kinetic energies of the electrons produced by photoionization determined. The latter distribution reflects the distribution of ion internal states produced. Photoelectron spectroscopy has been applied to determine vibrational-state distributions at the same photon wavelengths at which an ion–molecule reaction was studied using photoionization methods (Fig. 4).[86d]

Photoionization has been most successfully applied to study of the effects of reactant vibrational energy on the cross sections of exothermic and endothermic ion–molecule reactions. Reactions studied by this method are summarized in Table II. In addition, photoionization methods

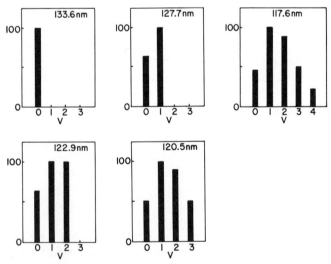

Figure 4. Vibrational-intensity distribution for ground state of NO^+ as determined by photoelectron spectroscopy at five wavelengths: (a) 133.6 nm, (b) 127.7 nm, (c) 122.9 nm, (d) 120.5 nm, (e) 117.6 nm.[86d]

have been employed for correct identification of the nature of the excited electronic state of reactant ions in certain ion–molecule reactions— notably O^{+*} in the reaction $O_2^+(O_2,O)O_3^+$ (see Table I). The energies of the $O_2^+(^2\Pi_u, v=0)$ and $O_2^+(a^4\Pi_u, v=5)$ states are very close, and only by utilizing a high-energy-resolution technique such as photoionization could the onset of the ion–molecule reaction be shown to correspond to the energy of the fifth vibrational level of the $a^4\Pi_u$ state.[44,45] This reaction and other similar processes are discussed in greater detail in later sections of this review.

3. Charge Transfer; Ion–Molecule Reactions; Penning Ionization

Charge transfer can produce reactant ions in fairly well defined internal-energy states. Important work in this area was initially conducted by Lindholm and co-workers.[6] In an exoergic-charge transfer reaction of the general type

$$A^+ + M \rightarrow A + M^+ \tag{I.6}$$

no light particle is available to carry off the excess energy. If the reaction as shown is exoergic, the excess energy has to appear as internal energy of the products (electronic, vibrational, or rotational). An important quantity in discussing such reactions is the so-called recombination energy, RE, of the ion, A^+, which is that energy released when A^+ recombines with an electron. If A^+ is an atomic ion characterized by a single recombination energy, M^+ will be formed in a well-defined state of internal energy given by $RE(A^+) - IP(M)$. However, the situation is often complicated by the fact that several electronic states of A^+ may be formed, each of which possesses a different recombination energy. Furthermore, if A^+ is polyatomic, a continuous range of RE values is possible such that

$$RE(A^+) \leqslant IP(A) \tag{I.7}$$

since, depending on the relevant transition probabilities, A may be left in various excited vibrational states. Lindholm has compiled tables listing best estimates of recombination energies of various atomic and molecular ions.[6] In many cases these are not well known.

Most of the studies of ions formed by charge transfer have been concentrated on the unimolecular reactions of M^+ ions formed in well-defined internal-energy states (e.g., fragmentation patterns[6]) and more recently have been concerned with rate-coefficient measurements.[118] Some work has also been reported on consecutive ion–molecule reactions of M^+ ions produced in well-defined internal states (mostly electronically excited states) by charge transfer. Results of such experiments are included in Table I [reactions of $O^+(^2D)$;[4c] $CO^+(A^2\Pi$ and $B^2\Sigma_u^+)$;[79a] $C_2H_2^+(^2\Sigma_g^+)$[71,72,79b]]. In many instances it is possible to deduce the excitation energy of ionic reactants from the reactions observed. This holds also for vibrationally excited ions produced by ion–molecule reactions,[105–109] such as H_3^+, or by Penning ionization,[97] such as He_2^+ (see Table II).

One very important technique for probing excited states of products of charge transfer and for determining their populations is that of light emission or luminescence measurements. These methods are discussed in greater detail in Section IV.A.1.

4. Lifetimes of Excited Ionic States

Obviously, the various electronically excited states of an atomic or molecular ion vary in their respective radiative lifetime, τ. The probability distribution applicable to formation of such states is thus a function of the time that elapses following ionization. Ions in "metastable" states, which have no allowed transitions to the ground state, are most likely to contribute to ion–neutral interactions observed under *any* experimental conditions since these states have the longest lifetimes. In addition, the experimental time scale of a particular experiment may favor some states over others. In single-source experiments, short-lived excited states may be of greater relative importance than in ion-beam experiments, in which there is typically a time interval of a few microseconds between ion formation and the collision of that ion with a neutral species, so that most of the short-lived states will have decayed before collision. There are several recent compilations of lifetimes of excited ionic states.[1h,120,121]

For some very long-lived ionic states, there have been no spectroscopic determinations of the radiative lifetimes. Studies of ion–molecule reactions of the states have contributed to the knowledge of their lifetimes, as is exemplified for the $(N_2^+)^*$ case.[26,28,31,33,34] The reaction sequence by which N_3^+ is produced in a single-source mass-spectrometric experiment is as follows:

$$N_2 + e \xrightarrow{k_i} N_2^+ + 2e \qquad (I.8a)$$

$$N_2 + e \xrightarrow{k_e} (N_2^+)^* + 2e \qquad (I.8b)$$

$$(N_2^+)^* + N_2 \xrightarrow{k_r} N_3^+ + N \qquad (I.8c)$$

$$(N_2^+)^* \xrightarrow{\tau^{-1}} (N_3^+)^\dagger + h\nu \qquad (I.8d)$$

where $(N_2^+)\dagger$ represents an electronic state of N_2^+ that cannot undergo reaction with N_2 to produce N_3^+. For this reaction scheme, the appropriate kinetic expression is

$$\frac{I_{N_3^+}}{I_{N_2^+}} = \frac{(k_e/k_i)k_r[N_2]}{\tau^{-1}+k_r[N_2]}\left(1-\exp\left\{-\left(\tau^{-1}+k_r[N_2]\right)t\right\}\right) \qquad (I.9)$$

where t is the time between ion formation and ion detection and $[N_2]$ is the density of N_2. Experimentally, a pulsed mass-spectrometric method has been employed to study the kinetics of this system.[28] The ratio $I_{N_3^+}/I_{N_2^+}$ was determined as a function of the delay time between the electron-beam pulse and the ion-drawout pulse for a range of ion-source pressures. For a certain fixed pressure at short delay times, $I_{N_3^+}/I_{N_2^+}$ rises, and at longer delay times the ratio attains a constant level. A plot of the asymptotic values $(I_{N_2^+}/I_{N_3^+})_{t=\infty}$ versus $[N_2]^{-1}$ yields, according to (I.9), a straight line, the intercept and slope of which are (k_i/k_e) and $(k_i/k_e)(1/\tau k_r)$, respectively. Furthermore, since

$$\lim_{t\to 0}\frac{d(I_{N_3^+}/I_{N_2^+})}{dt}=\left(\frac{k_e}{k_i}\right)k_r[N_2] \qquad (I.10)$$

$(k_e/k_i)k_r$ is obtained. These experiments thus yield values of $k_e/k_i, k_r$, and τ. The value of τ obtained for (N_2^{+*}) reacting under these specified conditions is on the order of several microseconds.[28,31,34a]

Utilizing ionization efficiency curves to determine relative populations of vibrationally excited states (as in the photoionization experiments) is a quite valid procedure in view of the long radiative lifetime that characterizes vibrational transitions within an electronic state (several milliseconds). However, use of any ionization efficiency curve (electron impact, photon impact, or photoelectron spectroscopic) to obtain relative populations of electronically excited states requires great care. A more direct experimental determination using a procedure such as the attenuation method is to be preferred. If the latter is not feasible, accurate knowledge of the lifetimes of the states is necessary for calculation of the fraction that has decayed within the time scale of the experiment. Accurate Franck–Condon factors for the transitions from these radiating states to the various lower vibronic states are also required for calculation of the modified distribution of internal states relevant to the experiment.[99b,102]

Radiative lifetimes have been included in Table I wherever available; some of these are to be found in individual articles cited, for example, $\tau(H_2O^+, {}^2A_1)$ is from Möhlmann and Deheer.[122]

B. Excitation of Neutral Reactants

Comparatively few studies have been conducted to assess the effects of internal excitation of neutrals on ion–molecule reactions. Some experiments of this type have been reported by Schmeltekopf et al,[123] who investigated reactions of various ions with vibrationally excited N_2, produced by passing it through a microwave discharge. The mechanism producing vibrationally excited N_2 in this case probably involves a temporary negative ion state N_2^- that is known to autoionize. The vibrational temperature, T_v, was varied between 300°K and 6000°K in these experiments by varying the discharge power or by changing the position of the microwave cavity on the flow tube, thus changing the extent of vibrational relaxation of the excited N_2 molecules on the walls of the tube before they entered the main reaction region. The temperature T_v was determined by vertical ionization of the $N_2(v)$, a process governed by known Franck–Condon factors, to a radiating state and observation of the light emitted. Two ionization methods were employed, electron impact and Penning ionization. Microwave discharge excitation has also been employed to produce electronically excited oxygen molecules, $O_2(a^1\Delta_g)$ ($\tau = 45$ min).[129] The discharge products were passed over a glass-wool plug containing mercuric oxide to remove atomic oxygen prior to reaction with ions.

It is also possible to obtain excited neutral species by heating the molecules in a furnance. This method was employed to obtain a vibrationally excited N_2 beam that was reacted with O^+ ions.[127] Since the molecules undergo a large number of collisions with the walls of the furnace before escaping into the beam, a Boltzmann distribution of internal-energy states is established. With such an apparatus, the source temperature is measured by an optical pyrometer and is typically in the range 1000–3000°K. Several reactions of ions with excited neutrals are listed in Table III.

C. Methods for Studying Excited Ion–Neutral Interactions

A considerable array of instrumentation has been developed in the past 10–15 years for the study of ion–neutral collisions. Such instruments can be classified into several categories, with each type having features that make it suitable for investigating particular aspects of the ion–neutral interaction. The present discussion is limited to those experimental techniques that have been specifically applied for examining the role of excited states in these processes.

1. Beam Techniques

Ion-beam apparatuses have been utilized extensively to study the reactions of excited ions with neutrals. The term *beam apparatus* as used in this

Table III.
Reactions of Ions with Excited Neutrals

REACTANT	EXCITATION	METHOD	REACTION STUDIED	COMMENTS	REFERENCE
H_2	Rotational $J=0; 1$	Low-energy ion beam–collision cell	$Ar^+(H_2,H)ArH^+$	Inverse dependence of cross section on rotational energy	124
H_2	Vibrational	Flowing afterglow–microwave discharge	$NH_3^+(H_2,H)NH_4^+$	Vibrational excitation of H_2 promotes reaction	125
$N_2(X^1\Sigma_g)$	Vibrational	Flowing afterglow–microwave discharge excitation of N_2	$He^+(N_2,He)N_2^+$ $He^+(N_2;He,N)N^+$	Rate constant for disappearance of He^+ is insensitive to N_2 vibrational temperature; branching ratio favors N^+ at higher vibrational temperatures	123
$N_2(X^1\Sigma_g)$	Vibrational	Flowing afterglow–microwave discharge excitation of N_2	$O^+(N_2,N)NO^+$	Although being exothermic, rate constant increases by a factor of $\sim$40 with increasing N_2 vibrational temperature, with a steep rise at $v=2$	123, 126
$N_2(X^1\Sigma_g)$	Vibrational	Crossed molecular beams–furnace heated N_2	$O^+(N_2,N)NO^+$	At high relative translational energies, reaction cross section is hardly dependent on N_2 vibrational temperature	127

Table III. *continued*

Reactions of Ions with Excited Neutrals

REACTANT	EXCITATION	METHOD	REACTION STUDIED	COMMENTS	REFERENCE
$N_2(X^1\Sigma_g)$	Vibrational	Merging molecular beams	$O^+,(^2D)(N_2,N)NO^+$	—	4d
$N_2(X^1\Sigma_g)$	Vibrational	Flowing afterglow–microwave discharge	$Ne^+(N_2,Ne)N_2^+$	Although being exothermic, rate constant increases by several orders of magnitude with increasing N_2 vibrational temperature, with a steep rise at $v=2$	128
$O_2(a^1\Delta_g)$	Electronic	Flowing afterglow–microwave discharge	$O^-(O_2,O_3)e$ $O_2^-(O_2,2O_2)e$	—	129
$O_2(a^1\Delta_g)$	Electronic	Crossed molecular beams–microwave discharge	$O^-(O_2,O)O_2^-$	Excitation function measured and compared with that for ground state	130

section is intended to indicate a device in which a well-defined ion beam is produced, undergoes collision with a neutral target species in a region separated and isolated from the region of formation, and the resultant product ions are detected and their mass and abundance measured. A typical beam apparatus may include the following major components:

1. Ion source. Ionization and formation of excited ionic states can be accomplished by any of the methods described earlier in this section (e.g., electron impact).
2. Reactant-ion mass analyzer. Mass analysis is usually achieved in a magnetic sector, although in some apparatuses quadrupole mass filters have been used.
3. Reactant-ion energy analyzer. An electric sector or a retarding lens (or some combination of these) is generally employed to define the translational energy of the ion beam.
4. Collision region. The neutral reactant can be introduced into the collision region in the form of a neutral beam, having well-defined velocity and orientation (e.g., produced by a nozzle), or neutrals may be admitted to a collision chamber, where their orientation is random and the velocity is characterized by a Boltzmann distribution.
5. Product-ion mass and energy analyzer. These are of the same type as those used for the reactant ion beam. Some beam instruments provide only for mass analysis of the product ions.
6. Ion detector. Ions are usually detected by an electron multiplier, using appropriate counting and recording devices.

Beam apparatuses incorporating some or all of these features have been developed in several laboratories.[4b, 5a, 38c, 48, 50, 53, 62, 66, 108a, 127, 130–137] One such apparatus, which has been employed by Rutherford and co-workers[53] to study a variety of excited ionic reactions, is illustrated schematically in Fig. 5. In this apparatus the neutral beam is modulated at 100 Hz by a mechanical chopper, and a phase-sensitive detection system is used to monitor the product ions. A very weak electric field (2 V/cm) is employed to extract ions from the collision region, and great care is taken to shield this region from the large accelerating fields used in the apparatus so that the ion–neutral interaction energy can be defined with reasonable certainty. The ion beam is also collimated to ensure that all reactant ions pass through the neutral beam. The reactant ion-beam intensity can be measured at the interaction region by using a movable Faraday cup, which facilitates cross-section determinations. A major experimental uncertainty associated with the apparatus of Rutherford et al.,[53] as well as with all beam apparatuses, is the collection efficiency for product ions. Because minimal extraction fields are used in such instruments, to avoid perturbing

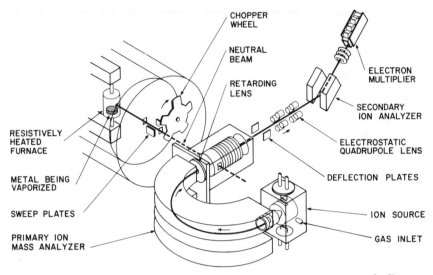

Figure 5. Schematic diagram of crossed ion–neutral beam apparatus.[53]

the ion–neutral interaction energy, it cannot be assured that all product ions are collected by the analyzer stage of the apparatus. Generally, an attempt is made to assess the collection efficiency, as Rutherford et al. have described,[53] but this is difficult because of the many experimental factors that control ion collection. Other major uncertainties associated with such beam experiments are the neutral beam density and the effective collision volume, both of which also affect the accuracy of cross-section measurements. More comprehensive discussions of the limitations of beam experiments with respect to kinetic determinations have been presented elsewhere.[138, 139]

Beam techniques are quite effective for studying the translational energy dependences of ion–neutral cross sections, and such methods have been the principal source of such data that have been reported in the literature. The translational energy regime accessible to beam apparatuses generally ranges from a lower limit of several tenths of an electron volt (incident ion energy in the laboratory scale) to energies as high as several kiloelectron volts. Relatively good definition of reactant ion-beam energies over a wide energy range is thus possible. Much lower interaction energies are possible with the merging-beam technique developed by Neynaber et al.[135, 136]

In studying the reactions of internally excited ions using beam instrumentation, an important consideration is the time required for the reactant ion to move from the region of formation (the ion source) to the region where collision with a neutral occurs. As already mentioned, this transit time is typically on the order of several microseconds, and thus excited states that have shorter lifetimes will decay prior to arrival at the collision region. Obviously this limits the ionic states that can be studied with these techniques. On the other hand, because regions of ion formation and interaction are separated in a beam apparatus and both regions are generally at relatively low pressures, collisional deactivation of internally excited ions is minimized. The latter is often a problem with other experimental methods used for studying ion–neutral collisions. Beam instruments of the type discussed here are most commonly used to study bimolecular ion–neutral events. Other techniques are more suitable for higher-order processes.

2. Afterglows

The flowing afterglow technique, which has been used in numerous ion–neutral reaction studies, has been described in great detail by Ferguson et al.[140] A buffer gas, normally helium, is admitted to a flow tube fitted with a high-capacity pumping system, and a fast gas flow ($v \sim 10^4$ cm/sec) is established. The buffer-gas pressure in the tube can be varied, typically from 0.1 to 5 torr by varying the helium flow or the pumping speed. The gas passes through an excitation region where ions are produced by electron impact (usually from a filament source, although microwave and DC discharges have also been used) and are carried down the tube in the flowing gas. A small amount of a suitable reagent gas is injected either with the helium or into the helium afterglow immediately after the ionization region. Reactant-ion species are formed and carried past another injection port, through which the neutral reactant is introduced. Reactions between reactant ions and neutrals occur in the region following the neutral injection port. At the end of the reaction zone is a sampling orifice and a quadrupole mass filter, with which ions in the flowing gas are sampled, mass analyzed, and counted. Variation of an ion intensity as a function of reactant concentration yields the reaction-rate constant. Reaction rates can also be deduced from experiments in which the neutral reactant flow is fixed and the gas velocity varied. Accurate rate-constant measurements require the application of a detailed mathematical model for the hydrodynamics and diffusion in the flowing gas stream.[140]

Reactant ions are produced in the flowing afterglow apparatus by either direct electron impact or secondary ion–neutral or helium metastable reactions. These reactant ions undergo many collisions with the buffer gas prior to reaching the region where they react with neutral species and are thus effectively thermalized. Therefore, the flowing afterglow experiment is designed exclusively for the study of thermal-energy ion–neutral processes, and there is no means for translational-energy variation. Apparently, under the usual conditions prevailing in the flowing afterglow, most reactant ions are in the ground electronic state because of the superelastic electron–ion collisions, which result in deexcitation of electronically excited ionic states. This has been demonstrated for atomic ions such as O^+ and can be checked experimentally by studying the dependence of reaction-rate constants on electron density. The vibrational-state distributions of molecular reactant ions in the flowing afterglow are not well known, although in some cases (e.g., N_2^+) it has been possible to determine that the reactant ions are in the ground vibrational state. The flowing afterglow technique, then, is not generally applicable for investigating reactions of excited ions. As mentioned in an earlier section, however, some afterglow experiments have been conducted in which the vibrational temperature of the neutral reactant was varied by using microwave excitation.[123, 128] Spectroscopic measurements were made to determine the vibrational-state distributions in this case. The same excitation technique was also employed to study reactions of selected ions with electronically excited oxygen molecules.[140]

A modification of the conventional flowing afterglow apparatus, in which a drift section is incorporated, is shown schematically in Fig. 6.[46a, 141] In the so-called flow-drift apparatus reactant ions are produced in the upstream section just as in the conventional afterglow system, but the downstream section, where reactions with neutrals occur, is a drift tube, in which a uniform electric drift field is applied. In the latter section ions can be accelerated from thermal kinetic energies to several electron volts. The two sections of the apparatus are separated by an electronic ion shutter, which makes it possible to admit narrow pulses of ions into the drift region at specified times. This permits measurements of ion-drift velocity and, in

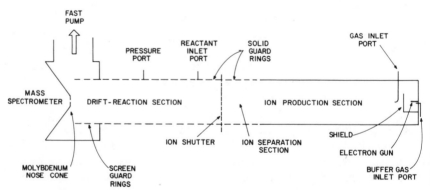

Figure 6.   Schematic diagram of flow-drift tube apparatus.[46]

turn, determinations of the ion–neutral scattering cross section. The flow-drift apparatus has recently been utilized by Ferguson et al.[46a] to study reactions of metastable O_2^+ ions with various molecules that do not react with ground-state O_2^+.

3. Mass Spectrometers—Single-chamber Experiments

Many of the initial experiments for study of ion–neutral interactions utilized slightly modified conventional mass spectrometers. Such experiments, which have been described in previous reviews,[1b] mainly involve elevating the ion source pressure so that ions formed in the initial ionization event undergo collisions with neutral molecules prior to leaving the source and entering the mass analyzer. Reactant ions are thus formed and engage in ion-neutral interactions in the same chamber. Thus it is possible, in principle, to study reactions of excited states having lifetimes shorter than those accessible to beam instruments. Complications often arise with such techniques because the neutral species from which reactant ions are formed and the species that serve as the neutral partner for the ion-neutral collision are simultaneously present in the interaction chamber. Since there is no mass selection of reactant-ion species, all ions formed in the initial ionization event may undergo reactions with neutral species present, and thus a multitude of ion–neutral reactions may occur simultaneously.

The complications just described can be minimized if there is greater selectivity in the ionization process, as is sometime possible when photo-ionization is used as the excitation mechanism. Because the ionization energy can be more precisely controlled, it is possible in selected cases to produce only the desired reactant-ion species, or at least to minimize production of other ions. As already noted in the earlier section on formation of excited ions, it is also possible to populate specific internal-energy states of some reactant ions by using a photoionization source. One of the earliest photoionization mass spectrometers used to study interaction of internally excited ions with neutrals was that constructed by Chupka et al.[91] Such apparatuses typically incorporate a photon source (either a line or a continuum source) and an optical monochromator, which are coupled to the reaction chamber. Various types of mass analyzer, including sector type, time-of-flight (TOF), and quadrupole mass filters, have been used with these apparatuses. Chupka has described the basic instrumental configuration in some detail.[85a] Photoionization mass spectrometers employed to study interactions of excited ions with neutral species have also been constructed in several other laboratories.[80, 114a, 142, 143] The apparatus recently developed by LeBreton et al.[80] is illustrated schematically in Fig. 7 and is typical of such instrumentation.

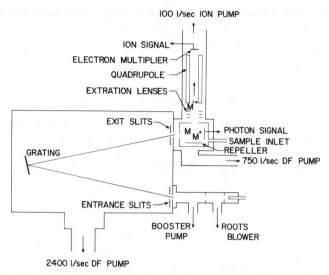

Figure 7. Schematic diagram of photoionization mass spectrometer. Axis of quadrupole mass spectrometer is parallel to exit slit of monochromator; mass spectrometer has been rotated 90° for purposes of illustration here.[80]

Specialized adaptations of the conventional electron-impact mass-spectrometric ion source have also been utilized to study internal-energy effects on the reactions of selected ions. For example, Ryan and co-workers[23e, 144, 145] have employed the ion-trapping technique, in which ions are produced by a short pulse (10^{-6} sec) of energetic electrons and are then trapped in the space-charge well created by an electron beam whose energy is below that required to cause further ionization ($\sim$4 eV). Typically, primary ions can be retained in the trapping field for a few milliseconds, and during this time they collide and react with neutral molecular species in the chamber. The reaction is finally terminated by application of an ion-ejection pulse that both destroys the space charge and accelerates the reactant and product ions into the mass analyzer.

The apparatuses described thus far were all designed to detect the ionic products from ion–neutral interactions, whereas the TOF apparatus described by Moran and co-workers[99b, 146, 147] has been employed to detect the neutral products from these processes. In the latter device the ions formed by electron impact are accelerated out of the source by an extraction pulse and are further accelerated by a series of grids. The primary beam is then focused by deflection and enters the drift region or flight tube, where mass separation of the different velocity groups occurs.

As the ions move through these regions, charge-transfer reactions can occur, and the fast neutral products of these reactions are separated from the ion products by applying a postacceleration voltage to a grid immediately in front of the electron-multiplier detector. This postacceleration causes ionic species to reach the detector ahead of neutral reaction products having the same mass. Actually, this apparatus is neither a true beam instrument nor a single-chamber device, but rather a hybrid between the two. Another hybrid multisection reaction-cell apparatus applied for the study of charge-transfer processes of excited ions (which is quite different from that just discussed here, however) has been discussed by Hayden and Amme.[96]

4. Ion-cyclotron Resonance Spectrometers

The operating principles of ion-cyclotron resonance (ICR) spectrometers and their application for ion–neutral reaction studies have been discussed in some detail in a recent review.[148] A schematic of a commonly used ICR reaction cell is presented in Fig. 8. Ions produced by electron impact in the source region (A in Fig. 8) are caused to drift in the X direction (see Fig. 8) following a cycloidal trajectory, by the application of a uniform magnetic field oriented along the Z axis and a DC electric field in the Y direction. The characteristic frequency of revolution is ω_c. In region B of the cell a DC electric field is applied in the Y direction, which maintains ion migration in the X direction. In addition, an RF (radiofrequency) electric field is applied normal to the magnetic field. If the frequency of the RF field, ω_{RF}, is equal to ω_c, an ion will absorb energy from the RF field and be accelerated. Thus the circular component of the ion's cycloidal trajectory will increase. Since ω_c for any given ion depends only on its mass and the magnetic field strength, a mass spectrum can be obtained by measuring the power absorbed from the RF electric field while sweeping either the RF electric field or the magnetic field. Various modifications of the ICR cell itself and the mode of operation have been described elsewhere.[148] The ICR technique is particularly suited for investigation of ion–neutral reactions at very low pressures and low kinetic energy (thermal or near-thermal), which is a consequence of the relatively long ion path lengths and residence times in the cell, which result from the cycloidal ion trajectories. Anicich and Bowers[149] have conducted several investigations of excited-ion reactions using both the more conventional ICR drift cell[149] and the trapped-ion cell developed originally by McMahon and Beauchamp.[150] Experiments of this type have also been accomplished using an ICR spectrometer in which reactant ions are generated using a continuum photon source and a monochromator.[151] Smith and Futrell[152] have described a unique tandem-ICR instrument that has been applied to

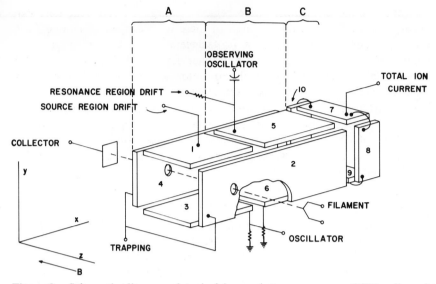

Figure 8. Schematic diagram of typical ion-cyclotron resonance (ICR) cell used for ion–molecule reaction studies. Regions A, B, and C designate ion source, analyzer, and ion collector regions, respectively. Electrodes 2 and 4 are used to apply trapping potential, 1 and 3 for source drift potential, 5 and 6 for analyzer drift and RF fields, and 7 to 10 for total ion collection.[148]

study a variety of internal-energy effects in ionic collision processes. The first stage of this latter apparatus is a 180° Dempster-type mass spectrometer, which produces a mass-analyzed ion beam. This beam is decelerated in energy to a few tenths of an electron volt and then injected into the cavity of an ICR spectrometer, where reactions with neutrals occur. Product ions are detected using the conventional ICR methods. This apparatus offers the advantage of superior control in ion preparation and a wider range of translational energy variation for the ion–neutral interaction than is normally possible in the ICR alone. At the same time, the advantages of the ICR method for product-ion detection (i.e., very little perturbation of the product species during measurement) are retained.

5. Luminescence Apparatuses

The instrumentation described in the foregoing sections is all based on mass analysis of the charged or neutral products of ion–neutral interactions. A different type of apparatus, which has been extensively utilized to characterize internally excited products from ion–neutral processes, is that in which an optical detector is employed to observe radiative emissions from the decay of excited-product states formed in those processes. Many of the earlier studies conducted with such apparatuses were concentrated on relatively high-energy ion–neutral collisions.[153–162] More recently, luminescence apparatuses have been developed that can be applied to study these processes at translational energies down to a few electron volts. Most of the instrumentation of the type that has been described in the literature consists of an ion-gun stage that produces a mass–energy-resolved beam of ions, which is directed into a collision chamber containing the neutral target molecules (or in some caes, the ion beam is crossed by a molecular beam), and a detector stage in which the optical emissions resulting from the collisions are resolved and detected using a monochromator and a photomultiplier. A representative example of such an apparatus developed by Hughes et al.[163] for lower-energy reaction experiments is shown schematically in Fig. 9. Similar devices have been described by several other authors,[9c, 12b, 164, 165] and that developed by Brandt et al.[9c] also permits mass analysis of product-ion species.

Facilities for monitoring luminescence from ion–neutral interactions have also been incorporated into other types of experimental apparatus, including afterglows,[166–170] an ion-cyclotron resonance spectrometer,[171, 172] and a drift tube.[173]

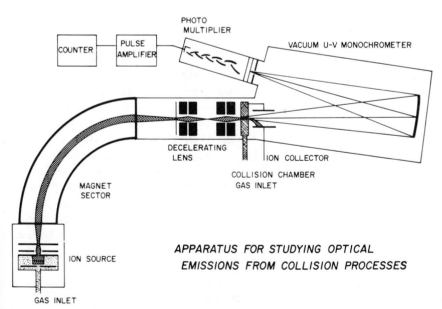

Figure 9.   Schematic diagram of apparatus at Wright State University for studying optical emissions from collision processes between ions and neutral molecules.[163]

II. INTERACTIONS OF EXCITED IONS WITH NEUTRALS

A. Reactive Scattering

1. Exoergic Reactions (Electronically, Vibrationally, and Rotationally Excited States)

Ion–neutral reactions that are exoergic for ground-state reactants do not require the excitation energy of the ion to overcome any endoergicity. Yet the reactivity of ions that are electronically and/or vibrationally excited is often different from that of the corresponding ground-state ions with the same neutral, even though the ground-state reaction is nominally exoergic. The excited state of the ion may be either more reactive or less reactive than the ground state, depending on the system and/or relative translational energy of the reactants. Among other factors influencing reactivity, the spin and symmetry of the reactants, products, and intermediate states appear to be important. For example, some (but by no means all) spin-forbidden reactions, such as

$$NO^+(^1\Sigma) + O_3(^1A) \rightarrow NO_2^+(^1\Sigma) + O_2(^3\Sigma) \tag{II.1}$$

are extremely slow despite the fact that they are exoergic.[174] Detailed understanding of these reactions requires knowledge of the potential surfaces on which the reactants interact, as discussed in more detail in a later section. Theoretical calculations of surfaces for systems involving as many as three interacting atoms are now feasible.

a. Charge Transfer

Several examples of exoergic charge-transfer reactions that proceed at different rates with ground-state and electronically excited ions are listed in Table I. In some cases the cross section for the excited-state reaction may be smaller than that for the ground state, as is the case for the reactions $Xe^+(O_2, Xe)O_2^+$; $Kr^+(N_2O, Kr)N_2O^+$; $Kr^+(CO_2, Kr)CO_2^+$, whereas in other instances the excited state is more reactive, as for the processes $N^+(Kr, N)Kr^+$, $N^+(CO, N)CO^+$, $O_2^+(Na, O_2)Na^+$, and $O_2^+(NO, O_2)NO^+$. The differences in reactivity are often more pronounced in the region of low ion translational energies[11b] (Fig. 10). The role of excited-state ions in charge-transfer reactions was reviewed by Hasted some time ago,[175] but much more experimental data has been obtained recently, as indicated by the data shown in Table I.

The factors that influence cross sections for charge-transfer reactions have not yet been completely assessed. Several theoretical models have been developed.[176–179] For asymmetric charge-transfer processes of the type

$$A^+ + B \rightarrow A + B^+ + \Delta E \tag{II.2}$$

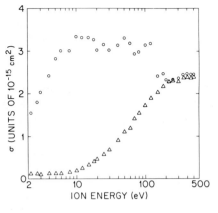

Figure 10. Cross sections for charge transfer from N^+ to Kr as function of reactant ion energy. Triangles represent data for ground state $N^+(^3P)$ and circles, for metastable $N^+(^1D)$.[11b]

where ΔE, the so-called energy defect, is the energy change in the collision, Massey's adiabatic hypothesis in its original form predicted very low cross sections for reactions of thermal and low-translational-energy ions. However, many thermal-energy charge-transfer reactions are observed to have very large cross sections, particularly when these involve molecular systems. In atom–atom charge exchange, factors such as accidental energy resonance,[157,180,181] which minimizes the energy defect ΔE, and crossings of potential curves, which cause ΔE to become very small at some relatively short internuclear distance, seem to be important. Accidental resonance may be achieved in some cases by production of one or both of the reaction products in an excited state. This is sometimes more easily realized with an excited-state ion reactant, as seems to be the case for the $N^+(^1D)$ reactions noted earlier,[11b] where the availability of excited product states that can carry away the exothermicity of the reaction is greater for the metastable ion reaction than for the ground-state reaction. Spin considerations dictate that production of excited $N(^2D)$ from reaction of $N^+(^1D)$ with krypton

$$N^+(^1D) + Kr(^1S) \rightarrow N(^2D) + Kr^+\left(^2P_{3/2}\right) \qquad (II.3)$$

is preferred over formation of ground-state $N(^4S)$. The recombination energy of $N^+(^1D)$ is thus 14.05 eV, which is nearly energy resonant with the ground-state ionization potential of krypton. On the other hand, the recombination energy of ground-state $N^+(^3P)$, 14.53 eV, is considerably higher than IP (Kr) if $N(^4S)$ is formed in the charge transfer, and the recombination energy is much lower (12.15 eV) if the final product state is $N(^2D)$.

The importance of energy resonance and the role of Franck–Condon factors in charge-exchange reactions of atomic ions with diatomic and simple polyatomic molecules have been discussed by several authors.[23c, e, 182–187] It has been argued that the magnitude of charge-transfer cross sections depends on both the availability of an energy level of the molecular ion resonant with the recombination energy of the atomic ion and the value of the Franck–Condon factor for the transition between this energy level of the molecular ion and the molecular neutral. The reactions of $N^+(^1D)$ again serve as an example.[11b] The recombination energy of this ionic state is nearly resonant with the ionization potential of CO in the reaction

$$N^+(^1D) + CO(^1\Sigma) \rightarrow N(^2D) + CO^+(X^2\Sigma, v=0) \qquad (II.4)$$

where CO^+ is formed in the ground vibronic state. This reaction has a large cross section over a wide translational-energy range, $\sigma = (4 + 0.5) \times 10^{-16}$ cm^2 from 2–100 eV. Resonance can also be achieved between the recombination energy of the ground-state ion, $N^+(^3P)$, and ionization of CO into a vibrationally excited level of CO^+; however, the Franck–Condon factor connecting this ionic CO^+ level with the ground-state neutral CO level is small, and thus the cross section for the latter charge-transfer reaction is low.

The role of incident-ion vibrational energy in molecular charge-transfer reactions has recently been studied from both the experimental and theoretical viewpoints by Moran and co-workers.[51,99b,103,188–190] The systems studied were of the general type AB^+–AB, where $AB = N_2$, CO, O_2, NO, and H_2. These reactions are included in the summary presented in Table II. The experimental results for these processes were shown to be consistent with theoretical predictions based on a multistate impact-parameter treatment of the variation of the total charge-transfer cross section with incident-ion vibrational quantum number. The behavior of the $N_2^+(N_2, N_2)N_2^+$ and $CO^+(CO, CO)CO^+$ reactions, for which the cross sections decrease with increasing population of higher vibrational levels of the reactant ions, may be contrasted with the corresponding NO^+ and O_2^+ symmetric charge-transfer reactions, which exhibit the opposite trend. These differences may be traced to the fact that the vibrational frequencies of the ion and neutral are similar in the former systems and dissimilar in the latter systems. One of these systems (viz., $NO^+(NO, NO)NO^+$) has been studied recently with vibrationally selected reactant ions. Squires and Baer[86c] determined the excitation functions for charge transfer from each of the levels $v' = 0$ to 5 separately (Fig. 11) and compared the results with the calculations by Moran et al.[103]

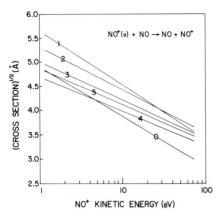

Figure 11. Photoion photoelectron coincidence studies of charge-transfer reactions of state-selected ions. Cross sections for nitric oxide symmetric charge-transfer reaction are plotted as function of reactant-ion kinetic energy and reactant-ion vibrational state ($v = 0, 1, 2, 3, 4, 5$). Solid lines are linear least-squares fits to experimental data (not shown).[86c]

Energy near-resonance and favorable overlap of vibrational states are the dominant factors affecting the magnitudes of the charge-transfer cross sections in the AB^+–AB systems. It was found[188] that an adequate theoretical treatment of the H_2^+–H_2 system necessitated inclusion of the effects of vibration–rotation interaction in calculating vibrational overlaps from accurate vibrational wave functions. Charge-transfer cross sections were thus computed as a function of different vibrational and rotational levels of the incident-ion species.

b. Heavy-particle Transfer

The cross sections as well as the dynamics of exothermic heavy-particle transfer reactions are affected by the electronic energy of the reactant ion, particularly at low relative translational energies. It is difficult to predict the behavior of a particular reaction system, however. Some hydrogen-atom transfer reactions, such as $F^+(H_2, H)FH^+$ and $Ar^+(H_2, H)ArH^+$ (Table I) exhibit larger cross sections for the reaction of the excited-state ion than for the ground-state species (Fig. 12), whereas for others, such as $Kr^+(H_2, H)KrH^+$, the excited state ion reaction has a lower cross section than does the ground state. These effects have been discussed previously from the points of view of statistical phase-space theory[13] and potential surface crossing[21, 191] (see Section V).

The dynamics of some hydrogen-atom transfer reactions are quite different for the ground- and excited-state reactants. The reaction $N^+(H_2, H)NH^+$ proceeds at low relative translational energies via a long-lived NH_2^+ transition state when the $N^+(^3P)$ ground state is the reactant ion, whereas at the same energies the $N^+(^1D)$ excited-state reaction involves a direct spectator stripping-type mechanism[12a] (Fig. 13). This behavior was explained on the basis of the appropriate electronic-state

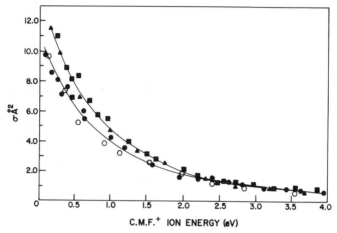

Figure 12.   Cross sections as function of F^+ ion energy for reactions $F^+(^3P)$ + $H_2, D_2, F^+(^1D) + H_2, D_2, F^+(^3P) + H_2(\bullet); \ F^+(^3P) + D_2(\bigcirc); \ F^+(^1D) + H_2(\blacksquare), F^+(^1D) + D_2(\blacktriangle).$[13]

correlation diagram, as discussed in more detail in Section V of this review. The velocity vector distributions show evidence of the metastable excited reactant at very low[12a] as well as very high initial relative energies.[192]

For several other heavy-particle transfer reactions that have been studied, the exact electronic states of the products have not been unequivocally determined. An example is the reaction $C^+(O_2, O)CO^+$, which is exoergic for both ground-state (^2P) and excited-state $(^4P)C^+$ reactants, provided that the products are formed in their ground states. The excitation functions for reactions of the two states are quite different, however, and only that for the excited reactant is "typical" of exothermic reactions, that is, the cross section monotonically decreasing with increasing relative translational energy. The excitation function for the ground state behaves as though the reaction possessed an energy barrier, with the cross section rising at low translational energy, passing through a maximum, and then declining at higher energies.[9a] The ground-state ion reaction has recently been demonstrated[193] to give excited CO^+ (see Section IV.A.2).

Several more complex reactions are also included in Table I, and these also exhibit pronounced energy effects. The cross section for the reaction $C_2H_4^+(C_2H_4, CH_3)C_3H_5^+$, for example, is about 13% lower when the $C_2H_4^+$ ion is in an electronically excited state than is the case for $C_2H_4^+$ in the lowest vibrational level of the electronic ground state.[80] The branching ratio between competing reaction channels is also affected by energy in some instances. Thus for the two reactions $C_2H_2^+(C_2H_2, H)C_4H_3^+$ and $C_2H_2^+(C_2H_2, H_2)C_4H_2^+$ of the acetylene ion, the former reaction is favored

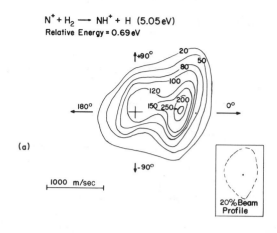

$N^+ + H_2 \longrightarrow NH^+ + H$ (5.05 eV)
Relative Energy = 0.69 eV

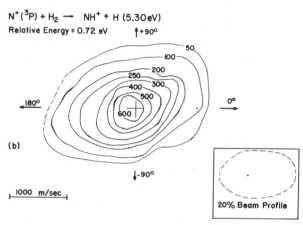

$N^+(^3P) + H_2 \longrightarrow NH^+ + H$ (5.30 eV)
Relative Energy = 0.72 eV

Figure 13. Cartesian [center-of-mass (CM)] contour diagrams for NH^+ produced from reaction of N^+ with H_2. Numbers indicate relative product intensity corresponding to each contour. Direction of N^+ reactant beam is $0°$ in center-of-mass system. For clarity, beam profiles have been displaced from their true positions (located by dots and $0°$). Tip of velocity vector of center of mass with respect to laboratory system is located at origin of coordinate system (+). Scale for production velocities in center-of-mass system is shown at bottom left of each diagram: (*a*) reactant N^+ ions formed by impact of 160-eV electrons on N_2; two components can be discerned, one approximately symmetric about the center of mass and the other ascribed to $N^+(^1D)$, forward scattered with its maximum intensity near spectator stripping velocity; (*b*) ground-state $N^+(^3P)$ reactant ions formed in a microwave discharge in N_2. Only one feature is apparent—contours are nearly symmetric about center-of-mass velocity.[12a]

for ground-state $C_2H_2^+$, whereas the latter channel is dominant for $C_2H_2^+$ in the electronically excited state.[71-78,79b]

The cross sections for heavy-particle transfer reactions, which are exoergic for ground-state reactants, either decrease with increasing vibrational quantum number of the reactant ion or remain unchanged (Table II). A decrease is observed for the reactions $H_2^+(H_2,H)H_3^+$, $NH_2^+(NH_3, NH_2)NH_4^+$ (Fig. 14), and $C_2H_4^+(C_2H_4,CH_3)C_3H_5^+$. This behavior may be explained by statistical theories such as the phase-space theory or the quasiequilibrium theory. There is good evidence to indicate that these reactions proceed at low relative translational energies via a long-lived intermediate collision complex or that they involve so-called strong coupling collisions. For such interactions the forward and reverse reaction channels are in competition. According to phase-space theory, the ratio of the volume in phase space of the exothermic product channel to that of the thermoneutral reactant channel thus decreases with increasing internal energy of the reactants.[21] Those reactions for which the cross sections are insensitive to the vibrational quantum number of the ionic reactant, such as $NH_3^+(H_2O,OH)NH_4^+$, presumably proceed via a direct atom-transfer mechanism.[21]

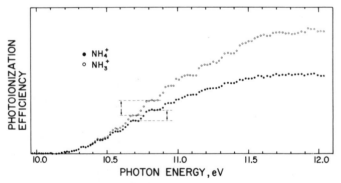

Figure 14. Photoionization cross section of NH_3 and apparent photoionization cross section for NH_4^+ produced by reaction of NH_3^+ with NH_3. Variation of reaction cross section as function of vibrational state of reactant NH_3^+ ion was determined by comparing relative step heights of curves for NH_3^+ and NH_4^+ after ordinate scales of both curves were adjusted so that data points of first plateau at about 10.2 eV coincide. Ratio of a pair of corresponding step heights is then proportional to ratio of cross section for reaction of vibrationally excited NH_3^+ to that for NH_3^+ in its ground vibrational state. Step heights used to determine relative cross section for reaction of NH_3^+ with $v=5$ are shown. Step ratio NH_4^+/NH_3^+ decreases with increasing v.[85a]

Several systems have been studied in which it has been observed that two or more competing product channels, all of which are exothermic, are affected differently by the internal vibrational energy of the reactant ion (Table II). For example,[112, 115a] the intensity ratios $C_4H_2^+/C_4H_3^+$ and $C_3H_3F_2^+/C_3H_5^+$ from the reactions of $C_2H_2^+$ and $C_2H_3F^+$ with acetylene and vinyl fluoride, respectively, decrease with increasing vibrational energy of the reactant ions. These results may again be explained on statistical grounds. Higher internal energy in a long-lived collision complex favors that product channel whose reaction has higher energy requirements (a high "activation energy") and a "loose" transition state (a high "entropy of activation").

c. Association

The association reaction involving excited states, which has been studied most extensively is the benzene dimer ion formation,[21, 194–203]

$$C_6H_6^+ + C_6H_6 \rightarrow C_{12}H_{12}^+ \tag{II.5}$$

Photoionization experiments have shown[21] that for benzene ions formed in the ground electronic state, the reaction is third order in pressure and involves almost exclusively the ground vibrational state of $C_6H_6^+$. Vibrationally excited monomer ions react in an approximately fourth-order reaction, requiring an initial deactivation to the vibrational ground state.[202] Highly excited benzene ions, produced 3.5 eV above the ground state (in either an electronically excited state or an isomeric $C_6H_6^+$ structure), dimerize in a bimolecular reaction to form a dimer that apparently has a different structure from the ground-state dimer and is much more strongly bound.[202] Some of the particle-transfer reactions discussed previously and included in Table II probably proceed via an intermediate association step. For example, in propylene, the probable reaction sequence is

$$C_3H_6^+ + C_3H_6 \rightleftharpoons (C_3H_6)_2^{+*} \tag{II.6}$$
$$(C_3H_6)_2^{+*} \rightarrow C_5H_9^+ + CH_3$$

All such reactions are characterized by a lowering of the cross section as the internal energy of the adduct is increased. The lifetime of the benzene dimer ion, as well as those of similar adduct ions, is very sensitive to its internal energy since it is very loosely bound (it has a binding energy of ~ 8 kcal/mole). Increasing the product-ion internal energy by vibrational excitation of the reactant readily promotes the dissociation back into reactants.

2. Endoergic Reactions (Electronically and Vibrationally Excited States)

As a general rule, the rates of those reactions that are endoergic for ground-state reactions are dramatically enhanced when excited-state reactants are employed. Electronic excitation and vibrational excitation may or may not have similar effects. In the case of vibrational excitation, the reactions may occur on a single electronic potential surface. In the case of electronic excitation, a single-surface path between electronically excited reactants and ground-state products may not exist because of the prevailing symmetries.

The utilization of various forms of energy to promote endoergic reactions is a topic of great general interest in chemical dynamics.[204] For neutral–neutral interactions, laser excitation of one of the reactants has been employed, among other methods, to increase the reaction rates of endoergic processes. The question arises as to which form of energy is more effective in promoting an endoergic reaction—relative translation of the reactants or internal energy. The answer is of considerable importance from both theoretical and practical standpoints. Consideration of the principle of microscopic reversibility leads to an even more interesting question with respect to the reverse exoergic reaction, namely, whether the excess energy will appear as translational, vibrational, rotational, or electronic energy of the products. The latter has important consequences for practical devices such as lasers that depend on highly exoergic reactions to create a population of excited states. Energy partitioning in neutral–neutral collision processes has thus been intensely investigated using both experimental and theoretical methods. It can be accurately stated, however, that some of the earliest and most important experiments that addressed such questions involved studies of ion–neutral interactions. Several of these are discussed in this section, and others are listed in the summaries given in Tables I to III.

a. Charge Transfer

Many charge-transfer reactions that are endoergic for ground-state ionic reactants and are thus not observed at translational energies below that corresponding to the reaction endoergicity exhibit measurable cross sections when the ionic reactants are electronically excited and the reaction becomes energetically feasible. Such differences in reactivity of the ground and excited states form the basis for detection and determination of the abundance of many excited states using the attenuation method described in a previous section. Most of the early experiments of this type involved species of interest in atmospheric chemistry.[4b, 10] However, reactions of many other excited ionic species have since been studied, including (see

Table I) $B^+(N_2O, B)N_2O^+$, $O^+(Ar, O)Ar^+$, $O^+(N_2, O)N_2^+$, $O_2^+(Ar, O_2)$ $Ar^+, O_2^+(N_2, O_2)N_2^+$, $O_2^+(H_2O, O_2)H_2O^+$, $NO^+(H_2O, NO)H_2O^+$, $H_2O^+(N_2O, H_2O)N_2O^+$, and $C^-(O_2, C)O_2^-$. Excitation functions (the variation of reaction cross section with translational energy) for the excited-state reactions are generally quite different from those of the ground-state ion reactions. The cross section for the reaction

$$O^+(^2D) + N_2 \rightarrow O + N_2^+ \tag{II.7}$$

is shown as a function of the ion translational energy in the laboratory system[4c] in Fig. 15. This reaction is exoergic and has a finite cross section at the lowest laboratory translational energy studied. The ground-state $O^+(^4S)$ does not react with N_2, even at higher translational energies. In some instances, of course, the reaction of the excited state also demonstrates a translational-energy threshold, but a lower one than that for the ground-state reaction, as is seen in Fig. 16 for the process

$$B^+(^3P), B^+(^1S) + N_2O \rightarrow B + N_2O^+ \tag{II.8}$$

Recently,[46a] the flow-drift-tube method has been employed to determine the dependence on translational energy of the rate coefficients for various charge-transfer reactions of $O_2^+(a^4\pi_u)$. The results for one such reaction are shown in Fig. 17, and the other processes studied exhibit similar behavior.

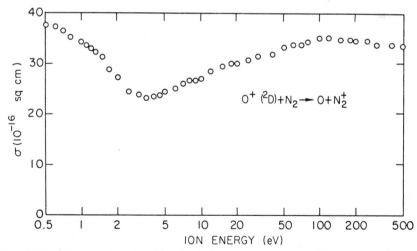

Figure 15.   Cross section for the charge-transfer reaction, $O^+(^2D) + N_2 \rightarrow O + N_2^+$ as function of ion energy in laboratory system.[4c]

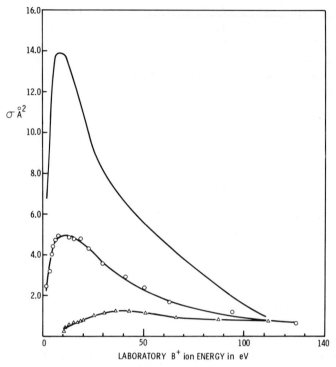

Figure 16. Cross section as function of ion kinetic energy for charge-transfer reaction $B^+(N_2O, B)N_2O^+$; ▲, cross section for reaction of $B^+(^1S)$ produced from BI_3; ○, cross section for reaction of B^+ produced from BF_3 (35.3% ^3P and 64.5% ^1S); solid line, cross section for reaction of $B^+(^3P)$ obtained by taking difference between two lower curves and correcting for appropriate abundance.[7a]

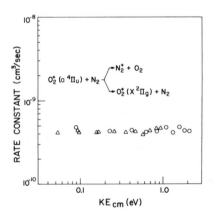

Figure 17. Rate constants for reactions of $O_2^+(a^4\Pi_u)$ with N_2 as function of kinetic energy in center-of-mass system; ▲ and ● indicate data taken under slightly different experimental conditions.[46a]

All these reactions are exothermic for the metastable $O_2^+(a^4\pi_u)$ state. As the corresponding ground-state O_2^+ reactions, are endoergic, all have negligibly small rate coefficients ($< 10^{-12}$ cm^3/sec) over the same translational-energy range.

Many dissociative charge-transfer reactions[6] that are endoergic for ground-state ions also become energetically allowed for some electronically excited state of the reactant ion. For example, the reaction of excited Kr^+ ions with methane has been observed[18,19]

$$Kr^+(^2P_{1/2}) + CH_4 \rightarrow Kr + CH_3^+ + H \qquad (II.9)$$

whereas the analogous ground-state $Kr^+(^2P_{3/2})$ reaction does not occur with a measurable cross section.

Still other charge-transfer reactions involving diatomic reactant ions, which are endoergic when the reactant ions are in the ground vibrational state, demonstrate enhanced cross sections for higher vibrational states, as is probably the case for the reaction $O_2^+(Xe,O_2)Xe^+$ (Table II).

b. Heavy-particle Transfer

The variations with translational energy of the cross sections for the reactions

$$B^+(^1S_g) + D_2(^1\Sigma_g^+) \rightarrow BD^+(^2\Sigma^+) + D(^2S_g) \qquad (II.10)$$

$$B^+(^3P_u) + D_2(^1\Sigma_g^+) \rightarrow BD^+(^2\Sigma^+) + D(^2S_g) \qquad (II.11)$$

are shown in Fig. 18.[7b] The ground-state-reaction cross section exhibits typical endoergic behavior, with onset occurring at the expected threshold (9.8 eV in the laboratory system and 2.6 eV in the center-of-mass system), then rising to a maximum, and finally declining. The reaction of the excited state shows behavior typical of exoergic processes, with the cross section decreasing monotonically with increasing translational energy.

Several other particle-transfer reactions have larger cross sections when the reactant ions are in electronically excited states. Two systems that have been studied extensively[24–37,58,79a] are

$$(N_2^+)^* + N_2 \rightarrow N_3^+ + N \qquad (I.8c)$$

$$(CO^+)^* + CO \rightarrow C_2O^+ + O \qquad (II.12)$$

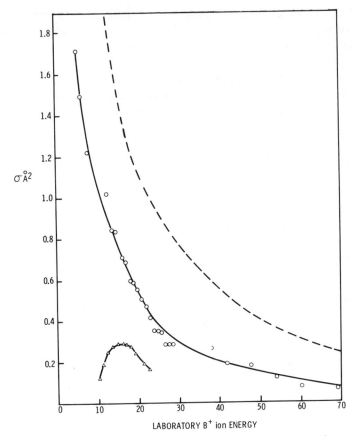

Figure 18. Cross sections as function of B^+ laboratory energy for reactions $\Delta - B^+(^1S) + D_2 \rightarrow BD^+ + D$; $\bigcirc$, $B^+(65\%\ ^1S,\ 35\%\ ^3P) + D_2 \rightarrow BD^+ + D$; dashed line, $B^+(^3P) + D_2 \rightarrow BD^+ + D$.[7b]

although the exact identity of the excited states responsible for these reactions is still not clear. Recent ICR[37] and ion-trapping[34b] studies of these reactions indicate that two excited states may be responsible in each case, the $A^2\Pi_u$, which is relatively short lived (see Section I.A.4) and a quartet state, probably $^4\Sigma_u^+$, having a much longer lifetime ($> 10^{-2}$ sec). Emission bands from a state, claimed to be the $N_2^+(^4\Sigma_u^+)$ state, have recently[205] been observed. Although there is considerable scatter in the reported data from various laboratories, the experimental rate coefficients that have been determined for reactions (I.8c) and (II.12) indicate that both reactions are quite fast, whereas ground-state N_2^+ and

CO^+ are nonreactive. These processes have rather high endoergicities (5.5 to 5.6 eV), based on calculations for the ground-state reactants. The energies required to form the excited reactant ions in these instances ($\sim$20 to 23 eV) are not readily accessible with ordinary photoionization sources;[85] therefore, these reactions have been studied only with electron-impact ionization. The identity of the excited state responsible for the analogous reaction

$$(O_2^+)^* + O_2 \rightarrow O_3^+ + O \tag{II.13}$$

which does occur in an energy range readily accessible with photoionization, has been accurately determined using photoionization methods (see Section I.A.4).

Several examples of endoergic particle-transfer reactions for which the cross sections are significantly enhanced by vibrational energy in the reactant ion are now known (Table II). The classical example of such a reaction and the most widely studied is

$$H_2^+ + He \rightarrow HeH^+ + H \tag{II.14}$$

Historically, this was the first triatomic reaction system for which a vibrational threshold was clearly demonstrated and for which a marked enhancement of the cross section with vibrational energy was observed. It is also one of the few ion–neutral interactions for which the relative importance of translational and vibrational energy in promoting reaction have been carefully determined. The two groups largely responsible for this important work were Friedman et al.,[99a] who conducted the early electron-impact work, and Chupka et al.,[92] who were responsible for the more accurate photoionization study. The status of our current understanding of this reaction has recently been reviewed by Chupka.[85] The importance of this reaction in Jovian ionospheric chemistry has also been noted in a review by Huntress.[95b] Several ion-beam[93,94] and ICR[95a] studies of the H_2^+–He reaction have also been reported. Because of the relative simplicity of the reaction, in terms of the number of electrons involved, several theoretical treatments have appeared (see also Section V). The important experimental findings may be summarized as follows: (1) the reaction has a vibrational-energy threshold corresponding to the H_2^+ ($v' = 3$) state when thermal translation-energy ions are employed; (2) both translational energy and vibrational energy of the reactants enhance the reaction cross section; and (3) at the same total energy, vibration is considerably more effective than relative translation in promoting the reaction. Some of these features, particularly the latter, are now known to be common to many

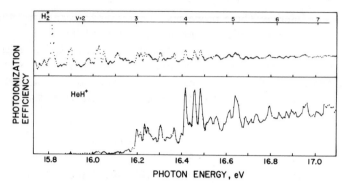

Figure 19. Photoionization efficiency curves for production of H_2^+ and HeH^+ from mixture of hydrogen and helium. Threshold energies for formation of H_2^+ in vibrationally excited states are indicated at top of figure.[85a]

neutral triatomic systems and may be explained, following J. Polanyi's work,[206] by a late barrier in the potential surface, as discussed in more detail in a later section. Some of the experimental results for this reaction are presented in Figs. 19–21.

Other endoergic reactions for which an increased cross section has been observed when the internal vibrational energy of the ionic reactant

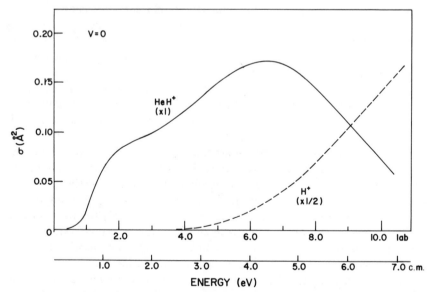

Figure 20. Microscopic cross section as function of kinetic energy for formation of HeH^+ and H^+ from reaction of H_2^+ ($v=0$) with helium.[85a]

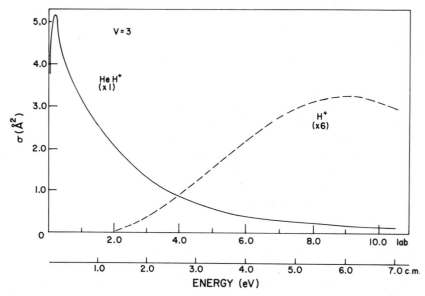

Figure 21.   Microscopic cross section as function of kinetic energy for formation of HeH^+ and H^+ from reaction of H_2^+ ($v=3$) with helium.[85a]

is increased are $H_2^+(Ne,H)NeH^+$, $O_2^+(O_2,O)O_3^+$, $O_2^+(CO,O)CO_2^+$, O_2^+ $(NO,O)NO_2^+$, $O_2^+(N_2,O)N_2O^+$, and $CH_2CO^+(CH_2CO,2CO)C_2H_4^+$.

The relative importance of endoergic and exoergic channels in the reactions of thermal H_3^+ has been shown to be strongly dependent on the extent of internal vibrational energy of the H_3^+ reactant (Table II), and the branching ratio has been used to monitor the changes in this internal energy as the H_3^+ ion undergoes deactivating collisions.[105–107] Similar effects have been observed[109] for other protonated species that are formed via exoergic ion–molecule reactions in highly excited (but poorly defined) vibrational-energy states. Reactions of this type, particularly of H_3^+, promise to be of considerable further interest as more detailed experiments[108a] and more rigorous theoretical calculations[207] are accomplished (see also Section II.B.1).

c. Collisional Dissociation

The energy dependences of the cross sections for collision-induced dissociation (CID) reactions of the form

$$AB^\pm + X \rightarrow A^\pm + B + X \qquad (II.15)$$

can provide a quantitative measure of the energy available for reaction in an ion beam. In an experiment of this type the relative translational energy of the reactants is transformed into internal energy of the ion, resulting in dissociation (see also Section II.B.2). If the reactant ion is initially electronically and/or vibrationally excited (because of the mode of formation), less relative translational energy will be necessary to overcome the endoergicity of the dissociation than is the case for the ground-state ion, provided that the products of the reaction are the same and that they are formed in the same energy states. Collision-induced dissociation reactions have thus served to detect metastable electronically excited states in ion beams [50] as well as vibrationally excited states. The presence of such states causes increased cross sections for collisional dissociation to be observed, as compared to those obtained for the ground-state ions, at the same relative translational energy.

The variable parameters in a collisional dissociation experiment are the ionizing (electron or photon) energy and the relative translational energy. The ionic reactant is normally in a well-collimated, energy-selected beam, whereas the neutral reactant can be located in a collision chamber or introduced as a neutral beam. Experiments have been performed at high translational energies (kiloelectron volt range), as well as at low energies (few electron volts). When a single mass spectrometer is employed for collisional dissociation studies, detection of the collision-induced product ion may be accomplished using the Aston band technique, that is, by observing the appearance of ions at an apparent mass m^*, which is given by the expression[98a]

$$m^* = \frac{(m_A)^2}{m_{AB}} \qquad (II.16)$$

Alternatively, double or "tandem" mass spectrometers may be employed for collisional dissociation experiments.[208] With the latter, the incident-ion kinetic energy can be much more effectively controlled and continuously varied. Such experiments, in which translational-energy thresholds for collision-induced dissociations have been measured,[50, 108c, 209–212] have yielded useful ionic thermochemical data.

Velocity and angular distribution measurements of the product ions from collision-induced dissociation have provided additional information concerning the mechanisms of these processes and the nature of the reactant and product states involved.[38b, 62, 208, 213–218] The experimental results obtained at relatively high incident-ion energies are generally explained by a mechanism involving vertical Franck–Condon-type transitions between electronic states of the reactant ion. For example, the first step in the high-energy collision-induced dissociation of H_2^+ apparently involves excitation from the electronic ground $(^2\Sigma_g^+)$ state to the $(^2\Sigma_u^+)$ repulsive state, followed by dissociation.[117] Similarly, collision-induced dissociation of $N_2^+(X\,^2\Sigma_g^+)$ and $N_2^+(A\,^2\Pi_u)$ involves initial excitation to the $(C\,^2\Sigma_u^+)$ state, followed by predissociation from vibrational levels, $v' \geqslant 3$.[218-220] Another mechanism for collisional dissociation entails direct excitation of the ion to the vibrational continuum of the ground state. Both of these mechanisms may operate in competition in the same system.[217]

Various theoretical treatments of collision-induced dissociation have appeared, covering both the high[221] (kiloelectron volt) and low[62] translational-energy regimes, as well as the threshold region.[222]

The effects of electronically excited reactant ions $(AB^+)^*$ on collision-induced dissociations have been studied in a number of systems, including those for which $AB^+ = O_2^+, N_2^+, NO^+, CO^+$, and for which the neutral target is either a rare-gas atom or another neutral molecule. Various reactions of this type that have been investigated are included in Table I.

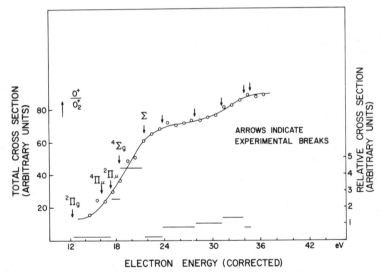

Figure 22. Cross section for collisional dissociation reaction, $O_2^+(O_2; O_2, O)O^+$, as function of energy of impacting electrons used to produce O_2^+ reactant. Solid curve represents cross section, bars indicate relative cross sections for various excited states, and arrows designate threshold energies for production of electronically excited states of O_2^+. Ion-beam energy in these experiments was 1.6 keV.[38a]

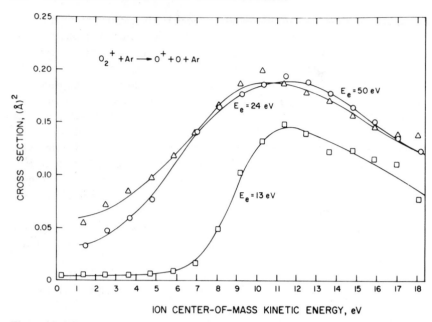

Figure 23.   Cross section for collisional dissociation reaction $O_2^+(Ar;Ar,O)O^+$, as function of reactant-ion kinetic energy at several different ionizing electron energies. Energy of impacting electrons used to produce O_2^+ from O_2 in each case is indicated.[50]

All reactant ions in these experiments were formed by electron-impact ionization. The cross sections for collision-induced dissociation sometimes rise sharply as the electron energies are increased, as a result of the formation and participation of excited electronic states in the reaction[38a] (see Fig. 22). The effect is particularly evident at low ion translational energies, below the dissociation energy for the respective ground states (see Figs. 23 and 24).[50,36a] The identity of the excited electronic state in the case of collisional dissociation of an O_2^+ beam has unequivocally been determined[50] to be $(a^4\Pi_u)$. The translational-energy threshold in the center-of-mass system (Fig. 25) was shown to correspond to the reaction

$$O_2^+(a^4\Pi_u) + Ne \rightarrow O^+(^4S^0) + O(^3P) + Ne \qquad (II.17)$$

The identity of the excited states contributing to the collision-induced dissociation in other systems studied, particularly N_2^+, is not as well known, however, and more than one reactant ion state may be present. Parenthetically, it may be noted that the identity of N_2^+ states producing

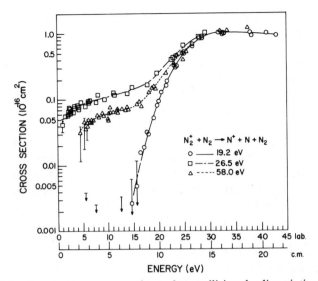

Figure 24. Apparent cross sections for collisional dissociation reaction, $N_2^+(N_2; N_2, N)N^+$, as function of reactant kinetic energy. Both laboratory (lab) and center-of-mass (CM) energy scales are shown. Energies of ionizing electrons producing N_2^+ in each case are indicated. Arrows indicate upper limits on cross section for reaction when N_2^+ is produced by 19.2-eV electrons.[36a]

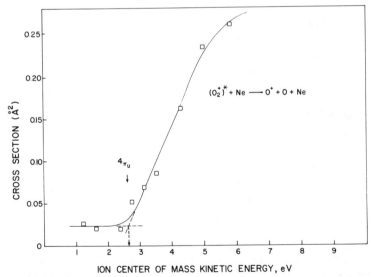

Figure 25. Kinetic-energy dependence of cross section for reaction $(O_2^+)^* + Ne \rightarrow O^+ + O + Ne$.[50]

139

optical emissions when these states are formed in certain ion–neutral collisions is also not well established, as is discussed in a subsequent section. It is also still unclear as to which metastable N_2^+ states are involved in the formation of N_3^+ via reaction (I.8c).[36b, 223] McGowan and Kerwin[38a] identified $N_2^+(A^2\Pi_u)$ and $N_2^+(^4\Sigma_u^+)$ as important nitrogen ion states in collisional dissociation. In addition, Maier[36a] (Fig. 26) showed that at low ion kinetic energies, the collision-induced dissociation of N_2^+ exhibits an electron-energy threshold of 24.1 ± 1.5 eV. In more recent experiments by Moran et al.[38c] in which the incident-ion-beam energy was varied over the range 0.65 to 5.0 keV, a sharp increase in the cross section for collision-induced dissociation of N_2^+ was observed when the ionizing electron energy was increased to 22.5 ± 1.5 V, at the lowest ion translational energy employed. The time interval between ion formation and the actual dissociative collision under the conditions of this experiment was such that the high vibrational levels of the $A^2\Pi_u$ state would have decayed radiatively prior to reaction. Therefore, the $A^2\Pi_u$ state could not contribute to the dissociative reaction that onsets at 22.5 eV. Apparently, then, the observed collision-induced dissociation involves N_2^+ in both the $(X^2\Sigma_g^+)$ and $(A^2\Pi_u)$ states, as well as some higher metastable state (or states) of

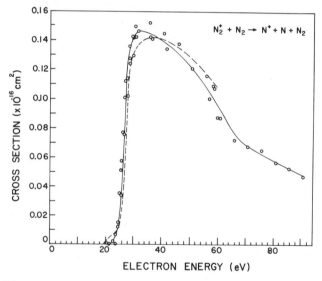

Figure 26. Apparent cross section for collisional dissociation reaction, $N_2^+(N_2; N_2, N)N^+$, as function of energy of electrons producing N_2^+ (solid curve and data points). Laboratory kinetic energy of primary ions was 10 eV. Cross section for radiative emissions from long-lived, excited states N_2^+ formed in electron impact on N_2^+ is also indicated (dashed line).[36a]

this ion. The long-lived state is possibly a quartet state such as $(^4\Sigma_u^+)$. The excited state in question contributes significantly to the total collisional dissociation cross section at low ion kinetic energies, despite its low abundance, because the excited state itself has a very large cross section for collisional dissociation (~ 30 Å^2). The energy dependences of the cross sections of the X and A states[38c] are shown in Fig. 27. A long-lived (possibly quartet) state also contributes to the collisional dissociation of CO^+.[38b] It thus seems that the metastable states that are involved in the collisional dissociation of O_2^+, N_2^+, and CO^+ are identical with the ones responsible for reactions (II.13), (I.8c), and (II.12), respectively. The measured threshold for dissociation of excited NO^+ ions was found to correspond most closely to the adiabatic dissociation energy of the long-lived $NO^+(^3\Delta)$ state[50] (Fig. 28). The possibility of contribution from lower excited states, such as the $^3\Sigma^+$ and $^3\Pi$, to the cross section above threshold could however not be entirely ruled out.[50]

The collision-induced dissociations of several triatomic and polyatomic ions have been studied and the participation of excited states such as $COS^+(A^2\Pi)$ has been detected.[224] Such experiments[225-227] have also provided insight into the structures of isomeric ions involved. All these studies were conducted with incident ions of relatively high (kiloelectron volt) kinetic energies and also entailed determination of the kinetic energies of the product ions and/or the fragmentation patterns.

In a related type of experiment, collisional ionization (or so-called charge stripping) of high-kinetic-energy ions is employed. The major reaction studied is of the general form

$$m^+ + N \rightarrow m^{+2} + N + e \qquad (\text{II.18})$$

and long-lived states of rare-gas ions have been detected with this method,[68b,98a] in conjunction with ion kinetic-energy measurements. When polyatomic ions are collisionally ionized, additional information is obtained concerning the structures of these species, which is complementary

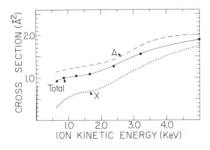

Figure 27. Cross section for collisional dissociation reactions of N_2^+ ions in $X\,^2\Sigma_g^+$ and $A\,^2\Pi_u$ states as function of incident-ion kinetic energy are indicated by dotted and dashed lines, respectively. Solid line is collisional dissociation cross section for incident-ion beam produced by impact of 50 eV electrons, which includes a mixture of the two states.[38c]

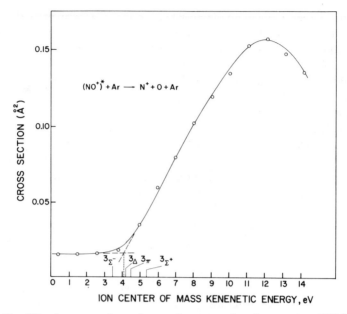

Figure 28.   Kinetic-energy dependence of cross section for reaction $(NO^+)^* + Ar$ $\rightarrow N^+ + O + Ar.$[50]

to that provided by collision-induced dissociation. In charge-stripping experiments, a more uniform sampling over a range of ion internal energies occurs than is the case for collision-induced dissociation.[228]

The effect on collision-induced dissociation of vibrational excitation in the reactant ion has been studied in several systems,[85a, 86b, 92, 97, 105, 108c, 117] including $H_2^+(X; X, H)H^+$ with $X \equiv H_2$, He, $He_2^+(He; 2He)He^+$, $D_3^+(X; D_2, X)D^+$ with $X \equiv He, Ne$, $H_3^+(X; X, H_2)H^+$ with $X \equiv He, H_2$, and $CH_2Br_2^+(CH_2Br_2; CH_2Br_2, Br)CH_2Br^+$. A marked increase in the reaction cross section is observed for most of these processes as the reactant-ion vibrational energy increases. This is true for the dissociation of H_2^+ at energies in the kiloelectron-volt range,[117] which is as expected from theory.[221] This effect of vibrational excitation in promoting collisional dissociation of high-kinetic-energy ions also causes the nonuniform sampling over the range of internal ion energies that occurs with electron-impact ionization when collision-induced dissociation (CID) spectra are monitored.[228]

The enhancement of collisional dissociation cross sections by reactant-ion vibrational excitation is also observed for interactions at relatively low translational energies, particularly in the energy region near threshold. This was demonstrated in studies of dissociation of H_2^+ ($v' = 0–5$) in which

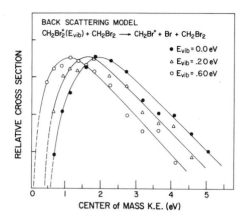

Figure 29. Relative cross sections for collisional dissociation of $CH_2Br_2^+$ in selected vibrational-energy states as function of kinetic energy. Data were obtained by photoion photoelectron coincidence technique and were analyzed by assuming only backscattering in center of mass. Maxima of curves were normalized to same relative cross section.[86b]

photoionization methods were utilized.[85a,92] The excitation functions for collisional dissociation reactions of H_2^+ ($v' = 0$) and H_2^+ ($v' = 3$) are shown in Figs. 20 and 21. It is seen that as the vibrational energy of H_2^+ is increased, the translational-energy threshold for the reaction is shifted to lower energies. This is generally the behavior encountered for all such processes studied. Sharp energy thresholds are typically observed, precisely at the energy required for reaction, which indicates that translational-to-vibrational energy transfer is easily accomplished at low kinetic energies[86b] (Fig. 29).

Since the velocities and angular distributions of products from collisional dissociations at low incident-ion energies have generally not been determined, the precise mechanism by which the products are formed is unknown. Thus in the collisional dissociation of H_2^+ with helium as the target gas, H^+ may result from dissociation of H_2^+ that has been directly excited to the vibrational continuum

$$H_2^+ + He \rightarrow H^+ + H + He \qquad (II.19)$$

or it may result from a dissociation of an intermediate HeH^+, with the

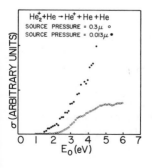

Figure 30. Relative cross section as function of nominal kinetic energy (center of mass), E_0, for collisional dissociation of He_2^+ on He. Data are presented for He_2^+ formed at two different ion-source pressures, as indicated.[97]

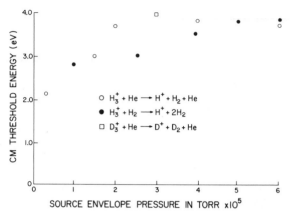

Figure 31. Variation of collision-induced dissociation thresholds (in center-of-mass system) for reactions indicated as function of source pressure.[108c]

latter produced in its vibrational continuum and undergoing dissociation before detection[229] (see also Section II.2.b),

$$H_2^+ + He \rightarrow [\, HeH_2^+ \,] \rightarrow HeH^{+*} + H \rightarrow He + H^+ + H \qquad (II.20)$$

Vibrational energy is more effective in promoting this dissociation reaction than is relative translational energy. That is, for the same total energy above threshold, the collisional dissociation cross section is greater if some of the energy was initially derived from vibrational excitation.[85a,92] This precludes a mechanism by which a $[HeH_2^+]^*$ intermediate is formed in which the energy is distributed statistically. On the other hand, for some collision-induced dissociation reactions, such as that of $CH_2Br_2^+$, the data are consistent with a model in which all the available energy, both internal and center-of-mass, is statistically distributed in the intermediate state.[86b]

The translational-energy thresholds for collision-induced dissociations of He_2^+, D_3^+, and H_3^+ have been measured to assess the extent of vibrational excitation in these ions. The initial study of this type was that of Leventhal and Friedman,[105] who observed that the kinetic-energy threshold for collision-induced dissociation of D_3^+ was approximately 2 eV lower when D_3^+ was formed in an ion source at low pressure than was the case for D_3^+ produced under higher source pressures. This suggested that the former D_3^+ species were formed with at least 2 eV of excess internal energy. At higher ion source pressure, deactivating collisions occur, reducing the level of internal excitation. The high-pressure limit of collisional dissociation thresholds can thus be employed to obtain true bond dissociation energies for ionic species. Examples of such results for $He_2^{+\,97}$ and $H_3^{+\,108c}$ are shown in Figs. 30 and 31.

B. Nonreactive Scattering—Energy Transfer

Energy transfer occurring in nonreactive neutral–neutral collisions is a very active field of investigation.[230] Important contributions to the understanding of collisional energy-transfer processes have also resulted from various studies of nonreactive ion–neutral collisions. The modes of energy transfer that have been investigated for the latter interactions include vibrational to relative translational (V–T), vibrational to vibrational (V–V), translational to vibrational (T–V), translational to rotational (T–R), vibrational to rotational (V–R), translational to electronic (T–E), and electronic to translational (E–T).

As has been discussed in previous sections, excited ions may be formed by electron or photon impact, and, in addition, excited ions may be the intermediates or products of ion–molecule reations. Such excited ionic species may be produced in environments where a relatively high pressure of the parent gas or of a foreign ("inert") gas is present and may thus undergo collisions that result in deexcitation. In discharges and ion lasers such collisional energy transfer is in competition with radiative decay. In some ion–neutral collisions, energy transfer occurs prior to reaction. Examples are the collision-induced dissociation processes discussed earlier, which require as an initial step the conversion of relative translational energy into internal energy. Understanding the mechanisms of nonreactive energy transfer in ion–neutral collisions is thus also crucial for the understanding of reactive systems.

1. Collisional Deactivation of Excited Ions.

Several cases in which excited ions are deactivated on collision have already been discussed in previous sections and are included in Tables I and II. Collisional deactivation of electronically excited ions (i.e., "quenching" of metastable states) quite likely occurs in competition with many of the reactive channels shown in Table I, although it has been specifically studied or discussed for only a few systems. Collisional deactivation is at least partly responsible for attenuation of the ion beam when the attenuation technique (described earlier) is employed to determine the abundance of electronically excited ionic states.

Very large rate coefficients for collisional deactivation of electronically excited states have been observed in systems in which no reactive channel is open, notably for the interaction[46a]

$$O_2^+ \left(a^4\Pi_u \right) + O_2 \rightarrow O_2^+ \left(X^2\Pi_g \right) + O_2 \qquad (II.21)$$

for which the rate coefficient is 3×10^{-10} cm^3/sec at room temperature and increases to approximately 7×10^{-10} cm^3/sec in the vicinity of 1-eV

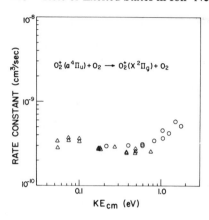

Figure 32. Rate constants for collisional deactivation of electronically excited $O_2^+(a^4\Pi_u)$ ions on collision with O_2 as function of kinetic energy in center-of-mass system.[46a]

center-of-mass translational energy, as shown in Fig. 32. This process probably involves E–T energy transfer and/or excitation of the neutral target.

When transitions from the excited to the ground ionic state are forbidden, as in the case of F^+, where the deactivation requires a singlet to triplet transition,[13] increased pressures of the parent gas are ineffective in collisionally deactivating the excited state. However, in such instances, addition of a paramagnetic impurity such as NO accomplishes the deactivation,[13, 64] just as is observed in deexcitation of analogous neutral systems.[231] This effect has been discussed by Chiu.[232] The F^+ ion and its neighboring NO molecule can be viewed as a composite molecule. Even though F^+ undergoes a spin nonconserving transition, the overall spin angular momentum of the composite pseudomolecule may still be conserved.

Electronically excited ionic states, for which the transitions to the ground state are allowed, normally have very short radiative lifetimes, typically on the order of 10 nsec to 1 μsec. Yet even these states are quite efficiently collisionally deactivated, particularly on interaction with the corresponding parent gases. Several such systems have been studied in detail, and the Stern–Volmer relation has been employed to determine rate coefficients for collisional deactivation.[233–239] Some of these reactions and the pertinent kinetic data are displayed in the reactions that follow.

$$N_2^+(B^2\Sigma_u^+, v=0) + N_2 \rightarrow N_2^+(X) + N_2 \qquad k = 7.5 \times 10^{-10} \text{ cm}^3/\text{sec} \qquad \text{(II.22)}$$
$$N_2^+(B^2\Sigma_u^+, v=0) + He \rightarrow N_2^+(X) + He \qquad k < 3 \times 10^{-11} \text{ cm}^3/\text{sec} \qquad \text{(II.23)}$$
$$CO_2^+(B^2\Sigma_u^+) + CO_2 \rightarrow CO_2^+(X) + CO_2 \qquad k = 7.5 \times 10^{-10} \text{ cm}^3/\text{sec} \qquad \text{(II.24)}$$
$$CO_2^+(A^2\Pi_u) + CO_2 \rightarrow CO_2^+(X) + CO_2 \qquad k = 8.6 \times 10^{-10} \text{ cm}^3/\text{sec} \qquad \text{(II.25)}$$
$$N_2O^+(A^2\Sigma^+) + N_2O \rightarrow N_2O^+(X) + N_2O \qquad k = 3.6 \times 10^{-10} \text{ cm}^3/\text{sec} \qquad \text{(II.26)}$$

Obviously charge transfer could also be the mechanism whereby collisional deactivation of these electronically excited ionic states is accomplished. If this is indeed the case, the low efficiency of helium in deactivating N_2^+* would be the result of its high ionization potential. However, helium was observed to be quite efficient in deactivating CO_2^+*. The identity of the products in some of these quenching reactions is actually unknown since only the intensity of radiation emitted on decay of the excited state was monitored in these experiments. Therefore, the relative role of reactive charge-transfer collisions as compared to pure nonreactive scattering in these processes has not been entirely established.

A somewhat more systematic study has been made of collisional deactivation of vibrationally excited ions. Some diatomic and triatomic systems that have been investigated are included in Table II. Vibrational-to-rotational transfer has been demonstrated[95a] for vibrationally excited H_2^+ colliding with helium:

$$(H_2^+)^v + He \xrightarrow{k_v} (H_2^+)^{v'} + \vec{He} \tag{II.27}$$

This process occurs in competition with the reactive channel. It was determined that the rate coefficient for this process is large at thermal kinetic energies and increases with increasing vibrational quantum number of the H_2^+, according to the approximate relation $k_v \cong (1.0 \pm 0.5)k_2^v$, where k_2^v is the rate coefficient for the reactive channel

$$(H_2^+)^v + He \xrightarrow{k_2^v} HeH^+ + H \tag{II.28}$$

These experiments did not resolve the question as to whether the deactivating collision of H_2^+ in $v > 1$ results in multiquantum transitions, single quantum transitions, or both. Since $k_2^v \sim 0.2$–0.3×10^{-9} cm^3/sec for reactions of H_2^+ in higher vibrational levels, this means that k_v, the rate coefficient for collisional deactivation, is on the same order of magnitude.

Vibrational-to-vibrational transfer tentatively has been identified in collisions of H_2^+ with H_2[85a, 91, 95a]

$$(H_2^+)^v + H_2 \rightarrow H_2^+ + H_2^v \tag{II.29}$$

and a lower limit of 0.25×10^{-9} cm^3/sec has been placed on the rate coefficient for this process.

An excited ionic species that has been widely studied from the standpoint of collisional deactivation is vibrationally excited H_3^+. The collisional deactivation process,

$$(H_3^+)^* + H_2 \xrightarrow{k_d} (H_3^+)^0 + H_2^* \tag{II.30}$$

where $(H_3^+)^*$ represents the initial distribution of excited H_3^+ ions formed in the exothermic reaction $H_2^+(H_2,H)H_3^+$, was found in ICR experiments by Kim et al.[107b] to have an "average" rate coefficient, $k_d = (2.7 \pm 0.6) \times$

10^{-10} cm³/sec, for deactivation to the ground state. Comparison of this value with the calculated gas kinetic-collision rate coefficient, reveals an average of one out of six collisions to be effective in removing the excess H_3^+ internal energy. More recent experiments employing beam techniques[108a] suggest a relaxation rate coefficient on the order of 10^{-11} cm³/sec for H_3^+ with about three quanta of vibrational excitation and a rate coefficient of about 10^{-12} cm³/sec for relaxation of H_3^+ in the lowest vibrationally excited states. The large average deactivation rate coefficient determined in ICR studies thus apparently indicates that the excited H_3^+ is initially formed in a high degree of vibrational excitation. The large rate coefficients for deactivation of higher vibrational levels of H_3^+ have been attributed to V–V transfer,[108a] whereas relaxation of the low-lying vibrational states of H_3^+, which are not near resonant with vibrational states of H_2, can probably only occur by relatively slow V–T transfer.[108a]

Other ICR experiments by Futrell and et al.[107a, c–f] also indicate that H_3^+ and D_3^+ formed in the $H_2^+–H_2$ and $D_2^+–D_2$ reactions, respectively, are initially in highly excited vibrational states. These studies show that a large number of collisions is necessary to quench these excited $H_3^+(D_3^+)$ states to the extent that they will not undergo reactions that are endoergic for ground-state ions. Typical results from the work of Futrell and co-workers are shown in Figs. 33 and 34. Similar results have been obtained for other vibrationally excited species,[109] including ArD^+, CD_5^+, and CO_2D^+.

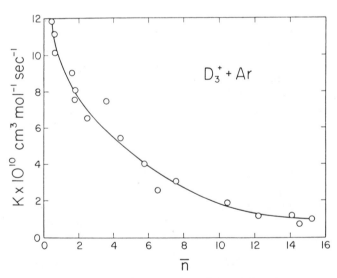

Figure 33.   Rate constant for reaction $D_3^+(Ar, D_2)ArD^+$ as function of average number ($\bar{n}$) of $D_3^+–D_2$ collisions occurring prior to reaction of D_3^+ with Ar. This reaction is endoergic by 0.4 eV for ground-state D_3^+.[107f]

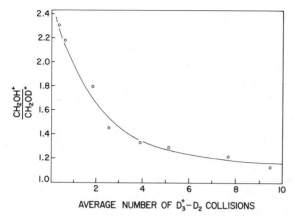

Figure 34. Ratio CH_2OH^+/CH_2OD^+ observed for reaction of D_3^+ with methanol as function of average number of $D_3^+-D_2$ collisions occurring prior to D_3^+ reaction with CH_3OH; CH_2OH^+ results from hydride ion-transfer reaction, $D_3^+(CH_3OH; D_2, HD)CH_2OH^+$, which is endoergic for ground-state D_3^+ reactant ions, whereas CH_2OD^+ results from a D^+ transfer reaction, $D_3^+(CH_3OH, D_2)$ CH_3OHD^+, which is exoergic for ground-state D_3^+, followed by H_2 elimination.[107e]

Up to this point we have discussed collisional deactivation of vibrationally excited ions formed by ionization or as products of exoergic particle-transfer ion–molecule reactions. A somewhat different situation prevails with larger vibrationally excited ions, such as those formed as intermediates in ion–molecule association reactions. Reactions in which such excited intermediates are formed generally demonstrate a third-order dependence of the rate on the concentrations of the reactants at relatively low pressures. The general reaction mechanism may be represented as

$$A^+ + B \underset{k_{-1}}{\overset{k_1}{\rightleftarrows}} (AB^+)^* \overset{k_{2i}}{\underset{M_i}{\rightarrow}} AB^+ \tag{II.31}$$

where M_i is a third body. Examples of such processes are displayed in the reactions that follow.

$$N_2^+ + N_2 \rightleftarrows (N_2^+ \cdot N_2)^* \overset{He,N_2}{\rightarrow} N_2^+ \cdot N_2 \tag{II.32}$$

$$O_2^+ + H_2O \rightleftarrows (O_2^+ \cdot H_2O)^* \overset{He, Ar}{\underset{N_2,O_2}{\rightarrow}} O_2^+ \cdot H_2O \tag{II.33}$$

$$CH_2CF_2^+ + CH_2CF_2 \rightleftarrows (C_4H_4F_4^+)^* \underset{M_i}{\rightarrow} C_4H_4F_4^+ \tag{II.34}$$

$$C_3H_5^+ + C_2H_4 \rightleftarrows (C_5H_9^+) \underset{M_i}{\rightarrow} C_5H_9^+ \tag{II.35}$$

In association reactions of this type, where a new bond is formed, the intermediate has excess vibrational energy equal to the bond energy of the newly formed bond and is thus unstable with respect to dissociation back to reactants unless stabilized by collision. The situation is very similar to that prevailing in neutral systems for atom–atom or radical–radical recombinations, as such larger systems are analogous to those studied by Rabinovitch and co-workers[241–243] by chemical-activation methods. Collisional stabilization or deactivation may result from V–T transfer if the third body, M_i, is monoatomic (a rare-gas atom) or from V–V transfer if it is polyatomic.

Ion–molecule association reactions and the collisional deactivation of excited ions have been the subjects of recent reviews.[244–246] Several systematic studies have been performed in which the relative deactivating efficiencies of various M_i species have been determined. By applying the usual kinetic formulations for the generalized reaction scheme of equation (II.31), and assuming steady-state conditions for $(AB^+)^*$, an expression for the low-pressure third-order rate coefficient can be derived:

$$k^{(3)} = \frac{k_1 k_{2i}}{k - 1} \qquad (\text{II.36})$$

Thus qualitative inferences concerning relative collisional deactivation efficiencies can be drawn from the magnitudes of experimentally measured termolecular ion–molecule rate coefficients.[247] The major systematic studies of the collisional deactivation process have involved the $(C_4H_8^+)^*$ dimer[248] and the $(C_5H_9^+)^*$ adduct[249] formed in ethylene and the dimer ions $(C_4H_4F_4^+)^*$ and $(C_{12}H_{12}^+)^*$ produced in 1,1-difluoroethylene and benzene, respectively.[201] The relative stabilizing effects of various inert bath gases in collisions with these species have been studied and indicate that polyatomic molecules are usually more efficient in stabilizing the excited intermediates than are diatomic or monoatomic species. In a study by Anicich and Bowers,[201] a mass effect was observed in the stabilizing efficiency of a series of rare gases as shown in Fig. 35, where the stabilization efficiency is plotted as a function of the square root of the reduced mass for a series of collision partners, M_i. These results have been interpreted by a model in which the probability of energy transfer is related to the duration of the collision, defined as the ion fly-by-time.[201] In other experiments Miasek and Harrison[249] observed no differences in the stabilization efficiencies of the rare gases for collisional deactivation of $(C_5H_9^+)^*$. These authors suggested that deactivation in this case involves formation of a complex in which energy redistribution occurs via the transitional modes of the complex[249, 242] (i.e., the number of translational and rotational degrees of freedom lost in forming the complex).

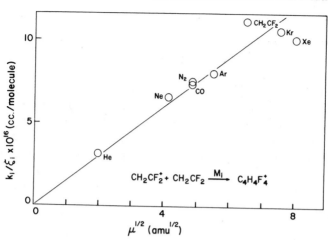

Figure 35. Stabilization efficiency for $C_4H_4F_4^+$ ion as function of square root of reduced mass for series of collision partners; M_i and k_i are the experimental third-order rate constants; ξ_i are the total collision frequencies for collisions of dimer species $(A_2^+)^*$ with nonreactive neutrals M_i.[201]

Additional experimental investigations and theoretical treatments of collisional deactivation processes have recently been reported from several laboratories.[250–253] Temperature effects on the lifetimes of intermediate adducts formed in the $O^- - CO_2$ interaction and in other relatively simple processes have been examined by Meisels and co-workers.[252–254] Here the theoretical treatment involves application of a modified RRKM approach to the unimolecular dissociation of the adduct and/or of the termolecular collision complex consisting of the adduct plus the deactivating species M_i.

In the preceding we have reviewed collisional deactivation of ions in thermal or low-translational-energy collisions. An interesting phenomenon, however, is the observation of E–T and V–T transfers in ion–neutral collisions at translational energies of several hundred electron volts. These transfers cause the so-called superelastic peaks observed in translational-energy measurements of the scattered ions. Electronic to translational energy transfer was observed in collisions of 3.5-keV N^+ with rare-gas atoms and with O_2.[255, 256]

$$N^+(^1S) + He \rightarrow N^+(^1D) + He + 2.15 \text{ eV (kinetic energy)} \quad (II.37)$$

Superelastic peaks caused by V–T energy transfer were also observed in scattering of 100-to-1500-eV H_2^+ from rare-gas atoms and N_2.[256, 257] The cross sections for these processes are quite large and increase monotonically with the increase in target-atom polarizability.

2. Energy Exchange

a. Electronic Excitation

Electronic excitation produced by collisions of medium-and high-translational-energy ions with neutrals has been studied in a large number of systems. If short-lived excited products are formed, such excitations may be conveniently monitored by observing the radiation emitted when the excited species decay radiatively. The wavelengths of the radiation emitted serve to identify the electronically excited states produced in the collision. An alternative method of investigating such processes entails measurement of ion-impact kinetic-energy-loss spectra. In such experiments a change in the internal energy of a particular atom or molecule is detected by measuring the kinetic energy loss, ΔE, by an ion that has been inelastically scattered from the atom or molecule.

Processes of this type can produce excitation of either the neutral target or the projectile ion. Direct excitation is accompanied in many cases by charge transfer (see also Section IV.A.1). The possible product channels are[258]

$$A^+ + B \begin{cases} \rightarrow A^+ + B^* \\ \rightarrow A^{+*} + B \\ \rightarrow A^* + B^+ \\ \rightarrow A + B^{+*} \end{cases} \qquad (II.38)$$

When A and B are identical atomic species (e.g., H^+–H^{259} or He^+–He^{260}), radiative emissions may result from neutral target excitation as well as from electron capture by the ionic projectile into the same excited state. Both processes yield Lyman α radiation in the case of H^+–H collisions,[259]

$$H^+ + H \begin{cases} \rightarrow H^+ + H(2p) & (II.39) \\ \rightarrow H(2p) + H^+ & (II.40) \end{cases}$$

The two processes illustrated here, direct excitation and charge transfer, are distinguishable because the radiative emission from the translationally thermal electronically excited target atom is not shifted in wavelength, whereas that produced by charge transfer exhibits a Doppler shift because of the high kinetic energy of the projectile.

Ion-impact excitation has been widely studied in the rare gases[260–270] and for alkali metal ion–atom collisions.[271–280] In many cases excitation functions have been measured (i.e., total cross sections as a function of initial relative translational energy), and in some instances the angular dependencies of the differential cross sections for inelastic scattering have been determined. The most striking feature of the results from these experiments is the oscillatory structure that is evident in many of the

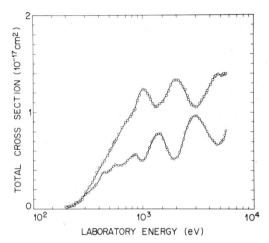

Figure 36. Absolute population cross sections for Ne*(3p) levels ($\square$) and Na*(3p) levels ($\bigcirc$) resulting from Na$^+$–Ne inelastic collisions, as function of laboratory energy.[279]

excitation functions. Figure 36 shows the absolute cross sections for population of the Ne*(3p) levels and Na*(3p) levels plotted as a function of laboratory energy. These results were derived from the analysis of the radiation observed from Na$^+$–Ne collisions, which arises from the excitation of both neon and sodium into 3p electronic states as a result of both direct excitation and charge transfer.

$$Na^+ + Ne \rightarrow \begin{cases} Na^+ + Ne^*(3p) \\ Na^*(3p) + Ne^+ \end{cases} \qquad (II.41)$$

The oscillatory structure just mentioned has been clearly demonstrated to result from quantum-mechanical phase-interference phenomena. The necessary condition[264, 265] for the occurrence of oscillatory structure in the total cross section is the existence in the internuclear potentials of an inner pseudocrossing, at short internuclear distance, as well as an outer pseudo-crossing, at long internuclear distance. A schematic illustration of this dual-interaction model, proposed by Rosenthal and Foley,[264] is shown in Fig. 37. The interaction can be considered to involve three separate phases, as discussed by Tolk and et al.:[279] (1) the primary excitation mechanism, in which, as the collision partners approach, a transition is made from the ground U_0 state to at least two inelastic channels U_1 and U_2 (the transition occurs at the internuclear separation R_i, the inner pseudocrossing, in Fig. 37), (2) development of a phase difference between the inelastic channels,

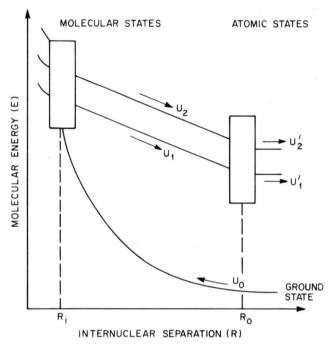

Figure 37. Schematic illustration of dual-interaction model for ion-impact excitation resulting in oscillatory structure in total cross section; see text.[279]

U_1 and U_2, as the collision partners separate ($R_i \leqslant R \leqslant R_0$ in Fig. 37), and (3) a second interaction (at the internuclear separation R_0, the outer pseudocrossing, in Fig. 37), at which interference occurs between the inelastic channels. In the Na^+–Ne system noted in (II.41) the quasimolecular states responsible for the oscillatory structure were uniquely determined.

The extent to which the Wigner spin rule is obeyed has been studied in collisions of 1.5-to-3.5-keV N^+ ions with various neutrals.[255] Inelastic and superelastic collisions involving both ground-state and metastable N^+ have been observed. Cross sections for those reactions in which total electron-spin angular momentum is conserved were observed to be larger than those in which it is not conserved, provided that the reactants are light mass species (e.g., interactions of N^+ with He, Ne, and Ar). On the other hand, spin angular momentum is not conserved in excitation and deexcitation of N^+ resulting from collisions with krypton or xenon. In the latter case (J, J) coupling is more important than (L, S) coupling.

Electronic excitations resulting from ionic collisions with diatomic and polyatomic systems have also been studied.[255, 256, 281-287] For example, excitations of the triplet ground state of O_2 to the excited $O_2(a^1\Delta_g)$ and $O_2(b^1\Sigma_g^+)$ states are induced by collisions of 1-keV H_2^+ ions with O_2. Such electronic transitions induced in neutral molecules by ion impact generally produce different states than do those excited by electron or photon impact. Whereas both H_2^+ and He^+ impact excite singlet–triplet transitions in neutral targets, H^+ does not. This difference in behavior is ascribed to the fact that because H^+ does not have an electron, it cannot cause electron exchange processes to occur. Most experiments of the type described have been accomplished at relatively high interaction energies. Collision-induced excitation processes at energies near threshold are normally masked by charge transfer and reactive scattering processes. To date, only two studies have been reported in which a diatomic molecule (viz,.NO) has been successfully excited by ion impact at low translational energies (a few electron volts). In these studies Rydberg excitation of NO was achieved in collisions with C^+, N^+, O^+, Ar^+, and H^+, at energies near threshold.[9c, 288] Figure 38 shows the NO emission spectrum resulting from 10-eV H^+–NO collisions, and Fig. 39 shows the excitation function for the process

$$H^+ + NO \rightarrow H^+ + NO(A^2\Sigma^+) \qquad (II.42)$$

as deduced from luminescence measurements. It was determined from these experiments that the electronic excitation occurs at relatively large impact parameters, with little distortion of the NO molecule. This is understandable in view of the fact that the excitations involve an outer Rydberg electron.

Several studies of direct excitations of diatomic ions (notably N_2^+)[160, 289, 290] and of polyatomic ions[291] have been reported. When N_2^+ is impacted on helium and neon target gases at laboratory energies in the range 0.3 to 5 keV, the charge-transfer channel is suppressed and direct excitation to the $N_2^+(B^2\Sigma_u^+)$ state occurs. The most interesting result of the latter experiments[160] was the observation of extensive vibrational and rotational excitation in the electronically excited N_2^+ species. This was somewhat unexpected in a reaction between a molecular ion and a neutral atom since the molecular ion should be barely affected by the weak field of the atom.[160] Similar results have recently[292] been reported for N_2^+ excitation on collision with argon. The emission spectrum of N_2^{+*} in this case is very similar to that resulting from Ar^+–N_2 charge-transfer reactions. Theoretical calculations for the three-atom intermediate system $(N-N-Ar)^+$ are in progress, and hopefully these results will help to rationalize the experimental observations (see also Sections II.B.2.b and IV.A.1).

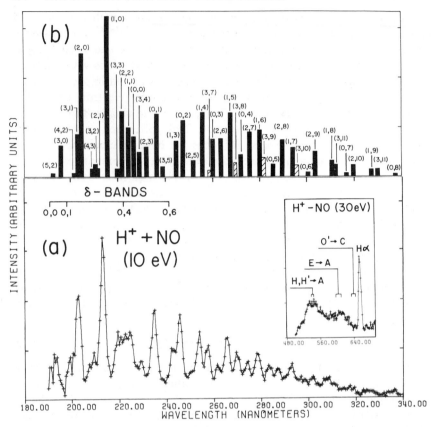

Figure 38. (a) Emission spectrum (190–340 nm) from excited species formed in collisions of 10-eV H^+ ions with NO (spectrum was taken with 1-nm resolution; major features are NO $A{\to}X$ γ bands, which are identified in diagram b); inset shows partial spectrum produced by impact of 30-eV H^+ ions with NO (6-nm resolution) and indicates Rydberg–Rydberg transitions; (b) model spectrum (vertical transitions) for the NO γ bands.[288]

In still other experiments, selective excitation of the H_2^+ molecular ion in collisions with rare gases at 0.1 to 10 keV was studied by determining the polarizations of the Balmer-α and Balmer-β lines resulting from the dissociative collision.[293]

The electronic excitation discussed up to this point involves translational to electronic energy transfer. Of great practical importance is near-resonant electronic excitation-energy transfer, which leads to ion lasers in the He–Kr and Ne–Xe systems.[294] The reactions involved are (He* and Ne*

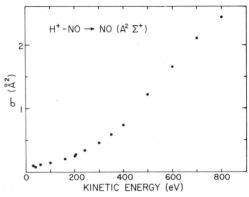

Figure 39. Absolute cross section for excitation of $NO[X(v=0)]\rightarrow NO(A)$ transitions in H^+–NO collisions as function of kinetic energy.[288]

are metastable atoms),

$$\mathrm{He}^*(2^3S_1) + \mathrm{Kr}^+(^2P) \rightarrow \mathrm{He}(^1S_0)\mathrm{Kr}^+(6s^4P_{3/2})_{,5/2} \qquad (\mathrm{II}.43)$$

and

$$\mathrm{Ne}^*(1s_5) + \mathrm{Xe}^+(^2P_{3/2})^0 \rightarrow \mathrm{Ne}(^1S_O) + \mathrm{Xe}^+(7s^4P_{3/2,5/2}) \qquad (\mathrm{II}.44)$$

These processes occur at thermal energies in the afterglow of a pulsed discharge and probably involve curve crossings. The resultant electronically excited states of Kr^+ and Xe^+ lead to several laser lines in the range of 512.8 to 431.9 nm and 609.5 to 486.4 nm, respectively.

b. Vibrational and Rotational Excitation

Translational-to-vibrational and -rotational energy-transfer processes have been studied in collisions between atomic ions and diatomic molecules, as well as in collisions between diatomic molecular ions and neutral atoms. Among the former studies are the experiments of Dittner and Datz,[295] Schöttler et al.,[296] and others,[297, 298] on excitation of H_2 in collisions with alkali ions (K^+, Li^+). Examples of diatomic ion–neutral atom experiments are those of Moran and co-workers,[134, 299, 300] in which excitations of O_2^+ and CO^+ in collisions with argon were investigated, and those of Cheng et al.,[301] which deal with excitations of NO^+ and O_2^+ in collisions with helium. The levels of vibrational and rotational excitation in these systems were determined by using beam methods and measuring ion-impact kinetic-energy-loss spectra. Two experimental techniques have been employed to determine the kinetic-energy distributions of inelastically scattered ions: (1) time-of-flight (TOF)[295, 296] measurements and (2) procedures utilizing electrostatic energy analyzers.[134, 283, 299–303] The early experiments employed incident ions of relatively high translational energies, whereas more recent studies have concentrated on low (few electron volts) energies. A survey of recent high-resolution, low-energy ($\lesssim 10$ eV) ion-scattering experiments, in which ion-impact kinetic-energy-loss spectra have been measured, is presented in a recent review by Toennies.[304] Here we discuss such experiments for only one system, the H^+–H_2 system, to illustrate the enormous progress made in this area.

Moore and Doering[303] were the first to achieve resolution of discrete vibrational levels excited in small-angle, high-energy scattering of H^+ and H_2^+ on H_2, D_2, and N_2. Their initial results are shown in Fig. 40. The cross sections for pure vibrational excitations were observed to be unusually

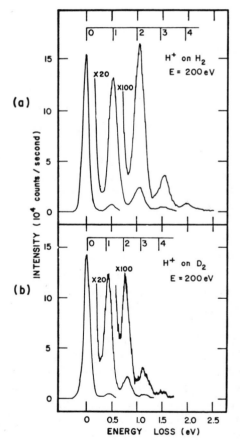

Figure 40. (*a*) Energy-loss spectrum of 200-eV protons scattered from molecular hydrogen (scattering angle 0° and target gas pressure 3.5 mtorr; positions of $v' = 1, 2, 3, 4$ vibrational levels of ground state of H_2 are indicated relative to $v' = 0$ level at 0-eV energy loss); (*b*) energy-loss spectrum of 200-eV protons scattered from deuterium at pressure of 3 mtorr.[303]

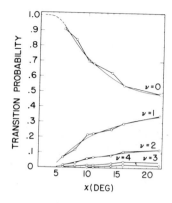

Figure 41. Transition probabilities for vibrational excitation as function of center-of-mass scattering angle for collisions of 10-eV H^+.[307]

large. More extensive experimental and theoretical investigations of the H^+-H_2 system have since been accomplished.[257, 305-308] The angular and energy dependences of the differential cross sections for resolved vibrational transitions in low-energy collisions of H^+ with H_2, HD, and D_2 have been determined by Udseth et al..[307] Figures 41 and 42 show vibrational transition probabilities as a function of center-of-mass scattering angle and energy, respectively, for these processes. The inelastic character of the collisions strongly increases with collision energy and scattering angle. The average energy transfer at a given collision energy and scattering angle was observed to be independent of the isotopic composition of the molecule. The dominant mode of excitation appears to be the perturbation of the electronic structure of the molecule by the proton, which results in a force tending to stretch the molecule.

Vibrational excitation is the dominant inelastic process in $H^+ + H_2$ scattering for collision energies above approximately 10 eV in the labora-

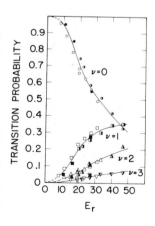

Figure 42. Transition probabilities for vibrational excitation as function of reduced energy $E/\hbar w$ [($\hbar = (h/2\pi)$, $w = 2\pi\nu$ and ν-vibrational frequency of neutral molecule) for $H^+ + H_2$ (O), $H^+ + HD$ (O), and $H^+ + D_2$ (●) collisions, at scattering angle of $11°$.[307]

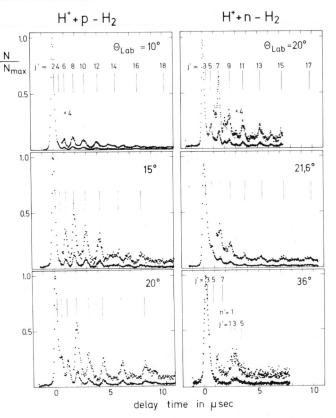

Figure 43. Time-of-flight spectra for H^+-p-H_2 (left) and H^+-n-H_2 (right) are shown at five different laboratory scattering angles. Number of ions normalized to maximum registered over entire measuring time (25 to 60 hr) in each TOF channel (40 nsec) is plotted as function of flight time (flight length = 1.25 m). Typically, N_{max} varies between 10,000 and 500. Vertical lines show absolute calculated locations of arrival times corresponding to indicated final rotational states. Center-of-mass scattering angles for elastic scattering corresponding to five laboratory angles are, in increasing order, $\vartheta_{CM} = 15.0°$, 22.4°, 29.9°, 32.1°, and 53.1°.[310]

tory system.[307] Pure rotational excitations are observed at lower energies.[309] Rotational quantum transitions corresponding to Δj up to 20 have been detected in experiments involving T–R excitation of H_2 by collision with protons at $E_{CM} = 3.7$ eV, where the subscript CM refers to center of mass. These results are shown in Fig. 43.[310] The measured probability distribution has a maximum for elastic scattering ($\Delta j = 0$) at all angles. It drops to 5–35% for $\Delta j = 2$ transitions and for most angles decreases only gradually with increasing Δj. These experimental findings are in agreement with the theoretical predictions of Giese and Gentry.[308] Rotational excitations were first observed by Moran, et. al.[299] in the CO^+–Ar interaction. Resolution of pure rotational transitions was first achieved by Toennies and co-workers in experiments with the Li^+–H_2 system.[311] The two scattering processes, H^+–H_2 and Li^+–H_2, show quite different behavior[309, 310] with respect to rotational and vibrational excitation, which must reflect differences in the potential surfaces for the two systems.

Both T–R and T–V energy transfer occurring in collisions between ions and polyatomic molecules have also been studied, and this work has been discussed by Toennies.[304] The increasing knowledge of T–R energy-transfer processes should also enhance the capability to study rotational excitation in reactive ion–neutral scattering, an area that has hardly been investigated as yet.

In addition to the processes just discussed that yield vibrationally and rotationally excited diatomic ions in the ground electronic state, vibrational and rotational excitations also accompany direct electronic excitation (see Section II.B.2.a) of diatomic ions as well as charge-transfer excitation of these species (see Section IV.A.1). Furthermore, direct vibrational excitation of ions and molecules can take place via charge transfer in symmetric ion molecule collisions, as the translational-to-internal-energy conversion is a sensitive function of energy defects and vibrational overlaps of the individual reactant systems.[312–314]

III. REACTIONS OF IONS WITH EXCITED NEUTRALS

Results of studies of reactions of ions with rotationally, vibrationally, and electronically excited neutrals are summarized in Table III. The limited number of such experiments that have been reported reflects the difficulty in preparing neutral reactants in selected states of internal excitation. It might be expected that increasing the internal energy of the neutral would have an effect on the reaction cross section equivalent to that of increasing the internal energy of the ionic reactant. In some instances, notably endoergic reactions, this has been found to be the case, insofar as the observed behavior is as expected for endoergic reactions. For example, electronic excitation of O_2 promotes endoergic charge transfer from O^-. However, most of the reactions involving neutral vibrationally excited N_2 behave in an "abnormal" manner. This includes the well-known and amply discussed exoergic reactions $O^+(N_2, N)NO^+, Ne^+(N_2, Ne)N_2^+$ for which the cross sections are significantly *enhanced* when the vibrational excitation of the N_2 reactant is increased.

The $O^+(N_2,N)NO^+$ reaction (Tables I and III) has been the subject of numerous investigations.[4c,d, 123, 126, 127, 315–317] It is one of the few examples of an exoergic ion–molecule particle-transfer reaction for which the cross section rises with increasing vibrational energy of the reactant. This reaction also exhibits several other unusual kinetic features, including a very low reaction-rate coefficient at room temperature, an inverse temperature dependence of the rate coefficient in the range 300 to 750°K, and a strong increase of k with translational energy above 750°K. This behavior has been explained on the basis of crossings between the potential energy surfaces on which the reaction occurs,[318–321] as discussed in more detail in the last section of this review. However, this model also predicts that vibrational excitation should affect the rate coefficient much more than translational excitation, which is not observed to be the case experimentally (Fig. 44).

The reaction $Ne^+(N_2,Ne)N_2^+$ (Table III) is the only exoergic charge-transfer process for which the cross section has been observed to increase with increasing vibrational quantum number of the neutral reactant.[128] A probable rationale for this phenomenon is that both criteria of efficient charge transfer, namely, energy resonance and favorable Franck–Condon overlap, are satisfied if the reaction entails formation of a quartet N_2^+ state from the $(v=2)$ vibrational level of N_2 in the ground electronic state.

The associative detachment reaction

$$O^- + N_2 \rightarrow N_2O + e \qquad (III.1)$$

even at relative translational energies well above the thermochemical

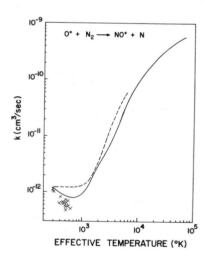

Figure 44. Rate constants for O^+–N_2 reaction as: ($\times$) function of effective temperature; (—), translational temperature data; (---), vibrational temperature data.[317]

threshold, does not occur with N_2 in the ground vibrational state.[322] However, it is predicted, on the basis of potential surface calculations,[323] to take place for $v \geqslant 1$. This has not yet been verified experimentally.

The effect of rotational energy on a reaction cross section has been studied experimentally in only one reaction, namely, $Ar^+(H_2, H)ArH^+$ (Table III). For this exoergic particle-transfer process, an inverse dependence of the cross section on rotational energy was observed.

It is evident from the foregoing discussion that although some progress has been made in defining the dependences of ion–neutral reaction cross sections on reactant internal energy, the data are still quite fragmentary, and the ultimate goal of obtaining state-to-state cross sections[324] has not yet been achieved.

IV. EXCITED PRODUCTS FROM ION–NEUTRAL COLLISIONS (ELECTRONIC, VIBRATIONAL AND ROTATIONAL EXCITATION)

The products of reactive ion–neutral collisions may be formed in a variety of excited states. Excited products from nonreactive collisions have already been discussed in a previous section. Theoretical calculations of vibrational excitation in the products of symmetric charge-transfer reactions have also been mentioned previously.[312–314] The present section deals with excited products from reactive ion–neutral scattering, with special emphasis on luminescence measurements.

Luminescence spectra resulting from pure vibrational or V–R transitions involving excited-product states formed in ion–neutral collisions have not yet been observed. However, vibrational and rotational excitation of the products of reactive ion–neutral collisions may be determined indirectly from measurements of Q, the translational exoergicity, which is defined as the difference between the translational energy of the products and that of the reactants. According to the energy-conservation principle, then,

$$Q = \Delta E_0^0 - U \qquad \text{(IV.1)}$$

where ΔE_0^0 is the internal energy change for the reaction and U is the internal-excitation energy of the products. Measurements of this type are usually made in beam experiments and involve determinations of both the angular and velocity distributions of the scattered products. In beam studies of neutral–neutral reactions, notably for the process $F + D_2 \rightarrow DF(v') + D$, vibrational structure has actually been resolved in the product-recoil translational-energy distribution.[325] Resolution of vibrational structure from kinetic-energy measurements on the products of ion–neutral reactions has been achieved only recently.[108a, 326] Many of the latter reactions have been studied by these methods, however, and in many instances it has been found that the products are formed with relatively large amounts of internal energy. Also, as the initial relative translational energy for these reactions is increased, the products tend to store increasing amounts of internal energy until the point is reached at which U exceeds the bond-dissociation energy and the product is no longer stable. In most reactions of the type $A^+(BC, C)AB^+$, and in particular for almost all exoergic hydrogen-transfer reactions, a strong maximum in the intensity of AB^+ is observed at a scattering angle of zero degrees. This product-intensity maximum exhibits a Q dependence on reactant translational energy that is consistent with the predictions of the spectator stripping model.[327] However, careful examination of intensity contour maps for such direct hydrogen-atom transfer processes reveals large-angle scattering as well as Q values in disagreement with the spectator stripping model. A sequential impulse model has been applied to explain these results.[328, 329]

In cases where short-lived electronically excited products are formed in ion–neutral reactions the excited states can be determined by monitoring the radiative emissions as these states decay. Several reactions in which metastable electronic states are formed have also been studied using special techniques. For example, charge transfer into the metastable $2S$ state of atomic hydrogen was observed for reactions of protons with cesium. The metastable atoms were detected by quenching them in an electric field and monitoring the resultant Lyman-α photons.[330] Penning ionization of acetylene has been employed as a detection technique for metastable argon atoms produced in Ar^+–Ar charge-transfer collisions.[331] More recently, ion–neutral reactions were employed to detect and estimate the fractional abundance of metastable rare-gas atoms. In the latter experiments rare-gas beams consisting partially of metastable atoms were obtained via charge-transfer reactions such as

$$He^+ + Cs \rightarrow He^* + Cs^+ \tag{IV.2}$$

$$Ne^+ + Na \rightarrow Ne^* + Na^+ \tag{IV.3}$$

$$Ar^+ + Rb \rightarrow Ar^* + Rb^+ \tag{IV.4}$$

The composition of these beams was estimated on the basis of the lower reactivity of the metastable atom as compared to that of the ground-state atoms in ion–molecule reactions with H_2^+, for example.[332]

The majority of the available information on electronically excited products formed in ion–neutral reactions has been derived from luminescence measurements. A comprehensive review of such data for the period up to 1970, which deals with ion interactions at translational energies of 10 eV and higher, is available.[258] More recent work is summarized here.

A. Chemiluminescent Reactions

Most of the work to date on chemiluminescence resulting from ion–neutral interactions has been concentrated on charge-transfer reactions. Very few particle-transfer reactions have been observed to produce radiative emissions. A large body of luminescence data has been accumulated for charge-transfer reactions occurring at high translational energies. McGowan and Stebbings[333] were among the first to suggest the potential role of charge-transfer processes in pumping lasers. In more recent experiments from several laboratories, luminescence has been detected from the products of thermal and low-energy ion–neutral reactions, and the importance of the reactions in selected laser systems has been confirmed. This is rather ironic, since most of the initial attempts at modeling lasers largely ignored the possibility of ion–neutral interactions. The feasibility of utilizing particular thermal-energy ion–neutral reactions in developing electronic transition lasers has been discussed in a recent review.[334] High-translational-energy ion–neutral reactions are generally of less interest for such applications since most laser systems operate at relatively high pressures under conditions where the energy of the reactant ions is near thermal. Studies of luminescence from low-energy ion–molecule interactions have also been reviewed by Marx.[335] Since it is difficult to specify precisely the boundary between the "low" and "high" translational-energy regimes with respect to such experiments, and since so much interesting information has been derived from experiments at high energies, we have attempted to include all the luminescence data reported for ion–neutral reactions in the compilation presented in Table IV.

In chemiluminescence experiments such as those described previously in the experimental section, emission spectra characteristic of the excited products of ion–neutral collisions are obtained, that is, intensities of the emitted radiation as a function of wavelength. This permits identification of the electronically excited states produced in the reaction as well as determination of the relative populations of these states. In addition if the luminescence measurements are made using beam techniques, excitation functions (intensity of a given transition as a function of the translational energy of the reactants) can be measured for certain transitions. As is discussed later, some of the observed transitions exhibit translational-energy thresholds. In the emission spectra from diatomic or polyatomic product molecules, band systems are sometimes observed from which the relative importance of vibrational and rotational excitation accompanying electronic excitation may be assessed.

1. Charge Transfer

The effects of reactant excitation on the cross sections for charge-transfer reactions have been briefly discussed in Section II.A.1.a. Bowers and

Table IV.A.
Luminescence in Charge Transfer and Ion–Molecule Reactions: Charge Transfer in Atomic Systems

REACTION	INCIDENT-ION TRANSLATIONAL ENERGY	EMITTING SPECIES	WAVELENGTHS AND/OR TRANSITIONS OBSERVED	REFERENCE
$H^+(H,H)H^+$	0.6 to 3.0 keV	$H(2p)$	Lymanα (121.57 nm)	259
$H^+(X,H)X^+$; X = He, Ne, Ar, Kr, Xe	1 to 25 keV	$H(2p)$	Lymanα	336
$He^+(He,He)He^+$	20 eV to 5 keV	He(I)	Visible $n=3 \rightarrow n=2$	262
$He^+(He,He)He^+$	5 to 100 keV	He(I)	Visible from $n=3$ to 6	260
$He^+(X,He)X^+$; X = He, Ne, Ar, Kr, Xe	1 eV to 10 keV	X(I); X(II); X = He, Ne, Ar, Kr, Xe	40 to 900 nm, X(I) resonance lines $(^1P, {}^3P \rightarrow {}^1S)$ XII lines arising From $^2S, {}^2P$, and ^2D states and other transitions	156, 163, 267, 270, 337–346
$He^+(X,He)X^+$ X = K, Rb	50 to 1600 eV	K(II), Rb(II), He(I)	Near-UV and visible	157
$He^+(Zn,He)Zn^+$	Thermal	Zn(II)	From levels $3d^{10}nx$ at 491 nm, 492 nm, 602 nm, 610 nm, 759 nm, and 776 nm	166, 351–354, 356, 357
$He^+(Cd,He)Cd^+$	Thermal	Cd(II)	From levels $4d^{10}nx$ at 538 nm, 636 nm, 724 nm, 728 nm, 807 nm, 853 nm, and 888 nm	166, 352, 353, 355–359
$He^+(Hg,He)Hg^+$	Thermal	Hg(II)	From $^2P^0$ states, 300–900 nm, especially 615 nm and 794.4 nm	173, 360

He⁺(I,He)I⁺	Thermal	I(II)	From $5s^25p^3(^2D^0)6p$	3a, 361
He⁺(Se,He)Se⁺	Thermal	Se(II)	447 to 1259 nm	353, 362, 363
He⁺(Au,He)Au⁺	Thermal	Au(II)	253 to 763 nm	364
Ne⁺(He,Ne)He⁺	10 eV to 5 keV	Ne(I)	$3^3S_1 \rightarrow 3^3P_2^0$	270
Ne⁺(Ne,Ne)Ne⁺	20 eV to 7 keV	Ne(I)	$3s \rightarrow 2p$ Resonance lines	365
Ne⁺(X,Ne)X⁺; X=Ar,Kr	5 to 300 eV	X(II)	350 to 550 nm	342
Ne⁺(Mg,Ne)Mg⁺	Thermal	Mg(II)	245 to 1009 nm	353
Ne⁺(Zn,Ne)Zn⁺	Thermal	Zn(II), Ne(I)	—	367
X⁺(Y,X)Y⁺ X=He,Ne,Ar; Y=Pb,Al,Mn	Thermal	Pb(II),Al(II),Mn(II)	—	353
Ar⁺(Ar,Ar)Ar⁺	20 eV to 7 keV	Ar(I),Ar(II)	36 to 720 nm	289, 366, 368
Na⁺(Ne,Na)Ne⁺	100 eV to 6 keV	Na(I),Ne(I)	$3p \rightarrow 3s$ NaD Lines	275, 278, 279
Na⁺(O⁻,Na)O	0.1 to 7 eV	Na(I)	$3p \rightarrow 3s$; $3d \rightarrow 3p$; $4p \rightarrow 3s$	369
X⁺(Ne,X)Ne⁺; X=Li,Be,B,C,N,O,F, Ne,Na,Mg,Al	0.1 to 15 keV	X(I)	Visible	277
K⁺(Ar,K)Ar⁺	0.2 to 3 keV	K(I),Ar(I)	—	280

Table IV.B.
Luminescence in Charge Transfer and Ion–Molecule Reactions: Charge Transfer in Molecular Systems

REACTION	INCIDENT-ION TRANSLATION ENERGY	EMITTING SPECIES	WAVELENGTHS AND/OR TRANSITIONS OBSERVED	VIBRATIONAL AND ROTATIONAL POPULATIONS OF EMITTINGS STATE	REFERENCE
		I. Atomic Ions on Diatomic Molecules			
$H^+(H_2,2H)H^+$	0.1 to 25 keV	$H(I)$	Lyman α	—	374,375
$H^+(H_2,2H)H^+$	5 to 140 keV	$H(I)$	Balmer $(H_\alpha, H_\beta, H_\gamma)$	—	376
$D^+(H_2;D,H)H^+$	1 to 25 keV	$H(I)$	Lyman α	—	375
$He^+(H_2;He,H)H^+$	~eV to 30 keV	$H(I)$ $(n), n=2,3,4$	Lyman$(L_\alpha, L_\beta, L_\gamma, L_\delta, L_\epsilon)$ Balmer $(H_\alpha, H_\beta, H_\gamma)$	—	266,374,375, 377–381
$He^+(H_2;He,H)H^+$	~eV to 30 keV	$He(I)$	447.1 nm, 587.6 nm, 667.8 nm	—	377–381
$X^+(H_2;X,H)H^+$ $X=Ne,Ar$	0.1 to 30 keV	$H(I)$ $(n), n > 2; Ne(I), Ar(I)$ $Ne(I), Ar(I)$	Balmer	—	377
$Li^+(H_2,Li)H_2^+$	0.2 to 3.0 keV	$Li(I)$	$ns \rightarrow 2p, n=3,4,5;$ $np \rightarrow 2s, n=2,3,4,5;$ $nd \rightarrow 2p, n=3,4,5$	—	382,383
$H^+(N_2,H)N_2^+$	1.5 to 4 keV	$H(I), N_2(II), N_2(II)$	H_β; 0,0 first negative band; 2,0 Meinel band		384
$H^+(N_2,H)N_2^+$ $D^+(N_2,D)N_2^+$	10 to 65 keV	$H(I)$ $D(I)$	Balmer		385
$He^+(N_2,He)N_2^+$	Thermal	$N_2(II)$	$C^2\Sigma_u^+ \rightarrow X^2\Sigma_g^+$ (Second Negative)	Non-Franck–Condon; resonant charge transfer to $v'=3,4$	168,171, 386–389
$He^+(N_2,He)N_2^+$	~eV to 25 keV	$N_2(II)$	$C^2\Sigma_u^+ \rightarrow X^2\Sigma_g^+$ (Second negative)	Near Franck–Condon modified by pre-dissociation from $v' > 3$	161,163,390–392

System	Energy	Emitter	Transition	Comments	References
He$^+$(N$_2$; He)N$_2^+$	Thermal	N$_2$ (II)	$B^2\Sigma_u^+ \rightarrow X^2\Sigma_g^+$ (First negative)	Franck–Condon modified by ion-induced dipole interaction	123,168,171,389
He$^+$(N$_2$; He)N$_2^+$	~eV to 450 eV	N$_2$ (II)	$B^2\Sigma_u^+ \rightarrow X^2\Sigma_g^+$ (First negative)	Franck–Condon at ion velocities $> 10^8$ cm/sec; at $v < 10^8$ cm/sec, relative population of upper vibrational states increases monotonically	155,385, 391–395
He$^+$(N$_2$; He)N$_2^+$	Thermal	N$_2$ (II)	$D^2\Pi_g \rightarrow A^2\Pi_u$ (Janin–d'Incan)	—	168
He$^+$(N$_2$; He)N$_2^+$	5 to 450 keV	N$_2$ (II)	$A^2\Pi_u \rightarrow X^2\Sigma_g^+$ (Meinel)	—	393
He$^+$(N$_2$; He)N$_2^+$	Thermal to 490 eV	N$_2$ (II)	164 to 320 nm (Unidentified)	—	389,390
He$^+$(N$_2$; He, N)N$^+$	23 to 450 eV	N (I)	120 nm, 124.3 nm, 131.5 nm, 141.2 nm, 149.3 nm, and 174.3 nm		390
He$^+$(N$_2$; He, N)N$^+$	500 eV to 25 keV	N (II)	$4d^3F^0 \rightarrow 3p^3D$ at 231.7 Å		161
Ne$^+$(N$_2$; Ne, N)N$^+$ X$^+$(N$_2$; X)N$_2^+$; X = Ne, Ar, H, N, D, Li, Na, K, Ca, Mg	2 eV to 25 keV	N (II) N$_2$ (II)	First negative	Non-Franck–Condon at low energies; strong energy-dependent deviations; rotational excitation	9c, 155,289, 290,368,391, 394–399

Table IV.B. *continued*

REACTION	INCIDENT ION TRANSLATIONAL ENERGY	EMITTING SPECIES	WAVELENGTHS AND/OR TRANSITIONS OBSERVED	VIBRATIONAL AND ROTATIONAL POPULATIONS OF EMITTING STATE	REFERENCE
$X^+(N_2;X,N)N^+$	—	N (II); X (I), X (II)	—	—	9c, 155, 289, 290, 368, 391, 394–399
$X^+(N_2,X)N_2^+$; $X^+=Kr^+,Xe^+$ in metastable excited states	Thermal	N_2 (II)	First negative	Non-Franck–Condon vibrational excitation; strong rotational excitation	400
$He^+(CO,He)CO^+$	200 to 1500 eV	CO (II)	$A^2\Pi \rightarrow X^2\Sigma^+$ (Comet tail) $B^2\Sigma^+ \rightarrow X^2\Sigma^+$ (First negative)	Franck–Condon	401
$X^+(CO,X)CO^+$; $X=He,Ne,Ar$	0.16 to 30 keV	CO(II)	Comet tail	Deviations from Franck–Condon at low energies	402
$Ar^+(CO,Ar)CO^+$	1000 eV	CO (II)	Comet tail	Distorted Franck–Condon	403
$Ar^+(CO,Ar)CO^+$	10 to 1000 eV	CO (II)	Comet tail	—	290
$X^+(CO,X)CO^+$; $X=Ar,Ne$	2 to 1000 eV	CO (II); C (II) (only with Ne)	Comet Tail $3p^2P^0 \rightarrow 2p^2S$	Distorted Franck–Condon	9c, 193
$X^+(CO,X)CO^+$; $X=Ar,Ne,He$	3 to 50 keV	CO (II); C(II), O(II), X (I)	—	—	404
$Ar^+(CO,Ar)CO^+$	2500 eV	CO (II)	Comet tail	Non-Franck–Condon	405
$Kr^+(CO,Kr)CO^+$	2000 eV	CO (II)	Comet tail	Non-Franck–Condon	406
$He^+(O_2,He)O_2^+$	Thermal	O_2 (II)	$c^4\Sigma_u^- \rightarrow b^4\Sigma_g^-$ (Hopfield)	—	407

Reaction	Target	Transition	Energy range	Comments	Ref.
$He^+(O_2, HeO_2^+)$	O_2 (II)	$A^2\Pi_u \rightarrow X^2\Pi_g$ (Second negative)	25 to 400 eV	Distorted Franck–Condon;	408
		$b^4\Sigma_g^- \rightarrow a^4\Pi_u$ (First negative) (Hopfield)	11 eV		391
$He^+(O_2; He, O)O^+$	O (I), O (II); He (I)	Many lines in range 300 to 850 nm	25 to 400 eV	—	408
$Ne^+(O_2, NeO_2^+)$	O_2 (II)	First negative; second negative	5 to 250 eV	—	398
$C^+(O_2, C)O_2^+$	O_2 (II)	Several band systems	>25 eV	—	193
$Ar^+(O_2; Ar, O)O^+$	O (I)	$3^3S^0 \rightarrow 4^3P$ (?)	900 eV	—	290
$Na^+(X, Na)X^+$; $X=O_2, N_2, CO$	Na (I)	268–819 nm	0.1 to 1 keV	—	162
$He^+(HCl, He)HCl^+$; $He^+(HBr, He)HBr^+$	HCl (II) HBr (II)	$A^2\Sigma^+ \rightarrow X^2\Pi$	0.3 to 4 keV	—	158,409,410
$He^+(HCl; He, H)Cl^+$; $He^+(HBr; He, H)Br^+$	H (I), Cl (II) H (I), Br (II)	—	0.3 to 4 keV	—	158,409
$X^+(HCl, X)HCl^+$; $Kr^+(HBr, Kr)HBr^+$; $X=Ar, Kr$	HCl (II) HBr (II)	$A^2\Sigma^+ \rightarrow X^2\Pi_i$	100 to 5000 eV	Deviations from Franck–Condon at low energies	405,406,410,41
$He^+(Cl_2; He, Cl)Cl^+$; $He^+(Br_2; He, Br)Br^+$; $He^+(I_2; He, I)I^+$	Cl (I); Cl (II); Br (I); Br (II); I (I), I (II)	60–870 nm (many lines)	21 to 170 eV	—	412

2. Atomic Ions on Triatomic and Polyatomic Molecules

Reaction	Target	Transition	Energy range	Comments	Ref.
$Ne^+(CO_2, Ne)CO_2^+$	CO_2 (II)	$\tilde{A}^2\Pi_u \rightarrow \tilde{X}^2\Pi_g$	2 to 200 eV	—	398
$He^+(CO_2, He)CO_2^+$	CO_2 (II); He(I)	$\tilde{A}^2\Pi_u \rightarrow \tilde{X}^2\Pi_g$; $\tilde{B}^2\Sigma_u^+ \rightarrow \tilde{X}^2\Pi_g$; triplet lines	200 to 1500 eV	—	159
$X^+(CO_2, X)CO_2^+$; $X=H, He, Ne$	CO_2 (II)	$A^2\Pi_u \rightarrow X^2\Pi_g$; $B^2\Sigma_u^+ \rightarrow X^2\Pi_g$	200 eV to 25 keV	—	420
$He^+(CO_2, He)CO_2^+$	CO_2 (II)	$A^2\Pi_u \rightarrow X^2\Pi_g$	100 to 5000 eV	Relative populations of low vibrational states of stretching mode increased monotonically with decreasing projectile-ion velocity	421

Table IV.B. *continued*

REACTION	INCIDENT ION TRANSLATIONAL ENERGY	EMITTING SPECIES	WAVELENGTHS AND/OR TRANSITIONS OBSERVED	VIBRATIONAL AND ROTATIONAL POPULATIONS OF EMITTING STATE	REFERENCE
$X^+(CO_2,X)CO_2^+$; H=He,Ne,Ar	500 to 5000 eV	CO_2^+(II), X(I)	$A^2\Pi_u \to X^2\Pi_g$ from high Rydberg states up to $n=9$	—	422
$X^+(CO_2,X)CO_2^+$; X=He,Ne,Ar,Kr	200 to 4000 eV	CO_2^+(II)	$\tilde{A}^2\Pi_u \to \tilde{X}^2\Pi_g$	Up to six quanta of bending vibrations are excited	423
		He(I)	$3^3P \to 2^3S$; $5^3D \to 2^3P$; $4^3D \to 2^3P$	—	
$Ar^+(CO_2,Ar)CO_2^+$	900 eV	CO_2^+(II)	$\tilde{A}^2\Pi_u \to \tilde{X}^2\Pi_g$; $\tilde{B}^2\Sigma_u^+ \to \tilde{X}^2\Pi_g$	—	422
$Ne^+(CO_2;Ne,O)CO^+$	2 to 200 eV	CO(II)	$A^2\Pi_u \to X^2\Sigma_g^+$ (Comet tail)	—	398
$He^+(CO_2;He,O)CO^+$	Thermal; 200 to 1500 eV	CO(II)	$A^2\Pi_u \to X^2\Sigma^+$ (Weak)	—	159,424
$X^+(CO_2,X,O)CO^+$; X=H,He,Ne	200 eV to 25 keV	CO(II)	$B^2\Sigma^+ \to X^2\Sigma^+$ (Weak)	—	420
$X^+(CO_2;X,O)CO^+$	500 to 5000 eV	CO(II)	Comet tail		
$X^+(CO_2;X,O)CO^+$	"	CO(I)	$a'^3\Sigma^+ \to a^3\Pi$; $d^2\Delta, a^2\Pi$; $e^3\Sigma^- \to a^3\Pi$		
$X^+(CO_2;X,CO)O^+$	"	O(I), O(II)	Strong emissions from high Rydberg states	—	422
$X^+(CO_2;X,2O)C^+$; X=He,Ne,Ar	"	C(I), C(II)	"		
$He^+(H_2O; He,H)OH^+$	<0.1 eV	OH(II)	$A^3\Pi_i \to X^3\Sigma^-$	$v' < 3$	172,389,425
$He^+(D_2O; He,D)OD^+$	"	OD(II)	"	"	
$He^+(H_2O; He,OH)H^+$	<0.1 eV	OH(I)	$A^2\Sigma^+ \to X^2\Pi$	$v' = 0$	424
	Thermal	OH(I)	"		
$X^+(N_2O,X)N_2O^+$; X=Ar,KR	2000 eV	N_2O(II)	$A^2\Sigma^+ \to X^2\Pi$	000 001	405,406

Reaction	Species	Energy	Transition / bands	Remarks	Ref.
$X^+(CS_2, X)CS_2^+$; X=Ar, Kr	CS_2 (II)	2000 eV	281.9 and 285.5 nm; $A^2\Sigma^+ \to X$	000	405, 406
$Ar^+(N_2O, Ar)N_2O^+$	N_2O (II)	900 eV	$A^2\Sigma^+ \to X^2\Pi$	—	290
$He^+(X; CH, ?)?$; X=$CH_4, C_2H_2, C_2H_4, C_2H_6$, acetone, acetonitrile	CH (I)	Thermal	$A^2\Delta \to X^2\Pi$; $B^2\Sigma \to X^2\Pi$	—	424
$Ar^+(CH_4; CH, H, ?)?$	CH (I); H (I)	900 eV	$A^2\Delta \to X^2\Pi$; $B^2\Sigma \to X^2\Pi$; Balmer	—	290
$He^+(C_2H_2; He, CH)CH^+$	CH (II), CD (I)	<0.1 eV	$A^2\Delta \to X^2\Pi$; $B^2\Sigma \to X^2\Pi$; $A^1\Pi \to X^1\Sigma^+$	$v' < 2$	389
$He^+(C_2D_2; H, CD)CD)^+$	CH (II), CD (II)	10 to 1000 eV	$A^2\Delta \to X^2\Pi$	—	290
$Ar^+(C_2H_2; CH, ?)?$	CH (I)	900 eV	$B^2\Sigma^- \to X^2\Pi$	—	290
$He^+(C_2H_2; C_2; ?)?$	C_2 (I)	900 eV	Balmer; Swan, high pressure, and Deslandres–D'Azambuja systems	—	424
$He^+(CH_3CN; CN, ?)?$	CN (I)	900 eV	—	—	424
$He^+(NH_3; NH, ?)?$	NH (I)	Thermal	$A^3\Pi \to X^3\Sigma$; $C^1\Pi \to a^1\Delta$	—	424
$He^+(X; ?)$; X=BF_3, SO_2, CS_2, COS	CS (I) and others	Thermal	—	—	424
3. Diatomic Ions and Monatomic, Diatomic, and Polyatomic Molecules					
$H_2^+(N_2, H_2)N_2^+$	N_2 (II)	0.4 to 3.0 keV	First negative system; rotational lines up to $k=50$	Rotational distribution grossly different from Boltzmann	426
$H_2^+(N_2, H_2)N_2^+$	N_2 (II)	100 eV to 13.5 keV	First negative system; vibrational excitation	Vibrational distribution non-Franck–Condon at low energies	155
$H_2^+(N_2, H_2)N_2^+$	N_2 (II)	30 keV	382–392 nm; $^4\Sigma_u^+ \to X^2\Sigma_g^+$	—	205
$H_2^+(CO_2, H_2)CO_2^+$	CO_2 (II)	100 to 5000 eV	$A^2\Pi_g \to X^2\Pi_g$	Stretching vibration excited; no excitation of bending vibrations	421–423
$N_2^+(N_2, N_2)N_2^+$	N_2 (II)	0.4 to 10 keV	Negative bands	—	289, 290 368, 414, 427

Table IV.B. *continued*

REACTION	INCIDENT ION TRANSLATIONAL ENERGY	EMITTING SPECIES	WAVELENGTHS AND/OR TRANSITIONS OBSERVED	VIBRATIONAL AND ROTATIONAL POPULATIONS OF EMITTING STATE	REFERENCE
$N_2^+(CO,N_2)CO^+$	10 to 250 eV	CO (II)	$A^2\Pi_u \to X^2\Sigma_g^+$ (Comet-tail bands)	Boltzmann distribution of vibrational states ($T \sim 2500°K$)	290,414
$N_2^+(O_2;N_2,O)O^+$	900 eV	O (I)	$3^5S^0 \to 4^5P$; $3^5S^0 \to 4^3P$	—	290
$N_2^+(CO_2;N_2)CO_2^+$	10 to 1000 eV; 500 to 5000 eV	CO₂ (II) "	$\tilde{A}^2\Pi_u \to \tilde{X}^2\Pi_g$; $\tilde{B}^2\Sigma_u^+ \to \tilde{X}^2\Pi_g$; $\tilde{A}^2\Sigma^+ \to \tilde{X}^2\Pi_i$	—	290,422
$N_2^+(N_2O,N_2)N_2O^+$	900 eV	N₂O (II)	$A^2\Delta \to X^2\Pi; B^2\Sigma^- \to X^2\Pi$ Balmer	—	290
$N_2^+(CH_4;N_2,CH,H,?)?$	900 eV	CH (I); H (I); N₂ (I)	$C^3\Pi_u \to B^3\Pi_g(?)$	—	290
$N_2^+(C_2H_2;N_2,CH,C_2,H,?)?$	10 to 1000 eV	CH (I); H (I); C₂ (I); N₂ (I)	$A^2\Delta \to X^2\Pi; B^2\Sigma^- \to X^2\Pi$ Balmer $A^3\Pi_g \to X^3\Pi_u; C^1\Pi_g \to b^1\Pi_u(?)$ $C^3\Pi_u \to B^3\Pi_g(?)$	—	290
$He_2^+(Zn;2He)Zn^+$[a]	Thermal	Zn (II)	589.4 nm, 747.8 nm; $3d^94S^2 \to 3d^{10}4p$	—	166,351
$He_2^+(Cd;2He)Cd^+$[a]	Thermal	Cd (II)	325.0 nm, 441.6 nm, 231.3 nm; $4d^95S^2 \to 4d^{10}5p$; $4d^{10}5d \to 4d^{10}5p$	—	166,169
$He_2^+(N_2;2He)N_2^+$[b]	Thermal	N₂ (II)	$B^2\Sigma_u^+ \to X^2\Sigma_g^+$ (First negative)	Franck-Condon	333,428–430
$He_2^+(N_2;2He)N_2^+$[c]	Thermal	N₂ (II)	First negative (0,0); (0,1); (0,2) and (0,3) vibrational components at 391.4; 427.8; 470.9 and 522.8 nm	—	3b,c,431,432
$He_2^+(N_2;2He)N_2^+$[d]	Thermal; 5 eV and 11 eV	N₂ (II)	First negative and (possibly) $A^2\Pi_u \to X^2\Sigma_g^+$ (Meinel)	Franck-Condon	168,391,433, 434

Reaction	Energy	Emitter	Transition (band system)	Distribution	Ref.
He₂⁺(CO,2He)CO⁺ᶜ	Thermal	CO (II)	(First negative) $B^2\Sigma^+ \to A^2\Pi$ (Baldet–Johnson)	Franck–Condon	430
			First negative (0,2) component at 247 nm, as well as (0,0) and (0,1) components of $B^2\Sigma^+ \to A^2\Pi$ (Baldet–Johnson) at 395.4 nm and 421 nm	—	435
He₂⁺(CO,2He)CO⁺ᵈ	Thermal; 11 eV	CO (II)	First negative ($B\to x$) comet tail ($A\to x$) Baldet–Johnson ($B\to A$)	Franck–Condon	168,391
He₂⁺(O₂;2He)O₂⁺ᵇ	Thermal	O₂ (II)	$b^4\Sigma^-_g \to a^4\Pi_u$ (First negative)	Franck–Condon	333,424,430
He₂⁺(O₂;2He)I₂⁺ᵈ	11 eV	O₂ (II)	First negative; $A^2\Pi_u \to X^2\Pi_g$ (Second negative)	Franck–Condon ?	391
He₂⁺(NO;2He)NO⁺ᵈ	11 eV	NO (II)	$A^1\Pi \to X^1\Sigma^+$ (Miescher–Baer) (tentative)	—	391
He₂⁺(CO₂;2He)CO₂⁺ᵈ	11 eV	CO₂ (II)	$\tilde{A}^2\Pi_u \to \tilde{X}^2\Pi_g$; $\tilde{B}^2\Sigma^+_u \to \tilde{X}^2\Pi_g$	Franck–Condon	436
He₂⁺(H₂); OH,H,?)?	Thermal	OH (I); H (I); N₂ (II)	$A^2\Sigma^+ \to X^2\Pi$; H-Lines	—	424
N₂⁺(N₂;2Ne)N₂⁺	5 eV	N₂ (II)	$B\to x$ (First Negative system)	Franck–Condon	437
N₂⁺(CO;2Ne)CO⁺	5 eV	CO(II)	$B\to x$ (First negative); $A\to x$ (comet tail)	Franck–Condon	437
N₂⁺(CO₂;2Ne)CO₂⁺	5 eV	CO₂ (II)	$B\to A$ (Baldet–Johnson)	000 Level of $\tilde{B}$ state; $\tilde{A}$ state population similar to He₂⁺ reaction	
CO²⁺(H₂;CO⁺)H₂⁺	7.5 eV (0.5 eV CMᵉ)	CO (II)	$B\to x$ (First negative); $A\to x$ (comet tail); $B\to A$ (Baldet–Johnson)	$v'=0$ Only of B state; $v'=0,1,2,3$ of A state	438
OD⁺(H₂,OD)H₂⁺	20 eV CM	OD (I)	$A^2\Sigma^+ \to X^2\Pi$	—	12b

ᵃ Reactions are probably fast; however, their contribution to the total chemiluminescence in He–Zn and He–Cd systems has not been demonstrated unequivocally.

ᵇ Early studies suggesting population inversion.

ᶜ Stimulated emission and/or laser development.

ᵈ Spectroscopic studies and cross-section determinations.

ᵉ Translational energy in center-of-mass system.

Table IV.C.
Luminescence in Charge-Transfer and Ion-Molecule Reactions: Ion–Molecule Reactions (Particle Transfer)

REACTION	INCIDENT-ION TRANSLATIONAL ENERGY	EMITTING SPECIES	WAVELENGTHS AND/OR TRANSITIONS OBSERVED	VIBRATIONAL AND ROTATIONAL POPULATIONS OF EMITTING STATE	REFERENCE
$N^+(NO,O)N_2^+$[a]	$3.4\ eV \leqslant E_{CM}^+ \leqslant 15\ eV$	N_2 (II)	$B^2\Sigma_u^+ \rightarrow X^2\Sigma_g^+$ (First negative system)	—	9c, 441
$C^+(NO,C)NO^+$ followed by $C(NO,O)CN$	$1.9\ eV \leqslant E_{CM} \leqslant 15\ eV$	CN (I)	$B^2\Sigma^+ \rightarrow X^2\Sigma^+$ (Violet system)	—	9c, 441
$C^+(H_2,H)CH^+$	$2\ eV \leqslant E_{CM} \leqslant 12\ eV;$ $7\ eV \leqslant E_{CM}$	CH (II)	$A^1\Pi \rightarrow X^1\Sigma^+;$ $B^1\Delta \rightarrow A^1\Pi;$ $b^3\Sigma \rightarrow a^3\Pi$	—	442, 443
$C^+(H_2,CH)H^+$	$6.3\ eV \leqslant E_{CM}$	CH (I)	$A^2\Delta \rightarrow X\Pi$	—	442, 443
$C^+(O_2,O)CO^+$	$3.6\ eV$ (CM) (Also $1.0\ eV$ (CM) and lower)	CO (II)	$A^2\Pi \rightarrow X^2\Sigma^+$ (Comet tail)	High degree ($fr = 0.31$) of rotational excitation	193
$O^+(H_2,H)OH^+$	1.5 to 35 eV (CM)	OH (II)	$A^3\Pi \rightarrow X^3\Sigma^-$	—	12b
$O^+(H_2,OH)H^+$	1.5 to 35 eV (CM)	OH (I)	$A^2\Sigma^+ \rightarrow X^2\Pi$	—	12b
$N^+(RH,NH)R^+;$ $R=H, CH_3, C_2H_3, C_2H_5, C_3H_7$	1 to 150 eV (CM)	NH (I)	$A^3\Pi \rightarrow X^3\Sigma^-$	Increasing population of $v'=1$ and 2 relative to $v'=0$ with increasing collision energy (high degree of rotational excitation)	444

Laudenslager[334] have stressed the importance of energy resonance and large Franck–Condon factors for fast charge-transfer reactions. Two distinct mechanisms have been suggested for charge transfer: (1) a long-range electron jump mechanism, in which transfer of the electron occurs between reactants that are essentially stationary isolated moieties on separated energy surfaces and (2) adiabatic transfer, which occurs on a single surface, possibly via a long-lived complex with (some) randomization of the energy among all the coordinates of the system. The second mechanism is only applicable at low relative translational energies, whereas the long-range mechanism is expected to be dominant at high energies. In the transition region both mechanisms are probably operative. Obviously, the factors that influence charge transfer in general also affect those reactions that produce luminescence.

a. Atomic Systems

Reported data for atomic charge-transfer reactions that produce luminescence are summarized in Table IV.A. Many of the initial studies were concerned with reactions of He^+ with rare-gas atoms. These were also the first processes yielding luminescence to be studied over a wide translational-energy range. Very large cross sections for radiative emissions were observed from reactions such as

$$He^+ + Ar \rightarrow He + Ar^+* \qquad (IV.5)$$

even at relatively low translational energies. These processes, therefore, were among the first to demonstrate[337] that Massey's qualitative adiabatic criterion for charge transfer[176] was not rigorously applicable. Figure 45 shows the excitation function reported for one of the Ar-II lines (probably misidentified)[338] in early experiments by Lipeles et al.[337] Many such excitation functions have since been determined, and in every case the cross section is very large at translational energies in the neighborhood of threshold. Other examples of excitation functions are shown for the He-I line from He^+–Ne charge exchange[345] and for Xe-II lines from He^+–Xe charge exchange[341] in Figs. 46 and 47, respectively.

In many of these atomic systems, emissions are observed from the products of the charge-transfer reaction channel as well as from those formed via a direct excitation channel (see also Section II.B.2). For example, the spectra shown in Fig. 48 indicate the occurrence of transitions attributable to both Ar-I, and Ar II, that result from direct excitation and charge-transfer excitation, respectively, in the He^+–Ar interaction.[344] More than 50% of the total charge-transfer cross section for this reaction is attributable to the production of excited Ar^+ states, rather than ground-state Ar^+.[347, 348] Similar results have been obtained for the other rare-gas

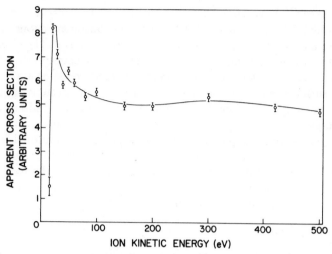

Figure 45. Kinetic-energy dependence of apparent cross section for production of 476.5-nm emission from Ar^{+*} produced by He^+ impact on argon.[337]

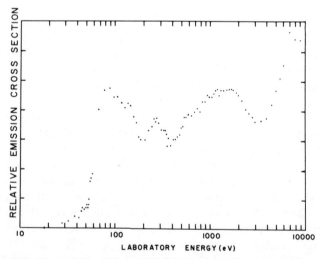

Figure 46. Emission cross section for He(I) resonance line, $1s2p$ $^1P \rightarrow 1S^2{}^1S$ (58.4 nm) observed from He^+–Ne collisions at translational energies ranging from threshold to 9 keV.[345]

178

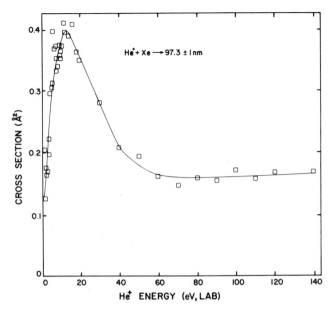

Figure 47. Emission cross section for Xe(II) line at 97.3 nm, observed from He$^+$–Xe collisions at translational energies ranging from threshold to 140 eV.[341]

systems. For example, the total cross section for excitation of radiating excited states is approximately 25% of the total He$^+$–Xe charge-transfer cross section at 100-eV translational energy.[341] This has been explained by considering the potential curves of the quasimolecule to consist of the two reactant atoms. The adiabatic elastic potential curve for the He$^+$ + X interaction (e.g., X = Ar) rises sharply as the internuclear separation decreases and initially crosses the potential curves that correlate to the (X$^+$)* + He products. In a somewhat higher energy region the same curve crosses the potential curve that correlates to the X$^+$ + He products.[347]

Quantum-mechanical oscillations in the excitation functions for reactions of this type have been mentioned earlier in the discussions relating to energy transfer. These oscillations are especially pronounced in the excitation functions for the He$^+$–Ne[163, 270, 340, 345] and Na$^+$–Ne reactions (see also Section II.B.2). The nonadiabatic interaction at large internuclear separation apparently involves quasiresonant charge transfer, as described by Lichten.[278, 349] This is deduced from the fact that oscillations in the cross sections for radiative emissions that arise from direct excitation are 180° out of phase with those arising from charge-transfer excitation (see Fig. 36).[279] Conservation of the total cross section for reaction and an approximate 1:1 branching ratio between the direct excitation and charge-transfer processes dictate this behavior.

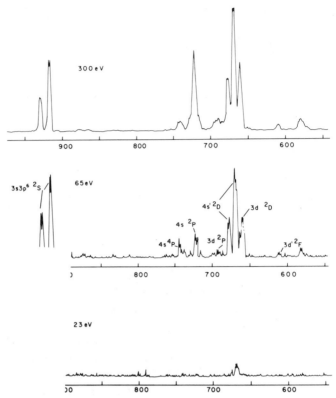

Figure 48. Far-ultraviolet spectral emissions produced by He^+–Ar collisions at translational energies of 300 eV, 65 eV, and 23 eV.[344]

Cross sections have been determined for the major emission lines appearing in the vacuum UV and visible spectral regions from the He^+–rare-gas atom reactions, and the importance of cascading as compared to direct excitation of the observed states had been assessed.[340, 341, 343–345] The cross sections for charge transfer from He^+ to rare-gas atoms are negligibly small at thermal translational energies.[350] Most of the observed emissions exhibit translational-energy thresholds of several electron volts or higher.[163, 341, 344, 345] These reactions are thus not likely candidates for populating excited states in a laser device. On the other hand, charge transfer from He^+ to the elements zinc, cadmium, mercury, iodine, selenium, and gold occurs readily at low energies, and these reactions are apparently the major pumping reactions in the respective lasers (see Table IV.A). These reactions have very large cross sections at thermal energies, as shown in Table V. Although the He^+ ion has a high recombination energy (24.586 eV) and

Table V.
Selected Cross Section and/or Rate Constants for Charge-transfer Reactions of He^+ and He_2^+ at Thermal Energies

IONIC REACTANT	NEUTRAL REACTANT(S)	$\langle\sigma\rangle$ OR k	LIGHT EMISSION	LASER ACTION OR STIMULATED EMISSION	REFERENCE
He^+	I	3×10^{-16} cm^2	Yes	Yes	361
He^+	Hg	1.3×10^{-14} cm^2; 1.6×10^{-9} cm^3/sec	Yes	Yes	360,370,371
He^+	Zn	2.1×10^{-15} cm^2	Yes	Yes	166
He^+	Cd	3.7×10^{-15} cm^2	Yes	Yes	355
He_2^+	Cd	2.7×10^{-14} cm^2	Not yet directly measured	Yes	166,169
He_2^+	N_2; He,N_2	1.1×10^{-9} cm^3/sec; 1.6×10^{-29} cm^6/sec	Yes	Yes	170,440;170
He_2^+	CO; He,CO	1.1×10^{-9} cm^3/sec; 3.6×10^{-29} cm^6/sec	Yes	Weak	170,440;170
He_2^+	CO_2; He,CO_2	1.6×10^{-9} cm^3/sec; 6.7×10^{-29} cm^6/sec	Yes	?	170,440;170
He_2^+	CH_4; He,CH_4	0.5×10^{-9} cm^3/sec; 0.5×10^{-29} cm^6/sec	Yes	?	170;170
He_2^+	Ar; He,Ar	0.2×10^{-9} cm^3/sec; 2.4×10^{-29} cm^6/sec	No	No	170;170

the reactant atoms have relatively low ionization potentials, the asymmetric charge-transfer reactions occur with small ΔE values because the product ions are formed in highly excited electronic states. In addition, a relatively small number of excited states is populated in each of these processes, so that the exoergicity of the reaction is efficiently channeled into a few emission lines in the visible or UV regions of the spectrum. The earliest experiments demonstrating ion laser action in these systems involved He–Hg[372] and He–I[3a] discharges. The He–I laser was the first ion laser for which thermal-energy charge transfer was postulated to be the dominant excitation mechanism.[3a] Subsequently, it was suggested[370] that the He^+–Hg charge-transfer process provides the inversion mechanism for lasing at 614.9 nm in He–Hg discharges. Direct experimental confirmation of this postulate has recently been obtained. It has been shown[360] that thermal-energy charge transfer is the dominant mechanism for production of the excited $7p\,^2P$ state of Hg-II, which supports cw laser oscillation at 615 nm and 794.4 nm (see Fig. 49). The specific reactions are

$$He^+\left(^2S_{1/2}\right) + Hg\left(^1S_0\right) \rightarrow He\left(^1S_0\right) + Hg^+\left(7p\,^2P_{3/2}\right) \qquad \Delta E = 0.27 \text{ eV}$$

(IV.6)

$$He^+\left(^2S_{1/2}\right) + Hg\left(^1S_0\right) \rightarrow He\left(^1S_0\right) + Hg^+\left(7p\,^2P_{1/2}\right) \qquad \Delta E = 0.72 \text{ eV}$$

(IV.7)

and the transitions producing the 614.9 nm and 794.4 nm lines are $Hg^+(7p\,^2P_{3/2} \rightarrow 7s\,^2S_{1/2})$ and $Hg^+(7p\,^2P_{1/2} \rightarrow 7s\,^2S_{1/2})$, respectively.

Spectroscopic studies of these charge-transfer reactions at near-thermal energies have also been accomplished using a drift-tube mass spectrometer[173] and a pulsed helium–metal vapor-discharge afterglow.[166] Partial charge-transfer cross sections for formation of discrete excited states of various product metal ions have been determined in these experiments. The branching ratios[173] do not indicate a preference for formation of those reaction channels having the lowest energy defects, ΔE. The charge-transfer mechanisms in these cases are proposed to involve curve crossings of molecular states formed during the collision, followed by fine-structure transitions into the various atomic J levels[373] (e.g., see Fig. 50).

The participation of both charge-transfer and Penning ionization excitation processes has been demonstrated in the He–Zn and He–Cd ion lasers, with charge transfer leading primarily to high-lying states of Zn-II and Cd-II, whereas Penning ionization collisions are the main source of lower-lying Zn-II and Cd-II levels.[354, 355]

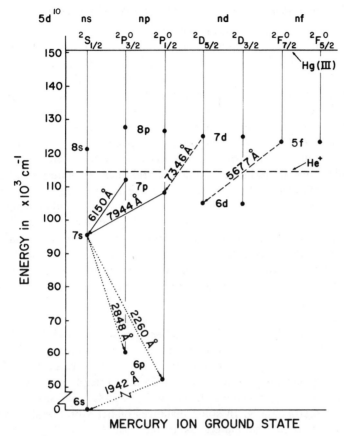

MERCURY ION GROUND STATE

Figure 49. Partial term diagram of Hg(II), indicating the two clockwise Hg(II) laser lines (solid lines), several pulsed Hg(II) laser lines (dashed lines), and other Hg(II) lines of interest. All wavelengths are in an angstroms. Energy available from He$^+$ is also shown.[360]

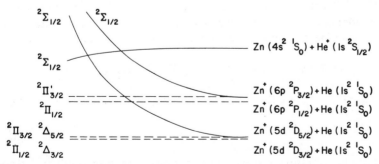

Figure 50. Schematic diagram of potential-energy curves for (He–Zn)$^+$ molecular system at large internuclear distance.[373]

b. Molecular Systems: Reactions of Atomic Ions with Various Molecules

Charge-transfer and dissociative-charge-transfer reactions of atomic ions with diatomic molecules that have been observed to produce luminescence are summarized in Table IV.B, part 1. Many of these reactions are nonspecific in terms of the energy states populated; that is, the reactions yield many states of the product ion(s). For example, at least four band systems have been observed in the radiative emissions resulting from the He^+–N_2 reaction, indicating formation of excited N_2^+ in the $A^2\Pi_u$, $B^2\Sigma_u^+$, $C^2\Sigma_u^+$, and $D^2\Pi g$ states. The emission spectra also demonstrate that various vibrational and rotational levels are populated in each of these electronic states. In addition, atomic spectral lines originating from excited N and N^+ are observed from the He^+–N_2 reaction. Hence this reaction is unsuitable as a pumping mechanism in a laser device. Other reactions of this type are even less suitable for producing population inversion in a low-energy plasma environment because radiative excited states are formed only at higher interaction energies. For example, the He^+–CO interaction leads primarily to dissociative charge transfer at thermal energies,[23c, 168] and CO^+ emissions are observed only from reactions of higher-energy He^+ ions.

Vibrational excitation resulting from reactions of various atomic ions with N_2, CO, and HX (X = Cl,Br) has been studied in some detail.[9c, 123, 155, 256, 390, 402, 403, 405, 411, 413–415] At high projectile-ion velocities ($\geqslant 10^8$ cm/sec) the vibrational-energy distribution in the product species can be rationalized by application of the Franck–Condon principle. At lower interaction energies strong deviations from the Franck–Condon distribution are observed experimentally. In some instances higher vibrational states tend to be preferentially populated, as is the case for the $N_2^+(B^2\Sigma_u^+)$ and $N_2^+(C^2\Sigma_u^+)$ electronic states that are formed in charge-transfer reactions with certain atomic ions. In the latter case this effect tends to become increasingly pronounced as the projectile velocity decreases (see Figs. 51–53). On the other hand, slow-ion excitation of the $CO^+(A^2\Pi)$ and $HCl^+(A^2\Sigma^+)$ states results in enhancement of the lower vibrational levels in abundances exceeding those expected for Franck–Condon transitions.

For reactions that have a small energy defect, such as[161, 386, 411]

$$He^+(^2S) + N_2(X^1\Sigma_g^+) \rightarrow He(^1S) + N_2^+(C^2\Sigma_u^+, v'=3,4) \qquad (IV.8)$$

$$Ar^+(^2P_{3/2,1/2}) + HCl(X^1\Sigma, v=0) \rightarrow Ar(^1S_0) + HCl^+(A^2\Sigma, v'=0)$$

$$(IV.9)$$

the product vibrational energy distributions have been explained using a model applicable to near-resonant charge-transfer processes, such as the

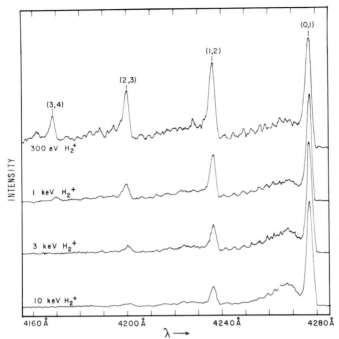

Figure 51.   Typical spectra of N_2^+ first negative system, $\Delta v = -1$ sequence, excited by collisions of 300-eV, 1.0-keV, 3.0-keV, and 10.0-keV H_2^+ ions with N_2.[155]

modified Rapp–Francis theory.[177, 416] A somewhat different approach, the so-called polarization model, was suggested by Lipeles[403] for the reactions

$$Ar^+ + CO \rightarrow Ar + CO^+(A\,^2\Pi) \qquad\qquad (IV.10)$$

$$X^+ + N_2 \rightarrow X + N_2^+(B\,^2\Sigma_u^+) \qquad\qquad (IV.11)$$

where X^+ represents a slow atomic-ion reactant. According to this model, the ground vibrational state of the neutral molecule is distorted as the projectile ion approaches. Vertical transitions then occur between the distorted neutral molecular state (which has an increased internuclear separation) and an unperturbed ionic state of the molecule. "Distorted" Franck–Condon factors are applied to characterize these transitions. Using this model, theoretically calculated vibrational distributions for $CO^+(A\,^2\Pi)$ and $N_2^+(B\,^2\Sigma_u^+)$ were obtained that are in good agreement with the experimentally measured distributions.[9c, 403] Less satisfactory agreement was achieved for other reactions.[411] Still another theoretical approach involves calculation of vibrational population distributions using the statistical phase-space model for ion–molecule reactions[397] (see also Section

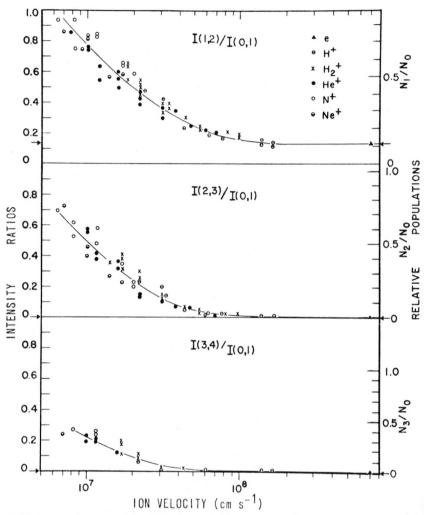

Figure 52. Ratios of intensities of vibrational transitions in N_2^+ first negative system as function of projectile-ion velocity (laboratory) for collisions of various ions with N_2. Intensity ratios and relative populations of upper vibrational states are indicated by left- and right-hand vertical axes, respectively. Arrows on vertical axes indicate relative populations and corresponding intensity ratios predicted by Franck–Condon principle.[155]

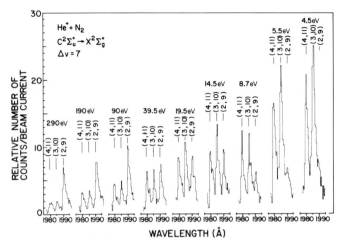

Figure 53.   Relative cross sections for production of $(2,9)$, $(3,10)$, and $(4,11)$ bands of the $N_2^+(2N)$ system from He^+–N_2 collisions at selected He^+ kinetic energies (laboratory). Bandhead wavelengths are indicated by vertical lines.[390]

V.B). Experimental distributions obtained for the Ne^+–N_2 system (Fig. 54) were shown to be intermediate between those predicted by application of the Franck–Condon principle and those predicted by such a statistical model. Finally, it has been suggested by Tomcho and Haugh[411] that the molecular orbitals of the transient triatomic system should be considered in predicting vibrational distributions of products from atomic-ion–diatomic-molecule reactions. Calculations based on this approach are in progress for the $(Ar–N_2)^+$ system.[417]

Rotational excitation accompanying the formation of electronically excited states in charge-transfer reactions has been investigated most intensively for reactions producing N_2^+. Both the first and second negative spectra for this ionic species have been studied.[395, 399, 400] As in the case of vibrational excitation, no conversion of translational energy into rotational energy occurs when the projectile-ion velocities are greater than about 10^8 cm/sec. At such high-impact velocities, the product rotational-energy distribution corresponds to that appropriate to a Boltzmann distribution at the ambient-gas temperature. At lower incident-ion velocities, excitation of high rotational states is found to increase monotonically as the projectile velocity decreases. The distribution of rotational states has been observed

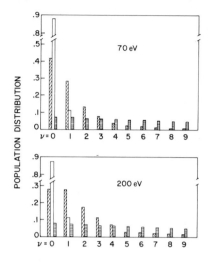

Figure 54. Relative vibrational population distributions of $N_2^+ B^2\Sigma(v)$ ions produced in 70-eV and 200-eV $Ne^+ - N_2$ collisions. Experimentally determined distributions are shown as diagonally striped bars. Distribution predicted from application of Franck–Condon principle is represented by open bars, whereas that computed from statistical phase-space model is given by horizontally striped bars. Franck–Condon distribution is essentially zero for $v > 2$.[397]

in some cases to be markedly different from a Boltzmann distribution,[395] so the designation of a "rotational temperature" for products of these low-velocity ion–neutral interactions is not meaningful. For the atomic-ion excitation processes discussed in this section, the extent of excitation of higher rotational levels is apparently solely a function of the incident ion velocity (see also Section IV.A1.c).

Several attempts have been made to develop a theoretical model to account for rotational excitation in the products of ion–neutral interactions. Liu has suggested that rotational excitation occurs as a consequence of conservation of angular momentum in inelastic collisions.[418] This model is applicable only for endoergic charge-transfer processes at incident-ion kinetic energies in the range of tens of electron volts or higher. A statistical phase-space model has also been applied to rationalize rotational excitations.[419] Although phase-space calculations were found to be inadequate in predicting vibrational distributions for products of ion–neutral reactions at kinetic energies in the neighborhood of 100 eV,[397] these calculations do yield a good representation of the vibrational–rotational distributions in the products of thermal-energy reactions of rare-gas ions with nitrogen. Agreement between phase-space theory and experiment was shown to be particularly good in the case of thermoneutral or slightly exoergic reactions (see Fig. 55). However, for moderately exoergic reactions, the computed vibrational–rotational distributions were observed to be significantly broader than those determined experimentally.

Internally excited products from interactions of atomic ions with tri-

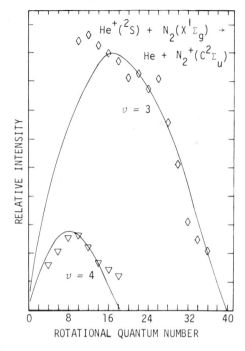

Figure 55. Comparison of relative intensity distributions of P-branch rotational lines of $(3,9)$ and $(4,10)$ bands of $N_2^+(C^2\Sigma_u, v, m) \rightarrow N_2^+(X^2\Sigma_g, v'', m'') + h\nu$ spontaneous radiative transitions. Points are experimental data and solid lines are distributions computed using statistical phase-space model to determine $N_2^+(C^2\Sigma_u, v, m)$ vibrational–rotational distribution resulting from reaction $He^+ + N_2(X^1\Sigma_g, v, m) \rightarrow N_2^+(C^2\Sigma_u, v, m) + He.$[419]

atomic and polyatomic molecules have been studied in relatively few instances. Experiments in which luminescence from such reactions has been observed are summarized in Table IV.B, part 2. Of particular interest here is the excitation of bending vibrations in the $A^2\Pi_u$ electronic state of CO_2^+ produced in charge-exchange collisions between He^+ and CO_2.[423] A "quasimolecule model" has been proposed for this reaction in which an ion complex $[He-CO_2]$ is formed that has a lifetime of approximately 10^{-15} sec. This $[He-CO_2]^+$ intermediate, which is isoelectronic with HCO_2, decomposes to a bent CO_2^+ ion, which then relaxes to the linear state having the bending vibrations excited.

c. Molecular Systems: Reactions of Diatomic Ions with Various Molecules

Charge-transfer and dissociative-charge-transfer reactions of diatomic ions with various molecules that yield luminescence spectra are summarized in Table IV.B, part 3. In some of these, for example, the $H_2^+-N_2$ reaction, vibrational and rotational excitation have again been observed to accompany electronic excitation.[155, 426] Molecular-ion reactions are generally accompanied by more extensive rotational excitation of the products than occurs with atomic-ion reactions.[439]

A particularly interesting group of reactions that fall in this category are those of the helium dimer ion, $He_2^+(X\,^2\Sigma_u^+)$. It was suggested some time ago that reaction of this species with nitrogen,

$$He_2^+ + N_2 \rightarrow 2He + N_2^+\left(B\,^2\Sigma_u^+\right) \qquad (IV.12)$$

might lead to an inverted population of N_2^+ excited states. This suggestion was prompted by still earlier studies by Collins and Robertson.[424] This reaction was shown to yield stimulated emission for several vibrational components of the first negative system,[431] and subsequently, the first nitrogen ion laser pumped by this charge-transfer reaction was reported by Collins et al.[3b] In this system a laser line is obtained in the violet region of the spectrum at 427.8 nm, originating from the vibrational transition:

$$N_2^+\left(B\,^2\Sigma_u^+, v'=0\right) \rightarrow N_2^+\left(X\,^2\Sigma_g^+, v=1\right) \qquad (IV.13)$$

Scaling of this laser to higher powers has now been achieved, and additional laser lines have been detected at 391.4 nm and 470.9 nm, arising from the (0,0) and (0,2) vibrational components.[432] The experiments of Collins and co-workers utilized a fast-pulsed electron beam gun that injects the beam into a mixture of helium and nitrogen at a pressure of several atmospheres. A similar nitrogen ion laser in which ionization is produced by an electric discharge has also been reported.[3c]

Several features of the $He_2^+(N_2, 2He)N_2^+$ reaction and the products thereof facilitate the use of this reaction as a laser-pumping process:

1. The reaction exhibits a very large cross section at thermal energies,[440] approaching the theoretical gas kinetic collisions limit (see Table V) and has an additional contribution from a termolecular component[170] at high helium pressures.
2. The reaction is highly specific in terms of energy states populated. Partial cross sections for charge transfer from He_2^+ into specific vibrational channels of the N_2^+ product have been determined.[391, 433] A comparison of the emission spectra from $He^+(N_2, He)N_2^+$ and $He_2^+(N_2, 2He)N_2^+$ at 11-eV projectile-ion translational energy is shown in Fig. 56. It can be seen that the He_2^+ reaction populates almost exclusively the $v'=0$ level of the $B\,^2\Sigma_u^+$ state of N_2^+. Similar results were obtained at incident-ion energies of 5 eV[433] and at thermal energies.[168,434]
3. The rate coefficient for quenching $N_2^+(B\,^2\Sigma_u^+, v'=0)$ in collisions with helium [reaction (II.23)] is relatively small.

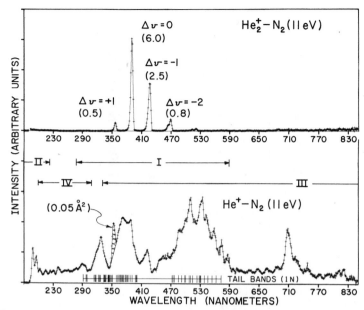

Figure 56. Emission spectra resulting from He_2^+–N_2 and He^+–N_2 collisions at 11 eV taken with 2-nm spectral resoultion. Known band systems of N_2^+ are indicated in lower spectrum (I, first negative; II, second negative; III, Meinel; IV, Janin-D'Incan). Emission cross sections in $\mathring{A}^2$ for some prominent sequences are indicated in parentheses.[391]

The large cross section observed for the $He_2^+(N_2, 2He)N_2^+$ reaction has been attributed to two factors:[391, 433] (1) the final state is a three-body state, which has available more phase space than does the two-body final state of most such charge-transfer processes; and (2) the recombination energy of He_2^+ and the ionization potential of N_2 leading to the B state of N_2^+ are energy resonant. The specificity of the interaction, in terms of the final electronic state produced, reflects this energy resonance. The vibrational specificity may be attributed to the Franck–Condon factors for this transition, which indicate that approximately 90% of the transitions should yield the $v' = 0$ state in vertical ionization of $N_2(X^1\Sigma_g^+, v = 0)$ to $N_2^+(B)$. The behavior of this reaction with respect to vibrational population is in direct contradiction to that observed for most slow-ion excitations of the B state in N_2^+, as discussed earlier (see also Fig. 57). The fact that transitions occurring in the course of thermal He_2^+ ion interactions with N_2 are in accordance with the Franck–Condon principle has been known since the earliest studies of this reaction.[430] This may be contrasted with the markedly different behavior of He^+, which excites high vibrational levels

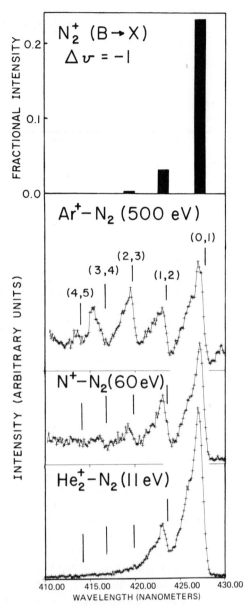

Figure 57. Spectra of $\Delta v = -1$ sequence of $N_2^+(1N)$ resulting from impact of Ar^+, N^+, and He_2^+ on N_2, taken at 0.4-nm spectral resolution. Also shown are fractional intensities calculated assuming occurrence of vertical transitions. Fractional intensity scale is chosen so that $(0,1)$ bar corresponds to $(0,1)$ band for $He_2^+ - N_2$.[391]

of the B state.[168, 391, 433] A long-range electron jump mechanism has been suggested for the helium-dimer ion reaction. This is also consistent with the large cross section for production of the B state. The effective impact parameter for the He_2^+ reaction is approximately 10 times as large as that for He^+.[433] At such interaction distances between the ion and the reactant molecule the ion-induced dipole interaction is too weak to significantly perturb the N_2 internuclear distance (according to the Lipeles polarization model).[403] Therefore, undistorted Franck–Condon factors are applicable to this reaction.[433]

Several other reactions of He_2^+ and of Ne_2^+ that yield luminescence are summarized in Tables IV and V. Stimulated emission has been observed at 247 nm, 395.4 nm, and 421 nm from the reaction $He_2^+(CO, 2He)CO^+$.[435] Lasing has not yet been reported from any of these processes, although the $He_2^+(Cd, 2He)Cd^+$ reaction is probably of importance in producing some of the Cd^{+*} excited states responsible for lasing in a He–Cd mixture.[169]

2. Heavy-particle Transfer

Relatively few ion–molecule reactions involving heavy-particle transfer have been observed to give luminescence. Those processes that have been reported are summarized in Table IV.C. The first chemiluminescent ion–molecule reaction of this type observed was reported by Brandt and co-workers.[9c, 441]

$$N^+ + NO \rightarrow N_2^+\left(B^2\Sigma_u^+\right) + O \tag{IV.14}$$

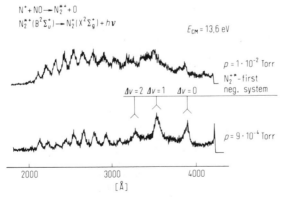

Figure 58. Spectral emissions from N^+–NO collisions. The N_2^+ first negative system, resulting from the indicated ion–molecule reaction, is observed at interaction energy of 13.6 eV, with 5-nm spectral resolution and indicated NO pressure.[441]

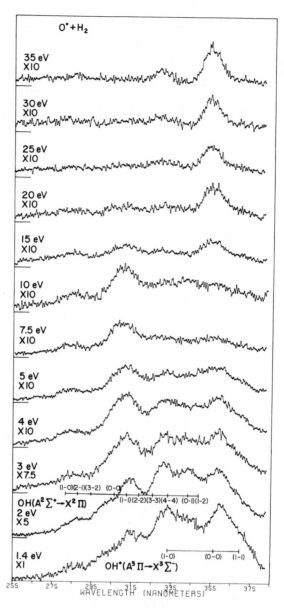

Figure 59. Emission spectra resulting from $O^+ - H_2$ collisions at various relative energies. Band heads for $A \to X$ transitions of OH and OH^+ are indicated.[12b]

In this case luminescence is atrributable to the first negative system of N_2^+ (Fig. 58).
The reaction

$$C^+(^2P^0) + H_2 \rightarrow CH^+(A\,^1\Pi) + H(^2S) \qquad \text{(IV.15)}$$

studied by both Ottinger et al.[442] and by Leventhal[443] provided the first direct evidence for an ion–molecule reaction yielding excited as well as ground-state products. (Similar behavior had previously been noted for the neutral–neutral reaction $Ba-Cl_2$.[445]) Reaction (IV.15) has been rationalized in terms of the electronic-state correlation diagram for the system. Correlation diagrams have also been employed to describe the O^+-H_2 system[12b] (see also Section V) (Fig. 59).
Some of these reactions, notably the process

$$C^+(^2P^0) + O_2 \rightarrow CO^+(A\,^2\Pi_u) + O \qquad \text{(IV.16)}$$

demonstrate a high degree of rotational excitation of the products (Fig. 60).[193] Vibrationally and rotationally resolved spectra were also obtained

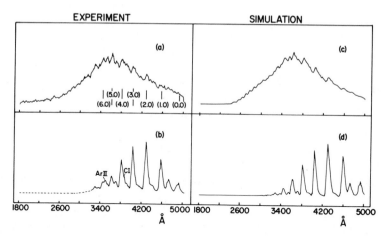

Figure 60. "Comet-tail" $CO^+(A\,^2\Pi \rightarrow X\,^2\Sigma^+)$ spectra from: (a, c) luminescent ion–molecule reaction $C^+ + O_2 \rightarrow CO^{+*} + O$ at $E_{lab} = 5$ eV; (b,d), charge-transfer reaction $Ar^+ + CO \rightarrow CO^{+*} + Ar$ at $E_{lab} = 1000$ eV. Experimental spectra (a, b) were obtained with 2-nm spectral resolution. Tabulated band heads for CO^+ ($A \rightarrow BX$) system are indicated. Spectral lines designated as Ar(II) and C(I) do not belong to CO^+ emission. Dashed portion of curves was not actually measured. Spectra simulated by computer calculations are given in diagrams (c and d). Rotational distributions assumed in simulation calculations were thermal with $T = 45,000°$K (c) and $1000°$K (d).[193]

for NH produced in the reaction

$$N^+ + RH \rightarrow R^+ + NH(A^3\Pi) \qquad (IV.17)$$

These reactions were also the first reported examples of nonadiabatic luminescent reactions exhibiting quite large cross sections ($\leq 10\%$ of the gas kinetic cross section).[444]

V. COLLISION MECHANISMS AND THEORETICAL IMPLICATIONS

A. General Effects of Internal Excitation

General effects of internal excitation on the kinetics, mechanisms, and collision-complex lifetimes of ion–neutral collisions have been discussed as an integral part of the presentation of accumulated data in previous sections. What one would hope to obtain from theoretical treatments is both a unifying generalized rationalization of such effects and some predictive capability. Many cases of selective energy consumption in endoergic ion–neutral reactions have been encountered, with the reaction $H_2^+(He, H)HeH^+$, which was discussed earlier, as a notable example. As noted in the foregoing discussion, several instances of specific energy disposal in exothermic reactions have also been observed, as exemplified by the reaction $He_2^+(N_2, 2He)N_2^+$. There are, however, other systems in which no such selectivity or specificity in energy utilization are observed. It seems evident that no existing theory would have predicted in advance the totally different behavior that is experimentally observed for the $He^+ - N_2$ and $He_2^+ - N_2$ reactions. Development of theory in this area has generally been of an interactive type, with the discovery of interesting experimental features prompting new theoretical calculations, with the objective to test the theories in light of the new experimental results.

The ion–neutral reaction that has received the greatest attention from a theoretical viewpoint is the $H_2^+ - He$ process. This is because of the relative simplicity of this reaction (a three-electron system), which facilitates accurate theoretical calculations and also to the fact that a wealth of accurate experimental data has been obtained for this interaction. Several different theoretical approaches have been applied to the $H_2^+ - He$ reaction, as indicated by the summary presented in Table VI. Most of these have treated the particle-transfer channel only, and few have considered the CID channel. Various theoretical methods applicable to ion–neutral interactions are discussed in the following sections. For the HeH_2^+ system, calculations using quasiclassical trajectory methods, employing an *ab initio* potential surface, have been shown to yield results that are in good agreement with the experimental results.

Table VI.

Theoretical Treatments of Reaction System $H_2^+(v') + He \rightarrow \begin{array}{l}\rightarrow HeH^+ + H \\ \rightarrow He + H^+ + H\end{array}$

METHOD	FEATURE	REFERENCE
Statistical phase-space theory	Calculations of total cross sections and vibrational enhancements	446
Statistical phase-space theory	Calculations of total cross sections and isotope effects	447
Statistical phase-space theory	Quantum-mechanical calculations for individual initial vibrational states; enhancement of cross section by vibrational excitation	448
Statistical phase-space theory	Inclusion of nuclear-spin considerations	449
Statistical theory of CID	Collision-induced-dissociation calculations for individual vibrational states; vibrational selectivity is achieved through a temperaturelike, energy-independent parameter, λ_v	222
Molecular-orbital correlation diagrams	Explanation for occurrence of endothermic reaction, $H_2^+(He, H)HeH^+$ and for failure to observe exothermic process $He^+(H_2, H)HeH^+$, indicating an extremely small cross section for this process	450
Electronic structure and chemical dynamics	In addition to molecular-orbital correlation diagrams, potential-energy curves for the diatoms in asymptotic reactant and product regions of the system are considered	451
Potential-energy surfaces	Calculation of adiabatic surface for linear $HeHH^+$ through superposition of configurations (SOC)	452
Potential-energy surfaces	Nonempirical LCAO–MO–SCF^a study for linear HeH_2^+; surface demonstrates a "late barrier," or "potential-energy step," characteristic for reactions showing vibrational enhancement (Fig. 61)	453

Table VI. *continued*

METHOD	FEATURE	REFERENCE
Potential-energy surfaces	Use of DIM method in fitting *ab initio* potential surfaces	454
Potential-energy surfaces	Collinear collision model combined with a square trough potential-energy surface	455
Quasiclassical trajectory calculations	Calculations of opacity functions and cross sections for particle transfer and CID in H_2^+ ($v'=0$)–He collisions above 2 eV; a classical model employing Lennard–Jones potentials was used	456
Quasiclassical trajectory calculations	Calculations of cross sections for particle transfer and CID for H_2^+ (v'), $v'=0-5$, employing a DIM representation of an *ab initio* potential function; vibrational energy was found to be far more effective than relative translational energy in promoting reaction	457
Quasiclassical trajectory calculations	Dramatic increase of σ with H_2^+ vibration at a given total energy; this effect decreases as E_{total} increases; both results are understandable from the shape of potential-energy surface	458
Quasiclassical trajectory calculations	Calculations on a spline-fitted *ab initio* surface	459
Quasiclassical trajectory calculations	Three-dimensional calculations on DIM surface of Kuntz	460
Collinear quantum-mechanical calculations	No vibrational enhancement observed	461
Collinear quantum-mechanical calculations	Qualitative agreement with experimental results	462
Collinear quantum-mechanical calculations	No vibrational enhancement	463
Collinear quantum-mechanical calculation		464
Distorted-wave Born approximation|calculations	Qualitative agreement with quasiclassical trajectory calculations and experimental data	465

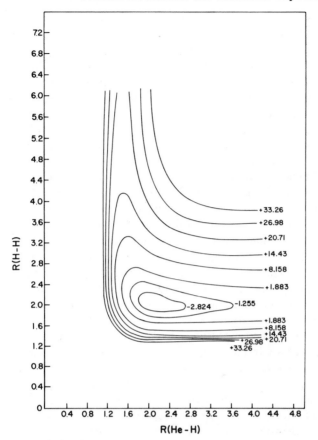

Figure 61. Contour plot of potential-energy surface for linear $(He-H-H)^+$ system. Indicated internuclear separations are in atomic units. Energies of equipotential lines are given in units of kcal/mol relative to $H_2^+ + He$ asymptotic limit.[453]

B. Theoretical Treatment of Energy Partitioning

Statistical theories treat the decomposition of the reaction complex of ion–molecule interactions in an analogous manner to that employed for unimolecular decomposition reactions.[466] One approach is that taken by the quasiequilibrium theory (QET).[467] Its basic assumptions are: (1) the rate of dissociation of the ion is slow relative to the rate of redistribution of energy among the internal degrees of freedom, both electronic and vibrational, of the ion; and (2) each dissociation process may be described as a motion along a reaction coordinate separable from all other internal coordinates through a critical "activated-complex" configuration. In applications of the QET to ion–molecule interactions, the assumption is generally made that a long-lived intermediate ion is formed in the interaction. There have been several recent extensions and reformulations of the QET. These were mainly developed as a result of the uncertainties concerning the properties of the activated complex. There has been a strong interaction between experiment and theory in QET development, with emphasis on the topics of ion lifetimes and energy disposal.[468]

One extension of the QET is the statistical phase-space theory.[477, 469, 470] It was originally developed for bimolecular interactions and later extended to unimolecular reactions.[471, 472] It assumes that the probability of formation of any given product in a "strong coupling" collision is proportional to the ratio of the phase space available to that product divided by the total phase space available with conservation of energy and angular momentum. This method has been found to have considerable value in predicting (from *ab initio* calculations) the total reactive cross sections for low-energy ion–molecule reactions involving three and four atoms. It is generally less successful in predicting the experimentally observed energy partitioning in such reactions. Such calculations do not predict vibrational population inversion for any reaction.

One system studied extensively by statistical theories, $C_4H_8^+$, is formed as an intermediate in the $C_2H_4^+ - C_2H_4$ interaction,[473] and also from charge transfer of various reactant ions with C_4H_8 isomers.[474] The QET was employed to calculate kinetic-energy distributions of the products of $C_4H_8^+$ decomposition.[475] Agreement with experimental results was achieved only by using an effective number of oscillators in the activated complex, which was smaller than the actual number of oscillators. In contrast to ordinary QET calculations, phase-space theory calculations[471g, 472a] enabled experimental distributions to be reproduced, indicating that energy equilibration within the $C_4H_8^+$ complex is complete. The QET does not provide for conservation of angular momentum and does not take into account the long-range potential of the system, two factors that can strongly affect product kinetic-energy distributions.

Techniques for rigorously calculating sums of states and densities of states of certain polyatomic systems under the constraints of conservation of both energy and angular momentum have recently been developed and employed by Chesnavich and Bowers[472a] in the statistical phase-space theoretical treatment of the $C_4H_8^+$ system. The experimental cross section for the reaction $C_2H_4^+ (C_2H_4, CH_3)C_3H_5^+$ decreases much more rapidly with the internal energy of the $C_2H_4^+$ ion [80] than is predicted by this theory, however. The phase-space calculations assume that the energy dependence of the cross section is solely a function of the properties of the products and reactants. Clearly, this assumption is not adequate for explanation of the data for this reaction. Additional calculations that will take into account the details of the potential surface in the region of the intermediate $C_4H_8^+$ ion are in progress. Apparently, some of the original ideas of QET must be retained when computing reaction cross sections for these bimolecular reactions, particularly the notion of an activated complex.

The effect of reactant internal vibrational energy on molecular: fragment ion ratios resulting from dissociative charge transfer of some large alkanes has also been accurately predicted by the QET.[476]

A third theoretical approach, which attempts to overcome the inadequacies of the previous two statistical theories to predict population inversion, is the so-called information-theory approach.[477] The latter has recently been applied to the problem of energy disposal and consumption in elementary chemical reactions. To the best of our knowledge, it has been applied to ion–neutral interactions only in the case of the collisional dissociation of H_2^+ (Table VI).

C. Calculations of Energy States, Correlation Diagrams, and Potential Surfaces

Statistical theories, such as those just described, are currently the only practical approach for many ion–neutral reactions because the fine details of the collision process are unknown; all the information concerning the dynamics of collision processes is, in principle, contained in the pertinent potential-energy surfaces. Although a number of theoretical groups are engaged in accurate *ab initio* calculations of potential surfaces (J. J. Kaufman, M. Krauss, R. N. Porter, H. F. Schaefer, I. Shavitt, A. C. Wahl, and others), this is an expensive and tedious task, and various approximate methods are also being applied. Some of these methods are listed in Table VI, for example, the diatomics-in-molecules method (DIM).

Valuable insight, particularly with regard to the effects of electronic excitation on reaction cross sections and reaction dynamics, has also been achieved without accurate knowledge of the actual potential surfaces, through the use of molecular-orbital correlation diagrams. Adiabatic correlation rules for neutral reactions involving polyatomic intermediates were developed by Shuler.[478] These were adapted and extended for ion–neutral interactions by Mahan and co-workers.[192, 451, 479, 480] Electronic-state correlation diagrams have been used to deduce the qualitative nature of the potential surfaces that control ion–neutral reaction dynamics. The dynamics of the reaction $N^+(H_2, H)NH^+$ and in particular the different behavior of the $N^+(^3P)$ and $N^+(^1D)$ states,[12a] for example, have been rationalized from such considerations (see Fig. 62). In this case the

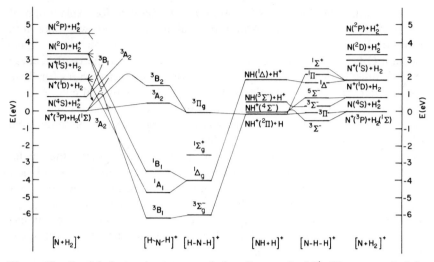

Figure 62. Partial electronic state correlation diagram for N^+–H_2 system. At left, N^+ is assumed to approach H_2 along perpendicular bisector of bond. At right, collinear approach is assumed. Crossings that are avoided in the more general conformations of C_S symmetry indicated by dotted lines.[480]

experimental findings indicate that the ^1D reaction involves a direct mechanism at all relative translational energies, whereas the ^3P reaction proceeds via a long-lived intermediate at relative energies of 1 eV and lower. This is true despite the fact that energetically, NH_2^+ lies a full 6 eV below the products. Apparently, this deep potential well path is not accessible to the ^1D state and is only available to the ^3P state at the lowest translational energies. A study of the correlation diagram suggests that this is indeed the case.[480, 192] The excited ^1D state leads to reaction via the $^1\Pi$ surface, which correlates with $N^+(^1D)$ and with the $^2\Pi$ ground state of NH^+. This surface exhibits no potential wells. The deep potential well of the $^3B_1$ state of NH_2^+ can be reached only when $N^+(^3P)$ approaches H_2 along a line that is somewhat off the perpendicular bisector of the hydrogen molecule. At low relative velocities the system proceeds adiabatically and passes from the $^3A_2$ surface at large N^+–H_2 separation, via an avoided crossing, to the $^3B_1$ surface. The avoided crossing has properties of a conical intersection such that, at somewhat higher initial relative velocities, the system moves adiabatically and transitions to the upper branch of the cone will occur with high probability. The system will then preferentially remain on the $^3A_2$ surface and not reach the deep potential well of the NH_2^+.

The important role of spin and symmetry restrictions in governing the course of ion–molecule reactions has been stressed by Kaufman and Koski and co-workers.[318, 321] The reaction $O^+(N_2,N)NO^+$ has played a central role in these arguments. *Ab initio* calculations have also been performed on the lower-state hypersurfaces of the N_2O^+ systems.[481, 482] There is general agreement that the reaction occurs at low energies via a multisurface mechanism, involving two spin-forbidden transitions $^4\Sigma^-\rightarrow^2\Pi\rightarrow^4\Sigma^-$. However, in disagreement with previous conclusions,[481] a single-surface treatment has recently been proposed[482–484] for the reaction at high energies, which involves the $1^4A''$ bent state of N_2O^+.

In some cases when detailed *ab initio* calculations of potential surfaces have become available, they have confirmed the major qualitative features of the surfaces deduced from experimental data and from preliminary data on correlation diagrams and the asymptotic properties of reactant, product, and intermediate states. One such case is the $C^+–H_2$ system,[451] which has aroused considerable interst. Angular and energy distributions were determined experimentally for the $CH^+ + H$ products from the reactions of the $C^+(^2P)$ ground state and the $C^+(^4P)$ excited state with H_2 (Table I, Jones et al.[9b] and references cited therein, and Jones et al.[326]). Chemiluminescence was also observed from this hydrogen-atom transfer reaction (Table IV).[442, 443] In addition, the reaction

$$C^+ + H_2 \rightarrow CH_2^+ + h\nu \qquad (V.1)$$

has been suggested as a first step toward "carbon fixation" and formation of carbon-containing interstellar molecules in dense H_2 clouds.[485] A theoretical estimation of the $C^+ + H_2$ radiative association (V.1) rate constant is quite important. First steps toward achieving that goal have been taken with the accomplishment of *ab initio* calculations of energy profiles along C_{2v} minimum energy paths,[486] and these have recently been improved.[487]

Calculations of potential surfaces are now in progress for a number of triatomic systems of atmospheric importance.[488, 489] These have direct pertinence to various atomic-ion–diatomic-molecule reactions.

The H_3^+ molecule ion is the simplest nonlinear molecule, of course, and its potential surface has been calculated in detail and with relatively high accuracy[490] (see Figs. 63–65). This *ab initio* surface, as well as that for $Li^+–H_2$,[491] have been employed in calculations of cross sections for vibrational excitation in nonreactive scattering collisions[492] (see also Section II.B.2.b). An *ab initio* calculation of the vibrational spectrum of H_3^+ has also been carried out,[207] and it has been suggested that vibrational chemiluminescence should be observable from the reaction $H_2^+(H_2,H)H_3^+$.

Potential-energy surfaces for a limited number of four-atomic systems have been calculated. Surfaces were constructed for H_4^+ using the DIM

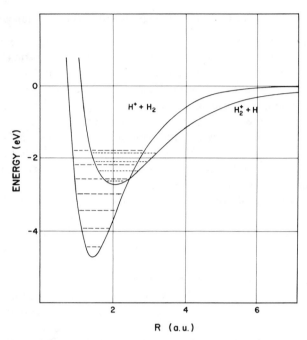

Figure 63. Potential curves of ground states of H_2 and H_2^+, drawn with same asymptote, showing a crossing at $R \simeq 2.5$ atomic units.[2]

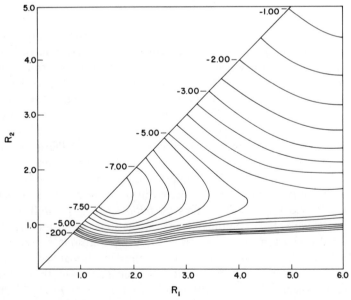

Figure 64. Contour plot of potential-energy surface for ground state of linear H_3^+. Values for R_1 and R_2 are in atomic units, and energies are in electron volts. Indicated energies are with respect to dissociated particles. Cut through surface at large R_1 coincides with potential curve of H_2 at small R_2 and with curve of H_2^+ at large R_2 (see Fig. 63).[2]

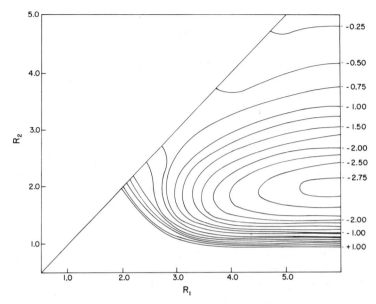

Figure 65. Potential-energy surface for first excited singlet state of linear H_3^+. Cut through surface at large R_1 coincides with potential curve of H_2^+ at small R_2 and with H_2 curve at large R_2 (see Fig. 63). There is a seam connecting this surface to that shown in Fig. 64, along which potential surface "hopping" can occur.[2]

method and representative geometries were investigated.[493] Some *ab initio* calculations are available as well.[494] Calculations of curves representing cuts through the $(He_2-N_2)^+$ potential surfaces have recently been discussed[495] in connection with the dynamics of the $He_2^+ - N_2$ reaction.

D. Quasiclassical and Collinear Quantum-mechanical Trajectory Calculations

When potential surfaces are available, quasiclassical trajectory calculations (first introduced by Karplus, et al.[496]) become possible. Such calculations are the theorist's analogue of experiments and have been quite successful in simulating molecular reactive collisions.[497] Opacity functions, excitation functions, and thermally averaged rate coefficients may be computed using such treatments. Since initial conditions may be varied in these calculations, state-to-state cross sections can be obtained, and problems such as vibrational specificity in the energy release of an exoergic reaction and vibrational selectivity in the energy requirement of an endoergic reaction can be treated. For example, trajectory calculations that have been accomplished to demonstrate the effectiveness of reagent vibrational excitation in the H_2^+–He interaction are summarized in Table VI. In this case trajectories were calculated on a single adiabatic potential surface (Fig. 61). Electronic excitation and electron exchange are energetically forbidden in this system at energies below about 13 eV and 9 eV, respectively. This is not the case in most other ion–molecule systems, where a number of closely lying and intersecting potential surface exist. Where there is an avoided crossing between two surfaces, the collision dynamics can be studied by following the classical trajectories on both surfaces. In this instance the trajectories start off on one surface and move on it until a region is encountered where the dynamical coupling between the two surfaces is large. At this point the trajectory either remains on the initial surface or "hops" to the new surface.

Trajectory surface-hopping (TSH) calculations have been accomplished for H_3^+ on DIM hypersurfaces.[2,498] A surface-hopping mechanism has also been suggested as applicable to several other systems, including various He_2^+ charge-transfer reactions,[495] the H_2^+–H_2 reaction,[499] and the C^+–H_2 radiative association process.[500]

In calculations utilizing the quasiclassical approach, only quantum-mechanically allowed states are accessible at the start of the trajectory. However, once the trajectory begins, all motion is classical and a continuum of states is allowed in the final state of the collision. In these calculations it is necessary to employ some arbitrary method for relating the final-state energies to quantum states. Exact quantum-mechanical calculations are obviously preferred, but the range of chemical systems and conditions that can be studied using current quantum-mechanical methods is limited. Only collinear calculations have been carried out for the H_2^+–He system and, in contrast to the predictions of quasiclassical trajectory calculations, no enhancement of the reaction probability from selective partitioning of the energy into initial H_2^+ vibration was observed.[461] This could be attributable to either the collinear nature of the calculations or slight inaccuracies in the potential-energy surface employed.[459]

ACKNOWLEDGMENT

The authors gratefully acknowledge the support of the Air Force Office of Scientific Research, Directorate of Chemical Sciences, under contract No. F44620-76-C-0007 with Wright State University, Fairborn, Ohio, during the period when this review was prepared.

REFERENCES

1. (a) R. F. Gould, Ed., *Ion–Molecule Reactions in the Gas Phase,* American Chemical Society, Washington, D. C., 1966; (b) J. H. Futrell and T. O. Tiernan, in P. Austoos, Ed., *Fundamental Process in Radiation Chemistry,* 1968,; Chapter 4, p. 171 (c) E. W. McDaniel, V. Čermák, A. Dalgarno, E. E. Ferguson, and L. Friedman, *Ion–Molecule Reactions,* (Wiley Interscience, New York, 1970; (d) J. L. Franklin, Ed., *Ion–Molecule Reactions,* Vols. 1 and 2, Plenum, New York, 1972; (e) J. Durup, *Adv. Mass Spectrom.* **6,** 691(1974); (f) P. Ausloos, Ed., *Interactions Between Ions and Molecules,* Plenum, New York, 1975, chapters by W. S. Koski, W. A. Chupka, R. Marx, and J. J. Kaufman; (g) E. E. Ferguson, *Annu. Rev. Phys. Chem,* **26,** 17 (1975); (h) J. W. McGowan, "The Excited State in Chemical Physics," *Advances in Chemical Physics,* Vol. **28,** 1975 Chapters II and VI; (i) R. D. Rundel and R. F. Stebbings, *Case Studies Atom. Collisions,* **2,** 549 (1972); (j) G. W. McClure and J. M. Peek, *Dissociation in Heavy Particle Collisions,* Wiley-Interscience, New York, 1972; (k) S. G. Lias and P. Ausloos, *Ion Molecule Reactions, Their Role in Radiation Chemistry,* ACS/ERDA Research Monograph, Washington, D.C., 1975.

2. R. K. Preston and J. C. Tully, *J. Chem. Phys.,* **54,** 4297 (1971).

3. (a) G. R. Fowles and R. C. Jensen, *Proc. IEEE,* **52,** 851 (1964); *Appl. Opt.,* **3,** 1191 (1964); (b) C. B. Collins, A. J. Cunningham, and M. Stockton, *Appl. Phys. Lett.,* **25,** 344 (1974); (c) J. B. Laudenslager, T. J. Pacala, and C. Wittig, *Appl. Phys. Lett.,* **29,** 580 (1976); (d) C. S. Willett, *Introduction to Gas Lasers: Population Inversion Mechanisms,* Pergamon, Oxford, 1974.

4. (a) B. R. Turner, J. A. Rutherford, and D. M. J. Compton, *J. Chem. Phys.,* **48,** 1602 (1968); (b) R. F. Stebbings, B. R. Turner, and J. A. Rutherford, *J. Geophys. Res.,* **71,** 771 (1966); (c) J. A. Rutherford and D. A. Vroom, *J. Chem. Phys.,* **55,** 5622 (1971); (d) R. H. Neynaber and G. D. Magnuson, *J. Chem. Phys.,* **58,** 4586 (1973).

5. (a) B. M. Hughes and T. O. Tiernan, *J. Chem. Phys.,* **55,** 3419 (1971); (b) B. M. Hughes and T. O. Tiernan, in *Ion–Neutral Reactions,* Chapter 18b, by D. A. Vroom, J. A. Rutherford, and R. L. Voigt, Defense Atomic Support Agency (DASA), *Reaction Rate Handbook,* Revision No. 5, June 1975, DNA 1948H, M. H. Bortner and T. Baurer, Eds.; (c) R. C. Amme and N. G. Utterback, in M. R. C. McDowell, Ed., *Atomic Collision Processes,* 1964, p. 847.

6. E. Lindholm, "Charge Exchange and Ion–Molecule Reactions Observed in Double Mass Spectrometers," in Ref. 1a, p. 1 and "Mass Spectra and Appearance Potentials Studied by Use of Charge Exchange in a Tandem Mass Spectrometer," in Ref. 1d, p. 457.

7. (a) K.-C. Lin, R. J. Cotter, and W. S. Koski, *J. Chem. Phys.,* **60,** 3412 (1974); (b) K.-C. Lin, H. P. Watkins, R. J. Cotter, and W. S. Koski, *J. Chem. Phys.,* **60,** 5134 (1974).

8. (a) Excitation energies of atomic ions are found in C. E. Moore, *Atomic Energy Levels,* National Bureau of Standards Circular 467, 1949; (b) excitation energies of some diatomic ions are found in B. Rosen, *Spectroscopic Data Relative to Diatomic Molecules,* Pergamon, Oxford, 1970; other energies are from the individual papers referenced in the table.

9. (a) R. C. C. Lao, R. W. Rozett, and W. S. Koski, *J. Chem. Phys.,* **49** 4202 (1968); (b) C. A. Jones, K. L. Wendell, J. J. Kaufman, and W. S. Koski, *J. Chem. Phys.,* **65,** 2345 (1976); (c) D. Brandt, C. Ottinger, and J. Simonis, *Ber. Bunsenges. Phys. Chem.,* **77,** 648 (1973).

10. B. R. Turner, J. A. Rutherford, and R. F. Stebbings, *J. Geophys. Res.*, **71**, 4521 (1966).

11. (a) H. B. Gilbody and J. B. Hasted, *Proc. Roy. Soc. (Lond.)*, **A238**, 334 (1957); (b) J. A. Rutherford and D. A. Vroom, *J. Chem. Phys.*, **62**, 1460 (1975); **65**, 1603 (1976).

12. (a) J. M. Farrar, S. G. Hansen, and B. H. Mahan, *J. Chem. Phys.*, **65**, 2908 (1976); (b) H. H. Harris and J. J. Leventhal, *J. Chem. Phys.*, **64**, 3185 (1976).

13. K.-C. Lin, R. J. Cotter, and W. S. Koski, *J. Chem Phys.*, **61**, 905 (1974).

14. W. A. Chupka and M. E. Russell, *J. Chem. Phys.*, **49**, 5426 (1968).

15. (a) P. Marmet and J. D. Morrison, *J. Chem. Phys.*, **36**, 1238 (1962); (b) J. W. McGowan and L. Kerwin, 2nd ICPEAC, Colorado, 1961, p. 36; (c) M. Hussain and L. Kerwin, *Can. J. Phys.*, **44**, 57 (1966).

16. (a) S. B. Karmohapatrd, *J. Chem. Phys.*, **35**, 1524 (1961); (b) H. Helm, *J. Phys.*, **B9**, 2931 (1976).

17. E. Lindholm, *Z. Naturforsch.*, **9a**, 535 (1954).

18. C. E. Melton, *J. Chem. Phys.*, **33**, 647 (1960).

19. (a) G. G. Meisels, *J. Chem. Phys.*, **31**, 284 (1959); (b) H. von Koch, *Arkiv. Fysik.*, **28**, 529 (1965).

20. R. D. Smith, D. L. Smith, and J. H. Futrell, *Internat. J. Mass Spectrom. Ion Phys.*, **19**, 395 (1976).

21. W. A. Chupka, in Ref. 1f, p. 259.

22. A. Galli, A. Giardini-Guidoni, and G. G. Volpi, *Nuovo Cimento*, **31**, 1145 (1964).

23. (a) G. K. Lavroskaza, M. I. Markin, and V. L. Tal'roze, *Kinet. and Catal. (U.S.S.R.)* (Engl. transl.), **2**, 18 (1961); (b) R. C. Amme and P. O. Haugsjaa, *Phys. Rev.*, **165**, 63 (1968); (c) J. B. Laudenslager, W. T. Huntress, Jr., and M. T. Bowers, *J. Chem. Phys.*, **61**, 4600 (1974); (d) J. M. Ajello and J. B. Laudenslager, *Chem. Phys. Lett.*, **44**, 344 (1976); (e) P. W. Harland and K. R. Ryan, *Internat. J. Mass Spectrom. Ion Phys.*, **18**, 215 (1975).

24. W. Kaul and R. Fuchs, *Z. Naturforsch.*, **15a**, 326 (1960).

25. M. Saporoschenko, *Phys. Rev.*, **111**, 1550 (1958).

26. V. Čermák and Z. Herman: (a) *Collect. Czech. Chem. Commun.*, **27**, 1493 (1962); (b) *J. Chim. Phys.*, **55**, 51 (1959).

27. M. S. B. Munson, F. H. Field, and J. L. Franklin, *J. Chem. Phys.*, **37**, 1790 (1962).

28. M. C. Cress, P. M. Becker, and F. W. Lampe, *J. Chem. Phys.*, **44**, 2212 (1966).

29. G. Junk and H. J. Svec, *J. Am. Chem. Soc.*, **80**, 2908 (1958).

30. R. K. Curran, *J. Chem. Phys.*, **38**, 2974 (1963).

31. V. Čermák and Z. Herman, *Collect. Czech. Chem. Commun.*, **30**, 1343 (1965).

32. J. J. Leventhal and L. Friedman, *J. Chem. Phys.*, **46**, 997 (1967).

33. R. K. Asundi, G. J. Schultz, and P. J. Chantry, *J. Chem. Phys.*, **47**, 1584 (1967).

34. (a) K. R. Ryan, *J. Chem. Phys.*, **51**, 570 (1969); (b) K. R. Ryan and H. M. Stock, *Internat. J. Mass Spectrom. Ion Phys.*, **21**, 389 (1976).

35. (a) S. Jaffe, Z. Karpas, and F. S. Klein, *J. Chem. Phys.*, **58**, 2190 (1973); (b) S. Jaffe and F. S. Klein, *Internat. J. Mass. Spectrom. Ion Phys.*, **14**, 459 (1974).

36. (a) W. B. Maier, II, *J. Chem. Phys.*, **55**, 2699 (1971); (b) R. F. Holland and W. B. Maier, II, *J. Chem. Phys.*, **57**, 4497 (1972).

37. M. T. Bowers, P. R. Kemper, and J. B. Laudenslager, *J. Chem. Phys.*, **61**, 4394 (1974).

38. (a) J. W. McGowan and L. Kerwin, *Can. J. Phys.*, **42**, 2086 (1964); (b) T. F. Moran, F. C. Petty, and A. F. Hedrick, *J. Chem. Phys.*, **51**, 2112 (1969); (c) T. F. Moran, J. B.

Wilcox, and L. E. Abbey, *J. Chem. Phys.*, **65**, 4540 (1976); (d) T. F. Moran and L. Friedman, *J. Chem. Phys.*, **42**, 2391 (1965).

39. J. T. Herron and H. I. Schiff, *Can. J. Chem.*, **36**, 1159 (1958).

40. Pham Dong and M. Cottin, *J. Chim. Phys.*, **57**, 557 (1960); **58**, 803 (1961).

41. V. Čermák and Z. Herman, *J. Chim. Phys.*, **57**, 717 (1960).

42. A. Henglein, Z. *Naturforsch.*, **11a**, 37 (1962).

43. J. L. Franklin and M. S. B. Munson, *Tenth Symposium International on Combustion*, The Combustion Institute, Pittsburgh, 1965, p. 561.

44. J. M. Ajello, K. D. Pang, and K. M. Monahan, *J. Chem. Phys.*, **61**, 3152 (1974).

45. P. M. Dehmer and W. A. Chupka, *J. Chem. Phys.*, **62**, 2228 (1975).

46. (a) W. Lindinger, D. L. Albritton, M. McFarland, F. C. Fehsenfeld, A. L. Schmeltekopf, and E. E. Ferguson, *J. Chem. Phys.*, **62**, 4101 (1975); (b) W. Lindinger, D. L. Albritton, F. C. Fehsenfeld, and E. E. Ferguson, *J. Geophys. Res.*, **80**, 3725 (1975).

47. M. H. Chiang, E. A. Gislason, B. H. Mahan, C. W. Tsao, and A. S. Werner, *J. Phys. Chem.*, **75**, 1426 (1971).

48. G. Bosse, A. Ding, and A. Henglein, *Ber. Bunsenges. Physik. Chem.*, **75**, 413 (1971).

49. J. M. Ajello, W. T. Huntress, A. L. Lane, P. R. LeBreton, and A. D. Williamson, *J. Chem. Phys.*, **60**, 1211 (1974).

50. T. O. Tiernan and R. E. Marcotte, *J. Chem. Phys.*, **53**, 2107 (1970).

51. P. C. Cosby, T. F. Moran, J. V. Hornstein, and M. R. Flannery, *Chem. Phys. Lett.*, **24**, 431 (1974).

52. (a) J. M. Ajello, *J. Chem. Phys.*, **62**, 1863 (1975); (b) J. M. Ajello and P. Rayermann, *J. Chem. Phys.*, **62**, 2917 (1975).

53. J. A. Rutherford, R. F. Mathis, B. R. Turner, and D. A. Vroom, *J. Chem. Phys.*, **55**, 3785 (1971).

54. J. A. Rutherford, R. F. Mathis, B. R. Turner, and D. A. Vroom, *J. Chem. Phys.*, **56**, 4654 (1972).

55. J. A. Rutherford, R. F. Mathis, B. R. Turner, and D. A. Vroom, *J. Chem. Phys.*, **57**, 3087 (1972).

56. (a) J. A. Rutherford and D. A. Vroom, *J. Chem. Phys.*, **57**, 3091 (1972); (b) B. R. Turner and J. A. Rutherford, *J. Geophys. Res.*, **73**, 6751 (1968).

57. S. E. Kuprianov, *Sov. Phys. JETP*, **3**, 390 (1969).

58. M. T. Bowers, M. Chau, and P. R. Kemper, *J. Chem. Phys.*, **63**, 3656 (1975).

59. R. F. Mathis, B. R. Turner, and J. A. Rutherford, *J. Chem. Phys.*, **49**, 2051 (1968).

60. T. F. Moran and J. R. Roberts, *J. Chem. Phys.*, **49**, 3411 (1968).

61. R. F. Mathis, B. R. Turner, and J. A. Rutherford, *J. Chem. Phys.*, **50**, 2270 (1969).

62. M. H. Cheng, M. Chiang, E. A. Gislason, B. H. Mahan, C. W. Tsao, and A. S. Werner, *J. Chem. Phys.*, **52**, 5518 (1970).

63. E. Lindemann, R. W. Rozett, and W. S. Koski, *J. Chem. Phys.*, **56**, 5490 (1972).

64. R. J. Cotter and W. S. Koski, *J. Chem. Phys.*, **59**, 784 (1973).

65. J. Collin, *J. Chim. Phys.*, **57**, 424 (1960).

66. P. Sullivan Wilson, R. W. Rozett, and W. S. Koski, *J. Chem. Phys.*, **52**, 5321 (1970).

67. D. W. Vance, *Phys. Rev.*, **169**, 263 (1968).

68. (a) F. C. Fehsenfeld, J. Appell, P. Fournier, and J. Durup, *J. Phys.*, **B6**, L268, (1973); (b) T. Ast, J. H. Beynon, and R. G. Cooks, *J. Am. Chem. Soc.*, **94**, 6611 (1972).

210   Role of Excited States in Ion–Neutral Collisions

69. (a) A. Tabché-Fouhaillé, J. Durup, J. T. Moseley, J.-B Ozenne, C. Pernot, and M. Tadjeddine, *Chem. Phys.*, **17**, 81 (1976); (b) J. T. Moseley, M. Tadjeddine, J. Durup, J.-B. Ozenne, C. Pernot, and A. Tabché-Fouhaillé, *Phys. Rev. Lett.*, **37**, 891 (1976).

70. (a) R. C. Dunbar and E. Fu, *J. Am. Chem. Soc.*, **95**, 2716 (1973); (b) B. S. Freiser and J. L. Beauchamp, *Chem. Phys. Lett.*, **35**, 35 (1975); (c) J. T. Moseley, P. C. Cosby, and J. R. Peterson, *J. Chem. Phys.*, **65**, 2512 (1976); (d) S. D. Rosner, T. D. Gaily, and R. A. Holt, *J. Phys.*, **B9**, L489 (1976).

71. E. Lindholm, I. Szabo, and P. Wilmenius, *Arkiv. Fysik*, **25**, 417 (1963).

72. J. H. Futrell and T. O. Tiernan, *J. Phys. Chem.*, **72**, 158 (1968).

73. C. E. Melton and W. H. Hamill, *J. Chem. Phys.*, **41**, 1469 (1964).

74. J. J. Myher and A. G. Harrison, *Can. J. Chem.*, **46**, 1755 (1968).

75. R. M. O'Malley and K. R. Jennings, *Internat. J. Mass. Spectrom. Ion Phys.*, **2**, 257 (1969).

76. A. A. Herod and A. G. Harrison, *Internat. J. Mass Spectrom. Ion Phys.*, **4**, 415 (1970).

77. P. G. Miasek and J. L. Beauchamp, *Internat. J. Mass. Spectrom. Ion Phys.*, **15**, 49 (1974).

78. M. S. B. Munson, *J. Phys. Chem.*, **60**, 572 (1965).

79. (a) P. J. Derrick and I. Szabo, *Internat. J. Mass Spectrom. Ion Phys.*, **7**, 71 (1971); (b) I. Szabo and P. J. Derrick, *Internat. J. Mass Spectrom. Ion Phys.*, **7**, 55 (1971).

80. P. R. LeBreton, A. D. Williamson, J. L. Beauchamp, and W. T. Huntress, *J. Chem. Phys.*, **62**, 1623 (1975).

81. P. Warneck. *Ber. Bunsen Gesellschaft*, **76**, 421 (1972).

82. J. M. Kramer and R. C. Dunbar, *J. Am. Chem. Soc.*, **94**, 4346 (1972).

83. E. G. Jones, A. K. Bhattacharya, and T. O. Tiernan, *Internat. J. Mass Spectrom. Ion Phys.*, **17**, 147 (1975).

84. (a) J. K. Layton, *J. Chem. Phys.*, **47**, 1869 (1967); (b) J. F. Paulson, *J. Chem. Phys.*, **52**, 5491 (1970).

85. W. A. Chupka: (a) in Ref. 1d, p. 33; (b) in Ref. 1f, p. 249.

86. (a) A. S. Werner and T. Baer, *J. Chem. Phys.*, **62**, 2900 (1975); T. Baer, B. P. Tsai, D. Smith, and P. T. Murray, *J. Chem. Phys.*, **64**, 2460 (1976); J. H. D. Eland and H. Schulte, *J. Chem. Phys.*, **62**, 3835 (1975); R. Stockbauer, *J. Chem. Phys.*, **58**, 3800 (1973); (b) T. Baer, L. Squires, and A. S. Werner, *Chem. Phys.*, **6**, 325 (1974); (c) L. Squires and T. Baer, *J. Chem. Phys.*, **65**, 4001 (1976); (d) K. Honma, K. Tanaka, I. Koyano, and I. Tanaka, *J. Chem. Phys.*, **65**, 678 (1976).

87. W. A. Chupka and M. E. Russell, *J. Chem. Phys.*, **48**, 1527 (1968).

88. V. H. Dibeler and R. M. Reese, *J. Chem. Phys.*, **40**, 2034 (1964); V. H. Dibeler and J. A. Walker, *Internat. J. Mass Spectrom. Ion Phys.*, **11**, 49 (1973).

89. W. A. Chupka and J. Berkowitz, *J. Chem. Phys.*, **48**, 5726 (1968); W. A. Chupka and J. Berkowitz, *J. Chem. Phys.*, **51**, 4244 (1969); P. M. Dehmer and W. A. Chupka, *J. Chem. Phys.*, **65**, 2243 (1976).

90. R. S. Berry, *Adv. Mass Spectrom.*, **6**, 1 (1974).

91. W. A. Chupka, M. E. Russell, and K. Refaey, *J. Chem. Phys.*, **48**, 1518 (1968).

92. (a) Ref. 14; (b) W. A. Chupka, J. Berkowtiz, and M. E. Russell, *Proceedings of the Sixth International Conference on the Physics of Electronic and Atomic Collisions*, MIT Press, Cambridge, Mass., 1969, p. 71.

93. J. A. Rutherford and D. A. Vroom, *J. Chem. Phys.*, **58**, 4076 (1973).

94. R. H. Neynaber and G. D. Magnuson, *J. Chem. Phys.*, **59**, 825 (1973).

95. (a) L. P. Theard and W. T. Huntress, Jr., *J. Chem. Phys.*, **60**, 2840 (1974); (b) W. T. Huntress, Jr., *Adv. Atom. Molec. Phys.*, **10**, 295 (1974).

96. H. C. Hayden and R. C. Amme, *Phys. Rev.*, **172**, 104 (1968).

97. T. R. Grossheim, J. J. Leventhal, and H. H. Harris, *Phys. Rev.*, **7A**, 1591 (1973).

98. (a) J. W. McGowan and L. Kerwin, *Can. J. Phys.*, **41**, 316, 1535 (1963); (b) J. W. McGowan, P. Marmet, and L. Kerwin, 3rd ICPEAC, North Holland, Amsterdam, 1964, p. 854.

99. (a) H. von Koch and L. Friedman, *J. Chem. Phys.*, **38**, 1115 (1963); T. F. Moran and L. Friedman, *J. Chem. Phys.*, **39**, 2491 (1963); F. S. Klein and L. Friedman, *J. Chem. Phys.*, **41**, 1789 (1964); (b) M. R. Flannery, P. C. Cosby, and T._F. Moran, *J. Chem. Phys.*, **59**, 5494 (1973); (c) J. W. McGowan, E. M. Clarke, H. D. Hanson, and R. F. Stebbings, *Phys. Rev. Lett.*, **14**, 620 (1964).

100. R. C. Amme and H. C. Hayden, *J. Chem. Phys.*, **42**, 2011 (1965).

101. P. O. Haugsjaa, R. C. Amme, and N. G. Utterback, *J. Chem. Phys.*, **49**, 4641 (1968).

102. J. M. Ajello and P. Rayermann, *J. Chem. Phys.*, **66**, 1372 (1977).

103. T. F. Moran, M. R. Flannery, and P. C. Cosby, *J. Chem. Phys.*, **61**, 1261 (1974).

104. W. A. Chupka and J. Berkowitz, *J. Chem. Phys.*, **54**, 4256 (1971).

105. J. J. Leventhal and L. Friedman, *J. Chem. Phys.*, **50**, 2928 (1969).

106. W. T. Huntress, Jr. and M. T. Bowers, *Int. J. Mass. Spectrom Ion Phys.*, **12**, 1 (1973).

107. (a) D. L. Smith and J. H. Futrell, *J. Phys.*, **B8**, 803 (1975); (b) J. K. Kim, L. P. Theard, and W. T. Huntress, Jr., *Internat. J. Mass Spectrom. Ion Phys.*, **15**, 223 (1974); (c) V. G. Anicich, J. H. Futrell, W. T. Huntress, Jr., and J. K. Kim, *Internat. J. Mass Spectrom. Ion Phys.*, **18**, 63 (1975); (d) D. L. Smith and J. H. Futrell, *Chem. Phys. Lett.*, **24**, 611 (1974); (e) A. Fiaux, D. L. Smith, and J. H. Futrell, *Internat. J. Mass Spectrom. Ion Phys.*, **20**, 223 (1976); (f) R. D. Smith and J. H. Futrell, *Internat. J. Mass Spectrom. Ion Phys.*, **20**, 33 (1976).

108. (a) M. L. Vestal, C. R. Blakeley, P. W. Ryan, and J. H. Futrell, Seventh International Mass Spectrometry Conference, Florence, 1976; *J. Chem. Phys.*, **64**, 2094 (1976); C. R. Blakeley, M. L. Vestal, and J. Futrell, *J. Chem. Phys.*, **66**, 2392 (1977); (b) D. L. Smith and J. H. Futrell, *Chem. Phys. Lett.*, **40**, 299 (1976); (c) R. P. Clow, T. O. Tiernan, and B. M. Hughes, 22nd ASMS Conference, Philadelphia, May 1974.

109. A. Fiaux, D. L. Smith, and J. H. Futrell, *Internat. J. Mass Spectrom. Ion Phys.*, **15**, 9 (1974); R. D. Smith and J. H. Futrell, *Internat. J. Mass Spectrom. Ion Phys.*, **19**, 201 (1976); **20**, 33, 43, 59, 71, 347 (1976).

110. W. T. Huntress, Jr., J. B. Laudenslager, and R. F. Pinizzotto, Jr., *Internat. J. Mass Spectrom. Ion Phys.*, **13**, 331 (1974).

111. S. E. Buttrill, Jr., J. K. Kim, W. T. Huntress, Jr., P. LeBreton, and A. Williamson, *J. Chem. Phys.*, **61**, 2122 (1974).

112. S. E. Buttrill, Jr., *J. Chem. Phys.*, **62**, 1834 (1975).

113. M. L. Gross and J. Norbeck, *J. Chem. Phys.*, **54**, 3651 (1971).

114. (a) L. W. Sieck and P. Ausloos, *J. Res. (N.B.S.)*, **76A**, 253 (1972); (b) L. Hellner and L. W. Sieck, *Internat. J. Chem. Kinet.*, **5**, 177 (1973).

115. (a) A. D. Williamson and J. L. Beauchamp, *J. Chem. Phys.*, **65**, 3196 (1976); (b) R. R. Corderman, P. R. LeBreton, S. E. Buttrill, Jr., A. D. Williamson, and J. L. Beauchamp, *J. Chem. Phys.*, **65**, 4929 (1976).

116. (a) J. A. D. Stockdale, R. N. Compton, and H. C. Schweinler, *J. Chem. Phys.*, **53**, 1502

(1970); (b) C. Lifshitz and R. Grajower, *Internat. J. Mass Spectrom. Ion Phys.*, **3**, App. 5 (1969).

117. J. W. McGowan and L. Kerwin, *Can. J. Phys.*, **42**, 972 (1964).

118. B. Andlauer and C. Ottinger, *J. Chem. Phys.*, **55**, 1471 (1971); *Z. Naturforsch.*, **27a**, 293 (1972).

119. A. Weingartshofer and E. M. Clarke, *Phys. Rev. Lett.*, **12**, 591 (1964).

120. W. L. Wiese, M. Smith, and B. M. Glennon, *Atomic Transition Probabilities*, Vol. I, National Standard Reference Data Systems (NSRDS) report no. NSRDS NBS-4, 1966; W. L. Wiese, M. Smith, and B. M. Miles, *Atomic Transition Probabilities*, Vol. II, NSRDS, NBS-22, 1969.

121. (a) A. Anderson, *Atom. Data*, **3**, 227 (1971); (b) P. H. Krupenie, *J. Phys. Chem. Ref. Data*, **1**, 423 (1972).

122. G. R. Möhlmann and F. J. DeHeer, *Chem. Phys. Lett.*, **43**, 170 (1976).

123. A. L. Schmeltekopf, E. E. Ferguson, and F. C. Fehsenfeld, *J. Chem. Phys.*, **48**, 2966 (1968).

124. N. Sbar and J. Dubrin, *J. Chem. Phys.*, **53**, 842 (1970).

125. F. C. Fehsenfeld, W. Lindinger, S. L. Schmeltekopf, D. L. Albritton, and E. E. Ferguson, *J. Chem. Phys.*, **62**, 2001 (1975).

126. A. L. Schmeltekopf, G. I. Gilman, F. C. Fehsenfeld, and E. E. Ferguson, *Planet. Space Sci.*, **15**, 401 (1967).

127. R. B. Cohen, *J. Chem. Phys.*, **57**, 676 (1972).

128. D. L. Albritton, Y. A. Bush, F. C. Fehsenfeld, E. E. Ferguson, T. R. Govers, M. McFarland, and A. L. Schmeltekopf, *J. Chem. Phys.*, **58**, 4036 (1973).

129. F. C. Fehsenfeld, D. L. Albritton, J. A. Burt, and H. I. Schiff, *Can. J. Chem.*, **47**, 1793 (1969).

130. R. F. Mathis and W. R. Snow, *J. Chem. Phys.*, **61**, 4274 (1974).

131. E. R. Weiner, G. R. Hertel, and W. S. Koski, *J. Am. Chem. Soc.*, **86**, 788 (1964).

132. M. A. Berta and W. S. Koski, *J. Am. Chem. Soc.*, **86**, 5098 (1964).

133. K. L. Wendell, C. A. Jones, J. J. Kaufman, and W. S. Koski, *J. Chem. Phys.*, **63**, 750 (1975).

134. P. C. Cosby and T. F. Moran *J. Chem. Phys.*, **52**, 6157 (1970).

135. R. H. Neynaber, G. D. Magnuson, and J. K. Layton, *J. Chem. Phys.*, **57**, 5128 (1972).

136. R. H. Neynaber, G. D. Magnuson, S. M. Trujillo, and B. F. Myers, *Phys. Rev.*, **A5**, 285 (1972).

137. M. L. Vestal, C. R. Blakeley, P. W. Ryan, and J. H. Futrell, *Rev. Sci. Instrum.* **47**, 15 (1976).

138. T. O. Tiernan, in Ref. 1f, p. 600.

139. M. Henchman, in Ref. 1d, p. 101.

140. E. E. Ferguson, F. C. Fehsenfeld, and A. L. Schmeltekopf, *Adv. Atom. Molec. Phys.*, **5**, 1 (1969).

141. M. McFarland, D. L. Albritton, F. C. Fehsenfeld, E. E. Ferguson, and A. L. Schmeltekopf, *J. Chem. Phys.*, **59**, 6610 (1973).

142. K. Tanaka and I. Tanaka, *J. Chem. Phys.*, **59**, 5042 (1973).

143. J. Yinon and F. S. Klein, *Adv. Mass Spectrom.*, **5**, 241 (1971).

144. K. R. Ryan and I. G. Graham, *J. Chem. Phys.*, **59**, 4260 (1973).

145. K. R. Ryan, *J. Chem. Phys.*, **61**, 1559 (1974).

146. R. J. Conrads, W. Pomerance, and T. F. Moran, *J. Chem. Phys.*, **57**, 2468 (1972).

147. T. F. Moran and R. J. Conrads, *J. Chem. Phys.*, **58**, 3793 (1973).

148. J. H. Futrell, in D. Price, Ed., *Dynamic Mass Spectrometry*, Vol. 2, Heyden, London, 1971.

149. V. G. Anicich and M. T. Bowers, *Internat. J. Mass Spectrom. Ion Phys.*, **14**, 171 (1974).

150. T. B. McMahon and J. L. Beauchamp, *Rev. Sci. Instrum.*, **43**, 509 (1972).

151. B. S. Freiser and J. L. Beauchamp, *J. Am. Chem. Soc.*, **96**, 6260 (1974).

152. D. L. Smith and J. H. Futrell, *Internat. J. Mass Spectrom. Ion Phys.*, **14**, 171 (1974).

153. M. Hollstein, D. C. Lorents, J. R. Peterson, and J. R. Sheridan, *Can. J. Chem.*, **47**, 1858 (1969).

154. L. Kurzweg, H. H. Lo, R. P. Brackmann, and W. L. Fite, *Phys. Rev.*, **179**, 55 (1969).

155. J. H. Moore, Jr., and J. P. Doering, *Phys. Rev.*, **177**, 218 (1969).

156. L. Wolterbeek Muller and F. J. deHeer, *Physica*, **48**, 345 (1970).

157. A. Salop, D. C. Lorents, and J. R. Peterson, *J. Chem. Phys.*, **54**, 1187 (1971).

158. M. J. Haugh, *J. Chem. Phys.*, **56**, 4001 (1972).

159. M. A. Coplan and J. E. Mentall, *J. Chem. Phys.*, **58**, 4912 (1973).

160. H. Bregman-Reisler and J. P. Doering, *Phys. Rev.*, **A9**, 1152 (1974).

161. T. R. Govers, C. A. van de Runstraat, and F. J. deHeer, *Chem. Phys.*, **9**, 285 (1975).

162. R. Odom, D. Siedler, and J. Weiner, *Phys. Rev.*, **A14**, 685 (1976).

163. B. J. Hughes, E. G. Jones, and T. O. Tiernan, *Proceedings of the Eighth International Conference on the Physics of Electronic and Atomic Collisions*, Vol. 1, Belgrade, 1973, p. 223.

164. S. Dworetsky, R. Novick, W. W. Smith, and N. Tolk, *Rev. Sci. Instrum.*, **39**, 1721 (1968).

165. R. C. Isler, *Rev. Sci. Instrum.*, **45**, 308 (1974).

166. G. J. Collins, *J. Appl. Phys.*, **44**, 4633 (1973).

167. R. S. Bergman and L. W. Chanin, *Phys. Rev.*, **A8**, 1076 (1973).

168. L. G. Piper, L. Gundel, J. E. Velazco, and D. W. Setser, *J. Chem. Phys.*, **62**, 3883 (1975).

169. M. Kamin and L. M. Chanin, *Appl. Phys. Lett.*, **29**, 756 (1976).

170. F. W. Lee, C. B. Collins, and R. A. Waller, *J. Chem. Phys.*, **65**, 1605 (1976).

171. G. Mauclaire, R. Marx, C. Sourisseau, C. van de Runstraat, and S. Fenistein, *Internat. J. Mass Spectrom. Ion Phys.*, **22**, 339 (1976).

172. T. R. Govers, M. Gérard, and R. Marx, *Chem. Phys.*, **15**, 185 (1976).

173. E. Graham, IV, M. A. Biondi, and R. Johnsen, *Phys. Rev.*, **A13**, 965 (1976).

174. F. C. Fehsenfeld, E. E. Ferguson, and C. J. Howard, *J. Geophys. Res.*, **78**, 327 (1973).

175. J. B. Hasted, *Adv. Atom. Molec. Phys.*, **4**, 237 (1968).

176. H. S. W. Massey, "Collisions Between Atoms and Molecules at Ordinary Temperatures," *Phys. Soc. (Lond.) Rep. on Progr. Phys.*, **12**, 248 (1949).

177. D. Rapp and W. E. Francis, *J. Chem. Phys.*, **37**, 2631 (1962), *Proc. Roy Soc. (Lond.)*, **A268**, 2349 (1964).

178. A. R. Lee and J. B. Hasted, *Proc. Roy Soc. (Lond.)*, **85**, 673 (1965).

179. F. T. Smith, in *Proceedings of Eighth International Conference on Phenomena in Ionized Gases*, Vienna, 1967 (International Atomic Energy Agency, Vienna, 1968), p. 75.

180. W. R. Henderson, J. E. Mentall, and W. L. Fite, *J. Chem. Phys.*, **46**, 3447 (1967).

214   The Role of Excited States in Ion–Neutral Collisions

182. D. K. Bohme, J. B. Hasted, and P. P. Ong, *Chem. Phys. Lett.*, **1**, 259 (1967); *J. Phys. B*, **1**, 879 (1968).

183. R. S. Hemsworth, R. C. Bolden, M. J. Shaw, and N. D. Twiddy, *Chem. Phys. Lett.*, **5**, 237 (1970).

184. D. L. Smith and L. Kevan, *J. Am. Chem. Soc.*, **93**, 2113 (1971); *J. Chem. Phys.*, **55**, 2290 (1971).

185. M. T. Bowers and D. D. Elleman, *Chem. Phys. Lett.*, **16**, 486 (1972).

186. A. F. Hedrick and T. F. Moran, *J. Chem. Phys.*, **64**, 1858 (1976).

187. S. G. Lias, *Internat. J. Mass. Spectrom. Ion Phys.*, **20**, 123 (1976).

188. T. F. Moran, M. R. Flannery, and D. L. Albritton, *J. Chem. Phys.*, **62**, 2869 (1975).

189. P. C. Cosby, T. F. Moran, and M. R. Flannery, *J. Chem. Phys.*, **61**, 1259 (1974).

190. M. R. Flannery, K. J. McCann, and T. F. Moran, *Chem. Phys. Lett.*, **39**, 374 (1976).

191. E. A. Gislason, *J. Chem. Phys.*, **57**, 3396 (1972).

192. B. H. Mahan and W. E. W. Ruska, *J. Chem. Phys.*, **65**, 5044 (1976).

193. C. Ottinger and J. Simonis, *Phys. Rev. Lett.*, **35**, 924 (1975).

194. F. H. Field, P. Hamlet, and W. F. Libby, *J. Am. Chem. Soc.*, **89**, 6035 (1967).

195. A. Giardini-Guidoni and F. Zocchi, *Transact. Faraday. Soc.*, **64**, 2342 (1968).

196. S. Wexler and R. P. Clow, *J. Am. Chem. Soc.*, **90**, 3940 (1968).

197. F. H. Field, P. Hamlet, and W. F. Libby, *J. Am. Chem. Soc.*, **91**, 2839 (1969).

198. S. Wexler and L. G. Pobo, *J. Phys., Chem.*, **74**, 257 (1970).

199. L. Friedman and B. G. Reuben, *Adv. Chem. Phys.*, **19**, 33 (1971).

200. J. A. D. Stockdale, *J. Chem. Phys.*, **58**, 3881 (1973).

201. V. G. Anicich and M. T. Bowers, *J. Am. Chem. Soc.*, **96**, 1279 (1974).

202. E. G. Jones, A. K. Bhattacharya, and T. O. Tiernan, *Internat. J. Mass Spectrom. Ion Phys.*, **17**, 147 (1975).

203. L. W. Sieck and R. Gorden, Jr., *Internat. J. Mass Spectrom. Ion Phys.*, **19**, 269 (1976).

204. K. G. Anlauf, D. H. Maylotte, J. C. Polanyi, and R. B. Bernstein, *J. Chem. Phys.*, **51**, 5716 (1969); J. B. Anderson, *J. Chem. Phys.*, **52**, 3849 (1970); R. J. Donovan and D. Husain, *Chem. Rev.*, **70**, 489 (1970); T. J. Odiorne, P. R. Brooks, and J. V. V. Kasper, *J. Chem. Phys.*, **55**, 1980 (1971); R. F. Heidner, III, and J. V. V. Kasper, *Chem. Phys. Lett.*, **15**, 179 (1972); T. M. Sloane, S. Y. Tang and J. Ross, *J. Chem. Phys.*, **57**, 2745 (1972); R. J. Gordon and M. C. Lin, *Chem. Phys. Lett.*, **22**, 262 (1973); Z. Karny, B. Katz, and A. Szöke, *Chem. Phys. Lett.*, **35**, 100 (1975); D. Arnoldi, K. Kaufmann, and J. Wolfrum, *Phys. Rev. Lett.*, **34**, 1597 (1975).

205. J. d'Incan and A. Topouzkhanian, *J. Chem. Phys.*, **63**, 2683 (1975).

206. J. C. Polanyi and W. H. Wong, *J. Chem. Phys.*, **51**, 1439 (1969).

207. G. D. Carney and R. N. Porter, *J. Chem. Phys.*, **65**, 3547 (1976).

208. R. L. Champion, L. D. Doverspike, and T. L. Bailey, *J. Chem. Phys.*, **45**, 4377 (1966).

209. R. L. C. Wu and T. O. Tiernan, 21st Annual Converence on Mass Spectrometry and Allied Topics (ASMS), San Francisco, May 1973.

210. T. O. Tiernan and R. P. Clow, *Adv. Mass Spectrom.*, **6**, 295 (1974).

211. D. G. Hopper, A. C. Wahl, R. L. C. Wu, and T. O. Tiernan, *J. Chem. Phys.*, **65**, 5474 (1976).

212. T. O. Tiernan and R. L. C. Wu, *Adv. Mass Spectrom.*, **7A**, 136 (1977).

213. G. W. McClure, *Phys. Rev.*, **140**, A769 (1965).

214. D. K. Gibson and J. Los, *Physica*, **35**, 258 (1967).

215. D. K. Gibson, J. Los, and J. Schopman, *Phys. Lett.*, **25A**, 634 (1967).

216. M. Vogler and W. Seibt, *Z. Phys.*, **210**, 337 (1968).

217. J. Durup, P. Fournier, and P. Dông, *Internat. J. Mass Spectrom. Ion Phys.*, **2**, 311 (1969).

218. P. Fournier, C. A. van de Runstraat, T. R. Govers, J. Schopman, F. J. deHeer, and J. Los, *Chem. Phys. Lett.*, **9**, 426 (1971).

219. J. Tellinghuisen and D. L. Albritton, *Chem. Phys. Lett.*, **31**, 91 (1975).

220. A. J. Lorquet and J. C. Lorquet, *Chem. Phys. Lett.*, **26**, 138 (1974).

221. J. M. Peek, *Phys. Rev.*, **140**, A11 (1965); J. M. Peek, T. A. Green, and W. H. Weihofen, *Phys. Rev.*, **160**, 117 (1967).

222. C. Rebick and R. D. Levine, *J. Chem. Phys.*, **58**, 3942 (1973).

223. F. R. Gilmore, *J. Quant. Spectrosc. Radiat. Transf.*, **5**, 369 (1965); D. C. Cartwright and T. H. Dunning, Jr., *J. Phys.*, **B8**, L100 (1975); E. W. Thulstrup and A. Anderson, *J. Phys.*, **B8**, 965 (1975).

224. K. C. Kim, M. Uckotter, J. H. Beynon, and R. G. Cooks, *Internat. J. Mass Spectrom. Ion Phys.*, **15**, 23 (1974).

225. K. R. Jennings, *Internat. J. Mass Spectrom. Ion Phys.*, **1**, 227 (1968).

226. F. W. McLafferty, P. F. Bente, III, R. Kornfeld, S.-C Tsai, and I. Howe, *J. Am. Chem. Soc.*, **95**, 2120 (1973).

227. K. Levsen and H. D. Beckey, *Org. Mass Spectrom.*, **9**, 570 (1974) and references cited therein.

228. R. G. Cooks, J. H. Beynon, and J. F. Litton, *Org. Mass Spectrom.*, **10**, 503 (1975).

229. R. D. Smith, D. L. Smith, and J. H. Futrell, *Internat. J. Mass Spectrom. Ion Phys.*, **19**, 369 (1976).

230. G. M. Burnett and A. M. North, *Transfer and Storage of Energy by Molecules*, Vols. 1–3, *Electronic, Vibrational, and Rotational Energy*, Wiley-Interscience, New York, 1970.

231. A. B. Callear and P. M. Wood, *Transact. Faraday Soc.*, **67**, 272 (1971).

232. Y.-N Chiu, *J. Chem. Phys.*, **56**, 4882 (1972).

233. B. Brocklehurst and F. A. Downing, *J. Chem. Phys.*, **46**, 2976 (1967).

234. F. J. Comes and F. Speier, *Chem. Phys. Lett.*, **4**, 13 (1969).

235. K. B. Mitchell, *J. Chem. Phys.*, **53**, 1795 (1970).

236. G. I. Mackay and R. E. March, *Can. J. Chem.*, **49**, 1268 (1971).

237. C. A. Winkler, J. B. Tellinghuisen, and L. F. Phillips, *J. Chem. Soc. Faraday II*, **68**, 121 (1972).

238. J. B. Tellinghuisen, C. A. Winkler, C. G. Freeman, M. J. McEwan, and L. F. Phillips, *J. Chem. Soc. Faraday II*, **68**, 833 (1972).

239. R. J. Alderson, B. Brocklehurst, and F. A. Downing, *J. Chem. Phys.*, **58**, 4041 (1973).

240. G. H. Dunn, in M. R. C. McDowell, Ed., *Atomic Collision Processes*, North-Holland, Amsterdam, 1964, p. 997.

241. J. H. Current and B. S. Rabinovitch, *J. Chem. Phys.*, **40**, 2742 (1964).

242. Y. N. Lin and B. S. Rabinovitch, *J. Phys. Chem*, **74**, 3151 (1970).

243. L. D. Spicer and B. S. Rabinovitch, *Annu. Rev. Phys. Chem.*, **21**, 349 (1970).

244. P. Kebarle, in Ref. 1d, Ch. 7 p. 327.

245. E. E. Ferguson, in Ref. 1d, Ch. 8, p. 370.

246. A. G. Harrison, in Ref. 1f, p. 263.

247. D. K. Bohme, D. B. Dunkin, F. C. Fehsenfeld, and E. E. Ferguson, *J. Chem. Phys.*, **51**, 863 (1969).

248. P. S. Gill, Y. Inel, and G. G. Meisels, *J. Chem. Phys.*, **54**, 2811 (1971).

249. P. B. Miasek and A. G. Harrison, *J. Am. Chem. Soc.*, **97**, 714 (1975).

250. R. C. Bhattacharjee and W. Forst, "Statistical Theory of Energy Transfer in Ion-Molecule Collisions," paper presented at Seventh International Mass Spectrometry Conference, Florence, 1976.

251. R. Houriet and J. H. Futrell, "Collisional Deactivation of Ions," paper presented at Seventh International Mass Spectrometry Conference, Florence, 1976.

252. G. G. Meisels, F. Botz, R. K. Mitchum, and E. F. Heckel, "Angular Momentum and the Temperature Dependence of Ion–Molecule Reactions," paper presented at Seventh International Mass Spectrometry Conference, Florence, 1976.

253. G. G. Meisels, F. Botz, and E. F. Heckel, "Lifetimes of Intermediate Complexes in Ion–Molecule Reactions," *Proceedings of 24th Annual Conference on Mass Spectrometry and Allied Topics*, ASMS, San Diego 1976, p. 373.

254. R. K. Mitchum, J. P. Freeman, and G. G. Meisels, *J. Chem. Phys.*, **62**, 2465 (1975).

255. J. H. Moore, Jr., *Phys. Rev.*, **A8**, 2359 (1973), **A10**, 724 (1974).

256. J. P. Doering, *Ber. Bunsenges. Phys. Chem.*, **77**, 593 (1973).

257. F. A. Herrero and J. P. Doering, *Phys. Rev. Lett.*, **29**, 609 (1972).

258. E. W. Thomas, *Excitation in Heavy Particle Collisions*, Wiley-Interscience, New York, 1972.

259. R. F. Stebbings, R. A. Young, C. L. Oxley, and H. Ehrhardt, *Phys. Rev.*, **A138**, 1312 (1965).

260. F. J. deHeer, L. Wolterbeek Muller, and R. Geballe, *Physica*, **31**, 1745 (1965).

261. D. C. Lorents, W. Aberth, and V. W. Hesterman, *Phys. Rev. Lett.*, **17**, 849 (1966).

262. S. H. Dworetsky, R. Novick, W. W. Smith, and N. Tolk, *Phys. Rev. Lett*, **18**, 939 (1967).

263. S. H. Dworetsky and R. Novick, *Phys. Rev. Lett.*, **23**, 1484 (1969).

264. H. Rosenthal and H. M. Foley, *Phys. Rev. Lett.*, **23**, 1480 (1969).

265. H. Rosenthal, *Phys. Rev.*, **A4**, 1030 (1971).

266. R. A. Young, R. F. Stebbings, and J. W. McGowan, *Phys. Rev.*, **171**, 85 (1968).

267. J. van Eck, F. J. deHeer, and J. Kistemaker, *Phys. Rev.*, **130**, 656 (1963).

268. J. van Den Bos, G. J. Winter, and F. J. deHeer, *Physica*, **40**, 357 (1968) and references cited therein.

269. D. Coffey, Jr., D. C. Lorents, and F. T. Smith, *Phys. Rev.*, **187**, 201 (1969).

270. N. H. Tolk, C. W. White, S. H. Dworetsky, and L. A. Farrow, *Phys. Rev. Lett.*, **25**, 1251 (1970).

271. S. V. Bobashev, *JETP Lett.*, **11**, 260 (1970).

272. S. V. Bobashev and V. A. Kritskii, *JETP Lett.*, **12**, 189 (1970).

273. V. A. Ankudinov, S. V. Bobashev, and V. I. Perel, *Sov. Phys. JETP*, **33**, 490 (1971).

274. S. V. Bobashev, V. I. Ogurtsov, and L. A. Razumovskii, *Sov. Phys. JETP*, **35**, 472 (1972).

275. N. H. Tolk, C. W. White, S. H. Neff, and W. Lichten, *Phys. Rev. Lett.*, **31**, 671 (1973).

276. A. N. Zavilopulo, I. P. Zapesochnyi, G. S. Panev, O. A. Skalko, and O. B. Shpenik, *JETP Lett.*, **18**, 245 (1973).

277. T. Andersen, A. K. Nielsen, and K. J. Olsen, *Phys. Rev. Lett.*, **31**, 739 (1973); *Phys. Rev.*, **A10**, 2174 (1974).

278. N. H. Tolk, J. C. Tully, C. W. White, J. Kraus, A. A. Monge, and S. H. Neff, *Phys. Rev. Lett.*, **35**, 1175 (1975).

279. N. H. Tolk, J. C. Tully, C. W. White, J. Kraus, A. A. Monge, D. L. Simms, M. F. Robbins, S. H. Neff, and W. Lichten, *Phys. Rev.*, **A13**, 969 (1976).

280. R. Odom, J. Caddick, and J. Weiner, *Phys. Rev.*, **A15**, 1414 (1977).

281. J. T. Park and F. D. Schowengerdt, *Phys. Rev.*, **185**, 152 (1969).

282. F. D. Schowengerdt and J. T. Park, *Phys. Rev.*, **A1**, 848 (1970).

283. J. H. Moore and J. P. Doering, *J. Chem. Phys.*, **52**, 1692 (1970).

284. J. H. Moore, Jr., *J. Chem. Phys.*, **55**, 2760 (1971).

285. F. D. Schowengerdt, J. T. Park, and D. R. Schoonover, *Phys. Rev.*, **A3**, 679 (1971).

286. J. H. Moore, Jr., *J. Phys. Chem.*, **76**, 1130 (1972); *J. Geophys. Res.*, **77**, 5567 (1972).

287. J. P. Doering and J. H. Moore, Jr., *J. Chem. Phys.*, **56**, 2176 (1972).

288. G. H. Bearman, J. D. Earl, H. H. Harris, P. B. James, and J. J. Leventhal, *Chem. Phys. Lett.*, **44**, 471 (1976).

289. S. H. Neff and N. P. Carleton, *Atomic Collision Processes*, North Holland, Amsterdam, 1964, p. 652.

290. C. Liu and H. P. Broida, *Phys. Rev.*, **A2**, 1824 (1970).

291. M. Medved, R. G. Cooks, and J. H. Beynon, *Chem. Phys.*, **15**, 295 (1976).

292. G. H. Bearman, H. H. Harris, P. B. James, and J. J. Leventhal, "Molecular Excitation in Low Energy Ion–Molecule Collisions," paper presented at 19th Annual Gaseous Electronic Conference, Cleveland, October 1976.

293. R. C. Isler, *Phys. Rev.*, **A9**, 1865 (1974).

294. L. Dana and P. Laures, *Proc. IEEE*, **53**, 78 (1965).

295. P. F. Dittner and S. Datz, *J. Chem. Phys.*, **49**, 1969 (1968); *J. Chem. Phys.*, **54**, 4228 (1971).

296. J. Shöttler and J. P. Toennies, *Z. Physik*, **214**, 472 (1968); W. D. Held, J. Shöttler, and J. P. Toennies, *Chem. Phys. Lett.*, **6**, 304 (1970).

297. H. Van Dop, A. J. H. Boerboom, and J. Los, *Physica*, **54**, 223 (1971).

298. W. L. Dimpfl and B. H. Mahan, *J. Chem. Phys.*, **60**, 3238 (1974).

299. T. F. Moran, F. Petty, and G. S. Turner, *Chem. Phys. Lett.*, **9**, 379 (1971).

300. F. Petty and T. F. Moran, *Phys. Rev.*, **A5**, 266 (1972).

301. M. H. Cheng, M. H. Chiang, E. A. Gislason, B. H. Mahan, C. W. Tsao, and A. S. Werner, *J. Chem. Phys.*, **52**, 6150 (1970).

302. T. F. Moran and P. C. Cosby, *J. Chem. Phys.*, **51**, 5724 (1969).

303. J. H. Moore and J. P. Doering, *Phys. Rev. Lett.*, **23**, 564 (1969).

304. J. P. Toennies, *Annu. Rev. Phys. Chem.*, **27**, 225 (1976).

305. F. A. Herrero and J. P. Doering, *Phys. Rev.*, **A5**, 702 (1972).

306. H. Udseth, C. F. Giese, and W. R. Gentry, *J. Chem. Phys.*, **54**, 3642 (1971).

307. H. Udseth, C. F. Giese, and W. R. Gentry, *Phys. Rev.*, **A8**, 2483 (1973).

308. C. F. Giese and W. R. Gentry, *Phys. Rev.*, **A10**, 2156 (1974).

218   The Role of Excited States in Ion–Neutral Collisions

309. H. Schmidt, V. Hermann, and F. Linder, *Chem. Phys. Lett.*, **41**, 365 (1976).

310. K. Rudolph and J. P. Toennies, *J. Chem. Phys.*, **65**, 4483 (1976).

311. H. E. Van den Bergh, M. Faubel, and J. P. Toennies, *Faraday. Discuss. Chem. Soc.* **55**, 203 (1973); R. David, M. Faubel and J. P. Toennies, *Chem. Phys. Lett.*, **18**, 87 (1973); M. Faubel, K. Rudolph, and J. P. Toennies, in J. S. Risley and R. Geballe, Eds., *Abstracts, ICPEAC (International Conference on the Physics of Electronic and Atomic Collisions)*, Vol. 1, Washington U. P., Seattle, 1975, p. 49; M. Faubel and J. P. Toennies, *J. Chem. Phys.* **71**, 3770 (1979).

312. M. R. Flannery, K. J. McCann, and T. F. Moran, *J. Chem. Phys.*, **63**, 1462 (1975).

313. T. F. Moran, K. J. McCann, and M. R. Flannery, *J. Chem. Phys.*, **63**, 3857 (1975).

314. M. R. Flannery and T. F. Moran, *J. Phys.*, **B9**, L509 (1976).

315. D. B. Dunkin, F. C. Fehsenfeld, A. L. Schmeltekopf, and E. E. Ferguson, *J. Chem. Phys.*, **49**, 1365 (1968).

316. R. Johnsen and M. A. Biondi, *J. Chem. Phys.*, **59**, 3504 (1973).

317. M. McFarland, D. L. Albritton, F. C. Fehsenfeld, E. E. Ferguson, and A. L. Schmeltekopf, *J. Chem. Phys.*, **59**, 6620 (1973).

318. J. J. Kaufman and W. S. Koski, *J. Chem. Phys.*, **50**, 1942 (1969).

319. T. F. O'Malley, *J. Chem. Phys.*, **52**, 3269 (1970).

320. T. E. Van Zandt and T. F. O'Malley, *J. Geophys. Res.*, **78**, 6818 (1973).

321. J. J. Kaufman, in Ref. lf, p. 185; in Ref. 1h, Chapter II and references cited therein.

322. W. Lindinger, D. L. Albritton, F. C. Fehsenfeld, and E. E. Ferguson, *J. Chem. Phys.*, **63**, 3238 (1975).

323. D. G. Hopper, A. C. Wahl, R. L. C. Wu, and T. O. Tiernan, "Theoretical and Experimental Studies of N_2O^- and N_2O Ground State Potential Surfaces. Implications for the $O^- + N_2 \rightarrow N_2O + e$ and Other Processes," paper presented at Annual Conference on Mass Spectrometry and Allied Topics, San Diego, May 1976, *J. Chem. Phys.*, **65**, 5474 (1976).

324. R. D. Levine and R. B. Bernstein, *Molecular Reaction Dynamics*, Clarendon, Oxford, 1974.

325. T. P. Schafer, P. E. Siska, J. M. Parson, F. P. Tully, Y. C. Wong, and Y. T. Lee, *J. Chem. Phys.*, **53**, 3385 (1970).

326. C. A. Jones, W. S. Koski, and K. L. Wendell, "Internal Energy of CH^+ Produced by the Reaction C^+ $(H_2, H)CH^+$," paper presented by the 173rd ACS National Meeting, New Orleans, March 1977.

327. A. Henglein, in C. Schlier, Ed., *Molecular Beams and Reaction Kinetics*, Academic, New York, 1970, p. 154; P. Hierl, Z. Herman, J. Kerstetter, and R. Wolfgang, *J. Chem. Phys.*, **48**, 4319 (1968); L. D. Doverspike, R. L. Champion, and T. L. Bailey, *J. Chem. Phys.*, **45**, 4385 (1966); E. A. Gislason, B. H. Mahan, C. W. Tsao, and A. S. Werner, *J. Chem. Phys.*, **54**, 3897 (1971); H. H. Harris and J. J. Leventhal, *J. Chem. Phys.*, **58**, 233 (1973); G. P. K. Smith and R. J. Cross, Jr., *J. Chem. Phys.*, **60**, 2125 (1974).

328. B. H. Mahan, in Ref. lf, p. 75.

329. B. H. Mahan, W. E. W. Ruska, and J. S. Winn, *J. Chem. Phys.*, **65**, 3888 (1976).

330. B. L. Donnally, T. Clapp, W. Sawyer, and M. Schultz, *Phys. Rev. Lett.*, **12**, 502 (1964).

331. P. O. Haugsjaa, R. C. Amme, and N. G. Utterback, *Phys. Rev. Lett.*, **22**, 322 (1969).

332. R. H. Neynaber and G. D. Magnuson, *J. Chem. Phys.*, **65**, 5239 (1976).

333. J. W. McGowan and R. F. Stebbings, *Appl. Opt. Suppl.*, **2**, 68 (1965).

334. M. T. Bowers and J. B. Laudenslager, in George Bekefi, Ed., *High Power Gas Lasers*, Wiley, New York, 1976.

335. R. Marx, in Ref. 1f, p. 563.

336. D. Pretzer, B. Van Zyl, and R. Geballe, *Phys. Rev. Lett.*, **10**, 340 (1963).

337. M. Lipeles, R. Novick, and N. Tolk, *Phys. Rev. Lett.*, **15**, 815 (1965).

338. M. Lipeles, R. D. Swift, M. S. Longmuire, and M. P. Weinreb, *Phys. Rev. Lett.*, **24**, 799 (1970); M. Lipeles, *Phys. Rev.*, **A4**, 140 (1971).

339. D. Jaecks, F. J. deHeer, and A. Salop, *Physica*, **36**, 606 (1967).

340. F. J. deHeer, B. F. J. Luyken, D. Jaecks, and L. Wolterbeek Muller, *Physika*, **41**, 588 (1969).

341. E. G. Jones, B. M. Hughes, D. C. Fee, and T. O. Tiernan, *Phys. Rev.*, **A15**, 1446 (1977).

342. H. Schlumbohm, *Z. Naturforsch.*, **23a**, 970 (1968).

343. R. C. Isler and R. D. Nathan, *Phys. Rev. Lett.*, **25**, 3 (1970).

344. R. C. Isler, *Phys. Rev.*, **A10**, 117 (1974).

345. R. C. Isler, *Phys. Rev.*, **A10**, 2093 (1974).

346. B. M. Hughes, E. G. Jones, and T. O. Tiernan, *Bull. Am. Phys. Soc.*, **19**, 156 (1974).

347. F. T. Smith, H. H. Fleischmann, and R. A. Young, *Phys. Rev.*, **A2**, 379 (1970).

348. R. L. Champion and L. D. Doverspike, *J. Phys.*, **B2**, 1353 (1969).

349. W. Lichten, *Phys. Rev.*, **139**, A27 (1965).

350. W. B. Maier, II, *Phys. Rev.*, **A5**, 1256 (1972).

351. R. C. Jensen, G. J. Collins, and W. R. Bennett, Jr., *Phys. Rev. Lett.*, **23**, 363 (1969).

352. C. E. Webb, A. R. Turner-Smith, and J. M. Green, *J. Phys.*, **B3**, L135 (1970).

353. A. R. Turner-Smith and J. M. Green, *J. Phys.*, **B6**, 114 (1973).

354. G. J. Collins, R. C. Jensen, and W. R. Bennett, Jr., *Appl. Phys. Lett.*, **18**, 282 (1971).

355. G. J. Collins, R. C. Jensen, and W. R. Bennett, Jr., *Appl. Phys. Lett.*, **19**, 125 (1971).

356. G. J. Collins, *J. Appl. Phys.*, **46**, 1412 (1975).

357. G. J. Collins, *Appl. Phys. Lett.*, **24**, 477 (1974).

358. W. K. Schuebel, *Appl. Phys. Lett.*, **16**, 470 (1970).

359. G. R. Fowles and W. T. Silfvast, *J. Quantum Electron.*, *QE*-1, 131 (1965).

360. H. Kano, T. Shay, and G. J. Collins, *Appl. Phys. Lett.*, **27**, 610 (1975).

361. T. Shay, H. Kano, and G. J. Collins, *Appl. Phys. Lett.*, **26**, 531 (1975).

362. W. T. Silfvast and M. B. Klein, *Appl. Phys. Lett.*, **17**, 400 (1970).

363. M. B. Klein and W. T. Silfvast, *Appl. Phys. Lett.*, **18**, 482 (1970).

364. R. D. Reid, J. R. McNeil, and G. J. Collins, *Appl. Phys. Lett.*, **29**, 666 (1976).

365. R. C. Isler and L. E. Murray, *Phys. Rev.*, **A13**, 2087 (1976).

366. R. C. Isler and L. E. Murray, private communication.

367. G. J. Collins, *J. Appl. Phys.*, **42**, 3812 (1971).

368. S. H. Neff, *Astrophys. J.*, **140**, 348 (1966).

369. J. Weiner, W. B. Peatman, and R. S. Berry, *Phys. Rev.*, **A4**, 1824 (1971).

370. D. J. Dyson, *Nature*, **207**, 361 (1965).

371. R. Johnsen, M. T. Leu, and M. A. Biondi, *Phys. Rev.*, **A8**, 1808 (1973).

372. W. E. Bell, *Appl. Phys. Lett.*, **4**, 34 (1964); A. L. Bloom, W. E. Bell, and F. O. Lopez, *Phys. Rev.*, **135**, A578 (1964).

373. C. F. Melius, *J. Phys.*, **B7**, 1692 (1974).

374. G. H. Dunn, R. Geballe, and D. Pretzer, *Phys. Rev.*, **128**, 2200 (1962).

375. B. Van Zyl, D. Jaecks, D. Pretzer, and R. Geballe, *Phys. Rev.*, **158**, 29 (1967).

376. R. H. Hughes, S. Lin, and L. L. Hatfield, *Phys. Rev.*, **130**, 2318 (1963).

377. G. N. Polyakova, V. A. Gusev, V. F. Erko, Ya. M. Fogel, and A. V. Zats, *Sov. Phys. JETP*, **31**, 637 (1970).

378. M. Hollstein, A. Salop, J. R. Peterson, and D. C. Lorents, *Phys. Lett.*, **32A**, 327 (1970).

379. R. C. Isler and R. D. Nathan, *Phys. Rev.*, **A6**, 1036 (1973).

380. R. D. Nathan and R. C. Isler, *Phys. Rev. Lett.*, **26**, 1091 (1971).

381. E. G. Jones, B. M. Hughes, D. G. Hopper, and T. O. Tiernan, American Society for Mass Spectrometry, Twenty-third Annual Conference, Houston, May 1975; D. G. Hopper, A. C. Wahl, E. G. Jones, B. M. Hughes, and T. O. Tiernan, 29th Gaseous Electronics Conference, Cleveland, October 1976.

382. B. L. Blaney and R. S. Berry, *Phys. Rev.*, **A13**, 1034 (1976).

383. R. Odom, J. Caddick, and J. Weiner, *Phys. Rev.*, **A14**, 965 (1976).

384. N. P. Carleton and T. R. Lawrence, *Phys. Rev.*, **109**, 1159 (1958).

385. J. R. Sheridan and K. C. Clark, *Phys. Rev.*, **A140**, 1033 (1965).

386. E. C. Y. Inn, *Planet, Sci.*, **15**, 19 (1967) and references cited therein.

387. D. L. Albritton, A. L. Schmeltekopf, and E. E. Ferguson, in I. Amdur, Ed., *VI ICPEAC*, Vol I, MIT Press, Cambridge, Mass., 1969, p. 331.

388. T. R. Govers, F. C. Fehsenfeld, D. L. Albritton, P. G. Fournier, and J. Fournier, *Chem. Phys. Lett.*, **26**, 134 (1974).

389. R. Marx, M. Gerard, T. R. Govers, and G. Mauclaire, "Luminescence in Near Thermal Charge Exchange," paper presented at Seventh International Mass Spectrometry Conference, Florence, 1976.

390. R. F. Holland and W. B. Maier, II, *J. Chem. Phys.*, **55**, 1299 (1971).

391. G. H. Bearman, J. D. Earl, R. J. Pieper, H. H. Harris, and J. J. Leventhal, *Phys. Rev.*, **A13**, 1734 (1976).

392. J. J. Leventhal, J. D. Earl, and H. H. Harris, *Phys. Rev. Lett.*, **35**, 719 (1975).

393. C. Y. Fan, *Phys. Rev.*, **103**, 1740 (1956).

394. G. N. Polyakova, Ya. M. Fogel, V. F. Erko, A. V. Zats, and A. G. Tolstolutskii, *Sov. Phys. JETP*, **27**, 201 (1968).

395. J. H. Moore, Jr., and J. P. Doering, *Phys. Rev.*, **182**, 176 (1969).

396. V. A. Gusev, G. N. Polyakova, and Ya. M. Fogel, *Sov. Phys. JETP*, **28**, 1126 (1969).

397. G. H. Saban and T. F. Moran, *J. Chem. Phys.*, **57**, 895 (1972); **57**, 5622 (1972).

398. H. Schlumbohm, *Z. Natuforsch.*, **23a**, 1386 (1968).

399. R. P. Lowe and H. I. S. Ferguson, *Can. J. Phys.*, **49**, 1680 (1971).

400. T. T. Kassal and E. S. Fishburne, *J. Chem. Phys.*, **54**, 1363 (1971).

401. M. A. Coplan and K. W. Ogilvie, *J. Chem. Phys.*, **61**, 2010 (1974).

402. G. N. Polyakova, V. F. Erko, A. V. Zats, Ya. M. Fogel, and G. D. Tolstolutskaya, *Sov. Phys. JETP Lett.*, **11**, 390 (1970).

403. M. Lipeles, *J. Chem. Phys.*, **51**, 1252 (1969).

404. V. A. Gusev, G. N. Polyakova, V. F. Erko, A. V. Zats, A. A. Oksyuk, and Ya. M. Fogel, *Sov. Phys. JETP*, **33**, 863 (1971).

405. M. J. Haugh and K. D. Bayes, *Phys. Rev.*, **A2**, 1778 (1970).

406. M. Haugh, T. G. Slanger, and K. D. Bayes, *J. Chem. Phys.*, **44**, 837 (1966).

407. F. J. LeBlanc, *J. Chem. Phys.*, **38**, 487 (1963).

408. H. H. Harris, M. G. Crowley, and J. J. Leventhal, *Chem. Phys. Lett.*, **29**, 540 (1974).

409. M. J. Haugh, *J. Chem. Phys.*, **59**, 37 (1973).

410. M. J. Haugh, B. S. Schneider, and A. L. Smith, *J. Molec. Spectrosc.*, **51**, 123 (1974).

411. L. Tomcho and M. J. Haugh, *J. Chem. Phys.*, **56**, 6089 (1972).

412. K. E. Siegenthaler, B. M. Hughes, E. G. Jones, and T. O. Tiernan, 24th Annual Conference on Mass Spectrometry and Allied Topics, ASMS, San Diego, May 1976.

413. J. P. Doering, in T. R. Govers and F. J. deHeer, Eds., *The Physics of Electronic and Atomic Collisions, VII ICPEAC*, North-Holland, Amsterdam, 1972, p. 341.

414. N. G. Utterback and H. P. Broida, *Phys. Rev. Lett.*, **15**, 608 (1965).

415. G. N. Polyakova, V. F. Erko, A. V. Zats, Ya. M. Fogel, and G. D. Tolstolutskaya, *Sov. Phys. JETP Lett.*, **12**, 204 (1970).

416. D. Rapp and W. E. Francis, *J. Chem. Phys.*, **33**, 179 (1960).

417. J. J. Leventhal, private communication, 1976; J. D. Kelley, G. H. Bearman, F. Ranjbar, H. H. Harris, and J. J. Leventhal, "Vibrational Excitation in $N_2^+(B\,^2\Sigma_u^+)$ Produced by $N_2^+(X\,^2\Sigma_g^+)$–Rare Gas Collisions," paper presented at the 173rd ACS National Meeting, New Orleans, March 1977.

418. C. Liu, *J. Chem. Phys.*, **53**, 1295 (1970) and references cited therein.

419. T. F. Moran and D. C. Fullerton, *J. Chem. Phys.*, **56**, 21 (1972).

420. M. J. Haugh and J. H. Birely, *J. Chem. Phys.*, **60**, 264 (1974).

421. H. Bregman-Reisler and J. P. Doering, *Chem. Phys. Lett.*, **27**, 199 (1974).

422. H. Bregman-Reisler and J. P. Doering, *J. Chem. Phys.*, **62**, 3109 (1975).

423. W. Sim and M. Haugh, *J. Chem. Phys.*, **65**, 1616 (1976).

424. C. B. Collins and W. W. Robertson, *J. Chem. Phys.*, **40**, 701 (1964).

425. M. Gérard, T. R. Govers, C. A. van de Runstraat, and R. Marx, *Chem. Phys. Lett.*, **44**, 154 (1976).

426. J. H. Moore, Jr., and J. P. Doering, *Phys. Rev.*, **174**, 178 (1968).

427. J. P. Doering, *Phys. Rev.*, **133**, A1537 (1964).

428. W. R. Bennett, Jr., *Ann. Phys.*, **18**, 367 (1962).

429. C. B. Collins, W. B. Hurt, and W. W. Robertson, in M. R. C. McDowell, Ed., *Atomic Collision Processes*, North Holland, Amsterdam, 1964, p. 517.

430. W. W. Robertson, *J. Chem. Phys.*, **44**, 2456 (1966).

431. C. B. Collins, A. J. Cunningham, S. M. Curry, B. M. Johnson, and M. Stockton, *Appl. Phys. Lett.*, **24**, 245, 477 (1974).

432. C. B. Collins and A. J. Cunningham, *Appl. Phys. Lett.*, **27**, 127 (1975).

433. J. J. Leventhal, J. D. Earl, and H. H. Harris, *Phys. Rev. Lett.*, **35**, 719 (1975).

434. E. Graham, IV, M. A. Biondi, and R. Johnsen, "Spectroscopic Studies of the Charge Transfer Reactions: $He^+ + Hg \rightarrow He + Hg^{+*}$ and $He_2^+ + N_2 \rightarrow 2He + N_2^{+*}$ at Thermal Energy," private communication, 1976.

435. R. A. Waller, C. B. Collins, and A. J. Cunningham, *Appl. Phys. Lett.*, **27**, 323 (1975).

436. G. H. Bearman, H. H. Harris, and J. J. Leventhal, *Appl. Phys. Lett.*, **28**, 345 (1976).

437. G. H. Bearman, J. D. Earl, H. H. Harris, and J. J. Leventhal, *Appl. Phys. Lett.*, **29**, 108 (1976).

438. G. H. Bearman, F. Ranjbar, H. H. Harris, and J. J. Leventhal, *Chem. Phys. Lett.*, **42**, 335 (1976).

439. J. H. Moore, Jr., and J. P. Doering, *Phys. Rev.*, **182**, 176 (1969).

440. D. K. Bohme, N. G. Adams, M. Mosesman, D. B. Dunkin, and E. E. Ferguson, *J. Chem. Phys.*, **52**, 5094 (1970).

441. D. Brandt and C. Ottinger, *Chem. Phys. Lett.*, **23**, 257 (1973).

442. J. Appell, D. Brandt, and C. Ottinger, *Chem. Phys. Lett.*, **33**, 131 (1975).

443. H. H. Harris, M. G. Crowley, and J. J. Leventhal, *Phys. Rev. Lett.*, **34**, 67 (1975).

444. I. Kusunoki, C. Ottinger, and J. Simonis, *Chem. Phys. Lett.*, **41**, 601 (1976).

445. M. Menzinger and D. J. Wren, *Chem. Phys. Lett.*, **18**, 431 (1973).

446. (a) J. C. Light, *J. Chem. Phys.*, **40**, 3221 (1964); (b) *Discuss. Faraday Soc.* **44**, 14 (1968).

447. J. C. Light and J. Lin, *J. Chem. Phys.*, **43**, 3209 (1965).

448. D. G. Truhlar, *J. Chem. Phys.*, **56**, 1481 (1972).

449. A. F. Wagner and D. G. Truhlar, *J. Chem. Phys.*, **57**, 4063 (1972).

450. B. H. Mahan, *J. Chem. Phys.*, **55**, 1436 (1971).

451. B. H. Mahan, *Acc. Chem. Res.*, **8**, 55 (1975).

452. C. Edmiston, J. Doolittle, K. Murphy, K. C. Tang, and W. Willson, *J. Chem. Phys.*, **52**, 3419 (1970).

453. P. J. Brown and E. F. Hayes, *J. Chem. Phys.*, **55**, 922 (1971).

454. P. J. Kuntz, *Chem. Phys. Lett.*, **16**, 581 (1972).

455. B. H. Mahan, *J. Chem. Educ.*, **51**, 377 (1974).

456. (a) G. R. North and J. J. Leventhal, *Chem. Phys. Lett.*, **23**, 600 (1973); (b) G. R. North, H. H. Harris, J. J. Leventhal, and P. B. James, *J. Chem. Phys.*, **61**, 5060 (1974).

457. P. J. Kuntz and W. N. Whitton, *Chem. Phys. Lett.*, **34**, 340 (1975).

458. W. N. Whitton and P. J. Kuntz. *J. Chem. Phys.*, **64**, 3624 (1976).

459. N. Sathyamurthy, R. Rangarajan, and L. M. Raff, *J. Chem. Phys.*, **64**, 4606 (1976).

460. (a) F. Schneider, U. Havemann, and L. Zülicke, *Z. Physik. Chem. (Leips.)*, **256**, 773 (1975); (b) F. Schneider, U. Havemann, L. Zülicke, V. Pacák, K. Birkinshaw, and Z. Herman, *IXth ICPEAC*, Seattle, 1975, p. 587; *Chem. Phys. Lett.*, **37**, 323 (1976).

461. D. J. Kouri and M. Baer, *Chem. Phys. Lett.*, **24**, 37 (1974).

462. W. G. Cooper, N. S. Evers, and D. J. Kouri, *Molec. Phys.*, **27**, 707 (1974).

463. J. T. Adams, *Chem. Phys. Lett.*, **33**, 275 (1975).

464. F. M. Chapman, Jr., and E. F. Hayes, *J. Chem. Phys.*, **62**, 4400 (1975).

465. C. Zuhrt, F. Schneider, and L. Zülicke, *Chem. Phys. Lett.*, **43**, 571 (1976).

466. P. J. Robinson and K. A. Holbrook, *Unimolecular Reactions*, Wiley-Interscience, London, 1972.

467. (a) H. M. Rosenstock, M. B. Wallenstein, A. L. Wahrhaftig, and H. Eyring, *Proc. Nat. Acad. Sci. (USA)*, **38**, 667 (1952); (b) H. M. Rosenstock and M. Krauss, in F. W. McLafferty Ed., *Mass Spectrometry of Organic Ions*, Academic, New York, 1963; (c) H. M. Rosenstock, *Adv. Mass Spectrom.* **2**, 251 (1963); (d) H. M. Rosenstock, *Adv. Mass Spectrom.*, **4**, 523 (1968); (e) M. L. Vestal, in P. Ausloos, Ed., *Fundamental Processes in Radiation Chemistry*, Wiley, New York, 1968; (f) A. L. Wahrhaftig, in A. Maccoll, Ed., *Mass Spectrometry, MTP International Review of Science*, Vol. 5, Butterworths, London, 1972, p. 1.

468. C. Lifshitz, *Adv. Mass Spectrom.*, **7** (1977).

469. F. A. Wolf, *J. Chem. Phys.*, **44**, 1619 (1966).

470. P. F. Knewstubb, in Ref. 1f, p. 139.

471. C. E. Klots; (a) *J. Chem. Phys.*, **41**, 117 (1964); (b) *J. Phys. Chem.*, **75**, 1526 (1971); (c) *Z. Naturforsch.*, **27a**, 553 (1972); (d) *J. Chem. Phys.*, **58**, 5364 (1973); (e) *Adv. Mass Spectrom.*, **6**, 969 (1973); (f) *Chem. Phys. Lett.*, **38**, 61 (1976); (g) *J. Chem. Phys.*, **64**, 4269 (1976).

472. J. Chesnavich and M. T. Bowers; (a) *J. Am. Chem. Soc.*, **98**, 8301 (1976); (b) *J. Am. Chem. Soc.*, **99**, 1705 (1977).

473. S. E. Buttrill, Jr., *J. Chem. Phys.*, **52**, 6174 (1970).

474. C. Lifshitz and T. O. Tiernan, *J. Chem. Phys.*, **51**, 1515 (1972).

475. (a) S. A. Safron, N. D. Weinstein, D. R. Herschbach, and J. C. Tully, *Chem. Phys. Lett.*, **12**, 564 (1972); (b) A. Lee, R. L. Leroy, Z. Herman, R. Wolfgang, and J. C. Tully, *Chem. Phys. Lett.*, **12**, 569 (1972).

476. C. Lifshitz and T. O. Tiernan; (a) *J. Chem. Phys.*, **57**, 1515 (1972); (b) *J. Chem. Phys.*, **59**, 6143 (1973).

477. R. D. Levine and R. B. Benstein, *Acc. Chem. Res.*, **7**, 393 (1974).

478. K. E. Shuler, *J. Chem. Phys.*, **21**, 624 (1953).

479. B. H. Mahan, *J. Chem. Phys.*, **55**, 1436 (1971).

480. J. A. Fair and B. H. Mahan, *J. Chem. Phys.*, **62**, 515 (1975).

481. A. Pipano and J. J. Kaufman, *J. Chem. Phys.*, **56**, 5258 (1972).

482. D. G. Hopper, *Chem. Phys. Lett.*, **31**, 446 (1975).

483. D. G. Hopper and T. O. Tiernan, paper No. E-4, 22nd ASMS Conference, Philadelphia, 1974.

484. D. G. Hopper, "Mechanisms of the Reaction of Positive Atomic Oxygen Ions with Nitrogen," private communication.

485. (a) J. H. Black and A. Dalgarno, *Astrophys. Lett.*, **15**, 79 (1973); (b) E. Herbst and W. Klemperer, *Physics Today*, **29**, 32 (1976).

486. D. H. Liskow, C. F. Bender, and H. F. Schaefer, III, *J. Chem. Phys.*, **61**, 2507 (1974).

487. P. K. Pearson and E. Roueff, *J. Chem. Phys.*, **64**, 1240 (1976).

488. M. Krauss, D. G. Hopper, P. J. Fortune, A. C. Wahl, B. J. Rosenberg, W. B. England, G. Das, and T. O. Tiernan, ARL report No. TR 75-0202, Vols. I and II, Air Force Systems Command, June 1975 (available from the National Technical Information Service, Springfield, Va. 22151).

489. P. J. Fortune, B. J. Rosenberg, and A. C. Wahl, *J. Chem. Phys.*, **65**, 2201 (1976).

490. (a) I. G. Csizmadia, R. E. Kari, J. C. Polanyi, A. C. Roach, and M. A. Robb, *J. Chem. Phys.*, **52**, 6205 (1970); (b) C. W. Bauschlicher, Jr., S. V. O'Neill, R. K. Preston, H. F. Schaefer, III, and C. F. Bender, *J. Chem. Phys.*, **59**, 1286 (1973); (c) G. D. Carney and R. N. Porter, *J. Chem. Phys.*, **60**, 4251 (1974).

491. J. Schaefer and W. A. Lester, Jr., *J. Chem. Phys.*, **62**, 1913 (1975).

492. R. T. Skodje, W. R. Gentry, and C. F. Giese, *J. Chem. Phys.*, **65**, 5532 (1976).

493. J. R. Krenos, K. K. Lehmann, J. C. Tully, P. M. Hierl, and G. P. Smith, *Chem. Phys.*, **16**, 109 (1976).

494. R. D. Poshusta and D. F. Zetik, *J. Chem. Phys.*, **58**, 118 (1973).

495. P. J. Kuntz and W. N. Whitton, *Chem. Phys.*, **16**, 301 (1976).

496. M. Karplus, R. N. Porter, and R. D. Sharma, *J. Chem. Phys.*, **43**, 3259 (1965).

497. (a) D. L. Bunker, *Meth. Comput. Phys.*, **10**, 287 (1971); (b) J. C. Polanyi, *Acc. Chem. Res.*, **5**, 161 (1972); (c) R. N. Porter, *Annu. Rev. Phys. Chem.*, **25**, 317 (1974); (d) D. L. Thompson, *Acc. Chem. Res.*, **9**, 338 (1976); (e) P. J. Kuntz, in Ref. 1f, p. 123.

498. (a) J. C. Tully and R. K. Preston, *J. Chem. Phys.*, **55**, 562 (1971); (b) J. C. Tully, *Ber. Bunsenges. Phys. Chem.*, **77**, 557 (1973).

499. J. R. Stine and J. T. Muckerman, *J. Chem. Phys.*, **65**, 3975 (1976).

500. E. Herbst and J. B. Delos, *Chem. Phys. Lett.*, **42**, 54 (1976).

CHAPTER THREE

ELECTRONIC STRUCTURE OF EXCITED STATES OF SELECTED ATMOSPHERIC SYSTEMS

H. Harvey Michels

United Technologies Research Center

Contents

I. INTRODUCTION   227

II. ELECTRONIC STRUCTURE CALCULATIONS   228
  A. Electronic States and Wave functions   229
  B. Born–Oppenheimer Separation   231
  C. Variational Methods   232
  D. Potential-energy Curves and Surfaces   239

III. ELECTRONIC STRUCTURE AND POTENTIAL-ENERGY CURVES   240
  A. Nitrogen Molecule   241
  B. Oxygen Molecule   265
  C. Nitric Oxide Molecule   288
  D. O_2^- Ion   301
  E. NO^+ Ion   318

I. INTRODUCTION

To evaluate the thermodynamic and radiation properties of a natural or perturbed state of the upper atmosphere or ionosphere, the thermal and transport properties of heated air are required. Such properties are also of particular interest in plasma physics, in gas laser systems, and in basic studies of airglow and the aurora. In the latter area the release of certain chemical species into the upper atmosphere results in luminous clouds that display the resonance electronic–vibrational–rotational spectrum of the released species. Such spectra are seen in rocket releases of chemicals for upper-atmosphere studies and on reentry into the atmosphere of artificial satellites. Of particular interest in this connection are the observed spectra of certain metallic oxides and air diatomic species. From band-intensity distribution of the spectra and knowledge of the f-values for electronic and vibrational transitions, the local conditions of the atmosphere can be determined.[1]

Present theoretical efforts that are directed toward a more complete and realistic analysis of the transport equations governing atmospheric relaxation and the propagation of artificial disturbances require detailed information of thermal opacities and long-wave infrared (LWIR) absorption in regions of temperature and pressure where molecular effects are important.[2,3] Although various experimental techniques have been employed for both atomic and molecular systems, theoretical studies have been largely confined to an analysis of the properties (bound–bound, bound–free, and free–free) of atomic systems.[4,5] This is mostly a consequence of the unavailability of reliable wave functions for diatomic molecular systems, and particularly for excited states or states of open-shell structures. More recently,[6-9] reliable theoretical procedures have been prescribed for such systems that have resulted in the development of practical computational programs.

The theoretical analysis of atmospheric reactions requires knowledge of the electronic structure of the ground and low-lying excited states of atoms, ions, and small molecular clusters of nitrogen and oxygen and, in certain regions, the interaction of water or other small molecules with these clusters. Because of the computational complexity for systems with large numbers of electrons, traditional *ab initio* theoretical methods are difficult and expensive to apply. However, no clear alternative is currently on the horizon, although progress continues in semiempirical and perturbative approaches to the calculation of electronic structure. Such approaches have met with limited success in applications to electronically excited states of molecules. *Ab initio* computational programs based on the variational theorem, incorporating analytic basis functions and coupled with a configuration interaction analysis, remain the backbone of our techniques, for studies to chemical accuracy, which are applicable to systems such as the atmospheric diatomic molecules and ions.

The potential curves derived from such calculations can often be empirically improved by comparison with so-called experimental curves derived from observed spectroscopic data, using Rydberg–Klein–Rees (RKR) or other inversion procedures. It is often found, particularly for the atmospheric systems, that the remaining correlation errors in a configuration interaction (CI) calculation are similar for many excited electronic states of the same symmetry or principal molecular-orbital description. Thus it is often possible to calibrate an entire family of calculated excited-state potential curves to near-spectroscopic accuracy. Such a procedure has been applied to the systems described here.

The particular choice of the atmospheric systems to be presented here proved to be somewhat problematic. Calculations for the N_2 molecule, including all low-lying valence and Rydberg states, had been performed several years ago,[10] as had a fairly extensive set of calculations for O_2^-.[11] Gilmore has described the known spectroscopic states for N_2, O_2, NO, and several states of their corresponding positive and negative ions.[12] A complete analysis of all the valence and low-lying Rydberg states for N_2, O_2, and NO including ions seemed too formidable (and expensive) a task, so the systems N_2, O_2, NO, O_2^-, and NO^+ have been selected for inclusion in this review. This choice includes the most important atmospheric positive ion, NO^+, and the chemically most interesting negative ion, O_2^-.

The general composition of this chapter is as follows. We first present a critical review of the current status of electronic-structure calculations for molecular systems. This is followed by a compilation of the potential-energy curves, derived spectroscopic analysis, and pertinent discussion of the selected atmospheric molecules mentioned in the preceding paragraph.

II. ELECTRONIC-STRUCTURE CALCULATIONS

The application of quantum-mechanical methods to the prediction of electronic structure has yielded much detailed information about atomic and molecular properties.[13] Particularly in the past few years, the availability of high-speed computers with large storage capacities has made it possible to examine both atomic and molecular systems using an *ab initio* variational approach wherein no empirical parameters are employed.[14] Variational calculations for molecules employ a Hamiltonian based on the nonrelativistic electrostatic nuclei–electron interaction and a wave function formed by antisymmetrizing a suitable many-electron function of spatial and spin coordinates. For most applications it is also necessary that the wave function represent a particular spin eigenstate and that it have appropriate geometric symmetry.

In addition, we assume, for the systems of interest here, that the electronic motion is fast relative to the kinetic motion of the nuclei and that the total wave functions can be separated into a product form, with one term depending on the electronic motion and parametric in the nuclear coordinates and a second term describing the nuclear motion in terms of adiabatic potential hypersurfaces. This separation, based on the relative mass and velocity of an electron as compared with the nucleus mass and velocity, is known as the Born–Oppenheimer approximation.

The specific computational methods for electronic-structure calculations may be broken down into three main classifications. First there are the variational methods, which can be grouped into two categories. The first category is comprised of the self-consistent-field (SCF) procedures, which are based on wave functions consisting of a single term. Self-consistent-field wave functions are characterized by the requirement that the spatial orbitals be chosen so as to optimize the energy, normally with certain occupancy and symmetry restrictions. The second category includes variational methods based on configuration interaction (CI) wave functions, that is, on wave functions that are a linear combination of terms differing in either spatial or spin occupancy, with each separate term referred to as a *configuration*. Among the CI methods we may distinguish between those designed to optimize the description of a single electronic state and those applicable to the description of manifolds of states. A second class of methods consists of those based on many-body perturbation theory or on the use of Green's functions. These approaches have not yet been developed to a degree comparable with that of the SCF and CI methods, but they hold promise for the future and are worth serious consideration. A third class of methods is derived either as an extension of the original Hückel type of approach or from approximate solid-state models. Here certain difficult integrals are only approximately evaluated and/or a nonrigorous Hamiltonian employed.

A brief comment on the accuracy of current approaches seems in order. For atomic systems in the first or second row of the periodic table, quantitative ($\sim 0.01 \, \text{eV}$) studies have been carried out. For diatomic systems, constructed from these atoms, an accuracy of ~ 1 kcal in the potential curves can be realized. For polyatomic systems the situation is less clear because of the great increase in computational difficulty. Accuracy of 5 to 10 kcal can probably be achieved for simple potential-energy surfaces, although very few surfaces have been examined in detail.

A. Electronic States and Wave Functions

Nearly all the electronic-structure calculations performed to date are based on the use of one-electron orbitals and are of two types: Hartree–Fock or configuration interaction.[15] Hartree–Fock calculations are based on a single assignment of electrons to spatial orbitals, following which the spatial orbitals are optimized, usually subject to certain restrictions. Almost all Hartree–Fock calculations have been subject to the assumption that the spatial orbitals are all doubly occupied as nearly as possible and are all of definite geometric symmetry. These restrictions define the conventional, or restricted, Hartree–Fock (RHF) method.[16,17] Restricted Hartree–Fock calculations can be made with relatively large Slater-type orbital (STO) basis sets for diatomic molecules with first- or second-row atoms, and the results are convergent in the sense that they are insensitive to basis enlargement. Polyatomic systems have for the most part been examined using Gaussian-type orbitals. The RHF model is adequate to give a qualitatively correct description of the electron interaction in many systems and in favorable cases can yield equilibrium interatomic separations and force constants. However, the double-occupancy restriction renders the RHF method inappropriate in a number of circumstances of practical interest. In particular, it cannot provide potential curves or surfaces for molecules dissociating into odd-electron atoms (e.g., NO at large internuclear separation) or into atoms having less electron pairing than the original molecule [e.g., $O_2\,^3\Sigma_g^- \rightarrow O(^3P)$]; it cannot handle excited states having unpaired electrons (e.g., the $^3\Sigma_u^-$ state of O_2 responsible for the Schumann–Runge bands); and, in general, it gives misleading results for molecules in which the extent of electron correlation changes with internuclear separation.

Configuration-interaction (CI) methods have the advantage of avoiding the limitations of the RHF calculations. If configurations not restricted to doubly-occupied orbitals are included, a CI can, in principle, converge to an exact wave function for the customary Hamiltonian. However, many CI calculations have in fact been based on a restriction to doubly-occupied orbitals and thus retain many of the disadvantages of the RHF method.[15] The use of general CI formulations involves three considerations, all of which have been satisfactorily investigated: the choice of basis orbitals, the choice of configurations (sets of orbital assignments), and the specific calculations needed to make wave functions describing pure spin states.[6] The first and second considerations are the art associated with quantum-mechanical electronic-structure calculations. Many methods [iterative natural spin orbitals (INSO), perturbation selection, first-order CI, etc.] have been advocated for the optimum choice of configurations. There are no firm rules at present, and the optimum choice is a strong function of the insight of the particular research investigator. This last consideration has proved difficult to implement into computer programs! The CI method has been found to be valuable in handling excited states and dissociation processes that cannot be treated with RHF techniques.

Either of the methods for *ab initio* calculations described in the preceding paragraph reduces in practice to a series of steps, the most important of which are the evaluation of molecular integrals, the construction of matrix elements of the Hamiltonian, and the optimization of molecular orbitals (RHF) or configuration coefficients (CI). For most molecules these steps are all comparable in their computing time, so that a point has been reached where there is no longer any one bottleneck determining computation speed. In short, the integral evaluation involves the use of ellipsoidal or spherical coordinates and the introduction of the Neumann or Laplace expansion for the interelectronic repulsion potential;[18] the matrix-element construction depends on an analysis of the algebra of spin eigenfunctions,[19] and the orbital or configuration optimization can be carried out by eigenvalue techniques.[20] All the steps have by now become relatively standard and can be performed efficiently on a computer having approximately 65,000 words of core storage, a cycle time in the microsecond range, and several hundred thousand words of peripheral storage.

Both the RHF and CI methods yield electronic wave functions and energies as a function of the internuclear separation, with the RHF method for one state and the CI method for all states considered. The electronic energies can be regarded as potential curves, from which equilibrium internuclear separations, dissociation energies, and constants describing vibrational and rotational motion (including anharmonic and rotational–vibrational effects) may be deduced. It is also possible to solve the Schrödinger equation for the motion of the nuclei subject to the potential curves to obtain vibrational wave functions for use in transition-probability calculations. The electronic wave functions themselves can be used to estimate dipole moments of individual electronic states, transition moments between different electronic states, and other properties. Although all of the calculations described here have been carried out on some systems, the unavailability of good electronic wave functions and potential curves has limited actual studies of most of these properties to a small number of molecules.

B. Born–Oppenheimer Separation

For a system of n electrons and N nuclei, and considering only electrostatic interactions between the particles, we have for the total Hamiltonian

$$\mathcal{H} = \mathcal{H}_{el} - \sum \frac{\hbar^2}{2m_\alpha} \nabla_\alpha^2$$

$$+ \frac{\hbar^2}{2M_{total}} \left[\sum_{\substack{\beta=1 \\ \alpha \neq \beta}}^{N} \sum_{\alpha=1}^{N} \nabla_\alpha \cdot \nabla_\beta + 2 \sum_{\alpha=1}^{N} \sum_{i=1}^{n} \nabla_\alpha \cdot \nabla_i + \sum_{i=1}^{n} \sum_{\substack{j=1 \\ i \neq j}}^{n} \nabla_i \cdot \nabla_j \right] \quad (II.1)$$

where

$$\mathcal{H}_{\text{el}} = -\frac{\hbar^2}{2m_e} \sum_{i=1}^{n} \nabla_i^2 + V^{\text{el}}(\mathbf{r}_n, \mathbf{R}_N) \qquad (\text{II.2})$$

where m_e, m_α, and M_{total} are the masses of the electron, atom α, and combined system mass, respectively. Now since the ratios m_e/m_α and m_e/M_{total} are both small ($2 \cdot 10^{-6} - 5 \cdot 10^{-4}$), we can effect a separation of the electronic and nuclear coordinates, treating the total wave function as a product of a nuclear and an electronic part. We have

$$\psi(\mathbf{r}_n, \mathbf{R}_N) = \sum_k \chi_k(\mathbf{R}_n) \psi_k(\mathbf{r}_n, \mathbf{R}_N) \qquad (\text{II.3})$$

where $\psi_k(\mathbf{r}_n, \mathbf{R}_N)$ is an electronic wave function parametric in the nuclear coordinates as given in equation (II.3) and $\chi_k(\mathbf{R}_N)$ are nuclear-motion wave functions that satisfy (neglecting terms of the order of m_e/M_α)

$$\left[-\sum_{\alpha=1}^{N} \frac{\hbar^2}{2m_\alpha} \nabla_\alpha^2 + \frac{\hbar^2}{2M_{\text{total}}} \sum_{\substack{\alpha=1 \\ \alpha \neq \beta}}^{N} \sum_{\beta=1}^{N} \nabla_\alpha \cdot \nabla_\beta + V^{\text{el}}(\mathbf{r}_n, \mathbf{R}_N) \right] \chi_k = i\hbar \frac{\partial X_k}{\partial t}$$

$$(\text{II.4})$$

The cross term in $\nabla_\alpha \cdot \nabla_\beta$ can be eliminated by a proper change of variables, and equation (II.4) then reduces to a $(3N-3)$-dimensional Schrödinger equation.

For most systems, where the velocity of motion of the nuclei is slow relative to the electron velocity, this decoupling of electronic and nuclear motion is valid and is referred to as the *adiabatic approximation*. Equation (II.3) thus defines an electronic eigenstate $\psi_k(\mathbf{r}_n, \mathbf{R}_N)$, parametric in the nuclear coordinates, and a corresponding eigenvalue $E_k(\mathbf{R}_N)$ that is taken to represent the potential-energy curve or surface corresponding to state k.

C. Variational Methods

1. General

An *ab initio* method is one that starts from a zero-order Hamiltonian that is exact except for relativistic and magnetic effects and that involves the evaluation of electronic energies and other relevant quantities for wave functions that are properly antisymmetrized in the coordinates of all the electrons. For a system containing n electrons and M nuclei, the zero-order Hamiltonian depends parametrically on the nuclear positions and is of the form

$$\mathfrak{K} = -\frac{1}{2} \sum_{i=1}^{n} \nabla_i^2 - \sum_{i=1}^{n} \sum_{j=1}^{M} \frac{Z_j}{|\mathbf{r}_i - \mathbf{R}_j|} + \sum_{1 \leq i < j}^{M} \frac{Z_i Z_j}{|\mathbf{R}_i - \mathbf{R}_j|} + \sum_{1 \leq i < j} \frac{1}{|\mathbf{r}_i - \mathbf{r}_j|}$$

$$(\text{II.5})$$

where Z_i and $\mathbf{R}_i$ are the charge and position of nucleus i, $\mathbf{r}_j$ is the position of electron j, and ∇_j^2 is the Laplacian operator for electron j. All quantities are in atomic units, that is, with lengths in bohrs and energies in hartrees (1 hartree = 2 rydbergs). The many-electron wave function consists of one, or a linear combination, of terms of the form

$$\Psi_\mu = \mathcal{C} \prod_{i=1}^{n} \phi_{i\mu}(\mathbf{r}_i) \theta_\mu(\sigma_1, \dots, \sigma_n) \qquad (\text{II.6})$$

where $\phi_{1\mu}$, $\phi_{2\mu}$, ..., are spatial orbitals occupied in Ψ_μ, $\theta_\mu(\sigma_1, \dots, \sigma_n)$ is an n-electron spin function, and $\mathcal{C}$ is an antisymmetrizing projection operator. The spin function θ_μ may be required to be an eigenfunction of the spin operators S^2 and S_z and may, therefore, be characterized by quantum numbers S and M_S, but these quantum numbers will not suffice completely to describe θ_μ except under certain conditions that cannot always be met. The spatial orbitals $\phi_{i\mu}$ may be whatever basis orbitals have been introduced or arbitrary linear combinations thereof, or they may be specific linear combinations determined pursuant to the particular calculational method in use.

The spatial orbitals $\phi_{i\mu}$, the spin functions θ_μ, and the coefficients of different Ψ_μ if a linear combination of Ψ_μ is used, may be explicitly determined by invoking the variational principle

$$\delta \left[\frac{\int \Psi^* \mathfrak{K} \Psi \, d\tau}{\int \Psi^* \Psi \, d\tau} \right] = 0 \qquad (\text{II.7})$$

or they may be determined implicitly by the application of perturbation or other nonvariational methods. Various specific methods are described in the following sections for determining wave functions. However, we should first observe that the adequacy of an *ab initio* calculation or, for that matter, any energy calculation, will depend crucially on the extent to which the wave function can be qualitatively appropriate. Some of the considerations surrounding the choice of a wave function are as follows: (1) the necessity that the wave function possess flexibility sufficient to describe dissociation to the correct atomic and molecular fragments as various internuclear separations are increased, (2) maintenance of equivalent quality of calculation for nuclear geometries differing in the nature or number of chemical bonds, (3) ability to describe degenerate or near-degenerate electronic states when they are pertinent, (4) ability to describe different electronic states to equivalent accuracy when their interrelation (e.g., crossing) is relevant, in particular, ability to describe Rydberg–valence-state mixing, and (5) ability to represent changes in the coupling of electron spins as bonds are broken or reformed. These considerations indicate that it will often be necessary to consider wave functions with more than a minimum number of singly occupied spatial orbitals and that there will be many potential curves or surfaces for which a wave function consisting of a single Ψ_μ cannot suffice. However, it will then be necessary to allow mixing of Ψ_μ with different degrees of orbital spatial occupancy so as to obtain smooth transitions from the occupancies characteristic of separated atoms or molecules to those characteristic of a compound system or a different fragmentation. For example, a wave function for the system $O + N_2$ must be capable of describing the ^3P state of oxygen when it is far from the N_2 molecule, and this will require two singly occupied spatial orbitals. But the compound system N_2O may be well described in certain states using doubly occupied orbitals only. Moreover, in regions where one nitrogen atom is separated from an NO molecule, as many as three singly occupied spatial orbitals will be needed for the nitrogen atom, and there must be at least one single occupied orbital in any description of the NO molecule. Any calculational scheme that does not permit smooth transitions between these various situations cannot be of comparable accuracy in all regions of space and thus may be misleading as to the shapes and sizes of features of the potential-energy surfaces.

Another implication of the considerations surrounding the choice of a wave function is related to the treatment of electron spin. Not only is it necessary to require that the wave function be an eigenfunction of S^2 and S_z, but it is also necessary to take account of the fact that under many conditions there will be more than one spin eigenfunction of given S and M_s. The different spin eigenfunctions correspond to different couplings among the individual spins. Since reactive processes involve the breaking and forming of electron-pair bonds, they must necessarily be accompanied by reorganizations of the spin coupling. A failure to take account of this will lead to qualitatively inappropriate wave functions.

2. Self-consistent-field (SCF) Methods

The simplest and most widely used SCF procedure is the RHF, where the spatial orbitals are assumed as far as possible to be doubly occupied, and if there is molecular symmetry, to be of a pure symmetry type. As a result of the occupancy assumption, the wave function usually takes the form of a single Slater determinant.

Restricted Hartree–Fock calculations are most easily done after introducing an atomic-orbital basis $\chi_1, \ldots, \chi_p$, in terms of which the orbitals take the form

$$\phi_{ji} = \sum_{j=1}^{p} c_{ji} \chi_j \qquad \text{(II.8)}$$

The variational condition determining the coefficients c_{ji} is cubic in the unknowns, but iterative techniques permit these coefficients to be determined by repeated use of matrix diagonalization methods. Under most conditions it is possible to choose an iterative process facilitating convergence; there is much RHF experience, and inordinate difficulties are not usually experienced. Because of the occupancy assumptions, it is possible without loss of generality to take the RHF spatial orbitals as orthogonal, and this is an important feature simplifying the calculations.

Restricted Hartree–Fock wave functions are frequently unsatisfactory. For many systems they will not give good descriptions of dissociation fragments; they may preclude satisfactory description of intermediate situations in which one bond is partially formed while another is partially broken. In such cases insistence on an RHF formulation may force a discontinuous transition as the nuclear positions are changed because of changes in the energy ordering of the highest-occupied RHF orbitals.

The great speed and known properties of RHF calculations are not sufficient justification for a limitation to RHF methods when they are inherently inappropriate. It is worth remarking that most potential-energy surfaces describing reactions, and many describing dissociations are inappropriate for RHF methods. Restricted Hartree–Fock methods are also of limited validity in many situations where two or more surfaces are at nearly the same energy.

Some of the disadvantages of the RHF method are avoided if a wave function consisting of simply an antisymmetrized spin-orbital product is used, not requiring orbitals of opposite spin to be pairwise spatially identical. This spin-polarized method has the disadvantage of failing to yield spin eigenfunctions and is thus a poor approach if it is planned to discuss relationships between states of definite spin multiplicities. The calculations are somewhat more cumbersome than RHF studies, and under most circumstances the advantages of the spin-polarized method are probably insufficient to make it a favored approach. The disadvantage of not yielding a spin eigenfunction can be circumvented by spin projecting to an eigenstate after the spin-polarized orbitals have been determined, but the result obtained thereby is inferior to an energy optimization of the projected wave function.

An SCF procedure avoiding all these conceptual difficulties is that of simultaneously optimizing the orbitals and the spin function. This will give good results only if the spatial orbitals are not required to be either identical or orthogonal, and the nonorthogonality greatly increases the computational complexity. Because of the free choice of spin eigenfunction, the spatial orbitals cannot be orthogonalized without serious loss of flexibility in the wave function. This simultaneous spin and spatial orbital optimization has been carried out for a few systems and gives good results. However, computational complexities are prohibitive if there are many electrons, and under most circumstances it is more practical to achieve comparable or better results through a CI with orthogonal orbitals.

3. Configuration-interaction (CI) Methods

The specific form for a CI wave function, ψ, may be written as

$$\psi = \sum_\mu c_\mu \psi_\mu \qquad (\text{II}.9)$$

where each ψ_μ is referred to as a *configuration* and has the general structure given in equation (II.6). States of different symmetries are studied by restricting ψ_μ to the appropriate form, and excited states of any symmetry can be handled by simultaneously determining ψ for the excited state of interest and for all lower-lying states of the same symmetry.

a. Single-state Optimization Methods

Variational optimization of equation (II.9), where we are concerned with only one projection of ψ corresponding to a particular electronic eigenstate, has been extensively studied. There are at least two well-developed techniques for such situations, namely, the multiconfiguration SCF (MCSCF) and iterative natural spin-orbital (INSO) approaches.

Multiconfiguration Self-consistent Field

In this approach the emphasis is on constructing the best possible spatial orbital $\phi_{i\mu}$ within a CI framework that contains only those configurations necessary for proper wave-function dissociation and those configurations necessary to handle, at least to first order, changes in pair-correlation error as a function of internuclear separation. Typically only two or three configurations are required to properly account for correct atomic-limit connections. A few tens of configurations, which mainly include single orbital excitations, are often added to the CI to handle the correlation problem. Various prescriptions have been proposed to handle the numerical aspects of obtaining optimum molecular orbitals within the chosen CI.

Wahl et al.[15] suggest an analytic procedure similar to analytic Hartree–Fock methods whereby each molecular orbital is expanded as a linear combination of some simple basis functions over which all the one- and two-electron integrals can be efficiently calculated. We have

$$\phi_{i\mu} = \sum_{ik} a_{ki} u_k \qquad (\text{II.10})$$

where u_k is an elementary basis function such as a Slater-type orbital (STO) or a Gaussian-type orbital (GTO). The u_k are usually symmetry restricted for a given $\phi_{i\mu}$ and the molecular-orbital expansion coefficients a_{ki}, are found by nonlinear matrix diagonalization techniques. An alternate and more direct procedure is to determine the a_{ki} using Newton–Raphson gradient techniques.[21] The entire procedure involves, in any event, a double iteration, first to estimate the molecular-orbital coefficients, a_{ki}, and then to solve the secular equation for the optimum CI coefficients.

The MCSCF procedure has yielded excellent results in cases where there is a single dominant configuration in ψ, where there are no nearby degeneracies, and where the principal interest has been the ground molecular state. The method fails, or is difficult to apply, when the total wave function ψ may have two or more leading terms for proper chemical sense. Convergence difficulties have also been encountered in applying MCSCF to excited electronic states of the same symmetry as the ground state. Nevertheless, MCSCF is among the best of techniques that are currently available for producing potential-energy surfaces accurate to within approximately 1 kcal.

Iterative Natural Spin Orbitals

The NSO approach is based on the observation of Löwdin[22] that there exists a unique orbital transformation that results in a CI with optimum convergence properties. Unfortunately, the procedure for defining this orbital transformation requires that a full CI be performed over an arbitrarily chosen basis set. The resulting CI coefficients can then be used to construct a unique one-particle density matrix (unique to the total Hilbert space represented by the basis functions), and this matrix can be diagonalized to yield the optimum orbital expansion coefficients for constructing the NSOs. Practically, this can never quite be accomplished because of the rapid expansion of the size of a full CI with modest increases in the basis function size. Therefore, approximate iterative procedures have been suggested. One such procedure[23] starts with a modest STO or GTO basis that is first converted into SCF form. All single and, possibly, double orbital excitations are then constructed from the SCF function using the unoccupied basis functions. The resulting CI is diagonalized and reduced to NSO form. These new orbitals now form a leading configuration, and the basis set is augmented by the addition of new functions, possibly of higher orbital angular momentum, and the single, double, and further excitations are constructed to form the CI. The procedure is cycled until some kind of internal consistency in the NSOs is achieved. Typically, several thousand configurations are handled in the CI, as contrasted with the tens of configurations employed in MCSCF.

The two largest deficiencies of INSO procedures are: (1) the magnitude of effort involved is very large, and only the largest and most powerful computers prove useful in implementing this analysis; and (2) the method may converge to some nonphysical state if the starting configuration is not a reasonable description of the total ψ. Care must be taken that the iterative procedure produce a set of functions that are localized in the space of the eigenstate under study. The most accurate CI calculations available to date have been constructed using this procedure. The costs associated with the INSO method, as applied to potential-energy hypersurfaces, would appear, at this state of development, to be prohibitive except for a few isolated prototype calculations.

b. Multistate Analysis

For many applications, a single potential-energy surface will not suffice, and information may be required for a whole series of possibly intersecting hypersurfaces. In this case ultimate wave-function accuracy must be compromised with a consistent description of all the eigenstates of interest. The most successful approach has been to choose a basis-set optimum for all the ground and excited separated atomic and molecular dissociation products for a given system that are accessible within the energy range of interest. Such a basis set yields reasonable descriptions of the separate atomic states and when mixed in a CI framework, yields a set of potential-energy curves of consistent quality for all the excited molecular states that can be constructed. This is the main idea of the valence configuration-interaction (VCI) approach,[6] where a complete or nearly complete CI is constructed from a few well-chosen atomic basis functions. The method applies without modification to studies of surfaces.

A key advantage of such an approach is that several eigenstates can be examined simultaneously with about the same overall accuracy. If the basis set is augmented by the addition of diffuse or Rydberg-type orbitals, Rydberg–valence-state mixing can be studied. An important problem in this area is concerned with IR radiation in the Rydberg bands of the NO molecule. Perturbations in the radiation characteristics of NO resulting from valence-state mixing have been quantitatively examined with a VCI approach.[24] The principal drawback to the approach is the reliance on a full CI expansion of a limited basis set. Modest extensions of the basis size render the method unwieldy because of the large number of new configurations that result. Even with a minimum basis set, a typical triatomic system could have on the order of 10^3–10^4 configurations in a full CI. It is therefore essential to identify and employ those configurations that describe the significant part of the wave function. There are several ways to accomplish this objective. First, atomic-orbital occupancies may be screened to eliminate those with excessive formal charge. Alternatively, in a molecular-orbital framework, those configurations with excessive numbers of antibonding orbitals may be eliminated. A third possibility is to carry out an initial screening of configurations, using perturbation theory based on interaction matrix elements between some ψ_0 (possibly the Hartree–Fock function) and each constructed ψ_μ, rejecting all ψ_μ that do not satisfy some energy criteria. This last procedure suffers from the fact that the total ψ will be constructed from a different set of ψ_μ at different geometries of the nuclei.

D. Potential-energy Curves and Surfaces

Calculations of the dynamics of molecular collisions are ordinarily carried out with the aid of the Born–Oppenheimer separation of the electronic and nuclear motion. The procedure is to calculate the electronic energy as a function of the positions of the nuclei, which are assumed to be stationary. This electronic energy, plus the electrostatic repulsion between the nuclei, defines a potential-energy hypersurface that is useful for describing the kinetic motion of the nuclei. A potential-energy hypersurface defined in this way is referred to as *adiabatic* and is appropriate for describing the nuclear motion in the limit of low velocity. There are many collisions for which an adiabatic potential-energy hypersurface provides an adequate description. However, many reactive collisions are inadequately described by an adiabatic potential-energy hypersurface. These collisions are characterized by velocities of nuclear motion sufficient to adversely affect the Born–Oppenheimer separation, with the result that the overall wave function must be described as a superposition of terms involving different electronic-energy states. Under these conditions it is useful to consider adiabatic potential-energy hypersurfaces corresponding to all electronic states relevant to the overall wave function.

When the different potential-energy hypersurfaces are well separated in energy, the nuclear motion can ordinarily be described in terms of motion on a single hypersurface. However, when two or more hypersurfaces are close in energy, they can be expected to mix appreciably in the overall wave function, and it will then be necessary to calculate not only the hypersurfaces, but also the quantities needed to discuss their mixing in the overall wave function.

The calculation of a point on a potential-energy hypersurface is equivalent to calculating the energy of a diatomic or polyatomic system for a specified nuclear configuration and thus presents considerable practical computational difficulty. For certain problems or nuclear configurations, the maximum possible accuracy is needed, and under these conditions relatively elaborate *ab initio* methods are indicated. For other problems, the description to a uniform accuracy of many electronically excited states of a given system is required. Such is the situation for the atmospheric systems described here, and thus most of our final potential curves are based on the analysis of VCI wave functions constructed to uniform quality for representation of the excited states.

III. ELECTRONIC STRUCTURE AND POTENTIAL-ENERGY CURVES

Ab initio calculations were carried out for all the low-lying non-Rydberg states of the systems N_2, O_2, NO, O_2^-, and NO^+. In N_2, for example, there are 102 molecular states that result from nitrogen atoms in the lowest ^4S, ^2D, and ^2P states. These states were all uniformly described using VCI wave functions constructed as described in Section II. Minimum basis, double-ζ basis and double-ζ-plus-polarization basis sets were employed for these studies. For the minimum basis-set calculations, which were always carried out first, the VCI wave functions represent full CI projections with the constraint that the K shells were kept frozen for all states. However, no constraint on the $2\sigma_g$ and $2\sigma_u$ orbitals was made since a CI among these orbitals is necessary to ensure proper description of the hole states in these molecules, such as $C^3\Pi_u$ of N_2. The calculations all have the property of asymptotically connecting with the correct atomic states. This computational method has previously been applied, with reliable results, to both closed- and open-shell systems.[6,9-11]

Most of the *ab initio* methods described in the preceding section, including the VCI approach, yield calculated molecular dissociation energies that are generally somewhat smaller than experimental values. In addition, the predicted equilibrium separations are usually a few percent too large. Both of these effects are a manifestation of core overlap and polarization correlation errors that are minimal for large internuclear separations but gradually increase as the nuclei are brought closer together. Attempts to uniformly and quantitatively represent these calculation errors have led to enormously large and expensive CI calculations,[9] and it is logical to ask whether a semiempirical correction exists that can yield improved estimates of potential-energy curves beyond those obtained from some uniform calculation procedure such as VCI.

Abandoning a purist point of view, we ask whether an empirical correction of the form

$$\Delta V(R, \Gamma, S) = V_{\text{calc}}(R, \Gamma, S) - V_{\text{exp}}(R, \Gamma, S) \qquad \text{(III.1)}$$

exists such that information derived from the spectroscopy of a few electronic states can be transferred to other states for which no information exists. In equation (III.1), R is the internuclear separation, Γ represents the collection of quantum numbers for angular momentum, g or u and $+$ or $-$ represent symmetry, and S represents the total spin. In previous studies on the N_2 molecule[10] ΔV was found to have a nearly universal shape, increasing smoothly with decreasing R and with a very weak dependence on Γ and S. This is the expected behavior if ΔV represents an empirical calibration of the inner-shell correlation error that is not fully accounted for in VCI calculations. The valence-shell correlation appears, to first order, to be properly accounted for within a full valence-shell CI calculation. The possibility of equation (III.1) having a uniform functional form can be expected with even more assurance for molecular states, all of which stem from the same occupancy molecular-orbital designation. In this case the ΔV values for different molecular symmetries, but the same MO description, differ on the order of several hundredths of an electron volt.

It thus appears that a judicious application of correction curves of the form of (III.1) can transform a uniform set of *ab initio* potential-energy curves into a set of corrected curves that are very representative of the actual system. Such curves must be of similar chemical character such as all valence states or all Rydberg states. Rydberg–valence mixing cannot be easily accounted for in *ab initio* calculations, and simple empirical corrections do not seem possible for such situations.

The calculations described here are all for excited valence states, and only in the case of N_2 are Rydberg states included in our compilation of potential-energy curves. This is because a great deal of experimental evidence for N_2 exists that permits the accurate location of these states relative to the valence states of N_2. A similar analysis could probably have been done for NO, based on the extensive work of Miescher[25] on Rydberg states. The situation is less clear in the case of O_2.

A. Nitrogen Molecule

Any attempt to describe the electronic structure of the nitrogen molecule must be judged against the enormity of the task as is evident in the pioneering work of Mulliken, Loftus, Dressler, Carroll, and others. The spectrum of the excited states of the nitrogen molecule is enormously rich and complex, and this molecule has received more attention from spectroscopists than any other system. An excellent bibliography of the literature of nitrogen up to 1967 is available in the book by Wright and Winkler.[26] Other excellent reviews are those of Loftus,[27] Mulliken,[28a] and the recent compendium of the nitrogen spectrum by Loftus and Krupenie.[28b]

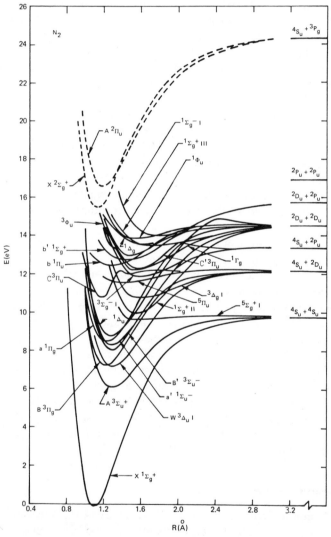

Figure 1. Potential-energy curves for bound states of N_2.

An extensive systematic study of the potential-energy curves for nitrogen is that of Gilmore.[12] These curves were calculated with a numerical RKR method,[29-31] using rotational and vibrational term values from the available literature, mainly the compilation of Wallace.[32] A less complete study of the potential-energy curves of nitrogen is the work of Vanderslice et al.[33] Mulliken (see Wilkinson[34]) has drawn the Rydberg potential curves be-

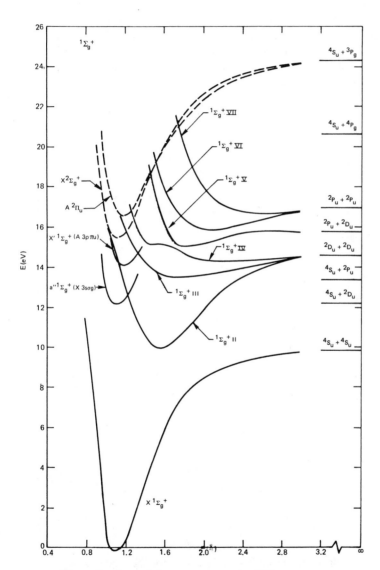

Figure 2. Potential-energy curves for $^1\Sigma_g^+$ states of N_2.

tween 12.0 eV and 14.5 eV, based on analysis of the spectroscopic data in this region and on analysis of the quantum defects for states Rydberg to the $X^2\Sigma_g^+$ and $A^2\Pi_u$ cores of N_2^+. Some of these states have been repositioned, based on the deperturbation analysis of Leoni and Dressler.[35]

Our compilation of the potential-energy curves for N_2, based on available spectroscopic data and our extensive calculations for the valence states, is given in Figs. 2 to 23 by symmetry type. A composite potential-energy curve for N_2 indicating only the bound states is shown in Fig. 1. The molecular correlations for the low-lying valence states of N_2 are shown in Table I. The dominant molecular-orbital configurations of nitrogen are given in Table II, and a listing of all known and predicted spectroscopic data is given in Table III.

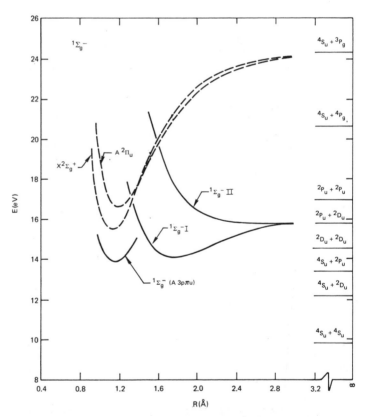

Figure 3.   Potential-energy curves for $^1\Sigma_g^-$ states of N_2.

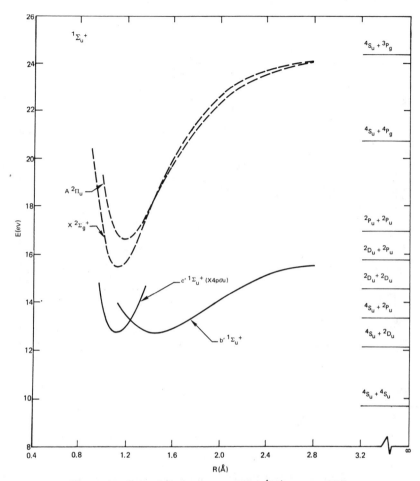

Figure 4. Potential-energy curves for $^1\Sigma_u^+$ states of N_2.

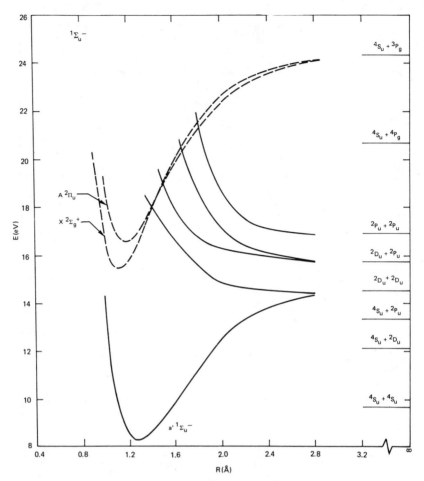

Figure 5. Potential-energy curves for $^1\Sigma_u^-$ states of N_2.

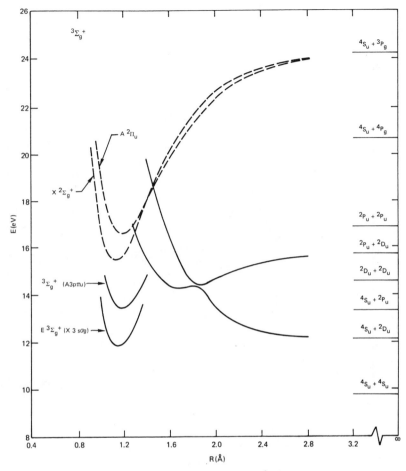

Figure 6. Potential-energy curves for $^3\Sigma_g^+$ states of N_2.

247

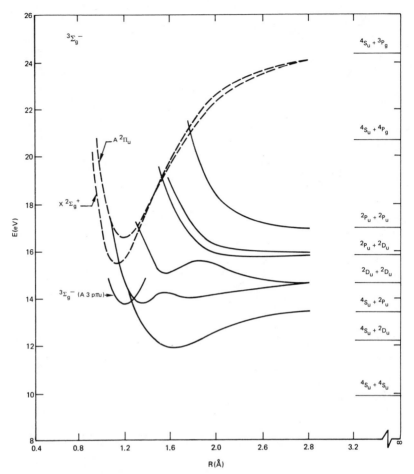

Figure 7. Potential-energy curves for $^3\Sigma_g^-$ states of N_2.

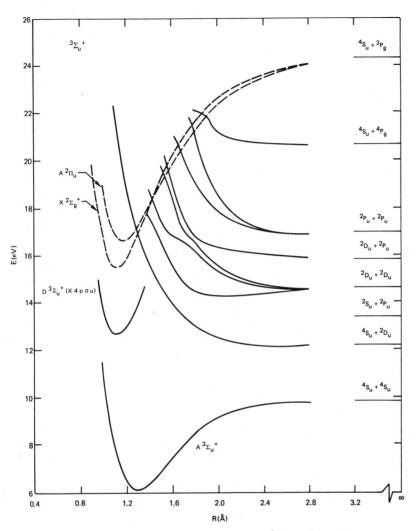

Figure 8. Potential-energy curves for $^3\Sigma_u^+$ states of N_2.

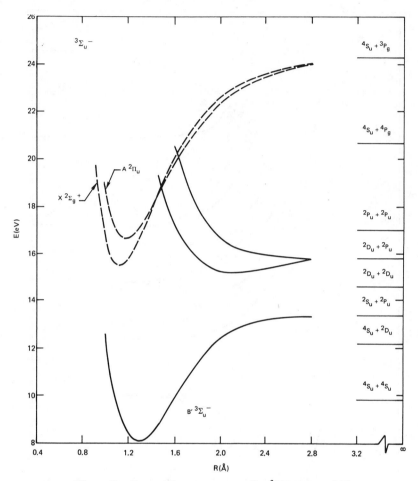

Figure 9. Potential-energy curves for $^3\Sigma_u^-$ states of N_2.

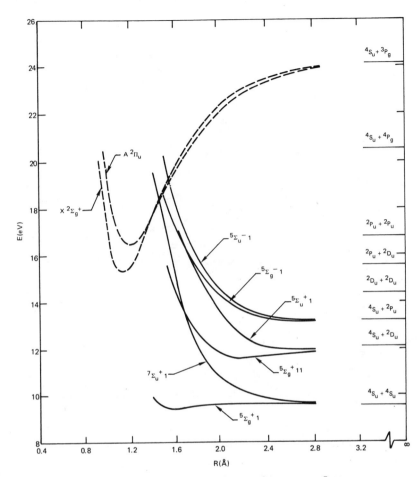

Figure 10. Potential-energy curves for $^5\Sigma_g^{+,-}$, $^5\Sigma_u^{+,-}$, and $^7\Sigma_u^+$ states of N_2.

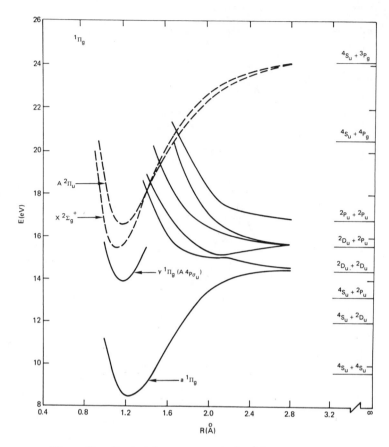

Figure 11. Potential-energy curves for $^1\Pi_g$ states of N_2.

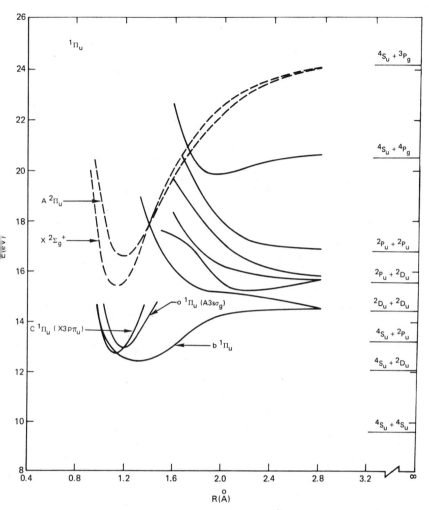

Figure 12. Potential-energy curves for $^1\Pi_u$ states of N_2.

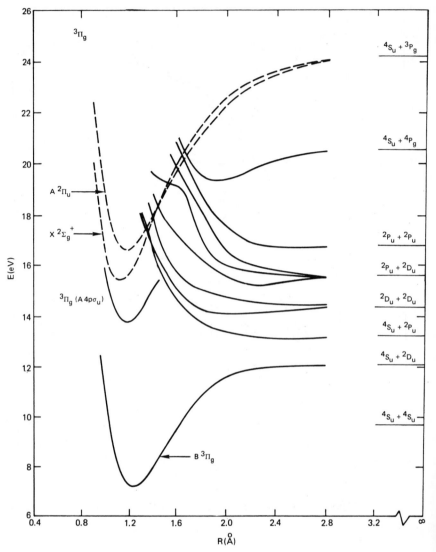

Figure 13. Potential-energy curves for $^3\Pi_g$ states of N_2.

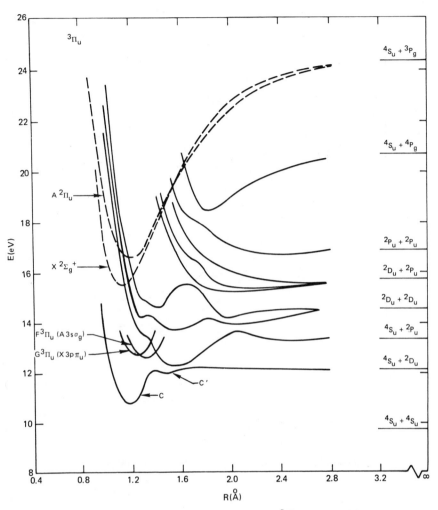

Figure 14. Potential-energy curves for $^3\Pi_u$ states of N_2.

255

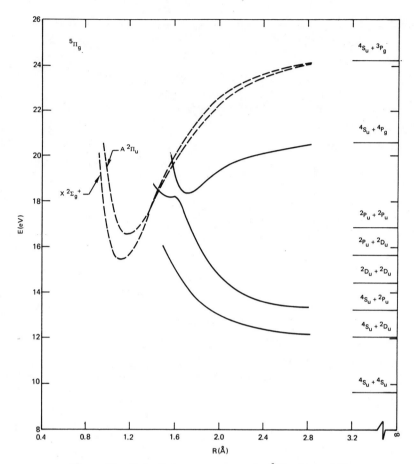

Figure 15. Potential-energy curves for $^5\Pi_g$ states of N_2.

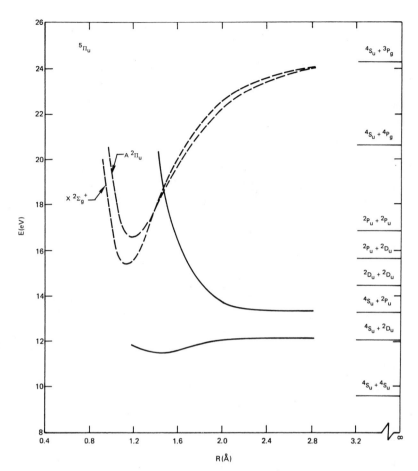

Figure 16. Potential-energy curves for $^5\Pi_u$ states of N_2.

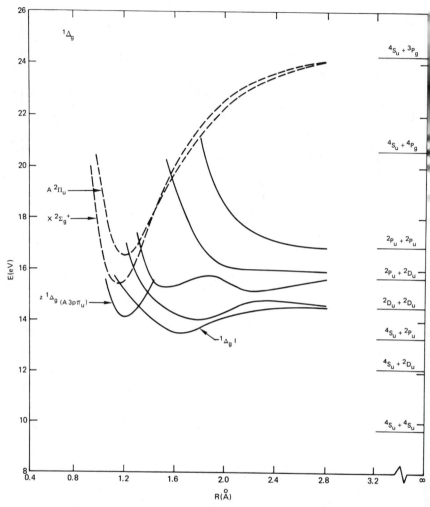

Figure 17. Potential-energy curves for $^1\Delta_g$ states of N_2.

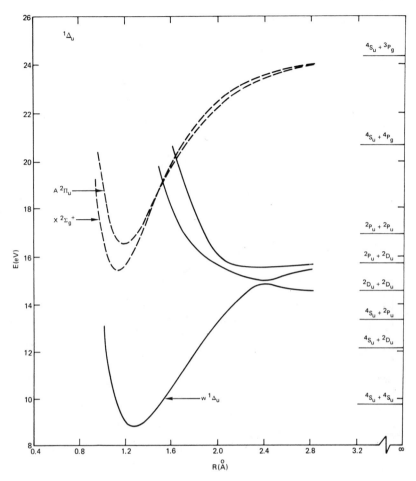

Figure 18. Potential-energy curves for $^1\Delta_u$ states of N_2.

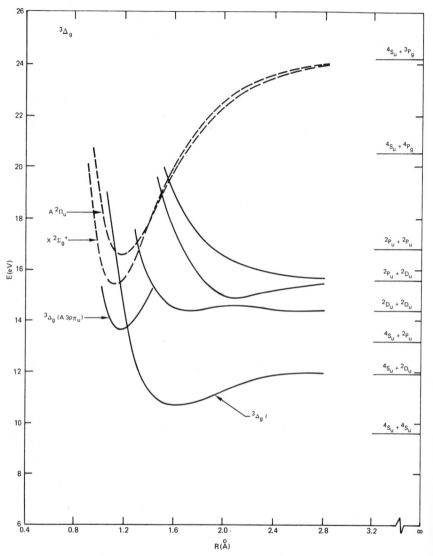

Figure 19. Potential-energy curves for $^3\Delta_g$ states of N_2.

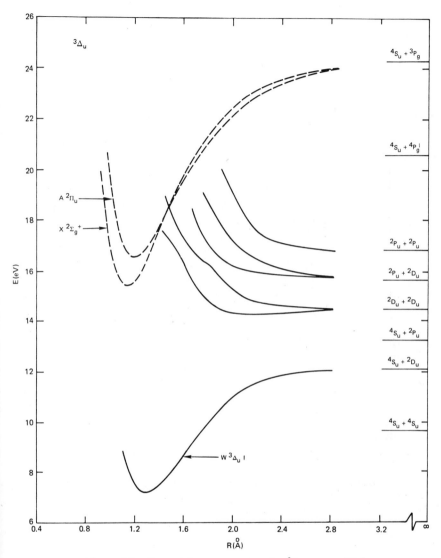

Figure 20. Potential-energy curves for $^3\Delta_u$ states of N_2.

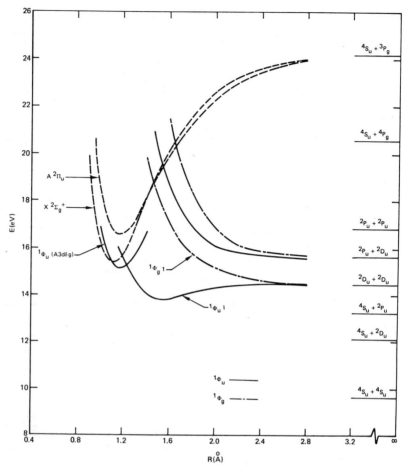

Figure 21. Potential-energy curves for $^1\Phi_g$ and $^1\Phi_u$ states of N_2.

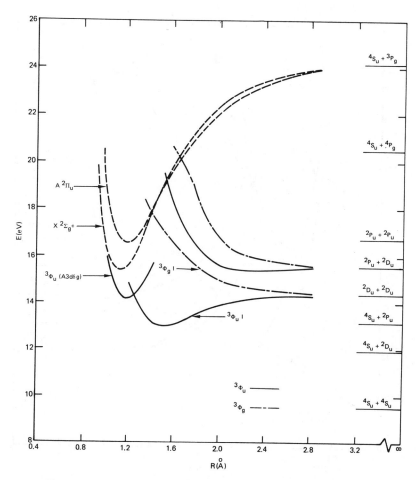

Figure 22. Potential-energy curves for $^3\Phi_g$ and $^3\Phi_u$ states of N_2.

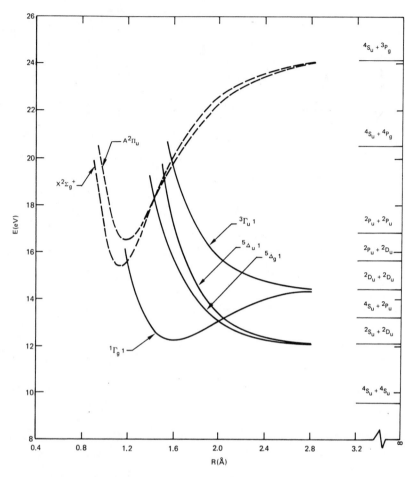

Figure 23. Potential-energy curves for $^5\Delta_g$, $^5\Delta_u$, $^1\Gamma_g$, and $^3\Gamma_u$ states of N_2.

Table I.

Low-lying Molecular States of N_2 and Their Dissociation Limits

DISSOCIATION LIMIT	MOLECULAR STATES
$N(^4S_u)+N(^4S_u)$	$^1\Sigma_g^+(1), {}^5\Sigma_g^+(1), {}^3\Sigma_u^+(1), {}^7\Sigma_u^+(1)$
$N(^4S_u)+N(^2D_u)$	$^3\Sigma_g^+(1), {}^3\Sigma_u^+(1), {}^5\Sigma_g^+(1), {}^5\Sigma_u^+(1), {}^3\Pi_g(1),$ $^3\Pi_u(1), {}^5\Pi_g(1), {}^5\Pi_u(1), {}^3\Delta_g(1), {}^3\Delta_u(1),$ $^5\Delta_g(1), {}^5\Delta_u(1)$
$N(^4S_u)+N(^2P_u)$	$^3\Sigma_g^-(1), {}^3\Sigma_u^-(1), {}^5\Sigma_g^-(1), {}^5\Sigma_u^-(1),$ $^3\Pi_g(1), {}^3\Pi_u(1), {}^5\Pi_g(1), {}^5\Pi_u(1)$
$N(^2D_u)+N(^2D_u)$	$^1\Sigma_g^+(3), {}^3\Sigma_g^-(2), {}^3\Sigma_u^+(3), {}^1\Sigma_u^-(2),$ $^1\Pi_g(2), {}^3\Pi_g(2), {}^1\Pi_u(2), {}^3\Pi_u(2), {}^1\Delta_g(2),$ $^3\Delta_g(1), {}^1\Delta_u(1), {}^3\Delta_u(2), {}^1\Phi_g(1), {}^3\Phi_g(1), {}^1\Phi_u(1),$ $^3\Phi_u(1), {}^1\Gamma_g(1), {}^3\Gamma_u(1)$
$N(^2D_u)+N(^2P_u)$	$^1\Sigma_g^+(1), {}^1\Sigma_u^+(1), {}^3\Sigma_g^+(1), {}^3\Sigma_u^+(1),$ $^1\Sigma_g^-(2), {}^1\Sigma_u^-(2), {}^3\Sigma_g^-(2), {}^3\Sigma_u^-(2),$ $^1\Pi_g(3), {}^1\Pi_u(3), {}^3\Pi_g(3), {}^3\Pi_u(3), {}^1\Delta_g(2),$ $^1\Delta_u(2), {}^3\Delta_g(2), {}^3\Delta_u(2), {}^1\Phi_g, {}^1\Phi_u, {}^3\Phi_g, {}^3\Phi_u$
$N(^2P_u)+N(^2P_u)$	$^1\Sigma_g^+(2), {}^3\Sigma_g^-(1), {}^3\Sigma_u^+(2), {}^1\Sigma_u^-(1),$ $^1\Pi_g(1), {}^3\Pi_g(1), {}^1\Pi_u(1), {}^3\Pi_u(1), {}^1\Delta_g(1),$ $^3\Delta_u(1)$

B. Oxygen Molecule

The spectra for the O_2 molecule is as sparse as that for nitrogen is rich, with the exception of Rydberg state transitions. Rydberg–Klein–Rees curves for several valence states of O_2 were first constructed by Vanderslice et al.[41] Gilmore[12] illustrates a much more complete set of potential-energy curves based on the spectroscopic data listed by Wallace.[32] An excellent bibliography of the literature and spectra of the oxygen molecule up to 1971 is the work of Krupenie.[42] A less complete set of potential-energy curves is included in his bibliography. Similar curves are available in the compilation by Jarmain.[43,44]

Calculations of potential-energy curves for O_2 include the valence-state studies of Schaefer and Harris[45] and the SCF studies of Pritchard et al.[46] Configuration-interaction calculations for the excited states of oxygen include those of Michels,[24] Schaefer,[47] and Schaefer and Miller.[48] The higher-lying $^1\Delta_u$ and $^1\Sigma_u^+$ states were examined by Itoh and Ohno[49] in an early computational study. Recently calculations for excited states of $^3\Sigma_u^-$ symmetry have been carried out by Yoshimine.[50] Buenker and Peyerimhoff[51] have recently reported some calculations for Rydberg states of O_2. The Schumann–Runge continuum in O_2 has recently been analyzed by Bunker et al.,[52] Julienne,[53] and Cartwright.[54]

Table II.
Important Molecular-orbital Electron Configurations for N_2

STATE	ELECTRON CONFIGURATION[a]			
	$1\pi_u$	$3\sigma_g$	$1\pi_g$	$3\sigma_u$
$X^1\Sigma_g^+$ (I)	4	2	0	0
$^1\Sigma_g^+$ (II)	2	2	2	0
$^1\Sigma_g^-$ (I)	2	2	2	0 }[b]
	3	1	1	1
$a'^1\Sigma_u^-$ (I)	3	2	1	0
$b'^1\Sigma_u^+$ (I)	4	1	0	1 }[c]
	3	2	1	0
$^3\Sigma_g^-$ (I)	2	2	2	0
$^3\Sigma_g^+$ (II)	2	2	2	0
$A^3\Sigma_u^+$ (I)	3	2	1	0
$B'^3\Sigma_u^-$ (I)	3	2	1	0
$^3\Sigma_u^-$ (II)	2	1	2	1
$^5\Sigma_g^+$ (II)	2	2	2	0
$a^1\Pi_g$ (I)	4	1	1	0
$^1\Pi_g$ (III)	3	2	0	1
$b^1\Pi_u$ (I)	3	1	2	0 }[b]
	4	2	1	$0\,(-2\sigma_u)$
$B^3\Pi_g$ (I)	4	1	1	0
$C^3\Pi_u$ (I)	4	2	1	$0\,(-2\sigma_u)$ }[d]
$C'^3\Pi_u$ (II)	3	1	2	0
$^1\Delta_g$ (I) $^1\Delta_g$ (II) $^1\Delta_g$ (III)	2	2	2	0^e
w { $^1\Delta_u$ (I) $^1\Delta_u$(II)	3	2	1	0^f
$^3\Delta_g$ (I)	2	2	2	0
$^3\Delta_g$ (II) $^3\Delta_g$ (III)	3	1	1	1^g
$^3\Delta_u$ (I)	3	2	1	0
$W^3\Delta_u$ (I)	3	1	2	0
$^1\Gamma_g$ (I)	2	2	2	0

[a] All states include $(1\sigma_g)^2$ $(1\sigma_u)^2$ $(2\sigma_g)^2$ $(2\sigma_u)^2$ electrons.

[b] The first configuration given is twice as important as the second at the equilibrium separation.

[c] These configurations are weighted about equally.

[d] There is an avoided crossing between these two states at $R = 1.56$ Å (see Fig. 14).

[e] There is an avoided crossing between these three states at $R = 1.9$ Å and 2.3 Å (see Fig. 17).

[f] There is an avoided crossing between these two states at $R = 2.4$ Å (see Fig. 18).

[g] There is an avoided crossing betwen these two states at $R = 2.1$ Å (see Fig. 19).

Table III.
Spectroscopic Constants for Bound States of N_2

STATE[a]		T_e (eV)	ω_e (cm^{-1})	$\omega_e X_e$ (cm^{-1})	α_e (cm^{-1})	r_e (Å)	B_e (cm^{-1})	D_e (eV)	D_0 (eV)
$^1\Sigma_g^-$ (I)	(C)	14.17	694.1	33.0	0.024	1.75	0.78	1.69	1.65
$^1\Sigma_g^+$ (III)	(C)	13.65	488.3	34.8	0.048	1.70	0.83	1.02	0.99
$^1\Phi_u$ (I)	(C)	13.99	736.9	18.9	0.021	1.47	1.11	0.68	0.64
$^1\Delta_g$ (I)	(C)	13.66	776.9	6.3	0.012	1.60	0.94	1.01	0.96
$H^3\Phi_u$ (I)	(E)	13.13	925.6	(17.2)	0.019	1.49	1.087	1.54	1.48
$b'^1\Sigma_u^+$	(E)	12.92	746.	(5.9)	0.0048	1.44	1.15	2.94	2.89
$b^1\Pi_u$	(E)	12.60	635.0	6.0	0.0048	1.23	1.45	2.07	2.03
$^1\Gamma_g$ (I)	(C)	12.49	856.2	9.7	0.011	1.60	0.94	2.18	2.13
$^3\Sigma_g^-$ (I)	(C)	12.04	792.0	7.4	0.0079	1.61	0.93	1.44	1.39
$^5\Sigma_g^+$ (II)	(C)	11.98	580.3	27.0	0.015	2.13	0.54	0.31	0.27
$C^3\Pi_u$	(E)	11.05	2047.2	28.4	0.019	1.15	1.82	1.23	1.10
$G^3\Delta_g$ (I)	(E)	10.91	765.9	(13.2)	0.016	1.61	0.93	1.37	1.33
$^1\Sigma_g^+$ (II)	(C)	10.09	1164.6	3.2	0.004	1.57	0.98	4.58	4.51
$^5\Pi_u$ (I)	(C)	11.69	475.4	14.6	(0.043)	1.45	1.15	0.59	0.56
$^5\Sigma_g^+$ (I)	(C)	9.74	694.0	24.0	0.025	1.61	0.92	0.16	0.12
$w^1\Delta_u$	(E)	8.94	1559.2	11.9	0.017	1.27	1.50	5.73	5.63
$a^1\Pi_g$	(E)	8.59	1694.2	13.9	0.018	1.22	1.62	6.08	5.98
$a'^1\Sigma_u^-$	(E)	8.45	1530.3	12.1	0.017	1.28	1.48	6.22	6.13
$B'^3\Sigma_u^-$	(E)	8.22	1516.9	12.2	0.017	1.28	1.47	5.26	5.17
$W^3\Delta_u$	(E)	7.56	1501.4	11.6	(0.016)	1.28	(1.47)	4.72	4.64
$B^3\Pi_g$	(E)	7.39	1733.4	14.1	0.018	1.21	1.64	4.89	4.78
$A^3\Sigma_u^+$	(E)	6.22	1460.5	13.8	0.018	1.29	1.45	3.68	3.59
$X^1\Sigma_g^+$	(E)	0.0	2358.0	14.1	0.018	1.10	2.00	9.90	9.75

[a]Code: ( ) Estimated, (C) calculated, (E) experimental.

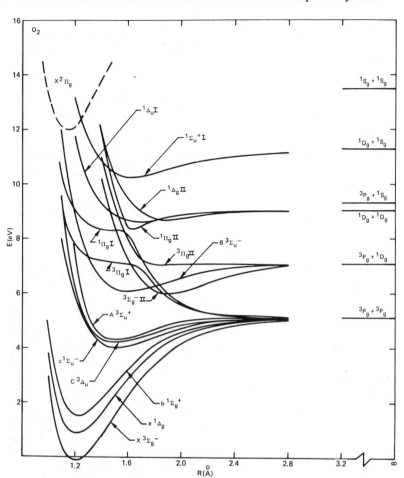

Figure 24. Potential-energy curves for bound states of O_2.

Our analysis of the valence states of O_2 is based on available spectroscopic data and on several extensive sets of CI calculations.[24] A more
recent study of the valence states is that of Beebe et al.[55] A composite
potential curve for the bound states of O_2 is shown in Fig. 24. Separate
curves showing all states for each symmetry, are given in Figs. 25 to 41.
The molecular correlations for the valence states of O_2 are given in Table
IV, and the dominant molecular-orbital configurations are given in Table
V. A summary of the available spectroscopic data, including calculated
curves, is given in Table VI.

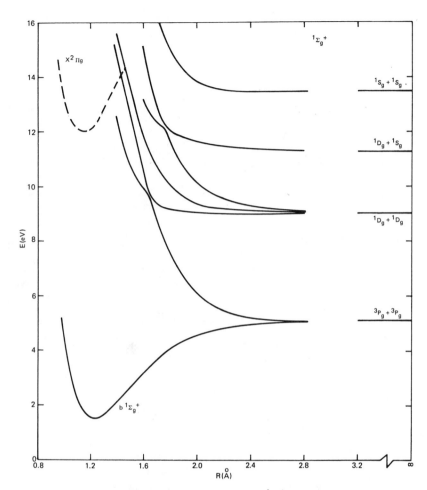

Figure 25. Potential-energy curves for $^1\Sigma_g^+$ states of O_2.

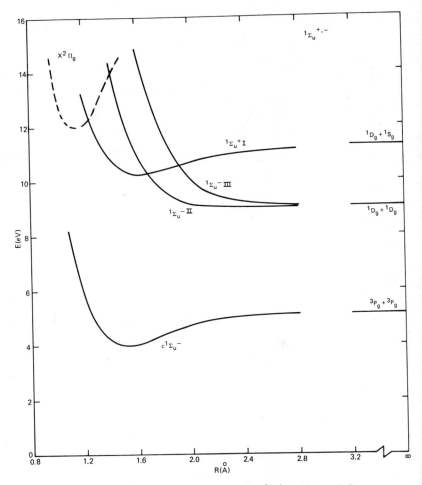

Figure 26. Potential-energy curves for $^1\Sigma_u^{+,-}$ states of O_2.

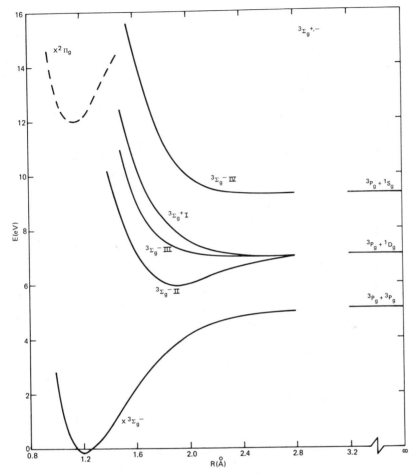

Figure 27. Potential-energy curves for $^3\Sigma_g^{+,-}$ states of O_2.

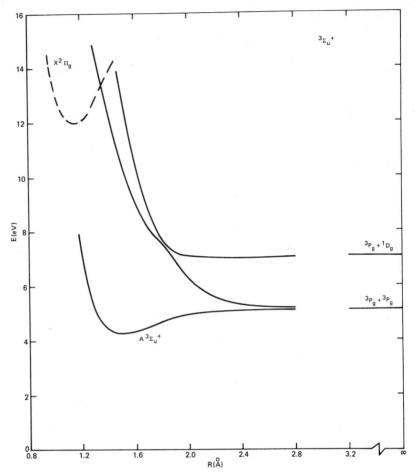

Figure 28. Potential-energy curves for $^3\Sigma_u^+$ states of O_2.

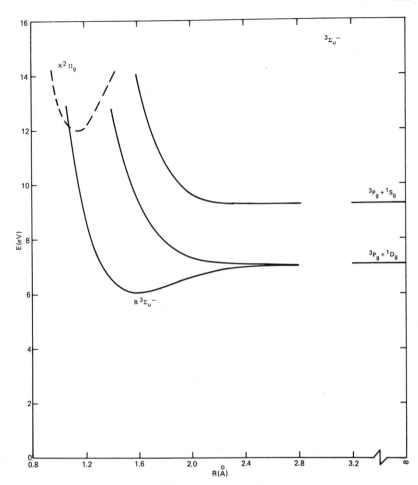

Figure 29. Potential-energy curves for $^3\Sigma_u^-$ states of O_2.

273

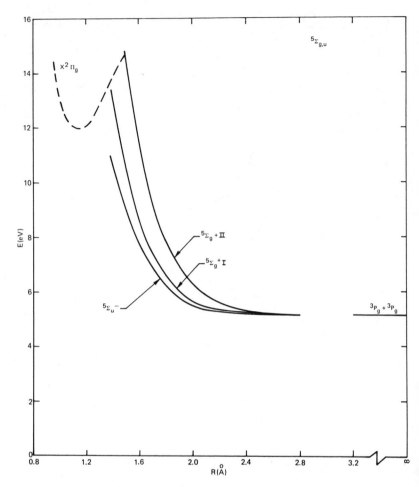

Figure 30. Potential-energy curves for $^5\Sigma_g^+$ and $^5\Sigma_u^-$ states of O_2.

274

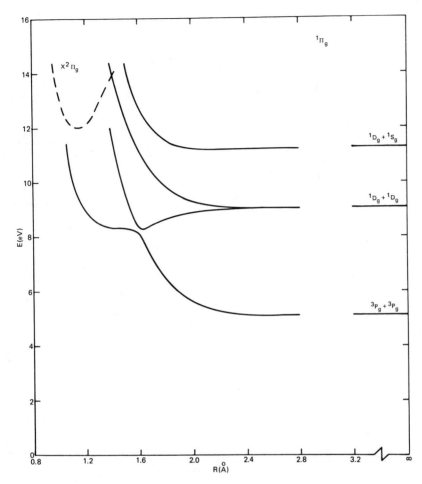

Figure 31. Potential-energy curves for $^1\Pi_g$ states of O_2.

275

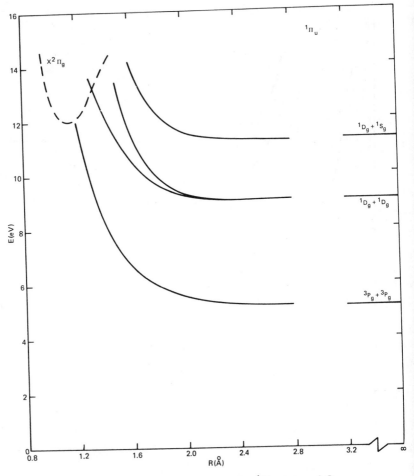

Figure 32. Potential-energy curves for $^1\Pi_u$ states of O_2.

276

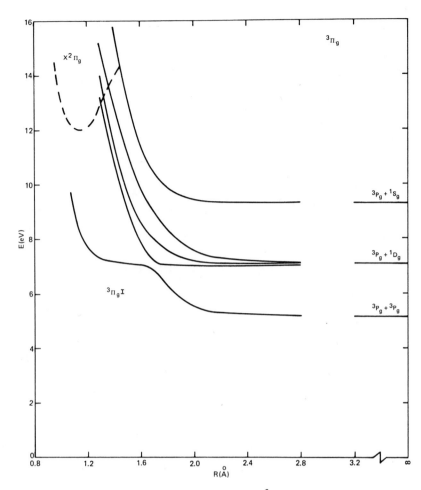

Figure 33. Potential-energy curves for $^3\Pi_g$ states of O_2.

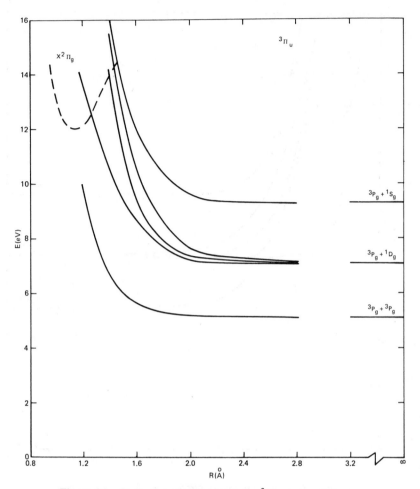

Figure 34. Potential-energy curves for $^3\Pi_u$ states of O_2.

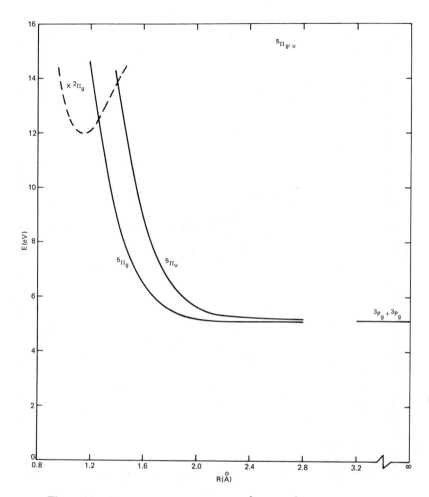

Figure 35. Potential-energy curves for $^5\Pi_g$ and $^5\Pi_u$ states of O_2.

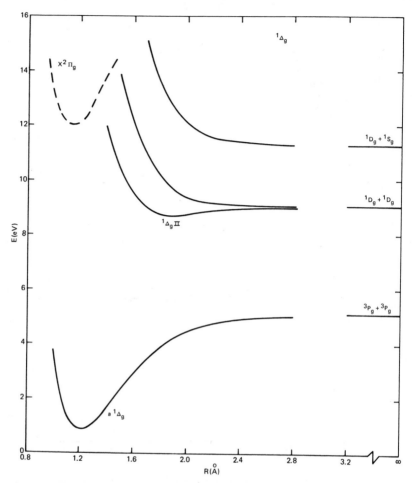

Figure 36. Potential-energy curves for $^1\Delta_g$ states of O_2.

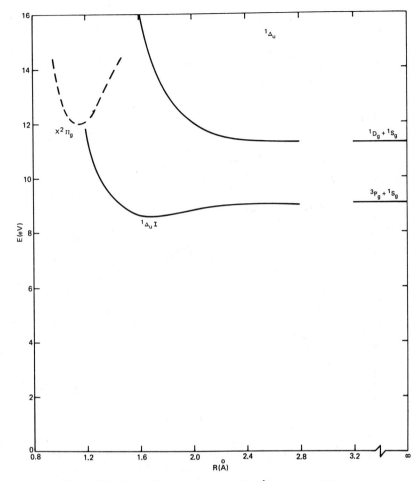

Figure 37. Potential-energy curves for $^1\Delta_u$ states of O_2.

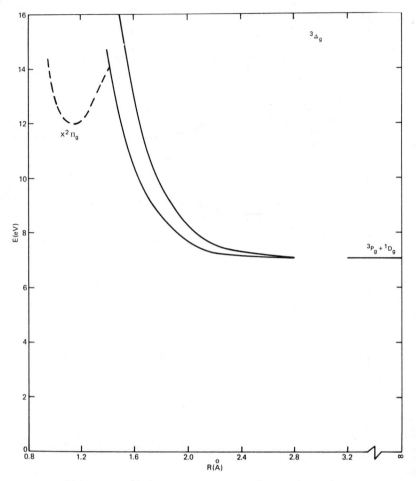

Figure 38. Potential-energy curves for $^3\Delta_g$ states of O_2.

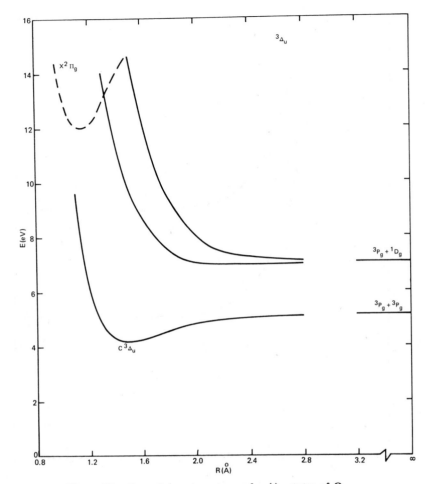

Figure 39. Potential-energy curves for $^3\Delta_u$ states of O_2.

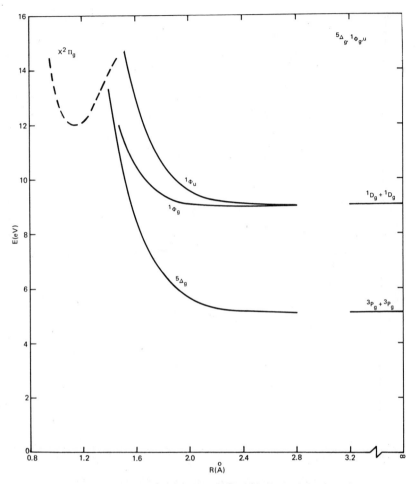

Figure 40. Potential-energy curves for $^5\Delta_g$, $^1\Phi_g$, and $^1\Phi_u$ states of O_2.

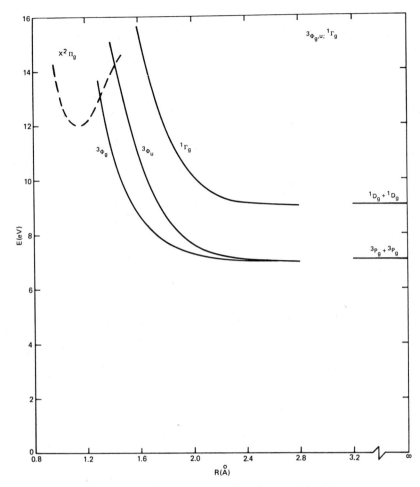

Figure 41. Potential-energy curves for $^3\Phi_g$, $^3\Phi_u$, and $^1\Gamma_g$ states of O_2.

Table IV.

Low-lying Molecular states of O_2 and Their Dissociation Limits

DISSOCIATION LIMIT	MOLECULAR STATES
$O(^3P_g)+O(^3P_g)$	$^1\Sigma_g^+(2), {}^3\Sigma_g^-(1), {}^3\Sigma_u^+(2), {}^5\Sigma_g^+(2),$ $^1\Sigma_u^-(1), {}^5\Sigma_u^-(1), {}^1\Pi_g(1), {}^3\Pi_g(1),$ $^5\Pi_g(1), {}^1\Pi_u(1), {}^3\Pi_u(1), {}^5\Pi_u(1), {}^1\Delta_g(1),$ $^5\Delta_g(1), {}^3\Delta_u(1)$
$O(^3P_g)+O(^1D_g)$	$^3\Sigma_g^+(1)\,{}^3\Sigma_u^+(1), {}^3\Sigma_g^-(2), {}^3\Sigma_u^-(2),$ $^3\Pi_g(3), {}^3\Pi_u(3), {}^3\Delta_g(2), {}^3\Delta_u(2), {}^3\Phi_g(1),$ $^3\Phi_u(1)$
$O(^1D_g)+O(^1D_g)$	$^1\Sigma_g^+(3), {}^1\Sigma_u^-(2), {}^1\Pi_g(2), {}^1\Pi_u(2),$ $^1\Delta_g(2), {}^1\Delta_u(1), {}^1\Phi_g(1), {}^1\Phi_u(1), {}^1\Gamma_g(1)$
$O(^3P_g)+O(^1S_g)$	$^3\Sigma_g^-(1), {}^3\Sigma_u^-(1), {}^3\Pi_g(1), {}^3\Pi_u(1)$
$O(^1D_g)+O(^1S_g)$	$^1\Sigma_g^+(1), {}^1\Sigma_u^+(1), {}^1\Pi_g(1), {}^1\Pi_u(1),$ $^1\Delta_g(1), {}^1\Delta_u(1)$
$O(^1S_g)+O(^1S_g)$	$^1\Sigma_g^+(1)$

Table V.

Important Molecular-orbital Electron Configurations for O_2

STATE	ELECTRON CONFIGURATION[a]			
	$3\sigma_g$	$1\pi_u$	$1\pi_g$	$3\sigma_u$
$b^1\Sigma_g^+$ (I)	2	4	2	0
$c^1\Sigma_u^-$ (I)	2	3	3	0
$^1\Sigma_u^+$ (I)	$\begin{cases} 2 \\ 1 \end{cases}$	$\begin{matrix} 3 \\ 4 \end{matrix}$	$\begin{matrix} 3 \\ 2 \end{matrix}$	$\left.\begin{matrix} 0 \\ 1 \end{matrix}\right\}c$
$X^3\Sigma_g^-$ (I)	2	4	2	0
$^3\Sigma_g^-$ (II)	1	3	3	1
$A^3\Sigma_u^+$ (I)	2	3	3	0
$B^3\Sigma_u^-$ (I)	$\begin{cases} 1 \\ 2 \end{cases}$	$\begin{matrix} 4 \\ 3 \end{matrix}$	$\begin{matrix} 2 \\ 3 \end{matrix}$	$\left.\begin{matrix} 1 \\ 0 \end{matrix}\right\}d$
$\left\{\begin{matrix}{}^1\Pi_g\ (I) \\ {}^1\Pi_g\ (II)\end{matrix}\right\}$	$\begin{cases} 1 \\ 2 \end{cases}$	$\begin{matrix} 4 \\ 3 \end{matrix}$	$\begin{matrix} 2 \\ 3 \end{matrix}$	$\left.\begin{matrix} 1 \\ 0 \end{matrix}\right\}b$
$a^1\Delta_g$ (I)	2	4	2	0
$^1\Delta_g$ (II)	1	3	3	1
$^1\Delta_u$ (I)	2	3	3	0
$C^3\Delta_u$ (I)	2	3	3	0

[a]All states include $(1\sigma_g)^2$, $(1\sigma_u)^2$, $(2\sigma_g)^2$, and $(2\sigma_u)^2$.
[b]There is an avoided crossing between these two states at $R = 1.6$ Å (see Fig. 31).
[c]The first configuration given is twice as important as the second at the equilibrium separation.
[d]These configurations are weighted about equally.

286

Table VI.
Spectroscopic Constants For Bound States of O_2

STATES		T_e (eV)	ω_e (cm^{-1})	$\omega_e X_e$ (cm^{-1})	α_e (cm^{-1})	r_e (Å)	B_e (cm^{-1})	D_e (eV)	D_0 (eV)
$^1\Sigma_u^+$ (I)	(C)	10.34	744.9	21.8	0.0184	1.62	0.80	1.03	0.98
$^1\Delta_g$ (II)	(C)	8.79	755.6	47.9	0.0170	1.88	0.60	0.36	0.31
$^1\Delta_u$ (I)	(C)	8.72	533.1	22.2	0.0261	1.68	0.75	0.68	0.64
$^1\Pi_g$ (I,II)[a]	(C)	8.37	1673.6	169.2	0.0282	1.62	0.80	0.78	0.68
$B^3\Sigma_u^-$	(E)	6.17	709.1	10.6	0.0119	1.60	0.82	1.01	0.96
$^3\Sigma_g^-$ (II)	(C)	6.05	786.2	10.7	0.0050	1.91	0.58	1.13	1.08
$A^3\Sigma_u^+$	(E)	4.39	799.1	12.2	0.0142	1.52	0.91	0.82	0.77
$C^3\Delta_u$	(E)	4.31	750.0	14.0	0.0201	1.50	0.94	0.91	0.86
$c^1\Sigma_u^-$	(E)	4.10	794.3	12.7	0.0139	1.52	0.92	1.11	1.07
$b^1\Sigma_g^+$	(E)	1.64	1432.7	13.9	0.0182	1.23	1.40	3.58	3.49
$a^1\Delta_g$	(E)	0.98	1509.3	12.9	0.0171	1.22	1.43	4.23	4.14
$X^3\Sigma_g^-$	(E)	0.0	1580.2	11.9	0.0159	1.21	1.45	5.21	5.11

[a]Avoided crossing for this symmetry.

C. Nitric Oxide Molecule

Potential curves for the NO molecule have been reported by Gilmore [12] and, in an earlier study, by Vanderslice et al.[56] Long-range interactions were studied by Meador[57] using a valence-bond framework. Studies of certain Rydberg states have been reported by Brion et al.[58] Spectroscopic data for NO are available in the compilation of Wallace[32] and the extensive studies by Miescher and co-workers.[59-67]

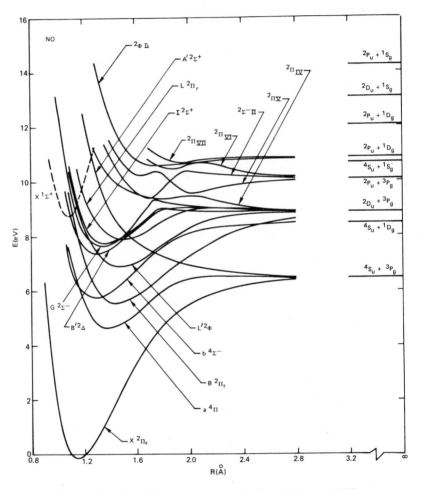

Figure 42.   Potential-energy curves for bound states of NO.

The ground-state potential curve and dipole-moment function for NO have been recently calculated within a CI framework.[68] Configuration-interaction studies of excited states have been carried out by Thulstrup.[69] An earlier CI study of the $X\,^2\Pi$ and $A\,^2\Sigma^+$ states has been reported by Green.[70]

A composite potential curve for the bound states of NO is shown in Fig. 42. Potential curves, including repulsive states, are shown in Figs. 43 to 52 for each symmetry type. The low-lying molecular states of NO and their dissociation limits are given in Table VII. The most important electronic

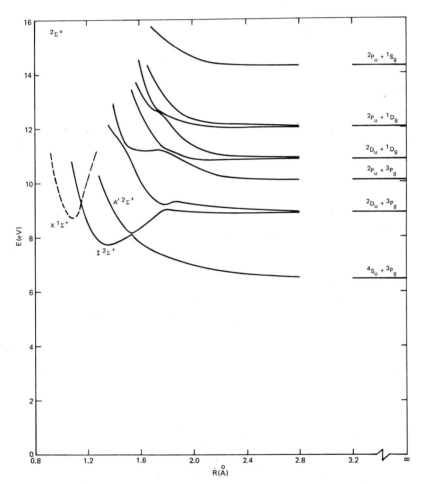

Figure 43. Potential-energy curves for $^2\Sigma^+$ states of NO.

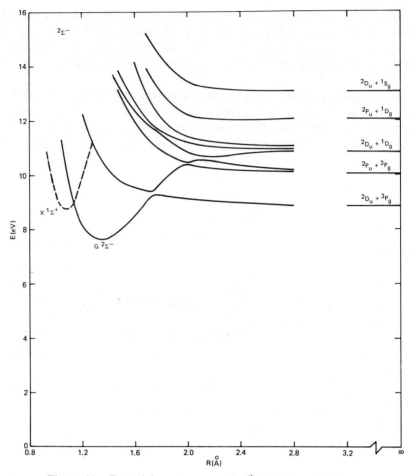

Figure 44. Potential-energy curves for $^2\Sigma^-$ states of NO.

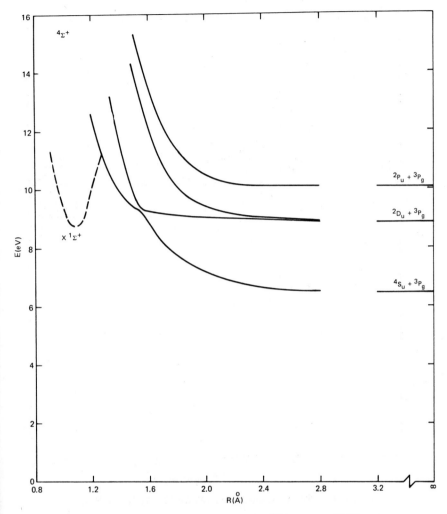

Figure 45. Potential-energy curves for $^4\Sigma^+$ states of NO.

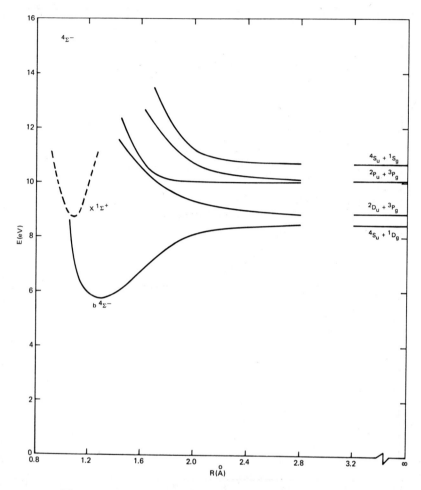

Figure 46. Potential-energy curves for $^4\Sigma^-$ states of NO.

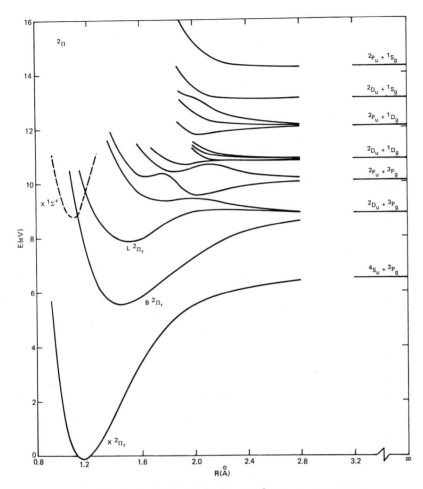

Figure 47. Potential-energy curves for $^2\Pi$ states of NO.

293

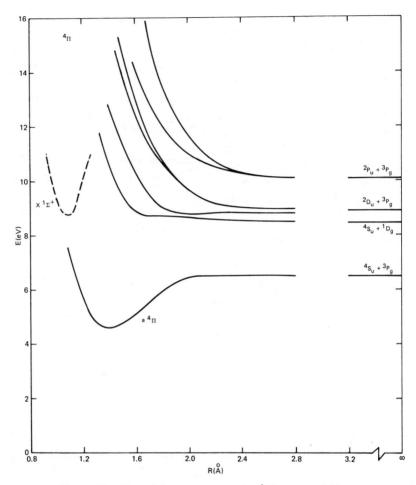

Figure 48. Potential-energy curves for $^4\Pi$ states of NO.

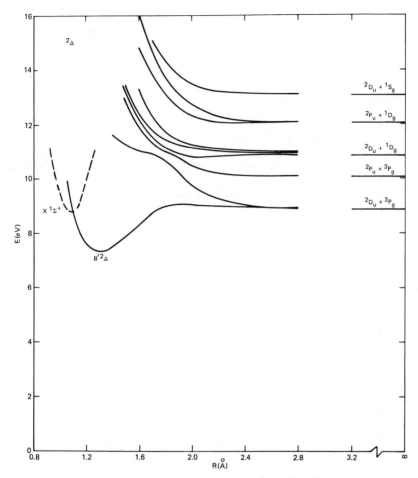

Figure 49. Potential-energy curves for $^2\Delta$ states of NO.

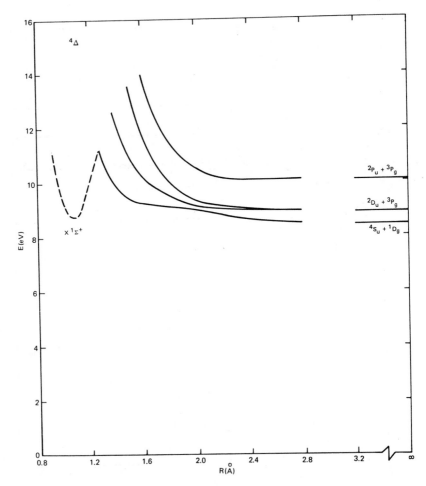

Figure 50. Potential-energy curves for $^4\Delta$ states of NO.

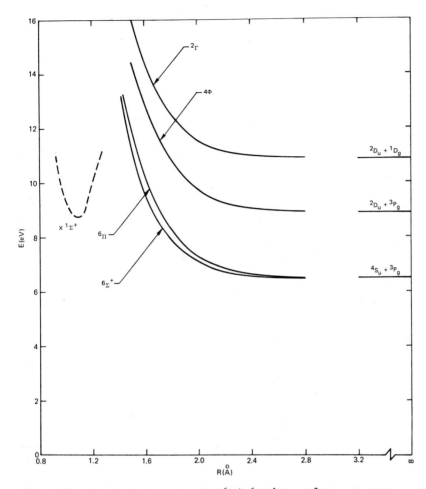

Figure 51. Potential-energy curves for $^6\Sigma^+$, $^6\Pi$, $^4\Phi$, and $^2\Gamma$ states of NO.

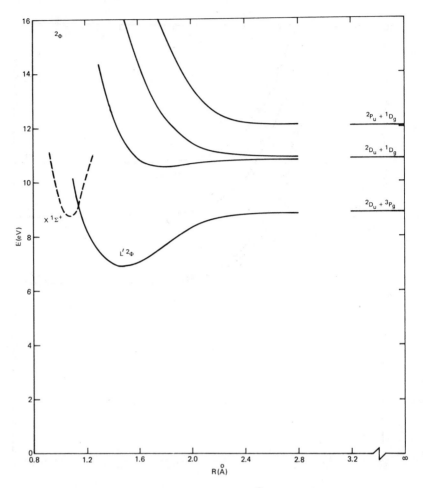

Figure 52. Potential-energy curves for $^2\Phi$ states of NO.

ow-lying Molecular States of No and Their Dissociation Limits

DISSOCIATION LIMIT	MOLECULAR STATES
$N(^4S_u) + O(^3P_g)$	$^2\Sigma^+(1), {}^2\Pi(1), {}^4\Sigma^+(1), {}^4\Pi(1), {}^6\Sigma^+(1),$ $^6\Pi(1)$
$N(^4S_u) + O(^1D_g)$	$^4\Sigma^-(1), {}^4\Pi(1), {}^4\Delta(1)$
$N(^2D_u) + O(^3P_g)$	$^2\Sigma^-(1), {}^4\Sigma^-(1), {}^2\Sigma^+(2), {}^4\Sigma^+(2),$ $^2\Pi(3), {}^4\Pi(3), {}^2\Delta(2), {}^4\Delta(2),$ $^2\Phi(1), {}^4\Phi(1)$
$N(^2P_u) + O(^3P_g)$	$^2\Sigma^-(2), {}^4\Sigma^-(2), {}^2\Sigma^+(1), {}^4\Sigma^+(1), {}^2\Pi(2),$ $^4\Pi(2), {}^2\Delta(1), {}^4\Delta(1)$
$N(^4S_u) + O(^1S_g)$	$^4\Sigma^-(1)$
$N(^2D_u) + O(^1D_g)$	$^2\Sigma^-(3), {}^2\Sigma^+(2), {}^2\Pi(4), {}^2\Delta(3), {}^2\Phi(2),$ $^2\Gamma(1)$
$N(^2P_u) + O(^1D_g)$	$^2\Sigma^-(1), {}^2\Sigma^+(2), {}^2\Pi(3), {}^2\Delta(2), {}^2\Phi(1)$
$N(^2D_u) + O(^1S_g)$	$^2\Sigma^-(1), {}^2\Pi(1), {}^2\Delta(1)$
$N(^2P_u) + O(^1S_g)$	$^2\Sigma^+(1), {}^2\Pi(1)$

Table VIII.
Important Molecular-orbital Electron Configurations for NO

STATE	ELECTRON CONFIGURATIONS[a]			
	$3\sigma_g$	$1\pi_u$	$1\bar\pi_g$	$3\bar\sigma_u$
$X^2\Pi$	2	4	1	0
$a^4\Pi$	2	3	2	0
$B^2\Pi$	2	3	2	0
$L^2\Pi$	2	3	2	0
$^2\Pi$ (IV)	2	3	2	0
$L'^2\Phi$	2	3	2	0
$b^4\Sigma^-$	1	4	2	0
$B'^2\Delta$	1	4	2	0
$I^2\Sigma^+$	1	4	2	0 |[b]
$A'^2\Sigma^+$	2	4	0	1	
$G^2\Sigma^-$	1	4	2	0
$^2\Sigma^-$ (II)	2	3	1	1
$^2\Pi$ (V)	0	4	3	0
$^2\Pi$ (VI)	2	2	3	0
$^2\Pi$ (VII)	2	2	3	0
$^2\Phi$ (II)	2	2	3	0

[a] All states include $(1\sigma)^2$, $(1\bar\sigma)^2$, $(2\sigma)^2$, and $(2\bar\sigma)^2$ and the near homopolar $g-u$ labels can be applied.

[b] There is a weak (~ 200 cm^{-1}) avoided crossing between these two $^2\Sigma^+$ states at $R \sim 1.5$ Å. The $I^2\Sigma^+$ state is nearly pure g and the $A'^2\Sigma^+$ respulsive state, near pure u in character.

Table IX.
Spectroscopic Constants For Bound States of NO

STATE		T_e(eV)	ω_e(cm^{-1})	$\omega_e X_e$(cm^{-1})	α_e(cm^{-1})	r_e(Å)	B_e(cm^{-1})	D_e(eV)	D_0(eV)
$^2\Phi$ (II)	(C)	10.69	553.7	50.3	0.038	1.79	0.71	0.27	0.24
$^2\Pi$ (V, VII)[a]	(C)	10.38	944.1	43.9	0.024	1.65	0.83	0.58	0.52
$^2\Pi$ (IV, V)[a]	(C)	9.50	544.2	4.1	0.011	1.80	0.69	0.69	0.66
$L\,^2\Pi$ (III)	(C)	8.00	957.1	6.7	0.006	1.50	1.00	1.01	0.95
$I\,^2\Sigma^+$ (I, II)[a]	(C)	7.87	1185.3	12.0	0.020	1.36	1.21	1.13	1.06
$G\,^2\Sigma^-$	(E)	7.80	1085.5	11.1	0.020	1.34	1.25	1.20	1.13
$B'\,^2\Delta$	(E)	7.48	1217.4	15.6	0.021	1.30	1.33	1.51	1.44
$L'\,^2\Phi$	(C)	7.01	1004.4	11.0	0.0055	1.49	1.02	1.99	1.93
$b\,^4\Sigma^-$	(E)	5.90	1168.0	13.3	0.020	1.30	1.34	2.68	2.61
$B\,^2\Pi_{1/2}$	(E)	5.69	1036.9	7.5	0.012	1.45	1.08	3.31	3.24
$a\,^4\Pi$	(E)	4.74	1019.0	12.8	0.020	1.41	1.14	1.87	1.80
$X\,^2\Pi_{1/2}$	(E)	0.00	1904.0	13.97	0.018	1.15	1.70	6.62	6.50

[a]Avoided crossing for this symmetry.

300

configurations for NO are shown in Table VIII, and our compilation of the available spectroscopic and calculated data is given in Table IX.

D. O_2^- Ion

The existence of a stable O_2^- molecular ion has been known for some time, and various experimental studies have been carried out to define the electron affinity and structure of the ground electronic state.[71-74] Relatively little is known about the excited electronic states of this ion. Bates and Massey[75] proposed that several excited state configurations, in addition to the $X\,{}^2\Pi_g$ ground state, should be stable. Boness et al.[76] suggested

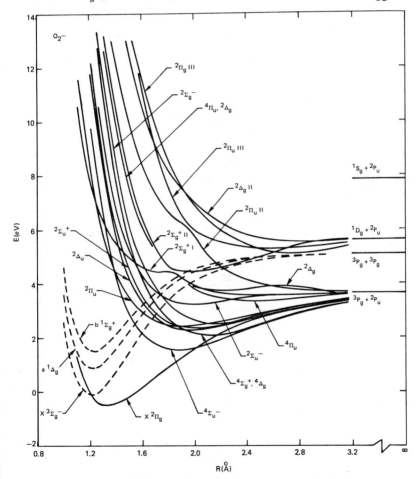

Figure 53. Potential-energy curves for bound states of O_2^-.

the possibility of a low-lying bound excited state for O_2^- to explain the discrepancy between their electron-attachment experiments and the evidence for a stable O_2^- in the solid phase.[77] Mulliken, however, has indicated that no low-lying electronic states of doublet symmetry should be stable.[78]

Ab initio calculations for the ground- and several excited-state configurations have been reported by Michels[11] and Krauss et al.[79] These early calculations indicated a large number of bound excited states, particularly of quartet symmetry, arising from the ground $^3P + {}^2P$ dissociation limits for

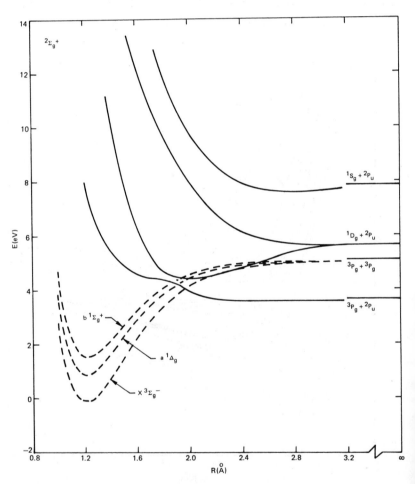

Figure 54. Potential-energy curves for $^2\Sigma_g^+$ states of O_2^-.

$O + O^-$. Additional MCSCF studies for several of these states have been reported by Zemke et al.[80a] and by Krauss et al.[80b]

The existence of a large number of bound electronic states, arising from $^3P + {}^2P$ atomic limits, has now been determined statistically by Ferguson.[81] This is now theoretically confirmed by our analysis of this system, which indicates 12 bound states of O_2^-, in addition to the $^2\Pi_g$ ground state, arising from the ground ^2P state of O^- and the ground ^3P and excited ^1D

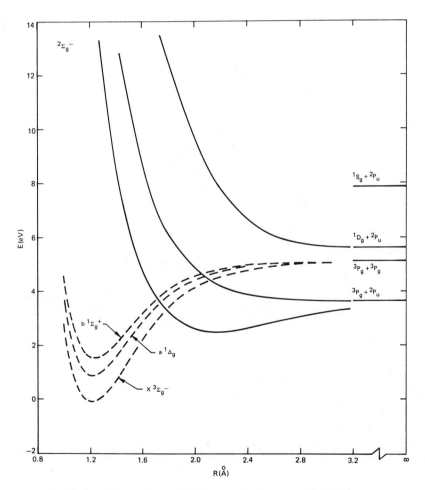

Figure 55. Potential-energy curves for $^2\Sigma_g^-$ states of O_2^-.

304   Electronic Structure of Excited States of Selected Atmospheric Systems

and ^1S state of O. A composite potential curve for the bound states of O_2^- is shown in Fig. 53. All the low-lying valence states arising from the preceding atomic limits are shown in Figs. 54 to 67. The three low-lying states of O_2 are indicated in these figures for comparison. All the possible low-lying molecular states of O_2^- and their dissociation limits are given in Table X. Our analysis of the most important molecular-orbital configurations is given in Table XI. Predicted spectroscopic constants for the ground and 12 bound excited states are given in Table XII.

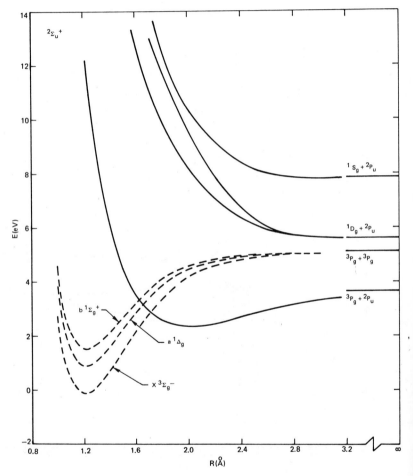

Figure 56.   Potential-energy curves for $^2\Sigma_u^+$ states of O_2^-.

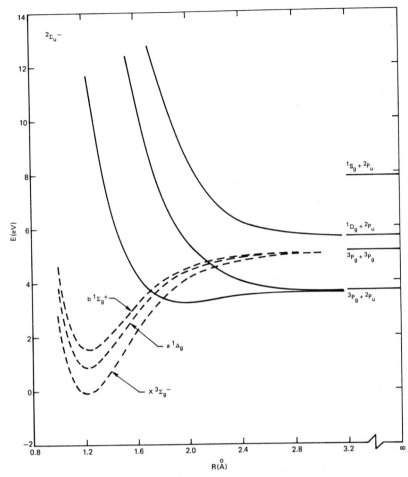

Figure 57. Potential-energy curves for $^2\Sigma_u^-$ states of O_2^-.

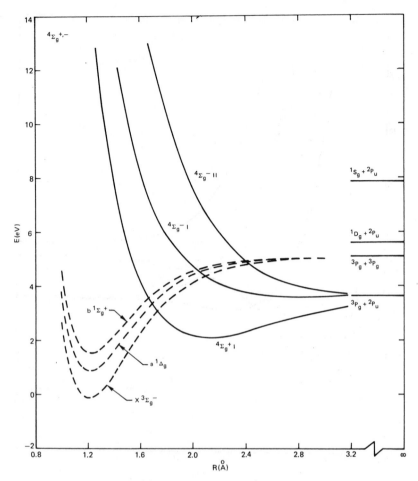

Figure 58. Potential-energy curves for $^4\Sigma_g^{+,-}$ states of O_2^-.

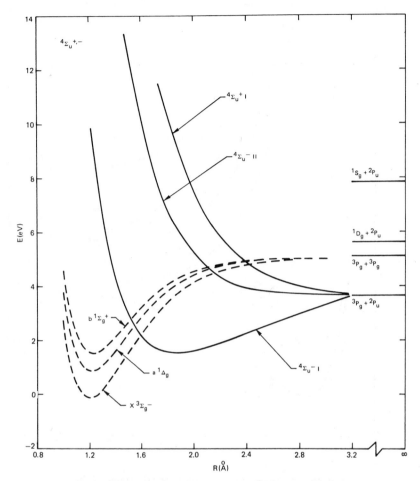

Figure 59. Potential-energy curves for $^4\Sigma_u^{+,-}$ states of O_2^-.

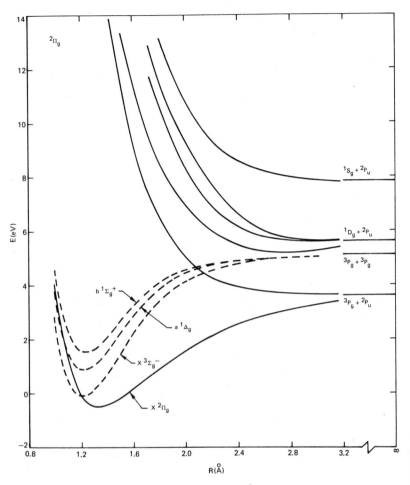

Figure 60. Potential-energy curves for $^2\Pi_g$ states of O_2^-.

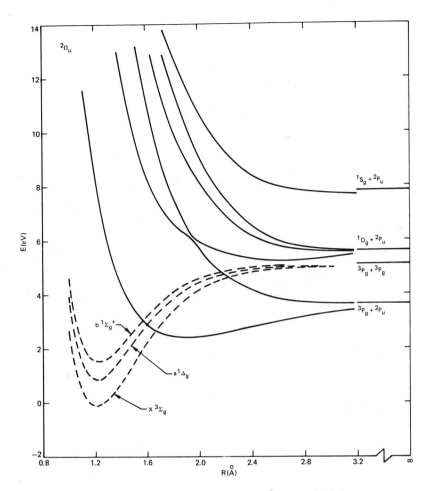

Figure 61. Potential-energy curves for $^2\Pi_u$ states of O_2^-.

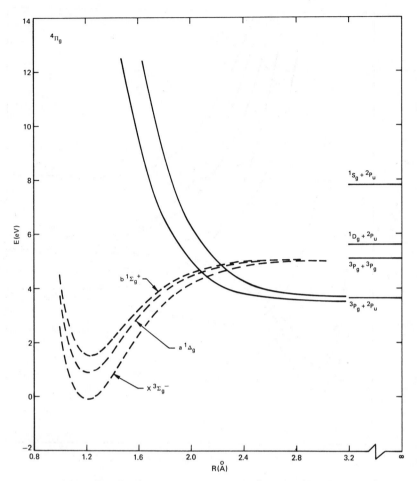

Figure 62. Potential-energy curves for $^4\Pi_g$ states of O_2^-.

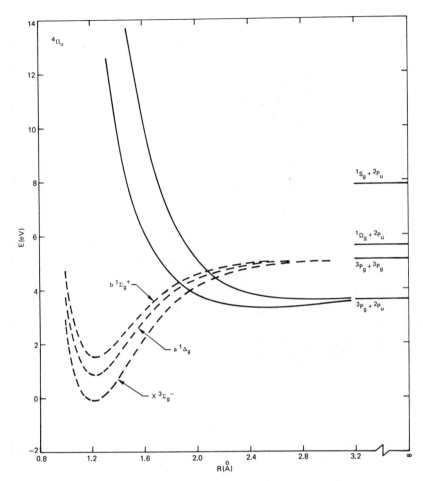

Figure 63. Potential-energy curves for $^4\Pi_u$ states of O_2^-.

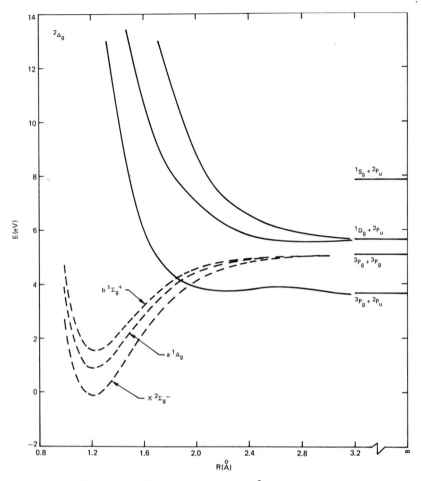

Figure 64. Potential-energy curves for $^2\Delta_g$ states of O_2^-.

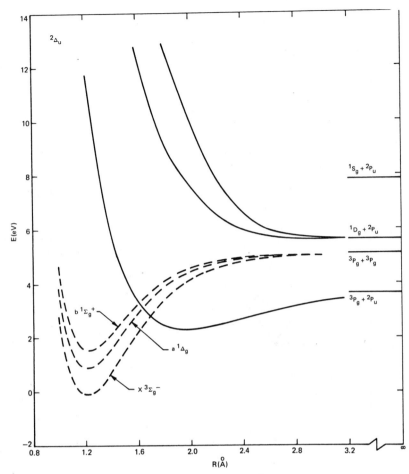

Figure 65. Potential-energy curves for $^2\Delta_u$ states of O_2^-.

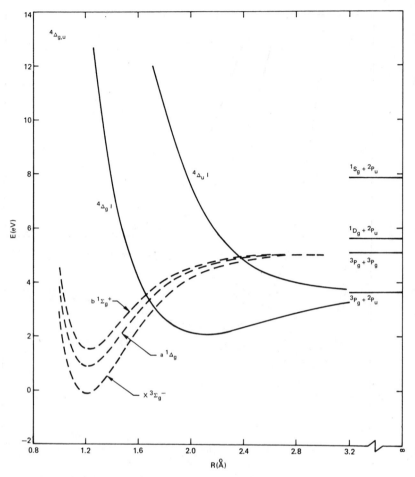

Figure 66. Potential-energy curves for $^4\Delta_g$ and $^4\Delta_u$ states of O_2^-.

314

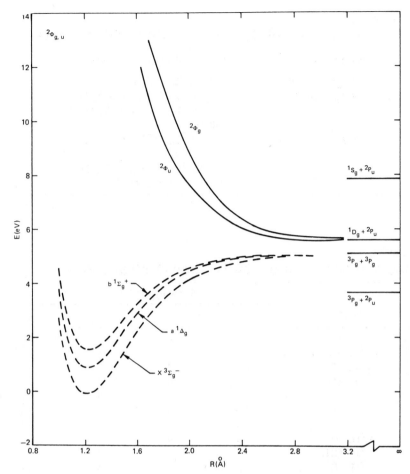

Figure 67. Potential-energy curves for $^2\Phi_g$ and $^2\Phi_u$ states of O_2^-.

315

Table X.

Low-lying States of O_2^- and Their Dissociation Limits

DISSOCIATION LIMIT	MOLECULAR STATES
$O(^1S_g)+O^-(^2P_u)$	$^2\Sigma_g^+(1), ^2\Sigma_u^+(1), ^2\Pi_g(1), ^2\Pi_u(1)$
$O(^1D_g)+O(^2P_u)$	$^2\Sigma_g^-(1), ^2\Sigma_u^-(1), ^2\Sigma_g^+(2), ^2\Sigma_u^+(2),$ $^2\Pi_g(3), ^2\Pi_u(3), ^2\Delta_g(2), ^2\Delta_u(2), ^2\Phi_g(1),$ $^2\Phi_u(1)$
$O(^3P_g)+O^-(^2P_u)$	$^2\Sigma_g^-(2), ^2\Sigma_u^-(2), ^4\Sigma_g^-(2), ^4\Sigma_u^-(2),$ $^2\Sigma_g^+(1), ^2\Sigma_u^+ +(1), ^4\Sigma_g^+(1), ^4\Sigma_u^+(1),$ $^2\Pi_g(2), ^2\Pi_u(2), ^4\Pi_g(2), ^4\Pi_u(2),$ $^2\Delta_g(1), ^2\Delta_u(1), ^4\Delta_g(1), ^4\Delta_u(1)$

Table XI.

Important Molecular-orbital Electron Configurations for O_2^-

STATE	ELECTRON CONFIGURATION[a]			
	$3\sigma_g$	$1\pi_u$	$1\pi_g$	$3\sigma_u$
$^2\Sigma_g^-$ (I)	2	3	3	1
$\left\{\begin{array}{l}{}^2\Sigma_g^+ \text{ (I)}\\{}^2\Sigma_g^+ \text{ (II)}\\{}^2\Sigma_u^+ \text{ (I)}\end{array}\right\}$	$\left\{\begin{array}{l}1\\2\end{array}\right.$	$\begin{array}{l}4\\3\end{array}$	$\begin{array}{l}4\\3\end{array}$	$\left.\begin{array}{l}0\\1\end{array}\right\}^b$
	$\left\{\begin{array}{l}2\\2\end{array}\right.$	$\begin{array}{l}4\\2\end{array}$	$\begin{array}{l}2\\4\end{array}$	$\left.\begin{array}{l}1\\1\end{array}\right\}^c$
$^2\Sigma_u^-$ (I)	$\left\{\begin{array}{l}2\\2\end{array}\right.$	$\begin{array}{l}4\\2\end{array}$	$\begin{array}{l}2\\4\end{array}$	$\left.\begin{array}{l}1\\1\end{array}\right\}^d$
$X^2\Pi_g$ (I)	2	4	3	0
$^2\Pi_g$ (III)	1	3	4	1
$^2\Pi_u$ (I)	2	3	4	0
$^2\Pi_u$ (III)	1	4	3	1
$^2\Delta_u$ (I)	$\left\{\begin{array}{l}2\\2\end{array}\right.$	$\begin{array}{l}4\\2\end{array}$	$\begin{array}{l}2\\4\end{array}$	$\left.\begin{array}{l}1\\1\end{array}\right\}^c$
$^4\Sigma_g^+$ (I)	2	3	3	1
$^4\Sigma_u^-$ (I)	$\left\{\begin{array}{l}2\\2\end{array}\right.$	$\begin{array}{l}4\\2\end{array}$	$\begin{array}{l}2\\4\end{array}$	$\left.\begin{array}{l}1\\1\end{array}\right\}^c$
$^4\Delta_g$ (I)	2	3	3	1

[a] All states include $(1\sigma_g)^2, (1\sigma_u)^2, (2\sigma_g)^2,$ and $(2\sigma_u)^2$ electrons.
[b] There is an avoided crossing between these two states at $R = 1.85$ Å (see Fig. 54).
[c] The first configuration given is twice as important as the second at the equilibrium separation.
[d] These configurations are weighted about equally.

316

Table XII.
Spectroscopic Constants For Bound States of O_2^-

STATES		T_e(eV)	ω_e(cm⁻¹)	$\omega_e X_e$(cm⁻¹)	α_e(cm⁻¹)	r_e(Å)	B_e(cm⁻¹)	D_e(eV)	D_0(eV)
$^2\Pi_u$ (II,III)[a]	(C)	5.74	417.5	40.3	0.0079	2.63	0.31	0.38	0.36
$^2\Pi_g$ (III)	(C)	5.63	442.0	9.5	0.0024	2.75	0.28	0.49	0.47
$^2\Sigma_g^+$ (I,II)[a]	(C)	4.92	603.2	24.1	0.0195	1.95	0.55	1.18	1.14
$^4\Pi_u$ (I)	(C)	3.85	345.6	8.9	0.0004	2.48	0.34	0.31	0.29
$^2\Sigma_u^-$ (I)	(C)	3.77	484.4	12.9	0.0104	1.99	0.53	0.39	0.36
$^2\Sigma_g^-$ (I)	(C)	3.02	451.5	3.5	0.0045	2.18	0.44	1.14	1.11
$^2\Pi_u$ (I)	(C)	2.93	452.1	4.0	0.0079	1.92	0.57	1.23	1.20
$^2\Sigma_u^+$ (I)	(C)	2.88	514.4	4.9	0.0063	2.00	0.52	1.28	1.25
$^2\Delta_u$ (I)	(C)	2.81	524.7	4.8	0.0065	1.98	0.54	1.35	1.32
$^4\Sigma_g^+$ (I)	(C)	2.61	504.1	3.0	0.0037	2.13	0.46	1.55	1.52
$^4\Delta_g$ (I)	(C)	2.60	503.7	2.9	0.0039	2.12	0.47	1.56	1.53
$^4\Sigma_u^-$ (I)	(C)	2.05	604.8	3.4	0.0061	1.88	0.60	2.11	2.07
$X^2\Pi_g$ (I)	(E)	0.00	1089.0	12.1	0.017	1.34	1.17	4.16	4.09

[a]Avoided crossing for this symmetry.

E. NO⁺ Ion

Since the dominant atmospheric molecular ion above 100 km is NO^+, its radiation characteristics and interactions with electrons and other atomic and molecular species are of primary importance in understanding the detailed chemistry of the atmosphere. Relatively little is known about the excited electronic states of the NO^+ ion, even though its spectrum must be nearly as rich as that for the isoelectronic nitrogen molecule. Accurate potential curves are available for the ground $X^1\Sigma^+$ state and for the $A^1\Pi$ and $a\,^3\Sigma^+$ excited states of NO^+.[12] Gilmore also gives some experimental estimates of other excited states of NO^+ based on the data of Wallace[32] and Miescher.[82] More recently, Thulstrup et al.[83,84] have assigned additional excited states of NO^+ based on photoelectron spectra. In addition to these valence excited states, there exists a large manifold of Rydberg states to the $N^+(^3P)+O^+(^4S)$ limit. It is obvious that only a small fraction of the low-lying excited states for this system have been experimentally investigated.

Few theoretical studies have been carried out for the excited states of this system. Lefebvre-Brion and Moser[85,86] have studied the lowest-lying $^1\Pi$ and $^3\Pi$ states in an SCF framework. More recently, Thulstrup and Öhrn[83,84] have examined low-lying singlet and triplet states of Σ^+, Σ^-, Π, and Δ symmetry. Their studies excluded higher spin and angular momentum states, which may be important in perturbation analysis. Their predicted locations for these states are in fair agreement with the best experimental estimates.

An extensive compilation of RKR potential curves for NO^+ has recently been reported by Albritton, et. al.[87] Their studies include all of the experimentally known electronic states of NO^+ below 24 eV. Potential energy curves and spectroscopic constants are tabulated and estimates of accuracy in the location of the excited states are reported.

A complete and quantitatively uniform *ab initio* CI study of all the low-lying valence states of NO^+ was carried out as part of the research conducted for this compendium. This study was similar to that carried out for the low-lying valence states of the nitrogen molecule.[10] Many of the states are quantitatively similar since NO^+ is isoelectronic with N_2.

A composite potential-energy curve for the bound states dissociating to the lowest two atomic limits is given in Fig. 68. Figures 69 to 83 illustrate the low-lying valence-state potential-energy curves for the NO^+ system. These calculated potential curves have the property of correct dissociation to atomic limits. The dissociation limits of the low-lying states of NO^+ are given in Table XIII, and the important molecular-orbital configurations are shown in XIV. Our summary of known experimental spectroscopic data for the low-lying bound valence states of NO^+ is given in Table XV.

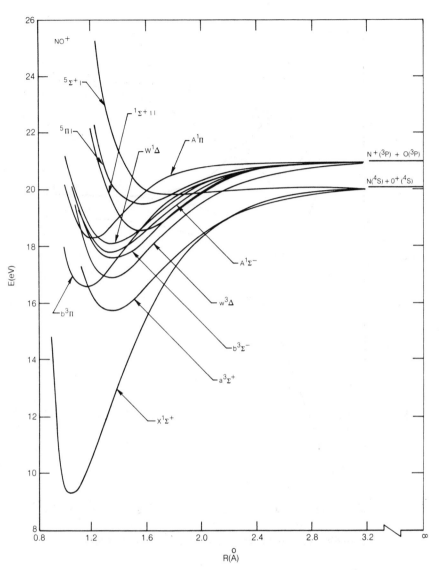

Figure 68. Potential-energy curves for bound states of NO⁺.

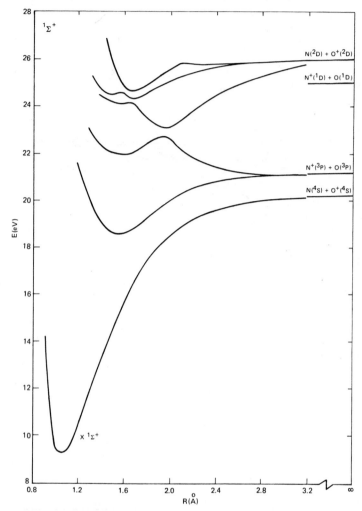

Figure 69. Potential-energy curves for $^1\Sigma^+$ states of NO^+.

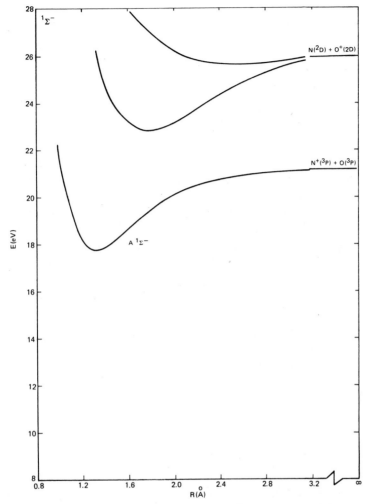

Figure 70. Potential-energy curves for $^1\Sigma^-$ states of NO^+.

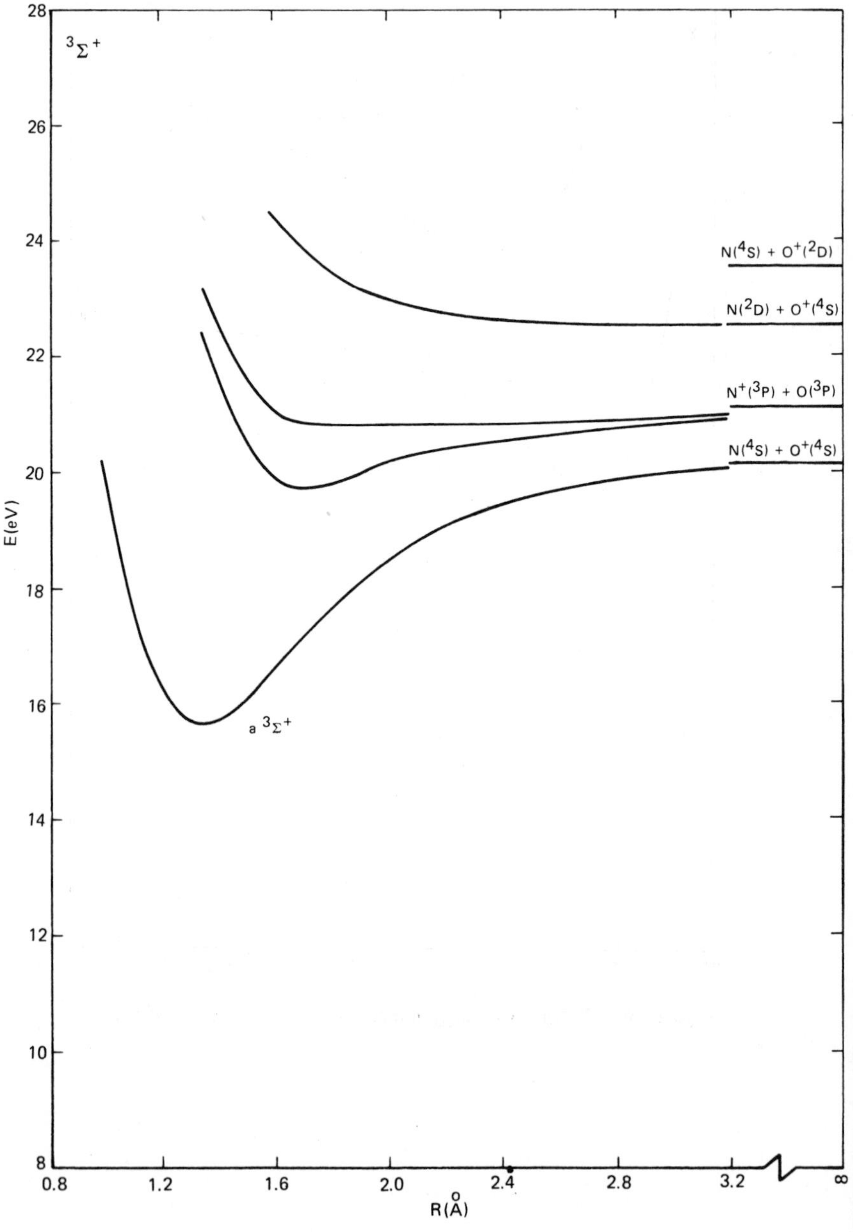

Figure 71. Potential-energy curves for $^3\Sigma^+$ states of NO^+.

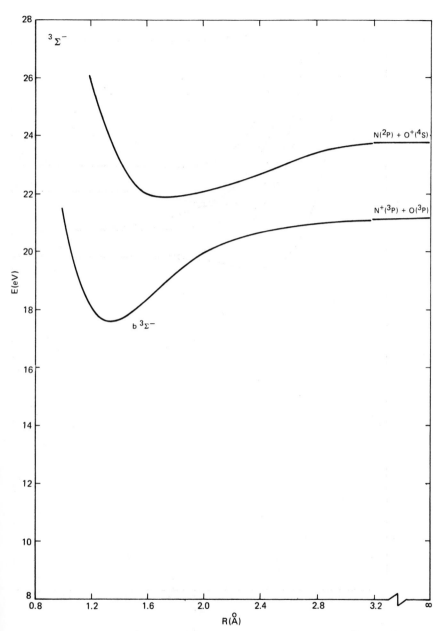

Figure 72. Potential-energy curves for $^3\Sigma^-$ states of NO^+.

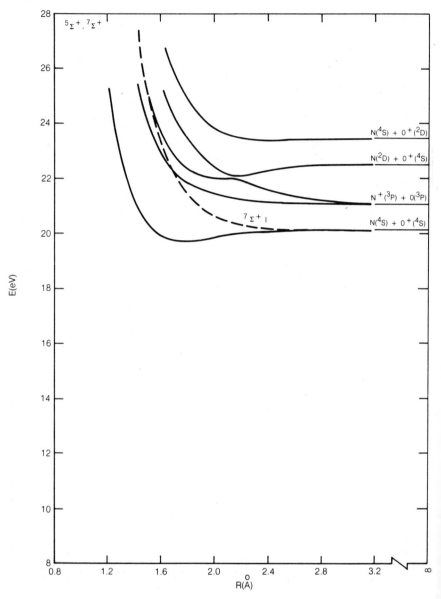

Figure 73. Potential-energy curves for $^5\Sigma^+$ and $^7\Sigma^+$ states of NO^+.

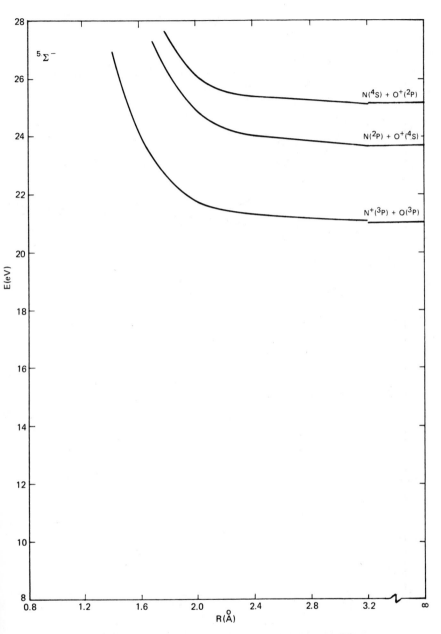

Figure 74. Potential-energy curves for $^5\Sigma^-$ states of NO$^+$.

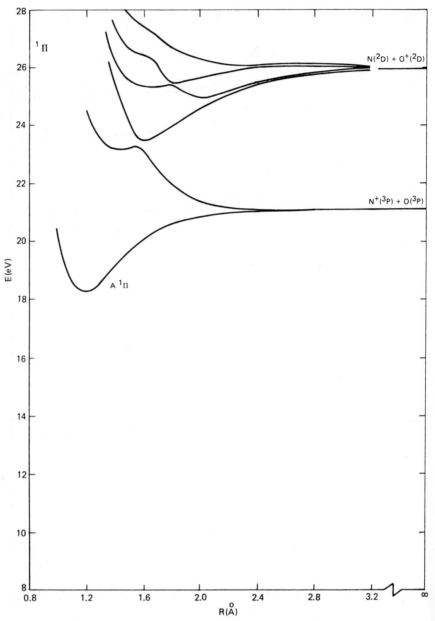

Figure 75. Potential-energy curves for $^1\Pi$ states of NO^+.

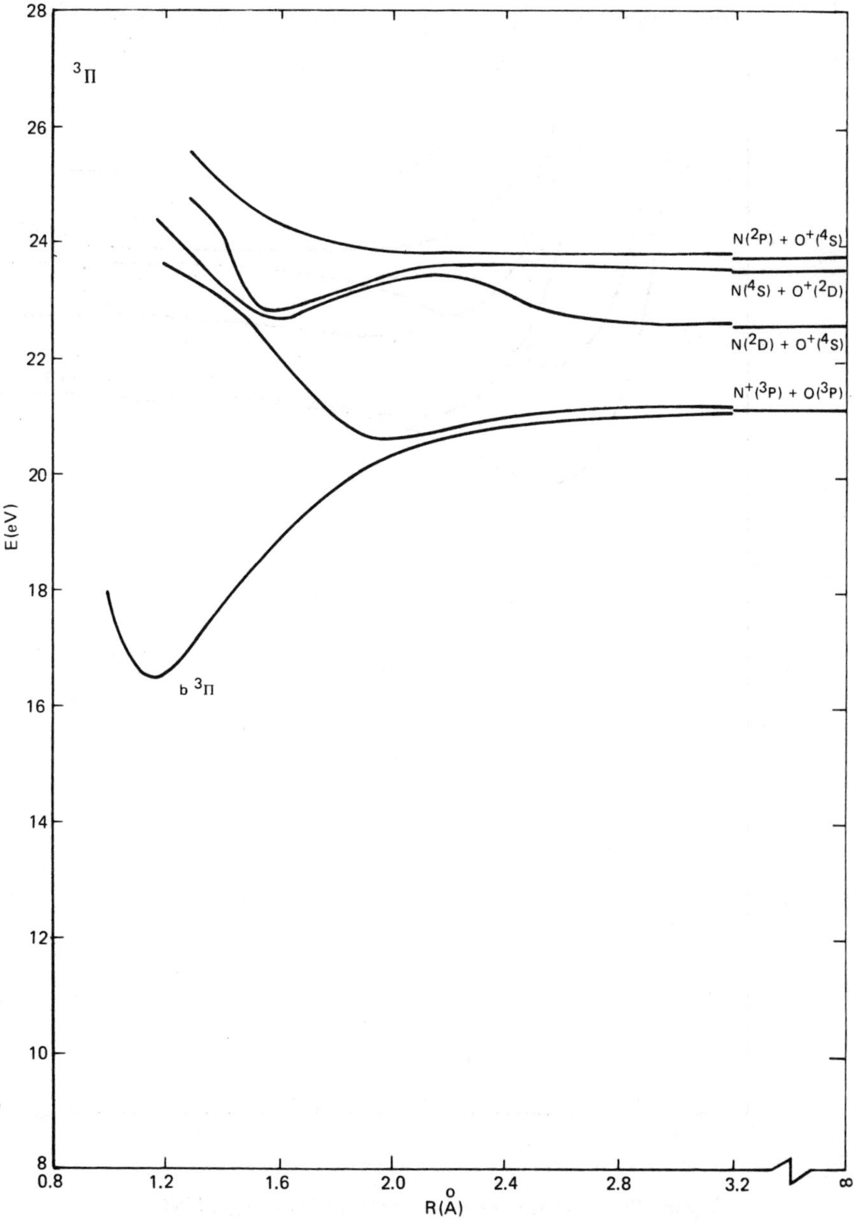

3Π
N(2P) + O+(4S)
N(4S) + O+(2D)
N(2D) + O+(4S)
N+(3P) + O(3P)
b 3Π
E(eV)
R(Å)
28
26
24
22
20
18
16
14
12
10
8
0.8
1.2
1.6
2.0
2.4
2.8
3.2
∞

Figure 76. Potential-energy curves for $^3\Pi$ states of NO^+.

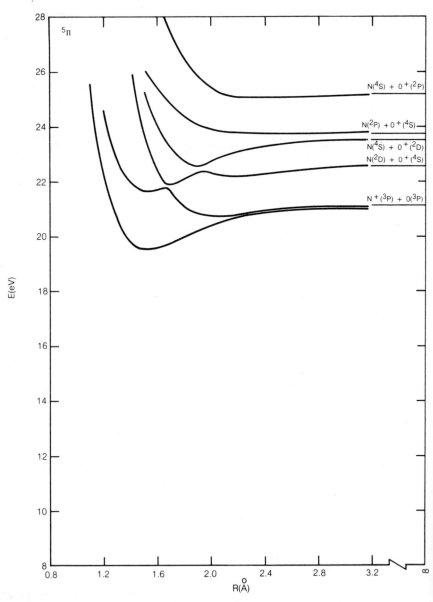

Figure 77. Potential-energy curves for $^5\Pi$ states of NO^+.

Labels in figure: $^5\Pi$

$N(^4S) + O^+(^2P)$
$N(^2P) + O^+(^4S)$
$N(^4S) + O^+(^2D)$
$N(^2D) + O^+(^4S)$
$N^+(^3P) + O(^3P)$

$E(eV)$
$R(\mathring{A})$

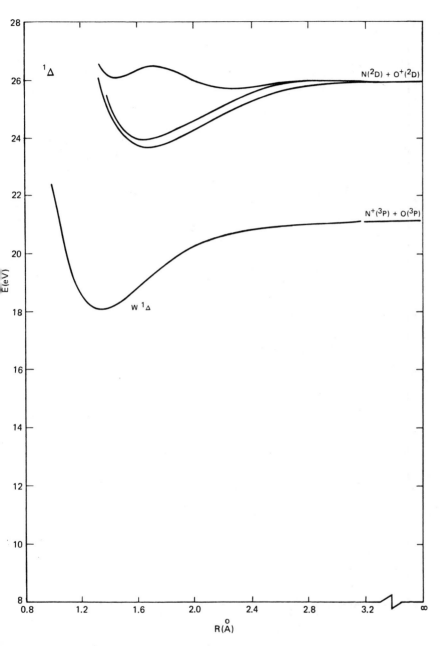

Figure 78. Potential energy curves for $^1\Delta$ states of NO^+.

329

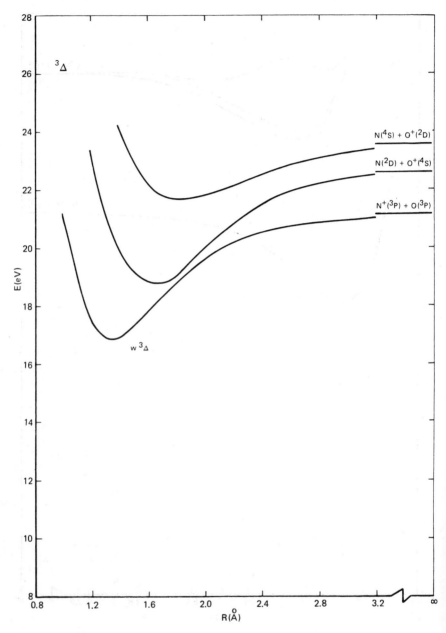

Figure 79. Potential-energy curves for $^3\Delta$ states of NO^+.

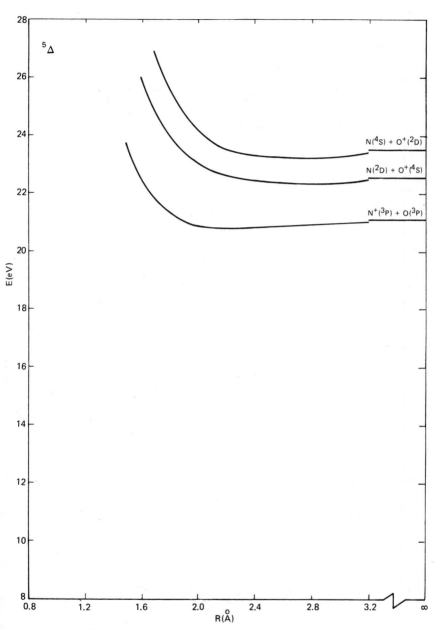

Figure 80. Potential-energy curves for $^5\Delta$ states of NO^+.

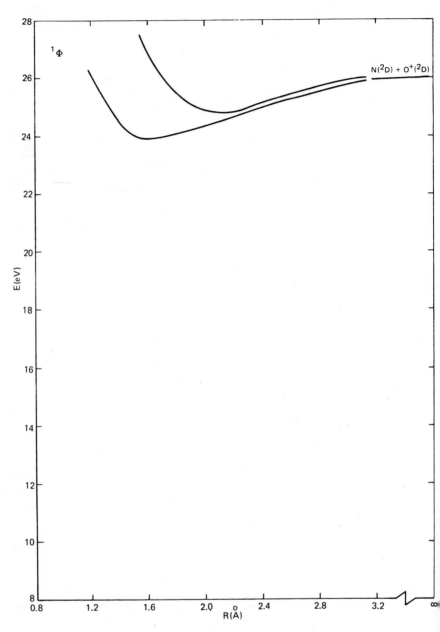

Figure 81. Potential-energy curves for $^1\Phi$ states of NO^+.

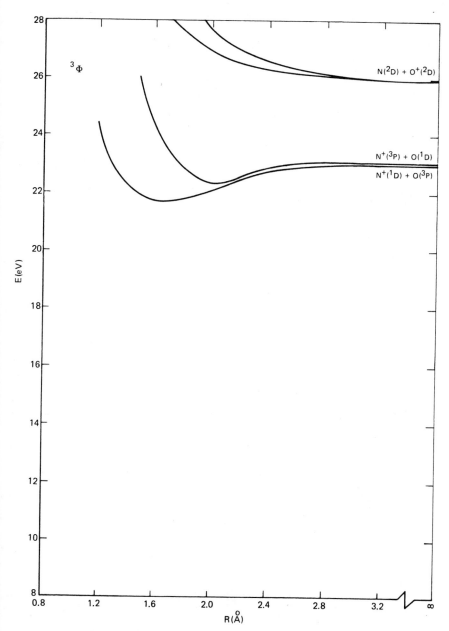

Figure 82. Potential-energy curves for $^3\Phi$ states of NO^+.

333

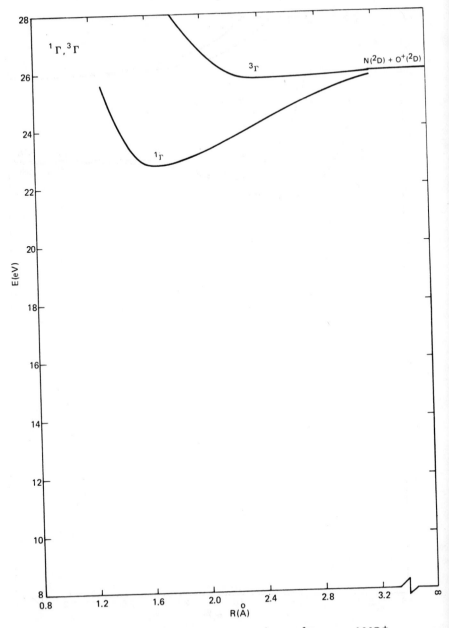

Figure 83. Potential-energy curves for $^1\Gamma$ and $^3\Gamma$ states of NO^+.

334

⌐ow-lying Molecular States of NO$^+$ and Their Dissociation Limits

DISSOCIATION LIMIT	MOLECULAR STATES
$N(^4S_u) + O^+(^4S_u)$	$^1\Sigma^+(1), {}^3\Sigma^+(1), {}^5\Sigma^+(1), {}^7\Sigma^+(1)$
$N^+(^3P_g) + O(^3P_g)$	$^1\Sigma^+(2), {}^3\Sigma^+(2), {}^5\Sigma^+(2), {}^1\Sigma^-(1), {}^3\Sigma^-(1),$
	$^5\Sigma^-(1), {}^1\Pi(2), {}^3\Pi(2), {}^5\Pi(2), {}^1\Delta(1),$
	$^3\Delta(1), {}^5\Delta(1)$
$N(^2D_u) + O^+(^4S_u)$	$^3\Sigma^+(1), {}^5\Sigma^+(1), {}^3\Pi(1), {}^5\Pi(1), {}^3\Delta(1),$
	$^5\Delta(1),$
$N^+(^1D_g) + O(^3P_g)$	$^3\Sigma^+(1), {}^3\Sigma^-(2), {}^3\Pi(3), {}^3\Delta(2), {}^3\Phi(1)$
$N^+(^3P_g) + O(^1D_g)$	$^3\Sigma^+(1), {}^3\Sigma^-(2), {}^3\Pi(3), {}^3\Delta(2), {}^3\Phi(1)$
$N(^4S_u) + O^+(^2D_u)$	$^3\Sigma^+(1), {}^5\Sigma^+(1), {}^3\Pi(1), {}^5\Pi(1), {}^3\Delta(1),$
	$^5\Delta(1)$
$N(^2P_u) + O^+(^4S_u)$	$^3\Sigma^-(1), {}^5\Sigma^-(1), {}^3\Pi(1), {}^5\Pi(1)$
$N^+(^1D_g) + O(^1D_g)$	$^1\Sigma^+(3), {}^1\Sigma^-(2), {}^1\Pi(4), {}^1\Delta(3), {}^1\Phi(2), {}^1\Gamma(1)$
$N^+(^1S_g) + O(^3P_g)$	$^3\Sigma^-(1), {}^3\Pi(1)$
$N(^4S_u) + O^+(^2P_u)$	$^3\Sigma^-(1), {}^5\Sigma^-(1), {}^3\Pi(1), {}^5\Pi(1)$
$N(^3P_g) + O(^1S_g)$	$^3\Sigma^-(1), {}^3\Pi(1)$
$N(^2D_u) + O^+(^2D_u)$	$^1\Sigma^+(3), {}^3\Sigma^+(3), {}^1\Sigma^-(2), {}^3\Sigma^-(2), {}^1\Pi(4),$
	$^3\Pi(4), {}^1\Delta(3), {}^3\Delta(3), {}^1\Phi(2), {}^3\Phi(2),$
	$^1\Gamma(1), {}^3\Gamma(1)$

Table XIV.
Important Molecular-orbital Electron Configurations for NO$^+$

STATES	ELECTRON CONFIGURATIONS^a			
	$3\sigma_g$	$1\pi_u$	$1\bar{\pi}_g$	$3\bar{\sigma}_u$
$X\,^1\Sigma_g^+$	2	4	0	0
$a\,^3\Sigma^+$	2	3	1	0
$b\,^3\Pi$	1	4	1	0
$w\,^3\Delta$	2	3	1	0
$b'\,^3\Sigma^-$	2	3	1	0
$A\,^1\Sigma^-$	2	3	1	0
$W\,^1\Delta$	2	3	1	0
$^5\Sigma^+\,(I)$	2	2	2	0
$A\,^1\Pi$	1	4	1	0
$^1\Sigma^+\,(II)$	2	2	2	0
$^5\Pi\,(I)$	1	3	2	0
$^5\Sigma^+\,(II)$	1	3	1	1

^{a}All states include $(1\sigma)^2$, $(1\bar{\sigma})^2$, $(2\sigma)^2$, and $(2\bar{\sigma})^2$, and the near-homopolar g–u labels can be applied.

335

Table XV.
Spectroscopic Constants For Bound States of NO$^+$

STATES		T_e (eV)	ω_e (cm^{-1})	$\omega_e X_e$ (cm^{-1})	α_e (cm^{-1})	r_e (Å)	B_e (cm^{-1})	D_e (eV)	D_0 (eV)
$^5\Sigma^+$ (I)	(C)	10.67	547.4	38.0	.0359	1.74	0.75	0.32	0.29
$^5\Pi$ (I)	(C)	10.38	900.8	10.2	.0112	1.56	0.92	1.53	1.47
$^1\Sigma^+$ (II)	(C)	9.44	1048.0	4.2	0.0034	1.57	0.92	2.47	2.40
$A^1\Pi$	(E)	9.11	1608.9	23.3	0.024	1.19	1.59	2.80	2.70
$W^1\Delta$	(E)	8.90	1205.1	18.2	0.014	1.35	1.24	3.01	2.94
$A^1\Sigma^-$	(E)	8.60	1260.0	14.9	0.015	1.34	1.26	3.31	3.23
$b'^3\Sigma^-$	(E)	8.41	1176.1	12.5	0.014	1.34	1.25	3.50	3.43
$w^3\Delta$	(E)	7.70	1360.4	17.3	0.015	1.34	1.27	4.21	4.13
$b^3\Pi$	(E)	7.28	1690.0	19.1	0.020	1.16	1.68	4.63	4.53
$a^3\Sigma^+$	(E)	6.50	1327.6	15.3	0.0120	1.34	1.26	4.49	4.41
$X^1\Sigma^+$	(E)	0.00	2377.1	16.3	0.020	1.06	2.00	10.99	10.84

336

ACKNOWLEDGMENTS

It is a pleasure to acknowledge the advice and assistance rendered by the author's associates. Special thanks are due to the Air Force Office of Scientific Research, Chemical Sciences Division, and to Lt. Colonel David S. Olson (AFWL), Lt. Colonel John T. Viola (AFOSR), and Drs. K. S. W. Champion and F. R. Innes (AFGL) for their support and interest in the author's research activities.

The author acknowledges the many fruitful discussions held with his associates, Drs. Gerald A. Peterson, H. J. Kolker, and R. H. Hobbs. Professor Frank E. Harris collaborated in the development of the computational techniques used in this research, and his advice and help are acknowledged with pleasure. The studies of the nitrogen molecule were aided by fruitful discussions with Professors A. Loftus, R. S. Mulliken, and F. Gilmore, whose early work served as a standard of accuracy and excellence.

Finally, we acknowledge the advice, constant encouragement, and tireless dedication to this research rendered by Judith B. Addison, who contributed to all aspects of this project. Without here continued support, this review could never have been accomplished.

REFERENCES

1. O. Harang, "A10 Resonant Spectrum for Upper Atmosphere Temperature Determination," AFCRL-66-314, Environmental Research Paper No. 192, 1966.
2. D. R. Churchill and R. E. Meyerott, *J. Quant. Spectrosc. Radiat. Transfer.*, **5**, 69 (1965).
3. E. B. Armstrong and A. Dalgarno, Eds., *The Airglow and the Aurorae*, Pergamon, New York, 1955.
4. B. H. Armstrong, R. R. Johnston, and P. S. Kelly, *J. Quant. Spectrosc. Radiat. Transf.*, **5**, 55 (1965).
5. R. R. Johnston, B. H. Armstrong, and O. R. Platas, *J. Quant. Spectrosc. Radiat. Transf.*, **5**, 49 (1965).
6. F. E. Harris and H. H. Michels, *Internat. J. Quantum Chem.*, **IS**, 329 (1967).
7. M. Krauss, "Compendium of *ab initio* Calculations of Molecular Energies and Properties," NBS Technical Note 438, 1967.
8. A. C. Wahl, P. J. Bertoncini, G. Das, and T. L. Gilbert, *Internat. J. Quantum Chem.*, **IS**, 123 (1967).
9. H. F. Schaefer, *Electronic Structure of Atoms and Molecules*, Addison-Wesley, Reading, Mass., 1972.
10. H. H. Michels, *J. Chem. Phys.*, **53**, 841 (1970).

11. H. H. Michels, in L. M. Branscomb, Ed., *Electronic and Atomic Collisions*, Vol. 2, North Holland, Amsterdam, 1971, p. 1170.

12. F. R. Gilmore, *J. Quant. Spectrosc. Radiat. Transf.*, **5**, 369 (1965).

13. J. C. Slater, *Quantum Theory of Molecules and Solids*, McGraw Hill, New York, 1963.

14. L. C. Allen, *Quantum Theory of Atoms, Molecules, and the Solid State*, Academic, New York, 1966.

15. A. C. Wahl, P. J. Bertoncini, G. Das, and T. L. Gilbert, *Internat. J. Quant. Chem.*, **IS**, 123 (1967).

16. C. C. J. Roothaan and P. S. Bagus, *Meth. Comp. Phys.*, **2**, 47 (1963).

17. C. C. J. Roothaan, *Rev. Mod. Phys.*, **23**, 69 (1951).

18. F. E. Harris, *J. Chem. Phys.*, **32**, 3 (1960).

19. F. E. Harris, *J. Chem. Phys.*, **46**, 2769 (1967).

20. W. Givens, "Eigenvalue–Eigenvector Technique," Oak Ridge report No. ORNL 1574 (Physics).

21. J. Todd, *Survey of Numerical Analysis*, McGraw-Hill, New York, 1962.

22. P. O. Löwdin, *J. Chem. Phys.*, **18**, 365 (1950).

23. E. R. Davidson, *Rev. Mod. Phys.*, **44**, 451 (1972).

24. H. H. Michels, "Theoretical Determination of Electronic Transition Probabilities for Diatomic Molecules", technical report No. AFWL-TR-72-1, 1972.

25. E. Miescher, *Can. J. Phys.*, **49**, 2350, (1971).

26. A. N. Wright and C. A. Winkler, *Active Nitrogen*, Academic, New York, 1968.

27. A. Loftus, "The Molecular Spectrum of Nitrogen," Spectroscopy report No. 2, Department of Physics, University of Oslo, 1960.

28. (a) R. S. Mulliken, "The Energy Leels of the Nitrogen Molecule," in *Threshold of Space*, Pergamon, New York, 1957; (b) A Loftus and P. H. Krupenie, *J. Phys. Chem. Ref. Data*, **6**, 113 (1977).

29. R. Rydberg, *Z. Phys.*, **73**, 376 (1931).

30. O. Klein, *Z. Phys.*, **76**, 226 (1932).

31. A. L. G. Rees, *Proc. Phys. Soc. (Lond.)*, **59**, 998 (1947).

32. L. Wallace, *Atrosphys. J. Suppl.*, **7**, 165, (1962).

33. J. T. Vanderslice, E. A. Mason, and E. R. Lippincott, *J. Chem. Phys.*, **30**, 129, (1959).

34. P. G. Wilkinson, *J. Molec. Spectrosc.*, **6**, 1 (1961) (see Fig. 8 due to R. S. Mulliken).

35. M. Leoni and K. Dressler, *Helv. Phys. Acta*, **45**, 959 (1972).

36. G. Herzberg, *Spectra of Diatomic Molecules*, 2nd ed., Van Nostrand, New York, 1950.

37. A. G. Gaydon, *Dissociation Energies and Spectra of Diatomic Molecules*, Chapman and Hall, London, 1968.

38. B. Rosen, *Spectroscopic Data Relative to Diatomic Molecules*, Pergamon, Oxford, 1970.

39. S. N. Suchard, "Spectroscopic Constants for Selected Heteronuclear Diatomic Molecules," Aerospace report No. TR-0074 (4641)-6, 1974.

40. S. N. Suchard and J. E. Melzer, "Spectroscopic Constants for Selected Homonuclear Diatomic Molecules," Aerospace report No. TR-0076 (6751)-1, 1976.

41. J. T. Vanderslice, E. A. Mason, and W. G. Maisch, *J. Chem. Phys.*, **32**, 515 (1960).

42. P. H. Krupenie, *J. Phys. Chem. Ref. Data*, **1**, 423 (1972).

43. W. R. Jarmain, *Can. J. Phys.*, **38**, 217 (1960).

44. W. R. Jarmain, "Transition Probabilities of Molecular Band Systems," University of Western Ontario report No. GRD-TN-60-498, 1960.

45. H. F. Schaefer and F. E. Harris, *J. Chem. Phys.*, **48**, 4946 (1968).

46. R. H. Pritchard, C. F. Bender, and C. W. Kern, *Chem. Phys. Lett.*, **5**, 529 (1970).

47. H. F. Schaefer, *J. Chem. Phys.*, **54**, 2207 (1971).

48. H. F. Schaefer and W. H. Miller, *J. Chem. Phys.*, **55**, 4107 (1971).

49. T. Itoh and K. Olnso, *J. Chem. Phys.*, **25**, 1098 (1956).

50. M. Yoshimine et al., *J. Chem. Phys.*, **64**, 2254 (1976).

51. R. J. Bunker and S. D. Peyerimhoff, *Chem. Phys. Lett.*, **34**, 225 (1975).

52. R. J. Bunker, S. D. Peyerimhoff, and S. Perie, *Chem. Phys. Lett.*, **42**, 383 (1976).

53. P. S. Julienne, *J. Molec. Spectrosc.*, **63**, 60 (1976).

54. D. C. Cartwright, *J. Phys. B, Atom. Molec. Phys.*, **9**, L419 (1976).

55. N. Beebe, E. W. Thulstrup, and A. Anderson, *J. Chem. Phys.*, **64**, 2080 (1976).

56. J. T. Vanderslice, E. A. Mason, and W. G. Maisch, *J. Chem. Phys.*, **31**, 738 (1959).

57. W. E. Meador, Jr., "The Interactions Between Nitrogen and Oxygen Molecule," NASA TR-R-68, 1960.

58. H. Brion, C. Moser, and M. Yamazaki, *J. Chem. Phys.*, **30**, 673 (1959).

59. K. Dressler and E. Miescher, *Astro. J.*, **141**, 1266 (1965).

60. E. Miescher, *J. Molec. Spectrosc.*, **20**, 130 (1966).

61. A. Lagerquist and E. Miescher, *Can. J. Phys.*, **44**, 1525 (1966).

62. C. Jungen, *Can. J. Phys.*, **44**, 3197 (1966).

63. C. Jungen and E. Miescher, *Can. J. Phys.*, **47**, 1769 (1969).

64. R. Suter, *Can. J. Phys.*, **47**, 881 (1969).

65. E. Miescher, *Can. J. Phys.*, **49**, 2350 (1971).

66. R. Field, L. Gottscho, and E. Miescher, *J. Molec. Spectrosc.*, **58**, 394 (1975).

67. E. Miescher and F. Alberti, *J. Phys. Chem. Ref. Data*, **5**, 309 (1976).

68. F. P. Billingsley, II, *J. Chem. Phys.*, **62**, 864 (1975).

69. P. W. Thulstrup, *J. Chem., Phys.*, **60**, 3975 (1974).

70. S. Green, *Chem. Phys. Lett.*, **23**, 115 (1973).

71. J. L. Pack and A. V. Phelps, *J. Chem. Phys.*, **44**, 1870 (1966).

72. M. J. W. Boness and J. B. Hasted, *Phys. Lett.*, **21**, 526 (1966).

73. J. B. Hasted and A. M. Awan, *J. Phys. B*, **2**, 367 (1969).

74. M. J. W. Boness and G. J. Schulz, *Phys. Rev.*, **A2**, 1802 (1970).

75. D. R. Bates and H. S. W. Massey, *Transact. Roy. Soc. (Lond.)*, **A239**, 269 (1943).

76. M. J. W. Boness, J. B. Hasted, and I. W. Larkin, *Proc. Roy. Soc.*, **A305**, 493 (1968).

77. J. Rolfe, *J. Chem. Phys.*, **40**, 1664 (1964).

78. R. S. Mulliken, *Phys. Rev.*, **115**, 1225 (1959).

79. M. Krauss, A. C. Wahl, and W. Zemke, in L. M. Branscomb, Ed., *Electronic and Atomic Collisions*, Vol. 2, North Holland, Amsterdam, 1971, p. 1168.

80. (a) W. T. Zemke, G. Das, and A. C. Wahl, *Chem. Phys. Lett.*, **14**, 310 (1972); (b) M. Krauss, D. Neumann, A. C. Wahl, G. Das, and W. T. Zemke, *Phys. Rev.*, **A7**, 69 (1973).

81. E. Ferguson, *Advances in Electronics and Electron Physics*, Vol. 24, Academic, New York, 1968.

82. E. Miescher, *Helv. Phys. Acta*, **29**, 135 (1956).
83. E. W. Thulstrup and Y. Öhrn, *J. Chem. Phys.*, **57**, 3716 (1971).
84. E. W. Thulstrup and A. Anderson, *J. Chem. Phys.*, **60**, 3975 (1974).
85. H. Lefebvre-Brion and C. M. Moser, *J. Chem. Phys.*, **44**, 2951 (1966).
86. H. Lefebvre-Brion, *Chem. Phys. Lett.*, **9**, 463 (1971).
87. D. L. Albritton, A. L. Schmeltekopf and R. N. Zare, *J. Chem. Phys.*, **71**, 3271 (1979).

CHAPTER FOUR

COLLISIONAL ENERGY-TRANSFER SPECTROSCOPY WITH LASER-EXCITED ATOMS IN CROSSED ATOM BEAMS: A NEW METHOD FOR INVESTIGATING THE QUENCHING OF ELECTRONICALLY EXCITED ATOMS BY MOLECULES

I. V. Hertel

Fachbereich Physik der Freien Universität Berlin, D 1000 Berlin 33

Contents

I INTRODUCTION   343

II EXPERIMENTAL TECHNIQUES FOR STUDYING
QUENCHING PROCESSES IN GASEOUS MIXTURES   346

III THEORETICAL QUENCHING MODELS   351

IV CROSSED-BEAM EXPERIMENTS WITH LASER-EXCITED
SODIUM ATOMS   358
  A General Aspects   358
  B Kinematics   362
  C Scattering Signal   364
  D Laser Optical Pumping of a Sodium-atom Beam   365
  E Experimental Setup   367

V DISCUSSION OF THE ENERGY-TRANSFER SPECTRA FOR
$Na(3^2 P_{3/2})$ QUENCHING BY SIMPLE MOLECULES   368
  A Diatomic Molecules N_2, CO, H_2, and D_2 369
  B Comparison of Experiments with Statistical State
    Populations   373
  C Linearly Forced Harmonic Oscillator Model   376
  D A More Complicated Case: E to E–V–R Transfer in Na*
    Quenching by O_2 377
  E Triatomic Molecules CO_2 and N_2O 377
  F Larger Polyatomic Molecules   379

VI POLARIZATION STUDIES IN QUENCHING PROCESSES
FROM LASER-EXCITED Na*(3p)   380
  A A Simple Example: $e + Na*(3p) \rightarrow e + Na(3s)$ 380
    1 General Aspects   380
    2 Linearly Polarized Light   382
    3 Circular Polarization   384
  B Polarization Effects in Quenching of Na(3p) by Simple
    Molecules   385
    1 Difference to Electron-scattering Processes   385
    2 Experimental Results   387
    3 Interpretation   389

VII OTHER BEAM EXPERIMENTS RELATED TO QUENCHING
OF ALKALI RESONANCE RADIATION   391

VIII CONCLUSION   393

I. INTRODUCTION

Electronic to vibrational and rotational (E–V–R) energy transfer certainly belongs to the elementary processes in chemical reaction dynamics and, in particular, in photochemistry. A fairly general type of photochemical reaction may be written as follows:

$$AB + CD + h\nu \rightarrow AB^* + CD \rightarrow (ABCD)^* \rightarrow AC^\# + BD^\# \qquad (I.1)$$

Conservation of energy is achieved by transferring excess energy into vibrational, rotational, and translational energy of the reaction products. Although it is a nearly intractable task to understand the preceding processes in all details and in full generality, a subset of reactions for which a complete experimental knowledge and at least a workable theoretical model may hopefully be gained are E–V–R transfer collision processes

$$A + BC + h\nu \rightarrow A^* + BC \rightarrow (ABC)^* \rightarrow A + BC^\# \qquad (I.2)$$

in which electronic excitation energy from an atom A^* is converted into vibrational and rotational internal energy of the molecule $BC^\#$ and into translational energy of the relative motion $A-BC^\#$.

This process is well known as quenching of resonance radiation, and the first measurements of quenching cross sections date back to 1926.[1] Besides the fundamental importance outlined in the preceding paragraph, quenching processes invoke interest for a number of more practical reasons:

1. Chemical reactivity of vibrationally excited molecules may be much higher than for the same species in thermal equilibrium. Although direct IR excitation is limited by the choice of lasers available and usually lends itself predominantly to the excitation of the first few vibrational levels and to IR active transitions,[2, 3] no such restriction applies to molecular excitation by E–V–R transfer.
2. In general, E–V–R transfer leads to a population inversion among vibrational levels and thus is of great importance for laser applications, and new types of laser such as those demonstrated recently[4, 5] may be built.
3. Complex reaction kinetics often incorporate processes of the preceding type and the inverse. Modeling the earth's atmosphere necessitates a detailed knowledge of its photochemistry, including the vibrational excitation and deexcitation of N_2, O_2, OH, and so on in E–V–R transitions with atoms and molecules. This has been reviewed by a number of authors,[6–9] and an informative survey is given in Chapter 6, of the first volume of this book.[10]

A vast amount of experimental material on quenching processes has been reported, especially within the last 20 years. Most of it is concerned with the determination of total quenching cross sections (or rate constants) without information about the final internal energy of the molecule $BC^{\#}$. Consequently, the discussion about the mechanism for the process [equation (I.2)] remained speculative for a long time. During the past few years experimental techniques have improved drastically. Thus, for instance, it has become possible to measure energy-dependent quenching cross sections from thermal energies up to 1 eV.[11, 12] And the development of laser technology, especially the availability of tunable dye lasers, has stimulated further progress. A crucial test for theoretical models is the vibrational–rotational-state population of the molecule after the collision.

The knowledge of the internal-energy distribution is of equal interest for the practical applications indicated in the preceding paragraphs. First spectroscopic obervations of the IR emission from the molecule $BC^{\#}$, which is related to the vibrational-state population, were reported by Karl and Polanyi[13] on the system $Hg^* + CO$. These measurements were subsequently improved and extended.[14-16] Recent time-resolved experiments with IR–laser absorption[17, 18] and emission techniques[19-21] yield more reliable results on the product-state distribution.

All these experiments are performed in a gaseous mixture of the reactants. The resulting state populations are the result of multiple scattering events, usually involving a number of collision processes for which the rate constants are also subject to investigation. Thus a set of rate equations has to be solved, which task in favorable cases is aided by time-dependent observations. The resulting rate constants $k(E_{\text{vib, rot}})$ are, at best, specific to the product-state population with an internal energy $E_{\text{vib, rot}}$ but are integrated over all scattering angles ϑ_{CM}, φ_{CM} and averaged over a broad initial thermal velocity (v_{CM}) distribution $f(v_{\text{CM}})$

$$k(E_{\text{vib, rot}}) = \int_0^\infty dv_{\text{CM}} v_{\text{CM}} f(v_{\text{CM}}) \int_{4\pi} d\Omega_{\text{CM}} \frac{d^2\sigma}{d\Omega_{\text{CM}} dE_{\text{vib, rot}}} \qquad (I.3)$$

The bulk of the experimental material involves an additional integration over all internal energies $E_{\text{vib, rot}}$ of the product-state distribution. As a consequence, no detailed theoretical prediction on differential cross sections $d^2\sigma/d\Omega_{\text{CM}} dE_{\text{vib, rot}}$ have emerged so far.

On the other hand, the physics of atomic and molecular collisions and its experimental technique have experienced a rapid progress during the last decades, allowing a very detailed investigation of binary collision processes and leading to a substantial amount of theoretical understanding. Among the most fruitful methods have been experiments with atomic and/or molecular beams, allowing the detailed investigation of differential scattering cross sections for binary collisions in single scattering events with relatively well-defined initial kinetic energies. An account of the possibilities of these methods in the field of chemical-reaction studies in general is given in an earlier work in this series.[22]

Thus it is tempting to investigate the quenching process [equation (I.2)] by such techniques to obtain a more detailed picture of the underlying dynamics. The progress in the applicability of dye lasers forms the necessary basis in studies of scattering processes involving short-lived excited species in a crossed-beam experiment. The necessary density of excited atoms may be prepared by irradiating the scattering volume with laser light tuned to the resonance transition of the atom under investigation. In 1973 we were able to report first experiments using this technique for inelastic collisions of electrons with sodium in the 3^2P excited state.[23] Related techniques were used subsequently to investigate elastic atom–atom scattering[24, 25] and fine-structure changing transitions,[26] and a first application to the quenching process [equation (I.2)] was performed recently in our laboratory.[27] In addition to the possibility of investigating differential scattering cross sections and to perform what may be called *energy-transfer spectroscopy*, the laser-excitation technique offers additional experimental possibilities as a result of the laser monochromacy and polarization. A detailed account of the principles, possibilities, and the current state of this technique is given in a recent review.[28]

It is the intent of this chapter to illustrate the application of these new possibilities to the problem of E–V–R transfer and to discuss the consequences of the results obtained so far for developing detailed theoretical models. We do not attempt in any way to give a comprehensive review on the work about E–V–R transfer. The earlier work with fluorescence cells and flames has been reviewed by Lijnse[29, 30], Krause[31] and Donovan.[32] An excellent survey of the work with product-state analysis in E–V–R transfer has been given by Lemont and Flynn,[33] and a bibliography on energy-transfer processes in general has been compiled by Beaty et al.[34] The appropriate theoretical approaches have been discussed by Nikitin on several occasions.[35] In the present chapter we give only some introduction to theory and illustrate by the experimental results where improvements of the models are indicated.

The results obtained so far with collisional energy-transfer spectroscopy are restricted to excited sodium atoms $A^* = Na(3^2P_{3/2})$ and quenching by a variety of simple polar and nonpolar molecules. The technique is applicable to any vaporizable molecule and will be available for a number of other atoms as well in due course with the progress of laser technology. The E–V–R transfer processes from and to sodium atoms have a number of interesting applications in laser physics and photochemistry. To mention just one prominent case, we recall the well-known sodium D-line emmission in the aurora that supposedly uses vibrationally excited N_2 as an energy reservoir[37, 38] that, in the light of our new results, could as well be metastable O_2. On the other hand, sodium, with its one valence electron only, may be sufficiently simple to serve as a model case in the development of improved theories, and it is thus important to learn as many details as possible about the quenching of this atom. Consequently, the discussion in this chapter is restricted mainly to $Na(3^2P)$, assuming that the results may be generalized in one way or another to different systems.

II. EXPERIMENTAL TECHNIQUES FOR STUDYING QUENCHING PROCESSES IN GASEOUS MIXTURES

Before discussing the theoretical models and the current state of the crossed-beam experiments, it seems appropriate to give a brief summary of the experimental techniques involving gaseous mixtures of the reactants. In contrast to the disadvantages mentioned in Section I, they usually allow absolute total quenching cross sections, which are difficult to obtain in crossed-beam experiments. The various possibilities to excite the atoms and to detect the quenching are listed schematically in Table I. The reference list given there is, of course, not intended to be complete. It refers to alkali atom quenching exclusively unless a particular technique has not been applied to it.

The alkali atoms may be excited by either being irradiated with resonance radiation

$$A + h\nu_{res} \rightarrow A^*$$

or by photodissociation of alkali halides[60]

$$AX + h\nu \rightarrow A^* + X + E_{kin}$$

The latter method has the advantage of low alkali-atom densities, thus avoiding radiation trapping and chemical reactions and allows selection of the initial kinetic energy of the A^* atom—subject, however, to some discussion about the velocity distribution and its relaxation before quenching. The excited atoms will loose their excitation energy be either spontaneous emission

$$A^* \overset{k_0}{\rightarrow} A + h\nu_{res} \tag{II.1a}$$

with the spontaneous rate constant $k_0 = (1/\tau_0)$, where τ_0 is the natural

Table I
Excitation and Quenching of Atoms

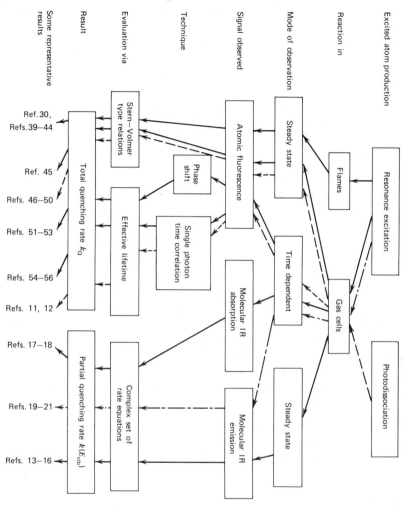

lifetime, or they will be deexcited by the process under discussion

$$A^* + BC \xrightarrow{k_Q} A + BC \qquad \text{(II.1b)}$$

where the total quenching rate constant

$$k_Q = \int k(E_{vib, rot}) dE_{vib, rot}$$

is a sum over all state selected rate constants $k(E_{vib, rot})$ [equation (II.3)]. The corresponding rate equations used are

$$-\frac{d[A^*]}{dt} = [A^*]_0 k_0 \qquad \text{(II.2a)}$$

without quenching and

$$-\frac{d[A^*]}{dt} = [A^*]_Q (k_0 + [BC] \cdot k_Q) \qquad \text{(II.2b)}$$

with quenching, $[A^*]$ and $[BC]$ refer to the number densities or concentrations in the reaction volume, which may be a gas cell or a shielded flame. The latter technique allows higher temperatures, and the determination of reactant concentrations seems to be well understood from flow measurements.[61]

The atomic fluorescence may be observed under either steady-state or time-dependent conditions. Under steady-state conditions the decrease of fluorescence intensity I is observed when the quenching gas is admixed, and from (II.2a, b) the well-known Stern–Volmer relation is obtained

$$\frac{I_0}{I_Q} = \frac{[A^*]_0}{[A^*]_Q} = \frac{k_0 + [BC] k_Q}{k_0} = 1 + [BC] \cdot \tau \cdot k_Q \qquad \text{(II.3)}$$

(details are discussed by Krause in an earlier volume of this series[31]). In the flame experiments a somewhat modified technique is used. Effectively, the fluorescence yield I_Q/I_0 is determined by measuring the absorption and the fluorescence when the flame is irradiated with resonance radiation. The alkali atoms are sprayed into the flame as metal salt. In time-dependent measurements the effective lifetime of the atom in the presence of the quenching reagent is determined. From equation (II.2b) we find

$$\frac{1}{\tau_{eff}} = k_0 + [BC] k_Q \qquad \text{(II.4)}$$

which again allows determination of the quenching rate constant k_Q. The lifetime measurement may be performed by modulating the exciting light intensity with a frequency ω_M and observing the fluorescence signal that is phase shifted by φ with tan $\varphi = \tau_{eff} \cdot \omega_M$.[51] Alternatively, in the single-photon time-correlation method (SPTC) the atoms are excited by a short pulse, and the exponential decay is observed by delayed coincidence techniques and signal averaging.[54] As an example, we show in Fig. 1 the experimental setup used by Barker and Weston[11] in an experiment with photodissociation of NaI by a flash lamp, an observation using the SPTC technique. Common to all these cell and flame experiments are the ambiguities discussed in Section I. In addition, a number of uncertainties impede the evaluation procedures and afford great care to produce reliable rate constants.

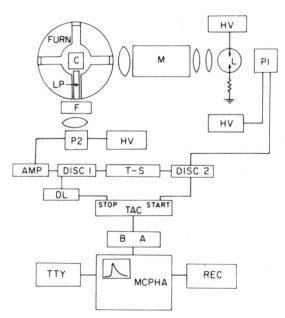

Figure 1. Block diagram of single-photon time-correlation apparatus from Barker and Weston:[11] HV, high-voltage supplies; L, lamp; Pl, photomultiplier; M, monochromator; FURN, furnace; C, sample cell; LP, light pipe; F, interference filter; P2, photomultiplier; AMP, amplifier; DISCl, discriminator; DISC2, discriminator; T-S, timer scaler; DL, delay line; TAC, time-to-amplitude converter; BA, biased amplifier; MCPHA, multichannel pulse–height analyzer; TTY, teletype printer and paper-tape punch; REC, strip-chart recorder.

The complexity of collision processes in a gaseous mixture becomes of even greater significance when the product-state population is to be analyzed. First steady-state experiments observing the IR emission in quenching collisions of Hg* with polar molecules by Polanyi and co-workers[13-16] still leave a considerable choice in the interpretation of experimental signals. Recent time-resolved experiments of IR emission [19-21] and IR absorption[17, 18] are the most promising among this type of experiment. Still, the experimental signals have to be interpreted in terms of a complex set of rate equations, which is, however, facilitated by the observed time dependence. For illustration, we show in Fig. 2 a schematic of the experimental setup of Hsu and Lin.[18] The sodium atoms are excited by a pulsed dye laser and quenched by CO molecules. The absorption signal for the CO laser light (turned to the various vibrational transitions in CO) is observed time dependently. In addition to the process to be observed, V–V, V–T, R–V, and R–T relaxion process may occur in the reaction tube. By suitable assumptions the observed signal is extrapolated back to a time at which only the E–V–R transfer is assumed to be of importance. In spite of these difficulties and the limitation to polar molecules, the method offers very attractive possibilities. In principle, the population of all rotational and vibrational states may be observed separately in E–V–R transfer. In practice, however, the laser lines available pose a limitation. Only high rotational states may be observed. In the system Na* + CO only $P(12)$ transitions have been investigated, except that the $P(13)$ absorption was used for the $v = 3 \rightarrow 4$ transition. However, it is by

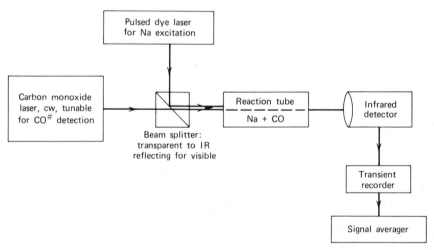

Figure 2. Schematic of Hsu and Lin's apparatus.[18]

no means *a priori* obvious that all final vibrational states should show the same rotational-state distribution. As we discuss later on, it is the total internal energy of the product molecule BC that is determined by the energy transfer, rather than the vibrational quantum number. Nevertheless, the time-resolved IR absorption technique, together with the time-resolved IR emission method,[19-21] may lead to very detailed information on the E–V–R transfer. If full advantage is taken from both these spectroscopic methods and crossed-beam techniques, a full picture on the processes under discussion will emerge in the near future.

Some more complex experiments in microwave discharges[62] and shock tubes[63, 64] are interesting but difficult to interpret quantitatively. A similar remark applies to mixed atom experiments, where fluorescence of the second atom indicates the combined transfer of energy from the first atom to the molecule and then to the second atom. Mixtures of Hg, N_2, Na[65] and Na, H_2, and Cs or Na, N_2, and Cs[66] have been used.

III. THEORETICAL QUENCHING MODELS

We wish to give a brief survey of the theoretical models that attempt to explain the quenching process to facilitate the discussion of the experimental results later on. We discuss the reaction

$$Na(3^2P) + BC \rightarrow Na(3^2S) + BC^{\#} \qquad (III.1)$$

at thermal initial energies.

A common result of all the experiments is that most molecules quench the alkali resonance radiation very effectively with total cross sections ranging from 10 Å^2 to over 200 Å^2. However, if the molecule BC is replaced by a rare-gas atom, the quenching cross sections become very small at thermal energies. They are probably below 10^{-2} Å^2 for quenching by helium, neon, argon, krypton, and xenon.[55] The latter result is easily understood in terms of Massey's adiabatic criterion.[67] If Δr is a characteristic interaction range, v the impact velocity, and ΔE the energy difference between initial and final electronic states $E(3p)$ and $E(3s)$, respectively, then we must have a Massey parameter

$$M = \frac{\Delta E \Delta r}{\hbar v} \gg 1 \qquad (III.2)$$

for the *adiabatic region*, where (i.e., at low impact energies) the characteristic "interaction time" $\Delta r / v$ is large compared to the transition time $\hbar / \Delta E$ that is needed by the atomic electron to adjust itself to changes in the internuclear distance. Thus electronic transitions should be negligible, as expected and observed for the alkali interaction with rare gases at thermal energies where $M \approx 300$. In contrast, the large quenching cross sections for atom–molecule collisions cannot be explained in that simple way. In fact, they are a prototype of *nonadiabatic transitions* for which

$$M = \frac{\Delta E \,\Delta r}{\hbar v} \lesssim 1 \qquad \text{(III.3)}$$

must be valid. At first sight it seems that this condition could be satisfied by taking account of the vibrational excitation of the molecule.

When the vibrational energy ΔE_{vib} transferred to the molecules is approximately equal to the electronic energy difference $E(3p) - E(3s)$, the total energy defect is small, and it could be assumed that relation (III.3) would indicate nonadiabaticity for such a near-resonant transition. On a second thought, however, it is realized that the internal motion of the molecule that is to be excited is also slow compared to the electronic movement. Thus there is no *a priori* reason why in such a process involving three particles (ABC) it should be more easy to transfer energy into the relative (vibrational) motion B–C rather than into the A–BC (translational) degree of freedom. As a matter of fact, there is no experimental evidence for any such electronic–vibrational resonance mechanism in alkali resonance quenching, unless the experiments are evaluated in a manner that explicitly assumes such a preference. In fact, Bästlein et al.[53] have compared the Na(3^2P) quenching by $^{14}N_2$ and $^{15}N_2$ and found no difference in the cross section, which had to be expected if a resonance mechanism would be of importance. The energy difference between the sodium excitation energy and the seventh vibrational level of $^{14}N_2$ and $^{15}N_2$ is 0.156 and 0.035 eV, respectively. Lijnse[30] has compared the quenching cross sections for various molecules for different alkali metals at different temperatures and found no correlation to the energy defect. And, of course, neither the IR absorption technique[18] nor the crossed-beam experiments to be reported in the following paragraphs have given any indication of a resonance mechanism.

Thus it seems clear that no direct transitions between essentially repulsive covalent potential surfaces Na*+BC and Na+BC are possible. This view is also supported by calculations.[68] Under such circumstances an additional ionic potential surface has been postulated,[69, 70] namely, Na$^+$ + BC$^-$, which was supposed to be strongly attractive and to couple with the covalent surfaces. All potentials depend on the molecular distance R_M, on the atom–molecule distance R_c during the collision, and on the molecular orientation relative to R_c measured by the angle γ. A two-dimensional cut through these surfaces along R_c is shown schematically in Fig. 3 for the

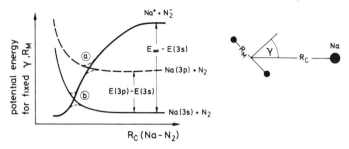

Figure 3.   Interaction potential of $Na + N_2$ for fixed λ, R_{N-N}.

most popular system $Na-N_2$. The ionic surface approaches asymptotically the energy $E_\infty = IP - E_{N_2}$, where IP is the sodium ionization potential and E_{N_2} is the electron affinity of the N_2 ($E_{N_2} < 0$ since there is no stable N_2^-). The dynamics on the potential surfaces may be understood by either using the adiabatic picture with avoided crossings and translational coupling or discussing the process in the diabatic picture with crossings and coupling by the electronic interaction. We do not emphasize these distinctions in our further discussion and use the term "crossing" freely for such regions where two surfaces come close to each other and interact. Plausible assumptions indicate that because of the Coulomb attraction $\propto 1/R_c$ of the ions, the surface crossings occur at large distances at 2 to 5 Å, which explains the large cross sections. In the crossing region equation (II.3) is valid. A number of assumptions may now be invoked for description of the potentials and the interaction dynamics that seem reasonable at these large interaction distances. One important assumption is implied in Fig. 3. Only one potential surface originating from the three $3p$ orbitals (π^+, π^-, and σ) is involved in the process. We discuss the implication in Section IV. The next assumption commonly made is to neglect the dependence of the potentials on the orientation angle γ. In view of the large interaction distances, this seems plausible at first sight. It leads, however, to a complete neglect of rotational transitions during the quenching process. As we see later, this is in contradiction to the experimental findings from our crossed-beam experiment.[†]

A further approximation adopted in theoretical computations is to omit all quantum-mechanical wave effects, which certainly is well justified for the large crossing radii and the small-particle wavelength of around 0.2 Å.

[†]*Note added in proof*: In view of recent ab initio surface calculations for the systems $Na - H_2$ and $Na - N_2$ the above discussion has to be revised substantially. There is no indication of an ionic intermediate state but sufficient attraction is found when the molecule is in a non equilibrium distance $R_M > R_M^{equil.}$. Details will be published in due course.[133–135]

The process may then be described by classical trajectories and by transition probabilities for changing the potential surfaces when these trajectories come close to a crossing. The transition probability for changing the diabatic curves at a crossing line R_c^x, R_M^x is given by a Landau–Zener formula

$$P_{12} = 1 - \exp\left\{ \frac{-(\pi/\hbar)U_{12}^2}{|[R_M(\partial/\partial R_M) + R_c(\partial/\partial R_c)](U_{11} - U_{12})|} \right\}_{R_c^*, R_M^*}$$

(III.4)

First calculations on this basis have been performed by Bjerre and Nikitin.[68] We may now understand the quenching process as follows. The excited atom and the molecule in, say, its vibrational ground state approach each other on the covalent surface Na* + N_2, with R_M assuming equilibrium distance. If the system reaches the crossing, it will more or less completely change onto the ionic surface Na$^+$ + N_2^- since $\dot R_M = 0$ and $\dot R_c$ is small [equation (III.4)]. Complementary to Fig. 3, we may visualize the following process on a one-dimensional cut through the surfaces along the molecular coordinate R_M at fixed R_c as illustrated schematically in Fig. 4. After crossing (Fig. 4a), the system finds itself on the ionic surface, and R_M is not in equilibrium position. Thus, when falling down the potential well, the collisional system will gain energy along the vibrational coordinate R_M as well as along the translational R_c. Now along the crossing line between the Na(3s) + N_2 potential and the ionic intermediate potential the system may change (as in Fig. 4b) to the electronic ground state. Precisely where that happens depends on the special trajectory and on the exact shape of the surfaces. The result will be translational *and* vibrational motion in the outgoing channel. Of course, on the way back, the energy may partially be transferred back into electronic energy. However, as

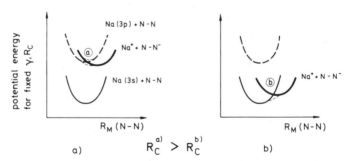

Figure 4. Interaction potential for fixed λ, R_{N_2-Na}.

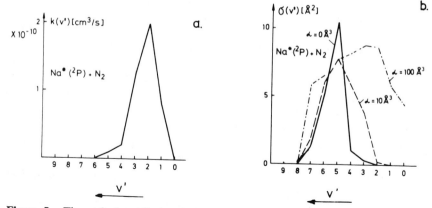

Figure 5. Theoretical predictions of partial quenching rate constants $k(v')$ and cross sections $\sigma\ (v')$ from Bjerre and Nikitin[68] (a) and Bauer et al.[71] (b) for different molecular electric polarization parameters.

calculations show,[68] this backward flux amounts to less than 10%, which may be understood in terms of the small probability for finding the system near the crossing On the way out. The vibrational-state population resulting from Bjerre and Nikitin's quenching model is illustrated in Fig. 5a. The potentials have been adjusted to yield a reasonable total quenching cross section and, lacking any detailed knowledge about the potential surfaces and coupling matrix elements, the authors understand their results as pure model calculations not to be compared in detail with experiments. The interesting outcome, however, is the clearly nonresonant character of the transition (v'=7 or 8 are the most resonant final vibrational levels). On the contrary, a pronounced maximum for v'=2 is found. Slightly more realistic potentials have been used by Bauer et al.[71] They introduced, however, an additional approximation [the Bauer–Fisher–Gilmore (BFG) model): to treat the problem as depending only on the collisional distance R_c, they parametrize the potentials according to the vibrational energy, and a whole grid of crossing potential curves μ, λ emerges, each one depending only on the distance between Na and N_2. At each crossing point a one-dimensional Landau–Zener formula is applied to find the transition probability for changing the diabatic curves

$$P_{\mu\lambda} = 1 - \exp\left[\frac{-(\pi/\hbar)U_{\mu\lambda}^2}{\left|\dot{R}_c\dfrac{d}{dR_c}(U_{\mu\mu}-U_{\lambda\lambda})\right|}\right]_{R_c^*} \tag{III.5}$$

where the coupling matrix element is given by the product of an electronic coupling element $V_{el}(R_c)$ and the Franck–Condon overlap factors between N_2^- and N_2 wave functions. This approximation is valid only to the extent that the wave function may be factorized into electronic, vibrational, and translational parts $\bar{\psi}(\mathbf{r}_{el}, R_M, R_c) = \psi_{el}(\mathbf{r}_{el}) \psi_{vib}(R_M) \cdot \psi_{trans}(R_c)$. Although the separation from the electron coordinate $\mathbf{r}_{el}$ is the usual Born–Oppenheimer approximation, the factorization of $\psi_{vib} \cdot \psi_{trans}$ is (under nonstationary conditions) valid only if $\dot{R}_c \ll \dot{R}_M$ which, at best, may be assumed during the initial time of the interaction but is no longer valid as the system gains translational energy during the interaction. No stationary vibrational state is allowed to build up. Nevertheless, the BFG model is a fair attempt at treating this complicated problem semiquantitatively in a simple way (results are shown in Fig. 5b).

The ionic potential has been adjusted by means of a polarization parameter α to give realistic total cross section. Again, a clearly nonresonant behavior is found. The relative populations of final vibrational states strongly depend, however, on the potential parameters chosen. This has been illustrated in some more detail by Fisher and Smith.[72–74] In fact, the most important handicap to all computations is that the potential surfaces are completely unknown, as are the coupling matrix elements for which semiempirical relations have been used.[75] Most disputable is the ionic intermediate state. Pseudopotential calculations by Bottcher[76] for Na(3p) + N$_2$ and Na(3s) + N$_2$, which give at least some guideline for the covalent states, fail to give any estimate of the Na$^+$ + N$_2^-$ surface. For diatomic systems such as NaH, KH,[77] and Zn$_2$,[78] it has been shown that such ionic configurations can have a strong influence even if they are very high lying for infinite separations. But no such calculations exist for triatomic systems. An additional complication arises for N$_2^-$, which as a free molecule does not exist for more than 10^{-15} sec and is only known as a resonance in the low-energy electron scattering.[79, 80] It could possibly be stabilized in the presence of an Na$^+$ atom that may be of interest in an *ab initio* calculation of the quenching process. More important, however, is the fact that the presence of an Na$^+$ almost certainly will change the shape of the N$_2^-$ potential.

In Fig. 6 the Na$^+$ core with the N$_2$ and the 3p electron surrounding both is illustrated in realistic dimensions at an internuclear distance that is typical for the crossing region. Characteristic is the large orbit of the 3p electron, and it is difficult to clearly distinguish an ionic and a covalent configuration. The N$_2^-$ potential differs from that for N$_2$ by its larger equilibrium distance and is also somewhat flatter. Clearly, when part of the electron cloud remains near the Na$^+$ core, the fictive N$_2^{-0.x}$ will be somewhere between the N$_2$ and N$_2^-$ potential. This would drastically

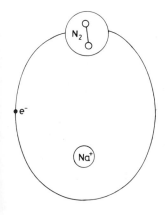

Figure 6. Schematic illustration of scattering dynamics seen as slow electron-collision process. Sodium ion core, N_2 molecule, and the $3p$ electron orbit are displayed in realistic dimensions.

change the quenching process in view of the previous discussion illustrated by Figs. 3 and 4. In contrast, all computations to date have used the undistorted N_2^- potential to evaluate the overlap between covalent and ionic wave functions.

Aside from the relative population of final vibrational states, the BFS model makes a prediction on the energy dependence of the cross section that is compared to the experiment in Fig. 7. The BFG model leads to a nearly constant quenching cross section up to 0.5 eV, which is in apparent contrast to the experimental findings that give a falling cross section with increasing energy. We later comment on the discrepancies between Lijnse's[30] flame experiments and Barker and Weston's[11] photodissociation results. With respect to theory, the constant cross section is obvious since essentially all collisions at thermal energies reach the ionic intermediate surface (R_c is small at the first crossing), and thus the initial energy is of little importance to the processes within the potential grid (which happen at much higher relative velocities). Lijnse[30] has claimed that by making the covalent potentials slightly attractive (which is certainly more realistic[72]), a centrifugal barrier may be constructed that may or may not be surmounted depending on the energy, thus leading to the observed energy dependence of the cross sections. A modified BFG curve crossing mechanism has been proposed recently by Barker[81] with essentially the same result. In view of the numerous other ambiguities of the BFG model, it is, however, difficult to decide which realistic significance these amendments have.

Before concluding this discussion on theoretical quenching models we wish to mention two more approaches. Andreev[82] proceeds along the BFG line for the $K(4p)$ quenching by N_2, investigating, however, more carefully the symmetries of the potentials, and uses interaction matrix elements deduced from $e + N_2$ scattering.[83] He finds the $v = 0$ level to be the most

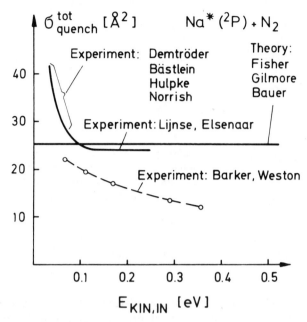

Figure 7. Total quenching cross section as function of incident kinetic energy from different experiments and theoretical predictions.

probable final vibrational state, which prediction unfortunately cannot be compared directly to the results for Na(3p) quenching. A very different and interesting idea has been tried by Bottcher and Sukumar.[84] They treat the quenching process as low-energy electron scattering (as emphasized by Fig. 6) and use empirical data from electron-scattering experiments by N_2, H_2, and O_2. Preliminary results are somewhat disappointing, but calculations of the N_2^- resonance in the presence of Na^+ would improve this model.

IV. CROSSED-BEAM EXPERIMENTS WITH LASER-EXCITED SODIUM ATOMS

A. General Aspects

The technique for investigating scattering processes in crossed-beam experiments is well developed. For example, elastic scattering experiments with neutral particles at thermal energies are well understood,[85] and the techniques for producing molecular and alkali atom beams and to detect them and interpret their kinematics has been reviewed on several occasions.[86, 87]. The new aspect of the present work is the technique for producing atoms in the excited state. Consequently, it allows investigation of electronically superelastic collisions in thermal crossed-beam experiments. Since the spontaneous lifetime of sodium in the 3^2P state is 1.6×10^{-8} sec, the atoms fly only some 10^{-3} cm during that time, and to maintain a substantial fraction of excited species, the scattering volume itself must be irradiated with mo.:ochromatic laser light tuned to the $3^2S_{1/2} - 3^2P_{3/2}$ resonance transition. The typical scattering geometry is illustrated in Fig. 8. To avoid Doppler broadening, the laser intersects the atom beam rectangular. Rectangular to both the atom and the laser beam is the molecular beam, which is somewhat wider. The scattering volume is defined by the intersect of the atom beam, laser beam, and detection region, with the latter given by limiting slits and other features in the detection system. For small laboratory scattering angles Θ_{lab}, the scattering volume is independent of Θ_{lab}.

One of the main goals of the crossed-beam experiment is to measure the internal energy $\Delta E_{vib, rot}$ transferred to the molecule. In principle, this is possible in either of two ways. First, the scattered molecules could be detected and their product-state population analyzed. Infrared emission or absorption techniques may be considered, similar to those used in cell experiments.[13-21] Although such studies would lead to the most detailed results (at least for polar molecules), under crossed-beam conditions they are impossible for intensity reasons, even if the possibility of measuring differential cross sections is renounced and the molecules in the scattering volume itself are detected. Detection via electronic molecular transitions may be invisaged. Unfortunately, the availability of tunable lasers limits this possibility to some exotic molecules such as alkali dimers. The future development of UV lasers could improve the situation. Hyper-Raman

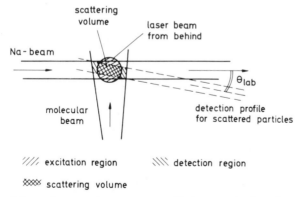

Figure 8. Schematic scattering geometry with laser excitation of atom beam.

scattering may be discarded for intensity reasons. Attempts by Burrow and Davidovits[88] to detect the vibrational product-state population by electron scattering near the $e-N_2$ resonance after $Rb^* + N_2$ collisions also suffered heavily from intensity limitations. Second, the energy transfer may be investigated by means of detecting the scattered sodium atoms and measuring their velocity. The reaction

$$Na(3^2P) + BC(v = 0, j) \rightarrow Na(3^2S) + BC(v', j') \tag{IV.1}$$

has the energy balance

$$E_{ex} + E_{rot}^0 + E_{IN} = E_{vib, rot} + E_{CM} \tag{IV.2}$$

where the 3^2P electronic excitation energy is $E_{ex} = 2.1$ eV, the initial (thermal) population of the rotational molecular states amounts to $E_{rot}^0 \approx$ 0.05 eV, the initial relative kinetic energy of the Na^*-BC system can be determined and is $E_{IN} \approx 0.1$ eV. Thus if the final relative kinetic energy E_{CM} (in the center-of-mass system $Na + BC$) is measured, $\Delta E_{vib, rot}$, the internal energy transferred into rotational and vibrational energy of the molecule ($E_{vib, rot}$), may be determined as follows:

$$\Delta E_{vib, rot} = E_{vib, rot} - E_{rot}^0 = E_{ex} + E_{IN} - E_{CM} \tag{IV.3}$$

The sodium velocity after collision v_{lab} may be measured and converted into relative kinetic energy E_{CM}, taking account of the scattering kinematics. The schematic of such an experiment[27] is illustrated in Fig. 9.

The supersonic sodium beam has a velocity distribution $\propto v^3 \exp[-m(v - u)/2kT]$ with a beam temperature of $T \approx 60°K$, a bulk velocity of u, a

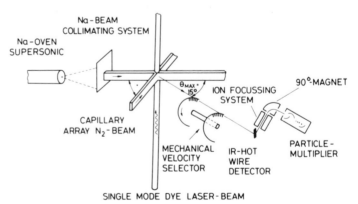

Figure 9. Schematic of energy-transfer experiment.

most probable velocity of $v_{Na} = 1370$ m/sec, and a full-width half maximum (FWHM) of 280 m/sec. Its angular divergence is less than 2^0, as its size is 2×6 mm. The sodium-atom density is typically $n_{Na} \approx 10^{11}$ atom/cm³ in the scattering center.

The molecular beam effuses from a capillary array 2×6 mm at a distance of 8 mm from the collision region. Its most probable velocity v_{mol} is given by the velocity distribution $\propto v^3 \exp(-mv^2/2kT_0)$ and is smaller than v_{Na}. The temperature T_0 may be chosen to $\approx 300°$K or $\approx 80°$K by cooling with liquid nitrogen. The angular beam divergence is below $15°$, and the density in the scattering volume is estimated to be $n_{mol} \approx 10^{13}$ molecules/cm³.

Scattered sodium atoms are observed in the scattering plane under small laboratory angles; they are velocity selected by a mechanical eight-disk selector of the Fizeau type with a resolution FWHM of 6%, and detected by a hot-wire detector followed by a dark-current-suppressing magnet and a particle multiplier. Details of the experimental arrangement are discussed by Hofmann[89] and Rost.[90] Alternatively, a Doppler shifted fluorescence (DSF) detector[91, 92-94], a time-of-flight (TOF) detector—with optical detection,[95, 96] or ionization on a very hot surface—could be used for simultaneous velocity selection and detection.[97] Each method has its own benefits and difficulties. The crossed-beam technique has a number of advantages over cell experiments, as discussed in Section II:

1. The processes investigated originate from pure single-collision events. This is obvious from the mean free path length $\lambda \approx 1/(\sigma_Q n_{mol})$. With $n_{mol} \approx 10^{13}$ cm⁻³ and the quenching cross section $\sigma_Q \approx 10^{-14}$ cm², we have $\lambda \approx 10$ cm, whereas the collision region extends over approximately 0.3 cm only. Thus the probability for one single quenching event per passage of an excited sodium atom is less than 0.03, and double collision events have a probability of less than 10^{-3}.
2. Any vaporizable quenching molecule may be investigated.
3. Differential rather than integrated cross sections are measured, although at present only small angle scattering results are available.
4. The polarization of the laser allows specific investigation about the symmetries of the potential surfaces involved in the quenching process and about the angular momentum transferred into molecular rotation.
5. The initial kinetic energy may be relatively well defined. For quenching by N_2, which is typical for the heavier molecules, we have the initial kinetic energy of the collision system defined to $E_{in} = 150$ meV ± 100 meV (FWHM) at $T_0 = 300°$K and (125 ± 60) meV at $T_0 = 80°$K, which is already better than that for thermal conditions and may be improved drastically by also using a supersonic molecular beam.

One limitation of the experiment is the impossibility of investigating near-resonant transitions. Purely elastic scattering from either excited or ground-state sodium has the trivial energy balance $E_{IN} = E_{CM}$, which has to be compared with equations (IV.2) and (IV.3). Thus, obviously, there can be no distinction of purely elastic scattering from an energy-transfer process where $\Delta E_{vib,\ rot} = E_{ex}$. This is of no significant importance as long as this ambiguity is limited to a small energy interval. The latter is determined by the experimental energy resolution and the magnitude of the elastic cross sections. We denote later the maximum $\Delta E_{vib,\ rot}$ (minimum E_{CM}) up to which the transfer processes may be investigated unambiguously as the "limit of elastic scattering." This is illustrated in Fig. 10 by a typical set of experimental scattering signals. The experimental signal with the laser light on, light off, and the difference (henceforth called *energy-transfer spectrum*) is shown. Below $v_{Na} = 2000$ m/sec the elastic scattering obviously dominates. The "light-off" signal originates from off ground-state elastic scattering in a scattering volume much larger than the excitation region from which the transfer processes originate exclusively. The merits (1 to 5 in the preceding list) of the crossed beam method must be paid for by a great complexity of the experimental setup. This is outlined in the text that follows.

B. Kinematics

Scattering intensities are measured in the laboratory; however, the physically relevant information has to be described in the center-of-mass

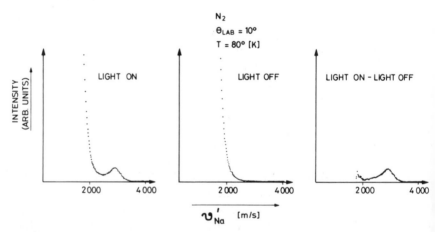

Figure 10. Experimental scattering signal as function of final sodium velocity: with light, without light, and difference signal.

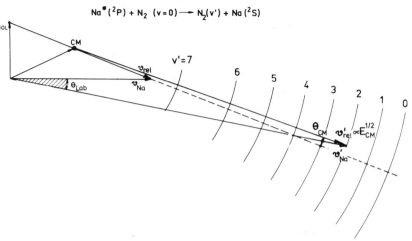

Figure 11.   Newton diagram for quenching process.

system. As discussed by Smith,[22] the kinematics of reactive heavy particle collision are best visualized in terms of a Newton diagram. A typical situation is illustrated in Fig. 11. The most probable molecular velocity v_{Mol} and atomic velocity v_{Na} is taken to represent the corresponding beams. After the collision, superelastically scattered atoms have a center-of-mass velocity given by circles around the center of mass (CM) as indicated in the Fig. 11. The line along the direction of the initial relative velocity $v_{Na} - v_{mol}$ indicates a center-of-mass scattering angle Θ_{CM}. The final sodium velocity in the laboratory v'_{Na} is indicated for the example of $\Theta_{CM} = 0$ and excitation of the $v' = 4$ vibrational level. Simple geometric relations allow transformation of the laboratory scattering angle Θ_{lab} and v'_{Na} into E_{CM} and to obtain ΔE_{vib} and Θ_{CM}.

These transformations are, however, exactly valid only if the velocity and angular distributions of atom and molecular beam and detection system are sharp. Since this is not the case, a finite-energy resolution and an uncertainty in the scattering angle is smearing out the experimental signal. Fortunately, these kinematic broadenings may be computed by Monte Carlo simulations of the scattering kinematics taking account of *all* energy, angular, and spatial distributions. The typical energy resolution computed in such a way is shown in Fig. 12 for the experimental conditions described in the preceeding paragraphs. Obviously, it is best for quenching by heavy molecules and low kinetic energy after collision.

The experiments are carried out at fixed laboratory angles Θ_{lab}, and the corresponding center-of-mass system scattering angle and its resolution is

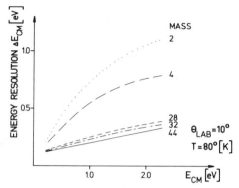

Figure 12. Energy resolution of crossed-beam E–V–R experiment, as determined by beam kinematics for different molecular masses.

illustrated for typical examples in Fig. 13. All these conditions may be improved by using better-defined beams, but the results reported later are obtained with these energy and angular spreads. The knowledge of their values allows us to a certain extent to deconvolute the experimental signal and to obtain the true relative differential cross sections.

C. Scattering Signal

A simpler first-order evaluation of the experiments may be obtained by using the most probable velocities and intersection angles as representative

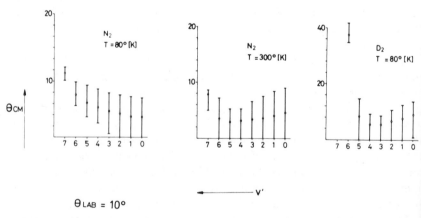

Figure 13. Range of scattering angles at fixed laboratory angle Θ_{lab} for three typical examples.

and by assuming the quenching cross sections to be constant over the range of the experimental resolution. Then the scattering signal in the laboratory system is

$$\dot{N}(v'_{Na},\Theta_{lab}) = P \cdot V_{sc} \cdot n_{ex} \cdot n_{mol} \cdot v_{IN} \cdot \frac{d^2\sigma}{d\Omega_{lab} \cdot dv_{Na}} \cdot v'_{Na} \cdot \Delta\Omega_{lab} \quad (IV.4)$$

where V_{sc} is the scattering volume, n_{ex} is the excited atom density, $\Delta\Omega_{lab}$ is the collected solid angle, and P is a selector and detector transmission constant. The final laboratory velocity of the sodium v'_{Na} accounts for the constant relative selector resolution $(\Delta v_{Na}/v_{Na}) = 0.06$ and thus linearly increasing transmittance.

The double differential cross section has to be transformed by the appropriate Jacobian $(v_{CM}/v'_{Na})^2$ (see e.g. Smith[22]) and the conversion of velocities into CM energies induces an additional factor $1/v_{CM}$. Thus finally

$$\frac{d^2\sigma}{d\Omega_{CM}dE_{CM}} \propto \frac{v_{CM}}{v'^3_{Na}} \cdot \dot{N}(\Theta_{lab},v'_{Na}) \quad (IV.5)$$

The deviation from numerical Monte Carlo evaluations is, as we see later, significant only for light molecules H_2 and D_2.

D. Laser Optical Pumping of a Sodium-atom Beam

A crucial part of these experiments is the preparation of the sodium atoms into the excited state by laser optical pumping. A commercial single-mode Rhodamin 6G continuous wave (cw)-dye laser (Spectra Physics model 580) is used, having 20–40-mW single mode output power when tuned to the sodium resonance line.

The optical pumping process has been described in detail elsewhere,[28, 98, 99] and we summarize only the most important points here. The dye laser has a bandwidth of $\delta_{\nu 1/2} < 50$ MHz, which is of the same order of magnitude as the hyperfine splitting of the excited sodium levels. Since Doppler broadening may be neglected for excitation of an atom beam, we are able to excite only one well-defined hyperfine level within one fine-structure level of the sodium atom in the 3^2P state. The term scheme of sodium is illustrated in Fig. 14. High number densities in the excited state may be achieved only by pumping into levels that decay spontaneously into only that ground-state hyperfine level out of which they were excited. Such a closed pumping cycle is possible between the $F=2$ level of the $3^2S_{1/2}$ ground state and the $F=3$ level of the $3^2P_{3/2}$ excited state. This

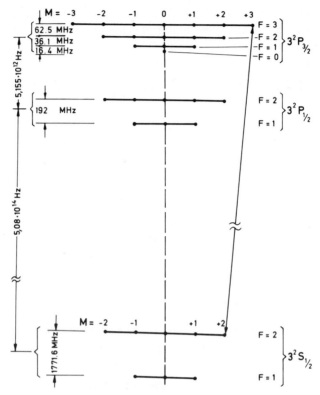

Figure 14. Hyperfine term scheme of sodium.

transition is used throughout the experiments described later, and up to 31.25% of the sodium atoms may be found in the excited state for steady-state conditions and sufficiently high laser pumping powers. In practice, this excited-state fraction is nearly reached in the center of the pumping region, whereas in the wings of the Gaussian laser beam the excited-state density rapidly decays. Fortunately, because of the saturation in the beam center, the excited-state density cutoff is very sharp, as is illustrated in Fig. 15. The scattering volume (see also Fig. 8) is defined in these experiments by the laser-excitation region to a much better degree than in usual crossed-beam experiments for scattering of ground-state atoms.

To give some feeling for the typical time scales, we note that the speed of the optical pumping process is determined by the spontaneous decay time $\tau \approx 1.6 \times 10^{-8}$ sec of the 3^2P level of sodium, which is two orders of magnitude less than the flight time $t_{fl} \approx 10^{-6}$ sec of the sodium atom through the excitation region. Thus each individual atom undergoes up to

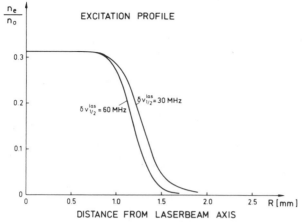

Figure 15. Estimated spatial distribution of excited-state density near scattering center. Atoms are excited by a Gaussian light beam.

100 pumping cycles, and the pumping process is essentially stationary. The time for induced optical transitions is $t_{ind} \approx 10^{-9}$ sec, and this, in turn, is long compared to the collisional interaction time $t_{col} \approx 10^{-12}$ sec for the quenching process to be investigated.

We finally wish to briefly mention the possibility of exploiting the laser polarization to obtain more detailed information on the scattering dynamics. Because of optical selection rules the linearly (or circularly) polarized laser light prepares the excited atoms in a known mixture of substates with defined angular momentum projection quantum number M. By rotating the plane of polarization with respect to the scattering coordinate system (center-of-mass system), the symmetry of the interaction potentials participating in the dynamics may be varied systematically. This is a completely new aspect in the field of experimental atomic crossed-beam collision, and it has been exploited quantitatively in electron–atom collisions[28, 100] using the language of state multipole moments.[101, 102] For the quenching process discussed at present, only qualitative first experiments investigating the polarization effects are available; nevertheless, these are informative and are discussed in Section VI. The energy-loss spectra presented in Section V have been recorded with fixed polarization angle parallel to the molecular-beam axis.

E. Experimental Setup

A somewhat more detailed schematic of the experimental arrangement is depicted in Fig. 16. The experiment is controlled by a minicomputer (MINICAL 621) via a CAMAC interface. Besides accepting the detector

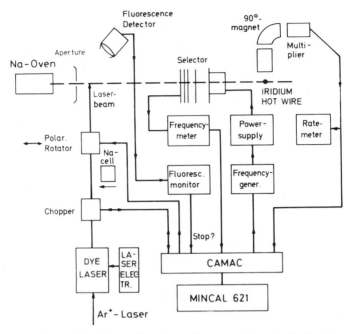

Figure 16. Schematic of experiment and electronic circuit.

signal via standard counting electronics and signal averaging, it performs a number of control functions. It monitors the atomic fluorescence for laser supervision, it controls the frequency generator for the selector and measures the actual selector frequency ($\propto v'_{Na}$), it allows opening and closing of the light chopper, thus allowing subtraction of the background noise from the signal, and it may be used to adjust the alignment of the laser-polarization vector.

Two modes of operation are used. When energy-transfer spectra are recorded, the selector frequency, that is, the final sodium velocity v'_{Na}, is swept several times with a period of some minutes from around 1500 m/sec to 4000 m/sec. Alternatively, at a fixed selector velocity the polarization direction may be rotated to observe the influence of atomic alignment on the scattering process.

V. DISCUSSION OF THE ENERGY-TRANSFER SPECTRA FOR Na($3^2P_{3/2}$) QUENCHING BY SIMPLE MOLECULES

In this section we report the results published so far[103] and some unpublished more recent data, and we give an appropriate discussion,[90] systematically summarizing the material.

A. Diatomic Molecules N_2, CO, H_2, and D_2

We first discuss some simple diatomic molecules H_2, D_2, N_2, and CO, which are known to be good quenching agents, having total quenching cross sections of around 10 to 40 Å^2 at the incident energies $E_{IN} \approx 0.05$ to 0.14 eV encountered here.[11] Figure 17 shows the experimental scattering signal (light on–light off) as a function of the sodium velocity v'_{Na} after the collisional quenching of the $3p$ state. From v'_{Na} we may obtain the final vibrational state v' of the molecule if pure vibrational excitation of the molecule is assumed. Monte Carlo computations for the experimental kinematics taking account of the angular and velocity distribution of the sodium and the molecular beam and of the properties of the detector allow determination of the experimental velocity resolution. If purely vibrational excitation is assumed, reconstruction of the experimental scattering signal can be attempted. The result is also shown in Fig. 17. Obviously, because of distinct excitation of the vibrational states, structures would be expected that clearly are not seen in the experimental data. Hence we conclude that rotational excitation plays an important role in the energy-transfer process and smears out the vibrational structure. Thus in the following we discuss the results only in terms of the total energy $\Delta E_{\text{vib, rot}}$ transferred to the molecule as we are unable to disentangle vibrational and rotational excitation. The vibrational level indicated in the following illustrations give a measure for $\Delta E_{\text{vib, rot}}$ only in terms of vibrational quanta. If equation (III.10) is applied, the experimental signal can be converted into differen-

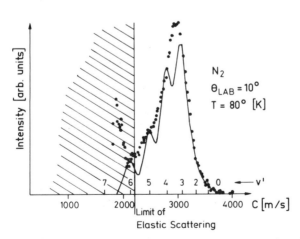

Figure 17. Experimental scattering signal for Na*+N_2 as function of final sodium velocity. Measured points do not exhibit structure that would be obtained for pure E–V energy transfer.

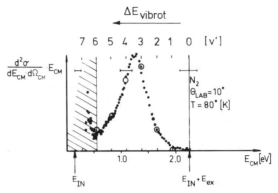

Figure 18. Energy-transfer spectrum for Na* + N$_2$. Energy transfer $\Delta E_{\text{vib, rot}}$ is measured in units of vibrational quanta v' after collision. Shaded area indicates strong superposition of elastic scattering processes. Horizontal bars illustrate experimental resolution. Kinematic deconvolution is indicated 0.

tial quenching cross sections per solid angle and per unit energy transferred to the molecule. More correctly, a kinematic Monte Carlo computation may be used to deconvolute the experimental data. Both results are given in Figs. 18 and 19 for N$_2$ and CO as a function of the relative center-of-mass energy after collision E_{CM} or, equivalently, as a function of $\Delta E_{\text{vib, rot}}$. For better visibility of structures, we have multiplied $d^2\sigma/d\Omega_{\text{CM}}\,dE_{\text{CM}}$ by E_{CM}. Obviously, for these heavier molecules the deconvolution agrees well with the simple transformation according to equation IV.5. This is different for the light molecules H$_2$ (Fig. 20) and D$_2$ (Fig. 21), which allow only a relatively poor energy resolution. The kinematic deconvolution is thus essential in the case of H$_2$ and D$_2$, and, of course, the

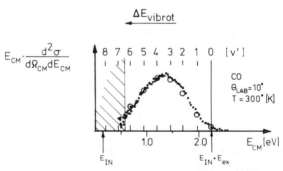

Figure 19. Energy-transfer spectrum for Na* + CO; otherwise as Fig. 18.

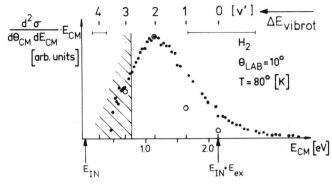

Figure 20.   Energy-transfer spectrum for Na* + H$_2$; otherwise as Fig. 18.

experiment cannot emphasize the importance of rotational excitation. Nevertheless, it can be expected to be as important as in the previous cases.

Before discussing these energy-transfer spectra, we recall two limiting aspects of these measurements. Because of elastic scattering, near $E_{CM} \approx E_{IN}$ and equivalently $\Delta E_{vib,\,rot} = E_{IN} + E_{ex}$, no reliable data may be taken. The spectra show, however, such a clear structure that only a very small quenching cross section may be expected under these near-resonant conditions. We also have to remember (see Fig. 13) that the experiment averages over a relatively large scattering angle (5–10°) that, in turn, varies with E_{CM}. Nevertheless, these results allow the most detailed qualitative interpretation thus far of the quenching mechanism and render a variety of new insights that will have to be taken into account for future theoretical developments.

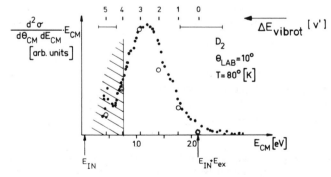

Figure 21.   Energy-transfer spectrum for Na* + D$_2$; otherwise as Fig. 18.

On inspection of the energy-loss spectra (Figs. 18 to 21), we observe as a common feature that for all these diatomic molecules the average energy transferred into vibrational and rotational excitation of the molecule is approximately half of that available $(E_{IN} + E_{ex})$. The maximum cross section is found for an energy transfer of around $\Delta E_{vib, rot} = 1$ eV for N_2, H_2, and D_2 and $\Delta E_{vib, rot} = 0.9$ eV for CO. The energy transfer is confined to a relatively well-defined energy region for the homonuclear case H_2, D_2, and N_2 (FWHM-0.5–0.8 eV), whereas the polar molecule CO has a broader distribution of internal energies after collision (FWHM-1.1 eV). Recent results for NO show even broader spectra, thus underlining the principal difference between polar and homonuclear molecules.[†]

We have already discussed the importance of rotational excitation. It could well be the reason for the observed difference between polar and nonpolar diatomic molecules. Polar molecules exposing the sodium to a strongly nonisotropic potential may be rotationally excited much more easily, thus leading to a broader energy distribution after the quenching process. We continue this argument in Section VI.

Obviously, the quantity in terms of which the quenching process may be described most adequately is the total energy transferred to the molecule $\Delta E_{vib, rot}$. The final vibrational state quantum number does not seem to be the adequate parameter. This is once more illustrated by plotting the H_2 and D_2 quenching cross sections as a function of the energy transfer (Fig. 22a) and, alternatively, as a function of the final vibrational state v' (Fig. 22b). Apparently, the shape of the spectra nearly coincide as a function of the energy transfer and disagree as a function of v'. The collisional systems $Na + H_2$ and $Na + D_2$ interact on the same potential surfaces, and obviously these give the relevant dynamic parameters for the process, despite the different vibrational spacings of H_2 and D_2.

The experimental findings described in the preceding paragraphs lead to a qualitative understanding of the E–V–R energy transfer in terms of the potential surface crossing model described in Section III (see Figs. 3 and 4). The relatively well-defined energy transfer to the homonuclear molecules could, very qualitatively, be interpreted as follows: When the system falls down on the ionic intermediate surface, it essentially gains relative kinetic energy. At the point of the crossing with the ground-state surface the remaining part of the electronic energy is released, and this relatively well-defined energy is then distributed into suitable rotational and vibrational states of the molecule. The much broader structure of the polar

---

[†]*Note added in proof*: Recently, Kwei and colloborators have investigated in a somewhat different crossed beam experiment $Na^* + N_2, CO, O_2$ and NO. Their results differ in some details from ours, but agree in general trends.[136]

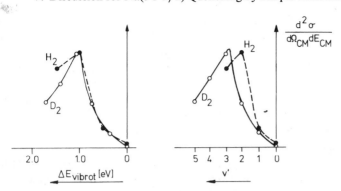

Figure 22. Partial quenching cross sections for Na*+H$_2$ and Na*+D$_2$ (from Figs. 20 and 21) plotted as function of energy transfer $\Delta E_{\text{vib, rot}}$ and alternatively as function of final vibrational state v'.

molecules may arise from the strongly anisotropic interaction potential that smears out the crossing region over a broader energy range depending on the molecular orientation in each individual collision. We gain additional support for this model in Section VI.

The parametrization according to vibrational quanta done in the BFG model[71–74] is, in view of the previous discussion, a somewhat arbitrary simplification since neither vibrational nor rotational states of the molecule have a higher *a priori* probability to be excited in the collision. In fact, a quantitative comparison[103] of the differential quenching energy transfer spectra with the numerical results of the BFG model yields only a qualitative agreement, clearly illustrating, however, that the BFG model gives broader energy distributions than observed in the experiment. Apparently, rather than to improve the details of BFG-type calculation, would have to be properly accounted for the rotational motion of the molecule would have to be properly accounted for and, most important, true interaction potentials would have to be used. An improved understanding would already be possible if the crossing regions were known, taking account of the deformation of the negative molecular ion in the presence of Na$^+$. In the meantime, a comparison of the experimental energy-transfer spectra with statistical *a priori* probabilities as described by Bernstein and Levine[104] might as well be attempted.

B. Comparison of Experiments with Statistical State Populations

In consequence of the complete lack of any reliable potential surfaces and bearing the preceding discussion in mind, it is tempting to compare the energy-transfer spectra to those obtained with the purely statistical population of all degrees of freedom. It has been shown in extensive studies by Levine and collaborators how to derive and apply such prior distributions for various collision problems. A comprehensive review has been provided by Bernstein and Levine.[104] Of course, such prior distributions do not give much physical insight into the collision dynamics. On the contrary, they merely describe the collisional energy-transfer process under the assumption that no dynamical preference to any of the vibrational, rotational, or translational degrees of freedom is given, and thus they allow only a first guess as to the shape of the energy-transfer spectra, assuming, in particular, spherical isotropy of the differential cross sections. Often experimental data are analyzed in terms of a so-called surprisal analysis. We do, however, prefer to compare our spectra directly to the original prior distributions.

The total energy available for distribution is $E = E_{ex} + E_{IN}$. The prior distribution $p(E_i)$ has to be classified according to the total internal energy of the molecule after collision $E_i \approx \Delta E_{vib, rot}$ or, alternatively, according to the translational energy $E_{CM} = E - \Delta E_{vib, rot}$. The probability of finding the molecule with a particular internal energy is proportional to the corresponding translational-state density $\rho_{CM} \propto E_{CM}^{1/2}$ and to the density of vibrational and rotational states energetically accessible $\rho_i(E - E_{CM})$, with the latter density zero for $E_{CM} = E$. Thus, prior distributions are expected with a maximum at intermediate energies between $E_{CM} = 0$ and $E_{CM} = E$. In Fig. 23a such distributions multiplied by E_{CM} are displayed for H_2

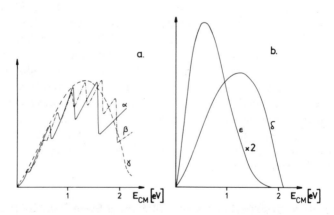

Figure 23. Statistical prior distribution for Na(3p) quenching by (α) H_2, (β) N_2, (γ) fictive diatomic molecule with 60-meV vibrational spacing, (δ) closely spaced vibrational levels, and (ε) triatomic quenching molecule.

(curve α) and N_2 (curve β), indicating steps at the onset of any vibrational channel. For simplicity, the rotational states are taken to be distributed continuously. As the vibrational spacing decreases, the prior probability becomes a smoother function (curve γ for 60-meV spacing) and in the limit of densely spaced vibrational levels may be written as[104]

$$p(E_i) \propto E_{CM}^{1/2} \cdot (E - E_{CM}) \qquad (V.1)$$

for atom diatom collisions and

$$p(E_i) \propto E_{CM}^{1/2} \cdot (E - E_{CM})^4 \qquad (V.2)$$

for atom triatom collision.

In Fig. 24 the energy-loss spectra for N_2 and CO are compared to the prior distribution normalized in height to the maximum of the experimental spectra. Clearly, a surprisingly good tendential agreement is seen, certainly as good as the BFG model computations. Carbon monoxide is populated more statistically than N_2. This finding is in good agreement

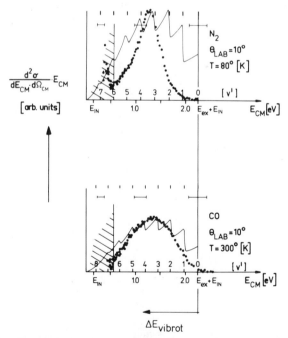

Figure 24. Comparison of prior distributions (—) with experimental Na* + N_2 and + CO, respectively.

with the previous discussion, since the strong anisotropy of the polar molecule spreads the surface crossing regions and thus incorporates the various vibrational and rotational states with more equal probability. For $Na^* + N_2$ quenching (as well as for H_2 and D_2), the collision dynamics become more dominating.

In conclusion, it can be stated that a good first estimate of the energy-transfer spectra is obtained from statistical distributions and that to obtain more quantitative theoretical results, certainly the knowledge of the potential surfaces is at least as important as the correct treatment of rotational *and* vibrational dynamics. Perhaps a suitable result may already be obtained by merely combining the statistical theory with some knowledge of the potential surface crossing regions.

C. Linearly Forced Harmonic Oscillator Model

Before turning to more complicated systems, we just wish to note that another simple model also yields curiously good agreements with the experimental differential quenching cross sections. Following Wilson and Levine, [105] the vibrational excitation is achieved by a sudden change in the equilibrium molecular distance at the crossing from intermediate ionic

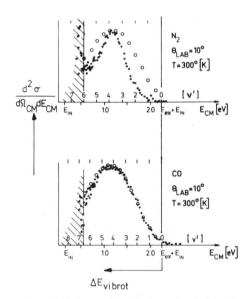

Figure 25. Comparison of Poisson distributions (○○) for vibrational levels with experiment (●●).

state to the ground state. This process may be treated in terms of the linearly forced harmonic oscillator, where the collisional perturbation of the assumed harmonic molecular potential is linearly approximated. Time-dependent perturbation theory leads to a Poisson distribution for the final vibrational state population[34]

$$p(0 \rightarrow v') = \frac{\varepsilon^{v'}}{v'!} \exp(-\varepsilon) \qquad (V.2)$$

where ε is the average energy transferred to the molecule and may be treated as a fitting parameter. As displayed in Fig. 25, excellent agreement is found again for CO (best $\varepsilon = 4.8$ in units of vibrational quanta), whereas the N_2 population is reproduced only tolerably well ($\varepsilon = 4.25$). Once more the importance of the actual collision dynamics are emphasized in the case of homonuclear diatomic molecules.

D. A More Complicated Case: E to E–V–R Transfer in Na* Quenching by O_2

Oxygen is an important molecule and is also known to be an efficient quenching agent for sodium resonance radiation. In fact, it seems to have an even larger quenching cross section ($\sigma_Q = 34$ Å^2) than $N_2 (\sigma_Q = 21$ Å^2) or $H_2 (\sigma_Q = 8$ Å^2).[106, 107] Although O_2 is also a diatomic molecule, it has a more complicated structure with respect to quenching of the Na ($3p$) state. Three electronic states are energetically accessible: the $X^3\Sigma_g^-$ ground state, the $a^1\Delta_g$ state at $E_{mol}^{el} \approx 1$ eV, and the $b^1\Sigma_g^+$ state at $E_{mol}^{el} \approx 1.6$ eV. The experimental energy-transfer spectrum given in Fig. 26, therefore, looks quite different from those in the previously discussed diatomic cases. The onset of the electronically excited channels may be seen at the appropriate energetic positions. Again, there can be a comparison with a prior statistical distribution, taking account of each electronically excited state individually with equal weight. The available total energy is $E = E_{ex} + E_{IN} - E_{mol}^{el}$. The full curve in Fig. 26 gives the so-obtained statistical distribution, normalized in height to match the experimental data. Again, we may be surprised about the excellent agreement with the experiment. Hopefully, useful predictions can be obtained for even more complicated processes of E to E–V–R energy transfer.

E. Triatomic Molecules CO_2 and N_2O

The statistical prior distribution predicts for triatomic molecules a more resonant behavior of energy-transfer processes than in the diatomic cases.

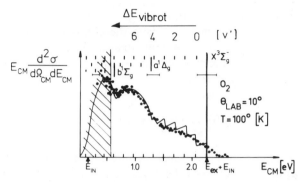

Figure 26. Experimental (●) energy-transfer spectrum for Na*+O₂ quenching and comparison with prior distributions (—).

The maximum in the $E_{CM} \times p(E_{CM})$ spectra is predicted at around 0.28 eV, as compared to 0.6 eV in the diatomic case. Unfortunately, the experimental observation of this maximum in the triatomic case is obscured by the elastic scattering. But clearly a much more resonant process is observed in the Na* quenching by CO_2 and N_2O illustrated in Fig. 27. A similar observation has been made for H_2O, however, the latter does not quench

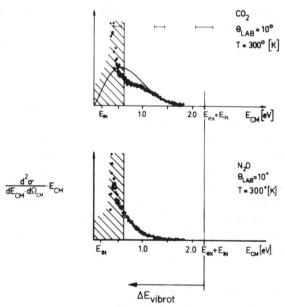

Figure 27. Experimental energy-transfer spectrum for Na*+CO_2 and +N_2O quenching and comparison with prior distribution for triatomic molecule.

Na* very efficiently ($\sigma_Q \approx 2$ Å^2) compared to CO_2 ($\sigma_Q \approx 50$ Å^2).[106, 107] Without going into detail, we state a qualitative agreement with the statistical predictions, although a quantitative comparison (see Fig. 27 for CO_2) is not very satisfactory.

F. Larger Polyatomic Molecules

When extending the quenching studies to larger molecules, a guideline hopefully may be obtained from statistical predictions since certainly *ab initio* calculations will not be possible in the near future. On the other hand, E–V–R transfer from Na* to larger molecules may be of interest for a number of practical reasons such as laser applications or the enhancement of chemical reactivities.

At first thought more and more resonant spectra (which are very difficult to investigate with our technique) might presumably be obtained since the number of atoms involved increases, as do the degrees of freedom. Experimental inspection of a typical example such as the quenching of Na($3p$) by ethylene ($H_2C–CH_2$) gives us a different understanding, as illustrated by Fig. 28. The energy-transfer spectrum looks much more like the spectrum for polar diatomic molecules, with a maximum at 1.3 eV (0.1 eV higher than for CO) and a half width of 1 eV (0.1 eV less than for CO). The explanation seems relatively obvious: As is well known, [29] organic molecules with double (and triple) bonds are very good quenching materials (for C_2H_4 $\sigma_Q = 130$ Å^2, [45]), whereas those with closed valence shells are very inefficient (e. g., methane $\sigma_Q \approx 0.35$ Å^2 [45]). Apparently, an extra electron is attached very easily to the double bond and leads to the attractiveness of the potential surface that is necessary for efficient quenching. For C_2H_4, the extra electron will be distributed in more or less cylindrical symmetry around the double bond, changing the equilibrium distance between the two CH_2 groups. Thus only the symmetric stretching

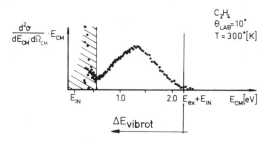

Figure 28.   Energy-transfer spectrum for C_2H_4, resembling $3p$ orbital quenching by diatomic molecule.

vibrational mode is excited in the collision. On the other hand, C_2H_4 is a large nonisotropic molecule, and rotational excitation rectangular to the axis connecting the CH_2 groups will be of importance. In total, it should act like a polar diatomic molecule that corresponds precisely to the experimental observation.

Most recent observations, [108] for even more complicated organic molecules with double and triple bonds allow an explanation along these lines. A careful discussion is necessary to distinguish those vibrational and rotational degrees of freedom that might be excited by attaching or not attaching the extra electron. Interesting examples are *cis*- and *trans*-buten. Whereas the first looks more like a triatomic atom, the latter gives a diatomic energy-transfer spectrum. Acetylen (C_2H_2), on the other hand, with its triple bond, has the behavior of a triatomic molecule. Many future studies are needed to improve our present understanding of the quenching process. Angular scattering distributions will give additional information, and an improved energy resolution will allow us to extend the spectra into the near-resonant region. Subsidiary information might be obtained by studying electron-scattering resonances of such molecules and their decay channels in a separate experiment.

VI. POLARIZATION STUDIES IN QUENCHING PROCESSES FROM LASER-EXCITED Na*(3p)

The polarization of the exciting laser light allows preparation of the excited atom in a well-defined mixture of states with specified angular-momentum projection quantum number m. Thus it is possible, in principle, to obtain direct information about phases and magnitudes of scattering amplitudes for the different m-levels, and critical tests of theoretical predictions prior to the usual averaging processes are also possible. The methods for preparing atoms in defined states, [98, 99] and disentangling geometric and dynamic effects, [102] have been described in detail elsewhere, as have the possibilities and difficulties for various collision problems. [28] In particular, the complicated hyperfine- or fine-structure coupling (necessary for describing the laser excitation process) must be factored out to retain only the dynamic interaction relevant to the collision process.

A. A Simple Example: $e + Na^*(3p) \rightarrow e + Na(3s)$

1. General Aspects

Before investigating the complicated quenching process it seems useful for illustration to discuss as a relatively simple example the dependence of the differential scattering cross section $e + Na^*(3p) \rightarrow e + Na(3s) + E_{ex}$ as a function of the polarization of the exciting laser light. This "test case" has been studied with considerable effort in our laboratory, [109-114] and the experimental determination of such detailed quantities as phases and ratios for scattering amplitudes has become possible.[100]

One of the simplifying aspects in electron collisions of the preceding type in the energy range of a few electron volts to some kiloelectron volts is that electrons interact essentially with the electric charge cloud of the atom only via the Coulomb potentials and charge exchange. Fine- or even hyperfine-structure interactions may be neglected for the collision dynamics.[115] Qualitatively, this is seen by comparing the collision time $t_{col} \approx 10^{-15}$ sec to the Larmor precession time for the atomic electron spin $t_{FS} \approx [1/(\hbar\,\Delta E_{FS})] \approx 10^{-12}$ sec and the nuclear spin $t_{HFS} \approx [1/(\hbar\,\Delta E_{HFS})] \approx 10^{-9}$ sec. Both electron spins and nuclear spin will stay untouched during the collision (an extremely nonadiabatic process) and enter into the scattering process only as statistical weight factors. Thus we may discuss the preceding scattering process by describing the sodium atom in terms of its electron orbital radial, angular and projection quantum numbers n, l, and m_l, with the interaction provided by the atomic charge-cloud distribution exclusively.

Thus because of laser optical pumping, the colliding electron initially finds the sodium atom in a mixture of the $3p$ states with $m_l = \pm 1$ or 0 and leaves it in the $3sm_l = 0$ state (Fig. 29). For simplicity, we neglect electron exchange in the following discussion. Then the scattering process may be described by three scattering amplitudes, $f_0, f_{\pm 1}$ (see Fig. 29), two of which are, for parity reasons, interrelated by $f_{-1} = -f_1$. Thus we need essentially three parameters to describe the scattering process completely, such as the inelastic differential cross section $\sigma_{3s \rightarrow 3p} \propto |f_0|^2 + 2|f_1|^2$, the amplitude ratio $\lambda = (|f_0|^2 / \sigma_{3s \rightarrow 3p})$, and a phase parameter χ defined by $f_0 f_1^* = |f_0||f_1|e^{i\chi}$. Alternatively, the collision-induced state multipole moments, that is to say, the orientation O_{1-} and alignment parameters A_0, A_{1+}, and A_{2+}, may be used for description. [101, 102]

It should be noted that in electron scattering the amplitudes (or collision parameters) are defined and computed in a laboratory space-fixed coordi-

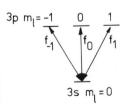

Figure 29. Angular momentum substates for Na($3p$) atom and scattering amplitudes for excitation–deexcitation.

nate system, the collision system (Z_{col}, Y_{col}, X_{col}). Conveniently, Z_{col} may be put parallel to the scattered electron (or to the incident e-beam direction). On the other hand, the atom excitation is conveniently described in a different frame, the photon frame (Z_{ph}, Y_{ph}, X_{ph}). The method for determining the collision parameters consists in a systemic variation of the relative orientation of the photon frame with respect to the collision frame, for example, by changing the direction of laser polarization or the incident light direction. Thus the scattering electron is exposed to a particular view of the charge of the excited atom.

2. Linearly Polarized Light

If we specialize to excitation of the atom with linearly polarized light, we prepare an aligned atom, that is, positive or negative m_l states are populated with equal weight. For symmetry reasons, the charge-cloud distribution has to exhibit cylindrical symmetry with respect to the electric vector (polarization vector) $\mathbf{E}_{h\nu}$ ($\| Z_{ph}$) of the exciting light. With respect to any space-fixed coordinate system, the excited atom may be seen as an incoherent mixture of $|\pi^+\rangle$, $|\pi^-\rangle$ and $|\sigma\rangle$ states (Fig. 30). Because of the selection rule ($\Delta m = 0$) the atom, is excited predominantly in the $|\sigma\rangle$ state with respect to $\mathbf{E}_{h\nu}$, but the $|\pi^+\rangle$ and $|\pi^-\rangle$ states are also admixed as a result of the fine and hyperfine coupling in the pumping process. The cigar-shaped charge cloud is thus aligned along the direction of light

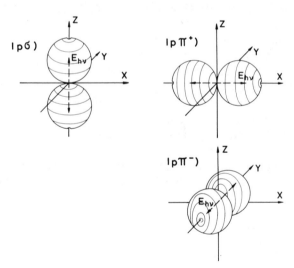

Figure 30. Schematic charge-cloud distribution of atomic p-state excited by linearly polarized light with electric vector $\mathbf{E}_{h\nu}$. Different relative alignments of $\mathbf{E}_{h\nu}$ lead to $|\sigma\rangle, |\pi^\pm\rangle$ states with respect to fixed coordinate frame.

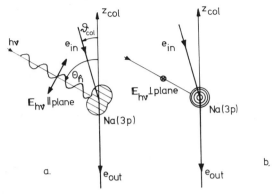

Figure 31. Schematic illustration of laser-excited atom and collision frame.

polarization. In Fig. 31 typical scattering geometries are illustrated, displaying the collision system ($Z_{col}-X_{col}$ plane). The atomic charge cloud is indicated for a polarization vector both in the plane (*a*) and rectangular to it (*b*).

The experimental observations and theoretical predictions show that the electron-scattering signal for the $3p \rightarrow 3s$ deexcitation in configuration *a* ($I_{\parallel}$) is larger (often substantially) than in situation *b* ($I_{\perp}$). This finding is immediately evident from Fig. 31, since in *a* configuration the electron has to cross through a much larger interaction region than in *b*. It is further obvious from Fig. 31 that the scattering intensity for *a* will critically depend on the angle Θ_n of light incidence, whereas it must be constant in *b* for symmetry reasons. We thus normalize the scattering intensity $I_{\parallel}$ (Θ_n, ϑ_{col}) to the latter signal ($I_{\perp}$). The measured normalized scattering signal *r* is shown in Fig. 32 as a function of the incident light angle for different scattering angles of the electron at an incident electron energy $E_{IN} = 10$ eV. The observed lobes follow a

$$I_{\parallel} = a + b \cos 2 \left(\Theta_n + \psi_{FIT} \right) \tag{VI.1}$$

law.[100, 102] They give a picture of the effective interaction that is experienced by the scattered electron. More precisely, the lobes of Fig. 32 depict the charge-cloud distribution that would be excited in the inversed collision process $e + Na(3s) \rightarrow e + Na(3p)$ at an incident energy $E_{IN}' = E_{IN} + E_{ex}$.[28]

Most easy to understand is the forward scattering $\vartheta_{col} = 0$ since in that case the selection rule $\Delta m_1 = 0$ holds for electron-impact-induced transitions, and the electron interacts only with the $|\sigma\rangle$ part (with respect to Z_{ph} $\|E_{h\nu}$) of the charge cloud. Of course, for $\vartheta_{col} = 0$, the lobe must be

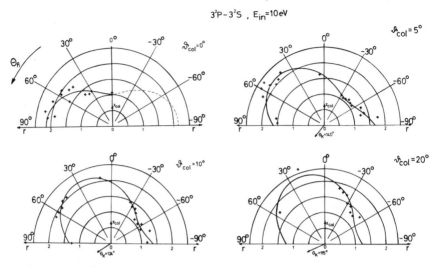

Figure 32. Scattering intensity $I_{\parallel}$ normalized to $I_{\perp}$ as function of light-incidence angle Θ_n for the $e + Na(3p) \rightarrow Na(3s) + e$ process at incident energy of $E_{IN} = 10$ eV for various collision angles ϑ_{col}.

symmetric to the Z_{col} axis, and the experimental signal directly reflects the thickness of the atomic electron distribution as transversed by the scattered electron. As Θ_n is varied (see Fig. 31), the scattering signal (Fig. 32 $\vartheta_{col} = 0$) is just the projection of the cigar shape on the Z_{col} axis. For $\Theta_n = 0$, it reaches a minimum that, however, is not equal to zero as it would be if the optical excitation would prepare the atom in the pure $|\sigma\rangle$ state.

As ϑ_{col} increases, the experimental scattering lobe is no longer symmetric to Z_{col} but turns its symmetry axis essentially into the direction of the momentum-transfer vector $\mathbf{K} = \mathbf{k}_{out} - \mathbf{k}_{in}$ as indicated by Θ_k in Fig. 32. In Born's approximation the lobe would be exactly symmetric to $\mathbf{K}$. This reflects the fact that electronic angular momentum of the atom is transferred into linear momentum of the scattered electron.

The conclusions to be drawn from series of measurements of this type, illuminating some new aspects of electron scattering theory, are interesting [28, 100] but completely beyond the scope of the present chapter.

3. Circular Polarization

When circular laser polarization is used to excite the atom, a somewhat different situation is experienced. The sodium $3p$ is prepared with a defined orientation rather than with alignment only. This implies a finite expectation value of the angular momentum projection $\langle m_1 \rangle \neq 0$ and toroidal rather than cigar shapes are prepared. The essential information may be obtained by performing a scattering experiment in the $X_{ph}-Y_{ph}$ plane, with Z_{ph} parallel to the incident light direction. Depending on which side of the torus the electron passes the atom, its absolute value of the orbital angular momentum decreases or increases during deexcitation of the $3pm_1 = +1$ (or -1) state. This leads to strong left–right asymmetries in the electron-scattering signal as a function of scattering angle $\vartheta_{col} > 0$.[114] In the quenching process no such effect has been observed yet: however, it could become important, for example when optically active molecules are investigated.

B. Polarization Effects in Quenching of Na(3p) by Simple Molecules

1. Differences from Electron-scattering Processes

It has been observed that the quenching cross section for the diatomic homonuclear molecules N_2, H_2, and D_2, clearly depends on the laser polarization, although to a lesser extent than the electron-scattering intensities.[116] Although, in principle, the same discussion may be applied as in electron scattering and the theory of the measurement[102] may be applied adequately, heavy-particle collisions, especially with molecules, bring a number of complications that have to be taken into consideration:

1. The collision system is defined by the relative particle velocities before and after collision (we choose $Z_{col}\|CM$ system before collision). The CM direction is, however, much less well defined than in electron-collision experiments, where it is essentially given by the incident electron beam (mass of electron ≪ mass of sodium). Therefore, larger experimental uncertainties arise in determination of quantities 1, 2, and especially ψ_{FIT} [equation (VI.1)]. Also, it is experimentally more difficult to change the incident light direction Θ_n without simultaneously changing the scattering volume. Therefore, only qualitative measurements have been carried out thus far, varying the polarization angle ψ rather than Θ_n while the laser light enters rectangular to the scattering plane (CM–CM′ system) as illustrated in Fig. 33. In this way nearly the same configuration is obtained as in the previous case (Fig. 31a); however, normalization of the scattering signal $I_\|$ is not possible. Otherwise, asymmetry lobes as shown in equation (VI.1) and Fig. 32 may be expected (Θ_n has to be replaced by $\psi \pm (\pi/2)$).
2. Since the electron-spin precession time $t_{FS} \approx 10^{-12}$ is of the same order of magnitude as the heavy-particle collision time, the influence of the spin on the dynamics may not be neglected *a priori* as previously. The influence of electron-spin rotation will almost certainly be to decrease

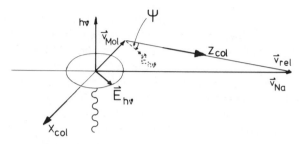

Figure 33. Experimental scattering geometry for observation of polarization effects in differential quenching process; Ψ may be varied to determine ratio of maximum (I_{max}) to minimum (I_{min}) partial quenching cross section.

the polarization effect. Nevertheless, we omit a discussion on the spin–orbit interaction for the tentative interpretation of the first experimental observations given in the text that follows, which in their qualitative essence can be described in terms of orbital angular momentum states $|\pi^{\pm}\rangle$, $|\sigma\rangle$.

3. In contrast to electron scattering, which is an extremely diabatic process, the quenching is certainly closer to the adiabatic picture. Therefore, scattering amplitudes are often described in a body-fixed system, rotating during the collision (Z_{body}, Y_{body}, X_{body}) as opposed to the previous case where f_0, $f_{\pm 1}$ (or f_σ, $f_{\pi\pm}$) referred to the space-fixed CM system. In the adiabatic limit (i. e., for Na* heavy-particle elastic collisions at low energies), neglecting electron spin and regarding the target molecule as spherical isotropic, the scattering process may be described in terms of three states with potentials Σ and the degenerate $\Pi^{\pm}$ (here and in the following, capital Σ and Π refer to the body-fixed system and lowercase σ and π, to the CM system). In the strictly adiabatic collision the three states do not mix in the body-fixed system (see Fig. 34), but obviously this usually implies $\Delta m_1 \neq 0$ (or $\sigma \leftrightarrow \pi^+$) transitions with respect to the CM system. Here, again, the atom preparation before collision as indicated in Fig. 34 is done with linearly polarized light, somewhat idealized by assuming pure $|\sigma\rangle$ or pure $|\pi^{\pm}\rangle$ state excitation (see Section VI.A.2). Obviously, $\pi^+ \leftrightarrow \pi^-$ transitions are not possible for symmetry reasons as long as the molecule is spherical symmetric. In the adiabatic limit we have for the amplitudes $f_{\pi+} = f_{\pi-}$. However, this picture certainly is oversimplified, and as the interaction becomes somewhat faster, Coriolis forces on the atomic electron are likely to induce $\Pi^+ \leftrightarrow \Sigma$ transitions. We recall this remark later when discussing the experimental results. Precisely as in electron

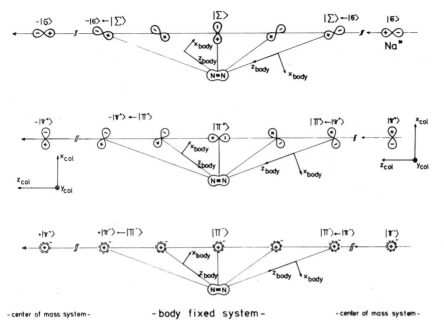

Figure 34. Schematic illustration of body-fixed adiabatic collisions of atomic $|\Sigma\rangle$, $|\Pi^+\rangle$, and $|\Pi^-\rangle$ states.

collisions, the final atomic state is always a state with $|\sigma\rangle = |\Sigma\rangle$ configuration ($3s$ $m_l = 0$). Thus anisotropies may be expected since the $|3p\ \Pi\rangle \rightarrow |3s\ \Sigma\rangle$ transition may well have a probability different from that of the $|3p\ \Sigma\rangle \rightarrow |3s\ \Sigma\rangle$ process.

4. The molecule is *not* spherical symmetric; thus the designation of the potentials at close distances is no longer simply Σ, $\Pi^\pm$ with respect to the body-fixed system as in assumed in Fig. 34. The states have to be designated according to the molecular symmetry. For a detailed discussion, the reader is referred to the literature.[34, 90] For the present subject it is, however, important to remember that the molecule may be rotationally excited and that to each final rotational state j, m_j specific scattering amplitudes exist. The great variety of scattering amplitudes that have to be taken into account when determining the differential cross section for the quenching process, will again tend to decrease the polarization anisotropy.

2. Experimental Results

The first experimental findings were reported for $Na^* + N_2$[116] and are currently extended:[117]

1. A definite but small anisotropy of the differential quenching cross section is observed when the electric vector $E_{h\nu}$ of the exciting laser light is rotated in the scattering plane. It follows equation (VI.1), where Θ_n has to be replaced by $\psi + (\pi/2)$.

2. The quenching cross section has its maximum when the $E_{h\nu}$ vector is approximately parallel to the CM system, that is, for the $|\sigma)$ configuration of the atom. This is relatively independent from the energy transfer, subject, however, to experimental errors as a result of uncertainties in the CM definition.

3. The anisotropy, that is, the maximum:minimum scattering intensity ratio I_{max}/I_{min}, seems to follow the energy-transfer spectrum as illustrated by Fig. 35. Taking account of some suitable corrections for the angular spread of the initial CM direction and for incomplete optical pumping, the maximum anisotropy is determined to be 26% for N_2 at 80°K, $\Theta_{lab} = 10°$, and $E_{CM} \approx 1.2$ eV.

4. For laboratory angles of $\Theta_{lab} = 7$ to 18°, the anisotropy is essentially independent of Θ_{lab}. It should be kept, in mind, however, that the range of CM scattering angles ϑ_{CM} does not change substantially for $\Theta_{lab} = 7$ to 18°. Preliminary results for larger Θ_{lab} indicate a rapid decrease of the anisotropy.

5. By a change in the temperature of the molecular beam, its distribution of rotational states is altered significantly, whereas, E_{IN} is determined mainly by the sodium velocity. At 300°K the anisotropy decreases to about half of the value at 80°K. One typical point is shown in Fig. 35.

6. Measurements on a circular asymmetry, that is, a change of the cross section for σ^+ and σ^- light excitation at $\Delta E_{vib,\ rot} = 1$ eV, $\Theta_{lab} = 18°$, $\vartheta_{CM} = 14.7°$, have shown no significant effect ($< 4\%$).

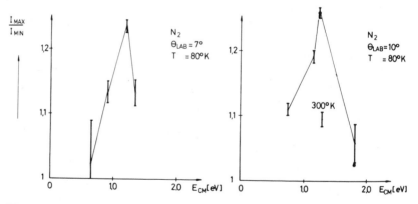

Figure 35. Experimentally observed anisotropy (see Fig. 33) for excitation with linearly polarized light. Note decrease in anisotropy for elevated temperatures.

7. Similar observations have been made for the quenching of Na* by D_2 and H_2. For the polar molecules CO and CO_2, measurements have only been carried out for 300K, where no anisotropy has been observed.

3. Interpretation

The experimental material available thus far does not allow a finally conclusive interpretation. The following discussion, therefore, has to be taken as tentative, joining, however, the experimental observations well to an improved qualitative understanding of the quenching process.

We start by recalling the observation of maximum quenching cross section for $E_{h\nu}$ parallel to the initial CM system, that is, for the atomic $|\sigma)$ configuration. For the small CM-scattering angles involved, the momentum transfer vector $\mathbf{K}$ is nearly parallel to the CM system within the limit of experimental uncertainties defining the latter. Thus the observation seems to be fairly equivalent to that in superelastic electron scattering (Section IV.A). However, in contrast to the simplified potential-energy diagram (Fig. 3), a somewhat refined curve crossing scheme (Fig. 36) indicates that three excited-state potentials, $\Pi^{\pm}$ and Σ, are involved in the quenching process.

Consequently, at second thought the quenching maximum for the $|\sigma)$ configuration should be surprising since in a body-fixed system $|\sigma)$ becomes a $|\Sigma)$ state (Fig. 34), and quenching could only happen if crossing ② is reached. It is not completely well known whether crossing ② is actually surmounted for thermal incident energies E_{IN}. The general belief is, however, that the Σ state may be neglected for quenching,[118] at thermal E_{IN}.* This view is supported by observing similar phenomena for N_2, H_2, and D_2, indicating that the energetic position of crossing ② is not critical. Thus we may assume that the quenching process has to proceed mainly via Π-state interaction and crossing ③. Then why does the atomic $|\Pi)$ preparation give a smaller section? The explanation that follows is essentially consistent with the arguments of Nikitin,[35] on coupling regions and $|\sigma) \rightarrow |\Pi)$ conversation.[118] As discussed in Section IV.B.1, we have to modify the strictly adiabatic picture. For large internuclear separation where the electrostatic interaction is still very weak, the atomic electron charge cloud will stay fixed to the CM system since no interaction there forces it to change its alignment. When Stark splitting becomes noticeable (i. e., larger than Coriolis forces), the atomic states will become fixed to the body frame. This may happen near crossing ① as illustrated in Fig. 37. For simplicity, a sudden change may be assumed from CM to body-fixed

---

* *Note added in proof*: This belief confirmed by the ab initio calculations.[133, 134]

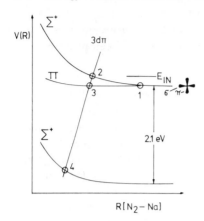

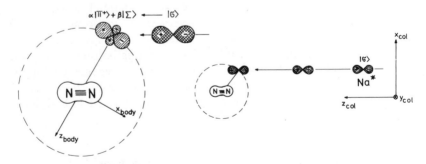

Figure 36.   Refined potential interaction scheme.

systems. For the large impact parameters relevant to small-angle scattering, it is obvious from Fig. 37 that the σ (CM) configuration becomes Π predominantly. Thus the larger quenching cross section for this case is explained by a more efficient way to reach crossing ③.

Following the quenching process further, the system is described best in the body-fixed frame, at least for the smaller internuclear distances. Then no orbital angular momentum can be transferred to the relative internuclear motion $l_Z = 0$, and the question arises as to how the electronic orbital angular momentum $L_Z = \pm 1$ of the $|3p\,\Pi\rangle$ state of Na* should be disposed of since the final state has to be $|3s\,\Sigma\rangle$. The obvious solution is to change the orientation of the molecular angular momentum j, that is, to induce a $\Delta j_Z = \pm 1$ transition in the molecule and thus maintain a constant

Figure 37.   Sudden transitions from center-of-mass-system to body-fixed system.

projection $J_Z = (L_Z + j_Z)$ of the total angular momentum onto the body-fixed internuclear axis. This discussion emphasizes the close connection between rotational transitions in the molecule and quenching as experimentally observed (Section V); no quenching seems to be possible when the molecule cannot undergo $\Delta j_Z = \pm 1$ transitions. It seems immediately obvious and has been proven for simpler systems[119] that such transitions with $\Delta j_Z = \pm 1$ are most probable for large rotational angular momentum j of the molecule, since j has to change its orientation only slightly. It is interesting to note that this interpretation would predict an increase of the quenching cross section with the rotational temperature of the molecule. This may explain the apparent discrepancies of the total quenching cross sections determined by photodissociation ("cold" molecules) and flame experiments ("hot" molecules) at similar incident energies E_{IN} (see Fig. 7). More refined measurements are needed to finally prove this hypothesis.

Next we have to explain the dependence of the quenching anisotropies on the molecular temperature. Taking molecular rotation into account, this becomes immediately obvious. The Σ and $\Pi^{\pm}$ potentials depend rather strongly[76] on the molecular orientation. Thus mixing of states occurs not only between crossings ③ and ④ as discussed earlier, but also between ① and ③,②. If the molecule rotates rapidly (high temperatures), the Σ and Π states are mixed thoroughly and the dependence of the quenching signal on the polarization decreases, as observed. Thus particularly high polarization effect should be observed for *para*-H_2, a crucial experiment that still remains to be carried out.

At once, the previous discussion explains why the polar molecules do not exhibit a polarization effect at all; the strong anisotropy of the CO–Na* potential leads to a complete mixing of Σ and Π states. Somewhat less easily explained is the dependence of the polarization effect on E_{CM} and its disappearance at larger scattering angles. One possibility is to ascribe small quenching cross sections to small collision parameters and thus to deeper penetration, where the molecular anisotropy is dominant and thus mixes the initial state preparation as discussed previously.

In summary, we note that important information may be gained from polarization studies about the nature of the interaction mechanism. The above discussion is still highly speculative and much more experimental data are needed before final conclusions can be drawn.

VII. OTHER BEAM EXPERIMENTS RELATED TO QUENCHING OF ALKALI RESONANCE RADIATION

Before concluding this chapter, we wish to mention very briefly a number of interesting beam experiments studying processes that are—to some extent—inverse quenching reactions. A first group of experiments studies the excitation of alkali atoms in fast collisions with molecules, for example,

$$\left.\begin{array}{c} \text{Na}(3s) \\ \text{K}(4s) \end{array}\right\} + \text{N}_2 \ (v=0\ ) \rightarrow \left.\begin{array}{c} \text{Na}(3p) \\ \text{K}(4p) \end{array}\right\} + \text{N}_2 \ ^{(v'\ )} \qquad \text{(VII.1)}$$

where the incident kinetic energy E_{IN} has to be larger than the excitation energy E_{ex}. Differential excitation cross sections for $K + N_2$ have been reported[120] for $E_{\text{IN}} \gg E_{\text{ex}}$. Most of the material is concerned with the observation of integrated cross sections for energies far above threshold. Usually the atomic fluorescence is detected and it is impossible to separate individual final vibrational states. The interested reader is referred to a review by Kempter.[121] The excitation mechanism at these energies from some 10 eV up to some keV is certainly quite different from the quenching process discussed in the present chapter. Recently, even coincidence experiments between photon and scattered atom have become possible[122] yielding information with a similar degree of detail as those discussed in Section VI.

Most directly related to the quenching process are recent studies of the differential excitation cross section for $K(4s) + N_2(v=0) \rightarrow K(4p) + N_2(v'=0, 1, 2)$.[123, 124] The seeded potassium beam had incident energies just above threshold. The potassium velocity after collision was measured by a time-of-flight (TOF) technique. The scattering kinematics (Newton diagram) is somewhat less favorable for clearly resolving the final v' states, since the energy scale is compressed in these inelastic collisions. Also, elastic scattering and vibrational transitions without electronic excitation in the backward direction partially obstruct the signal. Nevertheless, it may be concluded from these experiments that (1) the excited potassium atoms are dominantly scattered into the forward hemisphere with a sharp peak at $\Theta_{\text{CM}} = 0°$; (2) a peak of approximately $\frac{1}{3}$ the height of the one at $\Theta_{\text{CM}} = 0°$ appears in the backward direction; (3) the contour map is by no means symmetric with respect to $\Theta_{\text{CM}} = 90°$, supporting a direct excitation mechanism; and (4) the most probable amount of energy transferred into the internal degrees of freedom of N_2 is 10.5 kcal/mole, suggesting that the electronically inelastic encounters lead most probably to the first excited vibrational and some higher rotational states of N_2. This is in qualitative agreement with the curve-crossing model and computational results,[82] but no extrapolation to Na(3p) quenching is possible. A second type of experiment is concerned with excitation of the alkali resonance line by "hot" (i.e., vibrationally excited) N_2 in crossed beams

$$\left.\begin{array}{c} \text{Na}(3s) \\ \text{K}(4s) \end{array}\right\} + \text{N}_2 \ (v>0\ ) \rightarrow \left.\begin{array}{c} \text{N}(3p) \\ \text{K}(4p) \end{array}\right\} + \text{N}_2 \ (v'\ ) \qquad \text{(VII.2)}$$

Determination of the total excitation cross section is again done by observing the resonance-line emission. Both the initial relative kinetic energy and the N_2 temperature may be varied. First experiments of this type have studied the excitation of Na($3p$) by N_2, H_2, and D_2, vibrationally excited at 2000 to 3000°K for initial relative velocities from 1000 to 4000 m/sec [125, 126] as well as both potassium and sodium.[127] Recent studies with seeded beams [128-131] on $K + N_2$ and $K + CO$ involved lower initial vibrational states v. This type of experiment is very difficult to analyze and evaluation invokes a number of assumptions for fitting the experimental data. A special threshold law for the energy dependence of the partial cross sections is adopted:

$$\sigma(v \rightarrow v', E_{IN}) = \sigma_{v,v'} \cdot \left(\frac{E_{IN}}{E_{thresh}^{v,v'}} - 1 \right)^p \qquad (VII.3)$$

A Boltzmann distribution is assumed for the initial vibrational state population, the same order of magnitude is assumed for excitation cross sections at different v', and the angular beam spreads are neglected. Rotational-energy transfer is not discussed. A best fit to the experimental data[128] is found for $p = 1.5$, $(\sigma_{10}/\sigma_{00}) = 1.6$, and $(\sigma_{20}/\sigma_{00}) = 1$ for $K + N_2$. Markedly more different partial cross sections emerge for $K + CO$, again illustrating the obvious differences between homonuclear and polar molecules. All those experiments are in qualitative agreement with the curve-crossing model but do not warrant further discussion with respect to the quenching process at present. One may, however, look forward to future measurements along these lines.

Finally, we bring attention to an experiment by Schepper et al.[132] on vibrational excitation of N_2 and CO in collisions with ground-state potassium *without* electronic excitation, who observe a special type of oscillations in the differential cross section. This may be of importance in a detailed discussion of quenching processes in the outgoing channel.

VIII. CONCLUSION

The problem of quenching alkali resonance radiation in E–VR energy-transfer collisions with simple molecules is important as a model case for basic processes in photochemistry and serves its own right for a variety of practical applications, such as in laser physics. It has been studied for many years in the past, but only recent progress has led to information of the final internal energy of the molecule. In particular, crossed-beam experiments with laser-excited atoms allow a detailed measurement of energy-transfer spectra. There can be no doubt that the curve-crossing mechanism originally introduced to this problem by Bjerre and Nikitin is responsible for the large quenching cross section and for the nonresonant energy transfer. Many details have been revealed, and a useful simplifying qualitative understanding has become possible in consequence of the new experimental techniques.

However, at present there is insufficient information on the potential surfaces and interaction matrix elements for a quantitative interpretation. Among other improvements, theoretical computations will have to take account of rotational excitation of the molecule and be able to explain polarization effects. Experimentally, the range of scattering angles has to be widened, the energy and angular resolution must be improved, and a variation of incident energies as well as rotational temperatures is highly desirable. Thus much work remains to be done before quantitative agreement between theory and experiment will emerge.

ACKNOWLEDGEMENT

The author would like to acknowledge helpful discussions with Professor Nikitin and Dr. P. McGuire. The understanding of the quenching process as described in the present chapter has been developed during many exciting discussions with Dr. H. Hofmann and Dr. K. A. Rost. The experimental material presented here on quenching of laser-excited Na* in crossed-beam experiments is based mainly on their Ph.D. thesis [89,90] and is to be published elsewhere.[108,117] H. W. Hermann has been of great help by reading the manuscript.

The financial support from the Deutsche Forschungsgemeinschaft and from the Physics dept. of the University of Kaiserslautern is gratefully acknowledged.

REFERENCES[†]

1. R. Mannkopf, *Z. Phys.*, **36**, 22 (1926).
2. V. S. Letokhov, *Science*, **180**, 451 (1973).
3. K. L. Kompa, *Z. Naturforsch.* **276**, 89 (1972).
4. A. B. Peterson, C. Wittig, and S. R. Leone, *Appl. Phys. Lett.*, **27**, 305 (1975).
5. A. B. Peterson, C. Wittig, and S. R. Leone, *J. Appl. Phys.*, **47**, 1051 (1976).
6. E. C. Zipf, *Can. J. Chem.*, **47**, 1863 (1969).
7. R. L. Taylor, *Can. J. Chem.*, **52**, 1436 (1974).
8. R. S. Cventanovic, *Can. J. Chem.*, **52**, 1452 (1974).

[†]This article has been finished in 1977 and only a few recent publications have been added in proof.

9. M. N. Vlasov, *J. Atm. Terr. Phys.*, **38**, 807 (1976).

10. J. W. McGowan, R. H. Kummler, and F. R. Gilmore, in J. W. McGowan, ed., *The Excited State in Chemical Physics*, Vol. XXVIII, *Advances in Chemical Physics* Wiley, New York, 1975, p. 379.

11. J. R. Barker and R. E. Weston, Jr., *J. Chem. Phys.*, **65**, 1427 (1976).

12. S. M. Lin and R. E. Weston, Jr., *J. Chem. Phys.*, **65**, 1443 (1976).

13. G. Karl and J. C. Polanyi, *J. Chem. Phys.*, **38**, 271 (1963).

14. G. Karl, P. Kruus, and J. C. Polanyi, *J. Chem. Phys.*, **46**, 224 (1967).

15. G. Karl, P. Kruus, J. C. Polanyi, and I. W. M. Smith, *J. Chem. Phys.*, **46**, 244 (1967).

16. H. Heydtmann, J. C. Polanyi, and R. T. Taguchi, *Appl. Opt.*, **10**, 1755 (1971).

17. R. G. Shortrige and M. C. Lin, *J. Chem. Phys.*, **64**, 4076 (1976).

18. D. S. Y. Hsu and M. C. Lin, *Chem. Phys. Lett.*, **42**, 78 (1976).

19. S. R. Leone and F. J. Wodarczyk, *J. Chem. Phys.*, **60**, 314 (1974).

20. F. J. Wodarczyk and F. B. Sackett, *Chem. Phys.*, **12**, 1872 (1976).

21. A. Hariri, A. B. Petersen, and C. Wittig, *J. Chem. Phys.*, **65**, 1872 (1976).

22. I. W. M. Smith, in J. W. McGowan, Ed., *The Excited State in Chemical Physics 1*, Vol. XXVIII, *Advances in Chemical Physics*, Wiley, New York, 1975, p. 1.

23. I. V. Hertel and W. Stoll, in B. C. Cobic and M. V. Kurepa, Eds., *Abstracts, VIIIth ICPEAC*, Belgrade, 1975.

24. G. M. Carter, D. E. Pritchard, M. Kaplan, and T. W. Ducas, *Phys. Rev. Lett.* **35**, 1144 (1975).

25. R. Düren, H. O. Hoppe, and H. Pauly, *Phys. Rev. Lett.*, **37**, 743 (1976).

26. R. W. Anderson, T. P. Goddard, C. Parrano, and J. Warner, *J. Chem. Phys.*, **64**, 4037 (1976).

27. I. V. Hertel, H. Hofmann, and K. A. Rost, *Phys. Rev. Lett.*, **36**, 861 (1976).

28. I. V. Hertel and W. Stoll, *Adv. Atom. Molec. Phys.*, **13**, 113 (1978).

29. P. L. Lijnse, "Review of Literature on Quenching, Excitation and Mixing Cross Sections for the First Resonance Doublets of the Alkalies" Report i-398, Fysish Laboratorium, Utrecht University, 1972.

30. P. L. Lijnse, Ph.D. thesis, Utrecht University, 1973.

31. L. Krause, in J. W. McGowan, Ed., *The Excited State in Chemical Physics*, Vol. XXVIII, *Advances in Chemical Physics*, Wiley, New York, 1975 p. 267.

32. R. J. Donovan, *Prog. Reaction Kinetics*, **10**, 253 (1979).

33. S. Lemont and G. W. Flynn, *Annu. Rev. Phys. Chem.*, **28**, (1977).

34. E. C. Beaty, J. W. Gallagher, and J. R. Rumble. Jr., JILA Report, February 1977.

35. E. E. Nikitin, *Theory of Elementary Atomic and Molecular Processes in Gases*, Clarendon, Oxford 1974.

36. E. E. Nikitin, in J. W. McGowan, Ed., *The Excited State in Chemical Physics*, Vol. XXVIII, *Advances in Chemical Physics*, Wiley, New York, 1975, p. 317.

37. H. Derblom, *J. Atm. Terr. Phys.*, **26**, 791 (1964).

38. D. M. Hunten, *J. Atm. Terr. Phys.*, **27**, 583 (1965).

39. C. T. J. Alkemade and P. J. T. Zeegers, in J. D. Winefordner, Ed., *Excitation and Deexcitation Processes in Flames*, Wiley, New York, 1971.

40. H. P. Hooymayers and C. T. J. Alkemade, *Quant. Spectrosc. Radiat. Transf.*, **6**, 501 (1966).

41. H. P. Hooymayers and C. T. J. Alkemade, *J. Quant. Spectrosc. Radiat. Transf.*, **6**, 847 (1966).

42. D. R. Jenkins, *Proc. Roy. Soc. (Lond.)*, **A293**, 493 (1966).

43. H. P. Hooymayers and P. L. Ljinse, *J. Quant. Spectros. Radiat. Transf.*, **9**, 995 (1969).

44. P. L. Lijnse and R. J. Elsenaar, *J. Quant. Spectrosc. Radiat. Transf.*, **12**, 1115 (1972).

45. R. G. W. Norrish and W. M. Smith, *Proc. Roy. Soc. (Lond.)*, **A176**, 295 (1941).

46. B. L. Earl, R. R. Herm, S. M. Lin, and C. A. Mims, *J. Chem. Phys.*, **56**, 867 (1972).

47. B. L. Earl and R. R. Herm, *Chem. Phys. Lett.*, **22**, 95 (1973).

48. B. L. Earl and R. R. Herm, *J. Chem. Phys.*, **60**, 4568 (1974).

49. D. A. McGillis and L. Krause, *Can. J. Phys.*, **46**, 25 (1968).

50. D. A. McGillis and L. Krause, *Can. J. Phys.*, **46**, 1051 (1968).

51. W. Demtröder, *Z. Phys.*, **166**, 42 (1962).

52. E. Hulpe, E. Paul, and W. Paul, *Z. Phys.*, **177**, 257 (1964).

53. C. Bästlein, G. Baumgartner, and B. Brosa, *Z. Phys.*, **218**, 319 (1969).

54. B. P. Kibble, G. Copley, and L. Krause, *Phys. Rev.*, **159**, 11 (1967).

55. G. Copley, B. P. Kibble, and L. Krause, *Phys. Rev.*, **163**, 34 (1967).

56. G. Copley and L. Krause, *Can. J. Phys.*, **47**, 533 (1969).

57. J. N. Dodd, E. Enemark, and A. Gallagher, *J. Chem. Phys.*, **50**, 4838 (1969).

58. L. E. Brus, *J. Chem. Phys.*, **52**, 1716 (1970).

59. R. Bersohn and H. Horwitz, *J. Chem. Phys.*, **63**, 48 (1975).

60. A. Terenin, *Z. Phys.*, **37**, 98 (1926).

61. A. G. Gaydon and H. G. Wolfhard, *Flames, Their Structure, Radiation and Temperature*, Chapman and Hall, London, (1960).

62. C. M. Sadowski, H. I. Schiff, and G. K. Chow, *J. Photochem*, **1**, 23 (1972/73).

63. I. R. Hurle, *J. Chem. Phys.*, **41**, 3911 (1964).

64. S. Tsuchija and I. Suzuki, *J. Chem. Phys.*, **51**, 5725 (1969).

65. M. Czajakowski, L. Krause, and G. M. Skardis, *Can. J. Phys.*, **51**, 1582 (1973).

66. D. A. Jennings, W. Braun, and H. P. Broida, *J. Chem. Phys.*, **59**, 4305 (1973).

67. H. S. W. Massey, *Rep. Progr. Phys.*, **12**, 248 (1949).

68. A. Bjerre and E. E. Nikitin, *Chem. Phys. Lett.*, **1**, 179 (1967).

69. K. J. Laidler, *J. Chem. Phys.*, **10**, 43 (1942).

70. Y. Mori, *Bull. Chem. Soc. Jap.*, **35**, 1584 (1962).

71. E. Bauer, E. R. Fisher, and F. R. Gilmore, *J. Chem. Phys*, **51**, 4173 (1969).

72. E. R. Fisher and G. K. Smith, *Chem. Phys. Lett.*, **6**, 438 (1970).

73. E. R. Fisher and G. K. Smith, *Appl. Opt.*, **10**, 1083 (1971).

74. E. R. Fisher and G. K. Smith, *Chem. Phys. Lett.*, **13**, 448 (1972).

75. J. B. Hasted and A. Y. S. Chong, *Proc. Phys. Soc.*, **80**, 441 (1962).

76. C. Bottcher, *Chem. Phys. Lett.*, **35**, 367 (1975).

77. R. W. Numrich and D. G. Ruhlar, *J. Chem. Phys.*, **79**, 2745 (1975).

78. H. S. Taylor, F. W. Bobrowicz, P. J. Hay, and T. H. Dunning, Jr., *J. Chem. Phys.*, **65**, 1182 (1976).

79. H. Ehrhardt and K. Willmann, *Z. Phys.*, **204**, 462 (1967).

80. G. J. Schulz, *Rev. Mod. Phys.*, **45**, 378 (1973).

81. J. R. Barker, *Chem Phys.*, **18**, 175 (1976).

82. E. A. Andreev, *High Temp. Phys.*, **10**, 637 (1972).

83. J. N. Bardsley, F. Mandl and A. R. Wood, *Chem. Phys. Lett.*, **1**, 359 (1967).

84. C. Bottcher and C. V. Sukumar, *J. Phys.*, **B10**, (1977).

85. U. Buck, *Rev. Mod. Phys.*, **46-2**, 369 (1974).

86. H. Pauly and J. P. Toennis, *Meth. Exp. Phys.*, **7A**, 227 (1968).

87. H. Pauly and J. P. Toennis, *Adv. Atom. Molec. Phys.*, **1**, 201 (1965).

88. P. D. Burrow and P. Davidovits, *Phys. Rev. Lett.*, **21**, 1789 (1968).

89. H. Hofmann, Ph.D. thesis, University of Kaiserlautern, 1977.

90. K. A. Rost, Ph.D. thesis, University of Kaiserlautern, 1977.

91. I. V. Hertel, H. Hofmann, and K. A. Rost, *J. Phys. Sci. Instrum.*, **8**, 1023 (1975).

92. K. Bergmann, W. Demtröder, and P. Hering, *Appl. Phys.*, **8**, 65 (1975).

93. D. Hammer, E. Benes, P. Blum, and W. Husinsky, *Rev. Sci. Instrum.*, **47**, 1178 (1976).

94. W. Husinsky, R. Bruckmüller, P. Blum, F. Vieböck, D. Hammer, and E. Benes, manuscript in preparation.

95. K. Bergmann, U. Hefter, and P. Hering, *J. Chem. Phys.*, **65**, 488 (1976).

96. T. D. Gaily, S. D. Rosner, and R. A. Holt, *Rev. Sci. Instrum.*, **47**, 143 (1976).

97. H. J. Loesch, *Chem. Phys.*, **18**, 431 (1976).

98. I. V. Hertel and W. Stoll, *J. Phys.*, **B7**, 570 (1974).

99. I. V. Hertel and W. Stoll, *J. Appl. Phys.*, **47**, 214 (1976).

100. H. W. Hermann, I. Hertel, W. Reiland, A. Stamatovic, and W. Stoll, *J. Phys.*, **B10**, 251 (1977).

101. U. Fano and J. Macek, *Rev. Mod. Phys.*, **45**, 553 (1973).

102. J. Macek and I. V. Hertel, *J. Phys.*, **B7**, 2173 (1974).

103. I. V. Hertel, H. Hofmann, and K. A. Rost, *Chem. Phys. Lett.*, **47**, 163 (1977).

104. R. B. Bernstein and R. D. Levine, *Adv. Atom. Molec. Phys.*, **11**,

105. A. D. Wilson and R. D. Levine, *Molec. Phys.*, **27**, 1197 (1974).

106. H. P. Hooymayers and G. Nienhuis, *J. Quant. Spectrosc. Radiat. Transf.*, **8**, 955 (1968).

107. H. P. Hooymayers and P. L. Lijnse, *J. Quant. Spectrosc. Radiat. Transf.*, **9**, 995 (1969).

108. I. V. Hertel, H. Hofmann, and K. A. Rost, *J. Chem. Phys.*, **71**, 674 (1979).

109. I. V. Hertel and W. Stoll, *J. Phys.*, **B7**, 583 (1974).

110. W. Stoll, Ph.D. thesis, University of Kaiserslautern, 1974.

111. I. V. Hertel, in G. zu Putlitz, E. W. Weber, and A. Winnacker, Eds., *Atomic Physics 4*, 1974, p. 381.

112. I. V. Hertel, in H. Kleinpoppen and M. R. C. McDowell, Eds., *Electron and Photon Interactions with Atoms*, 1976, p. 375.

113. I. V. Hertel, H. W. Hermann, W.Reiland, A. Stamatovic, and W. Stoll, in J. Risely and R. Geballe, Eds., *Abstracts, lxth IPEAC*, Washington U. P., Seattle, 1975.

114. H. W. Hermann, I. V. Hertel, W. Reiland, and A. Stamatovic, *Abstracts, Xth ICPEAC*, Paris, 1977.

115. I. C. Percival and M. J. Seaton, *Proc. Phys. Soc.*, **53**, 644 (1957).

116. I. V. Hertel, H. Hofmann, and K. A. Rost, *Phys. Rev. Lett.*, **38**, 343 (1977).

117. W. Reiland and I. V. Hertel, manuscript in preparation.

118. E. E. Nikitin, private communication.

119. P. McGuire, *Chem. Phys.*, **13**, 81 (1976).

120. E. Gersing, H. Pauly, E. Schädlich, and M. Vonderschen, *Discuss. Faraday Soc.*, **55**, 260126 (1973).

121. V. Kempter, *Adv. Chem. Phys.*, **30**, 417 (1975).

122. V. Kempter, E. Clemens, P. J. Martin, and L. Zehnle, *Abstracts, Xth ICPEAC*, Paris, 1977.

123. M. Brieger and H. J. Loesch, *Abstracts, IVth International Symposium on Molecular Beams*, 1977, p. 298.

124. H. J. Loesch and M. Brieger, *Abstracts, Xth IPEAC*, Paris, 1977.

125. J. E. Mentall, H. F. Krause, and W. L. Fite, *Discuss. Faraday Soc.*, **44**, (1967).

126. H. F. Krause, J. Fricke, and W. L. Fite, *J. Chem. Phys.*, **56**, 4593 (1972).

127. P. J. Kalff, Ph.D. thesis, University of Utrecht, 1971.

128. H. J. Loesch, manuscript in preparation.

129. H. J. Loesch, *Abstracts, IVth International Conference on Molecular Beams*, 1977.

130. H. J. Loesch, manuscript in preparation.

131. U. Buck, E. Lessner, and D. Pust, *Verhandl. Deutsch. Phys. Ges.*, **2**, (1977).

132. W. Schepper, F. Pühl, and D. Beck, to be published.

133. P. Habitz, to be published.

134. P. Botschwina, W. Meyer and I. V. Hertel, manuscript in preparation.

135. I. V. Hertel, in *K. Lawley* ed. *The Dynamics of the Excited State, Adv. Chem. Phys.* 1981 in preparation.

136. J. A. Silvers, N. C. Blais and G. H. Kwei, *J. Chem. Phys.*, **71**, 3412 (1979).

CHAPTER FIVE

SPONTANEOUS IONIZATION IN SLOW COLLISIONS

A. Niehaus

Fysisch Laboratorium, Department of Atomic Fysics, Princetonplein 5, 3508 TA Utrecht, The Netherlands

Contents

I Introduction   401

II Penning Ionization—Simple Systems   402
    A Theoretical Background   403
    B Experimental Results and Their Evaluation   420

III Penning Ionization—Complications   460
    A Atomic Targets   460
    B Molecular Targets   463

IV Other Spontaneous Ionization Mechanisms   472
    A True Associative Ionization   472
    B Spontaneous Ionization by Electron Transfer   475

1. INTRODUCTION

Any atomic system that is excited to a state above the first ionization limit is unstable against the spontaneous ejection of an electron. The same is true, of course, for any molecular system consisting of two or more atomic systems at certain nuclear distances. The fact that in reality these nuclear distances change more or less rapidly in time can certainly not remove this possibility of spontaneous ionization. However, the expression "spontaneous ionization" should be limited to such molecular systems for which an ionization can in fact be "spontaneous" as compared to the relative motion of the nuclei, that is, to systems whose relative nuclear velocities are much smaller than the velocity of the ejected electron. Molecular systems formed temporarily in collisions at not too high collision energy are of this nature, and hence the phenomenon of spontaneous ionization in collisions exists. In this article we discuss this phenomenon for the case of "slow collisions," where by "slow" we mean that the electronic energy necessary for the spontaneous ionization cannot be gained from the collision energy but is carried into the system in the form of electronic energy of the collision partners.

Electronic energy can be carried into a collision system $(AB)^*$ in the form of excitation energy E_* of a collision partner A^*, but also into a system $(AB)^{+*}$ by A^+ in the form of the difference of ionization energies of A and B, $E_i(A) - E_i(B)$. If, in the case of an A^*/B collision, $E_* > E_i(B)$, the electronic state of the system $(AB)^*$ is already at large distances in the ionization continuum of $(AB)^+$. The spontaneous ionization process

$$A^* + B \rightarrow \text{Ionization} \tag{I.1}$$

is then called *Penning ionization* (PgI) after F. M. Penning, who first observed it in discharges.[1] If, in case of an A^+/B collision, $E_i(A) - E_i(B) > E_i(B^+)$, the state of the system $(AB)^{+*}$ is already at large distances in the ionization continuum of $(AB)^{++}$. The then-possible spontaneous ionization process

$$A^+ + B \rightarrow \text{Ionization} \tag{I.2}$$

we call *transfer ionization* (TI), because the electronic energy becomes available by transferring an electron from B to A^+. If the energy conditions for PgI and TI are not met, the state of the respective collision systems lies below the ionization continuum at large distances of the collision partners. But if the ionization energy of the collision system is lowered at smaller distances by an amount D, ionization is still possible if the respective conditions $E_* > E_i(B) - D$ and $E_i(A) - E_i(B) > E_i(B^+) - D$ are met. At sufficiently low collision energies—$E_k < E_* - E_i(B)$ and $E_k < E_i(A) - E_i(B) - E_i(B^+)$, respectively—this type of ionization leads to formation of a stable molecule. We call this process *true associative ionization* (AI). So far it has only been observed for systems of the type (AB)*.

Of the three spontaneous ionization processes mentioned in the preceding paragraphs, PgI has been studied most extensively, both theoretically and experimentally, and is now rather well understood, at least in its simplest form. Hence this article is largely devoted to PgI. To facilitate a good understanding of the experimental data from the physical point of view, we first present a short description of PgI theory. It follows a rather detailed discussion of experimental results and their interpretation. This discussion is divided into a first part that deals with simple systems and a second one dealing with complications. The other spontaneous ionization processes, AI and TI, on which detailed experimental data have become available only recently, are dealt with in Section IV. The theoretical background provided in the beginning for simple PgI systems is also very useful for the discussion of AI and TI.

The main goal in writing the article has been to present a sort of unified physical picture of a seemingly large variety of processes. A review of the literature in the field was not attempted, but rather such studies and results were selected that seemed most descriptive of the physics of the different processes.

II. PENNING IONIZATION—SIMPLE SYSTEMS

The first observation of reaction (I.1), PgI, was reported in 1927 by Penning,[1] who studied the influence of small amounts of impurities on the ignition voltage of rare-gas discharges and confirmed PgI for several systems A*/B, with A* as the rare-gas metastables and B as various atoms and simple molecules. First indirect determinations of cross sections from the Townsend ionization coefficient [2,3] showed PgI to occur in a large fraction of gas kinetic collisions. Schut and Smit,[4] who were the first to separate the excitation region of A from the reaction region of A* with B in a drift experiment, directly measured the ionizing effect of neon metastables on argon and hydrogen and of helium metastables on hydrogen. The rather large uncertainty in their cross-section data resulted from the difficulty in monitoring the absolute number of metastables by utilizing the effect of secondary-electron emission from surfaces on impact of metastables.[5]

As compared to, for instance, photoionization or electron-impact ionization, PgI is a rather complicated process, both theoretically and experimentally. Therefore, PgI remained poorly understood for a rather long time. Initiated by more detailed data obtained by mass spectrometry[6] and *Penning electron spectroscopy* (PgES),[7,8] the first models were proposed and discussed. Based on more mass-spectrometric studies using separated excitation and reaction regions[9] and on the first high-resolution PgES studies, [10-14] these models were further refined and led to a good qualitative understanding of PgI. Theoretical formulations of PgI were given by Nakamura [15] and by Miller,[16] who also discussed classical and semiclassical approximations to the problems, which made the theory more accessible for the experimentalist and also revealed the connection to the already available PgI models. The theory was then used mainly to carry out analyses of experimental data in terms of the theoretical quantities governing the PgI process. Such analyses are presented in detail in Section II, after the theoretical background has been provided. *Ab initio* calculations of experimentally observable quantities have been carried out for the systems $He(2^3S)-H^{17-19}$ and $He(2^3S)-H_2^{20}$.

A. Theoretical Background

1. Definition of Simple Systems

The PgI process can be complicated by other processes—for example, excitation transfer reactions. The available theory of PgI cannot describe these complications but is rather restricted to a class of simple systems. The conditions that such simple systems must fulfill are summarized as follows:

1. Particle A is an atom in a metastable excited state, and particle B is an atom.
2. The excitation energy of A* is larger than the lowest ionization potential of B, $E_*(A) > E_i(B)$.
3. The relative kinetic energy of the colliding particles is "small."

By condition 3 we want to ensure that the Born–Oppenheimer approximation can be applied to the description of the simple systems, allowing definition of adiabatic potential-energy curves for the different electronic states of the systems. Since the initial-state potential curve $\tilde{V}_*(R)$ (dissociating to A* + B) lies in the continuum of the potential curve $V_+(R)$ (dissociation to A + B⁺), spontaneous transitions $\tilde{V}_*(R) \to V_+(R) + e^-$ will generally occur. Within the Born–Oppenheimer approximation the corresponding transition rate $W(R)$—or energy width $\Gamma(R) = \hbar W(R)$ of $\tilde{V}_*(R)$

—is also a local quantity, calculated for fixed nuclei at distances R, like the potential curves themselves. It should, however, be noted that the validity of the Born–Oppenheimer approximation for potential curves lying in an ionization continuum is not as well established as for electronically stable molecular states. It seems possible that the influence of the relative motion of the nuclei on $\Gamma(R)$ becomes important at lower velocities than does the influence on the electronic energies. Up until present, however, PgI has always been described and experimental data interpreted in terms of the local quantities $V_*(R)$, $V_+(R)$, and $\Gamma(R)$.

A full theoretical treatment of PgI within the Born–Oppenheimer approximation involves two stages. First, the potentials $V_*(R)$, $V_+(R)$, and the width $\Gamma(R)$ of the initial state must be calculated in electronic-structure calculations at different values of the internuclear distance R—within the range relevant for an actual collision at a certain collision energy—and then, in the second stage the quantities observable in an experiment (e.g., cross sections, energy and angular distributions of electrons) must be calculated from the functions $V_*(R)$, $V_+(R)$, and $\Gamma(R)$, taking into account the dynamics of the heavy-particle collisions.

The electronic-structure calculations of the potentials and of the width have been carried out only for the especially simple systems $He(2^3S)/H, H_2^{17-20}$ and do not seem to be feasible at present for more complex cases. On the other hand, the theoretical connection of the functions $V_*(R)$, $V_+(R)$, and $\Gamma(R)$, with the quantities observable experimentally, is rather straightforward. Depending on whether the relative motion of the heavy particles is treated classically, semiclassically, or quantum mechanically, any observable quantity can be expressed in terms of $V_*(R)$, $V_+(R)$, and $\Gamma(R)$ by corresponding relations or at least can be calculated numerically from these functions. This is true, however, only for simple systems that, in addition to conditions 1 to 3, also fulfill the following condition:

4. The potential curves $V_*(R)$ and $V_+(R)$ are isolated; that is, at the collision energies considered there are no diabatic transitions possible from these curves to other adiabatic curves.

2. Potential Curve Diagram

A good basis for the qualitative understanding of the PgI process and its theoretical description is the "potential curve model of PgI,"[21] which was developed and applied[6-14] prior to the theoretical formulation of PgI (see Fig. 1). The spontaneous ionization occurring with probability $\Gamma(R_t)/\hbar$ at some distances R_t is the "vertical" transition $\tilde{V}_*(R_t) \rightarrow V_+(R_t)$, as indicated in the diagram. This "vertical" condition is a consequence of the Born–Oppenheimer approximation and has nothing to do with the approxima-

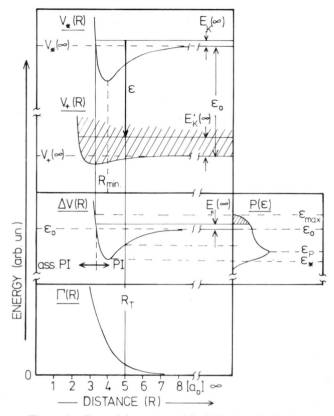

Figure 1.   Potential curve model of Penning ionization.

tions used to describe the heavy-particle motion. If we ask for a stationary state of the ionized system, characterized by the detection of an electron of energy ε, the diagram has to be read in the following different way for classical and quantum-mechanical treatment of the heavy-particle motion.

First, in the classical treatment a transition leading to ε can occur only at distances R_i, at which the vertical energy difference between the potential curves is equal to ε:

$$\varepsilon = V_*(R_i) - V_+(R_i); R_i = R_i(\varepsilon) \qquad (\text{II}.1)$$

This is so because no coupling between electronic and nuclear motion is assumed within the Born–Oppenheimer approximation, which in the classical limit leads to separate conservation of the instantaneous heavy-particle motion. Denoting by $E_k(R_i)$ and $E'_k(R_i)$ the instantaneous kinetic energy at the moment of transition in the upper- and lower potential curve, respectively, and by l and l' the corresponding angular momenta, we have

$$E_k(R_i) = E'_k(R_i); l = l' \tag{II.2}$$

Relations (II.1) and (II.2) neglect the recoil, which is transferred to the heavy particles on electron ejection.

Independent of the approximation used, we have from the energy-conservation principle the relation

$$E_0 = E_k(\infty) + V_*(\infty) = E'_k(\infty) + V_+(\infty) + \varepsilon \tag{II.3}$$

with E_0 the total energy and $E_k(\infty)$ and $E'_k(\infty)$ the asymptotic kinetic energies of the heavy-particle motion in the upper and lower electronic state. In case of collision, $E_k(\infty) > 0$. However, depending on ε, $E'_k(\infty)$ can be greater or smaller than zero. Denoting by $\varepsilon_0 = V_*(\infty) - V_+(\infty)$ the "nominal" electron energy, it follows from (II.3) that

$$E'_k(\infty) < 0 \qquad \text{if} \qquad (\varepsilon - \varepsilon_0) > E_k(\infty) \tag{II.4a}$$

$$E'_k(\infty) > 0 \qquad \text{if} \qquad (\varepsilon - \varepsilon_0) < E_k(\infty) \tag{II.4b}$$

Condition (II.4a) characterizes a bound heavy-particle system AB^+ and condition (II.4b), a free system $A + B^+$.

Second, in the quantum-mechanical treatment[15, 16, 22] the matrix element for a transition between the state of heavy-particle motion with kinetic energy $E_k(\infty)$ on the complex potential $\tilde{V}_*(R) = V_*(R) - \frac{1}{2}\Gamma(R)$ and the state of motion with kinetic energy $E'_k(\infty) = E_k(\infty) - (\varepsilon - \varepsilon_0)$ on the real potential $V_+(R)$ must be calculated. This matrix element is—with $\sqrt{\Gamma}(R)$ weighted— the overlap integral of the nuclear wave functions in the two states of motion. The main contributions to this matrix element will normally come from ranges of distances for which the classical relations (II.1) and (II.2) hold. This is so because at distances determined by (II.1) and (II.2) the relative phase of the oscillating wave functions in the overlap integral becomes stationary. Other than in the classical treatment, however, finite regions of distances contribute, and contributions from different regions can interfere in the probability for ejection of an electron of energy ε.

The potential curve diagram is especially instructive for the discussion of electron-energy distributions $P(\varepsilon)$ and of the corresponding distributions of relative kinetic energy after PgI, $P(E'_k(\varepsilon))$, which is connected to $P(\varepsilon)$ by the relation

$$E'_k(\infty) = E_k(\infty) - (\varepsilon - \varepsilon_0) \tag{II.5}$$

If $P(\varepsilon)$ is measured for a well-defined collision energy, from this distribution the ratio of associative PgI: normal PgI can be derived simply as the ratio of the areas for which conditions (II.4a) and (II.4b) are met, respectively. Further, as seen from Fig. 1, the width of $P(\varepsilon)$ reflects the variation of the difference potential $V_*(R) - V_+(R)$ within the range of distances accessible in the collision. Note that the difference potential is independent of impact parameter (or angular momentum l) in contrast to $V_*(R)$ and $V_+(R)$ themselves, if the effective potentials including the rotational energy are used. In the "normal" case $V_*(R)$ will have its minimum and its repulsive wall at larger distances than $V_+(R)$ because normally atom A^* is larger in its excited state than in its ground state, and in addition atom B is larger than ion B^+. The difference potential then has its minimum approximately at the position of the minimum of $V_*(R)$, $R_{\min}^{\mathrm{diff}} \approx R_{\min}$. A minimum of the difference potential shows up in $P(\varepsilon)$ as a low-energy edge, whose position ε_* with respect to the nominal energy ε_0 can then be used to estimate the well depth of $V_*(R), D_*$:

$$\varepsilon_0 - \varepsilon_* = V_*(\infty) - V_+(\infty) - V_*\left(R_{\min}^{\mathrm{diff}}\right) + V_+\left(R_{\min}^{\mathrm{diff}}\right)$$

$$\varepsilon_0 - \varepsilon_* \approx D_* - \left[V_+(\infty) - V_+(R_{\min}) \right]$$

(II.6)

As is outlined in the theoretical section, Section IIA5, the position ε_* is found in the measured spectrum as the electron energy at which $P(\varepsilon)$ has decreased to approximately 44% of the value of its low-energy peak.

If $P(\varepsilon)$ extends to values $\varepsilon > \varepsilon_0 + E_k(\infty)$, it can be concluded that in some part of the accessible R range, $V_+(R)$ is more attractive than $V_*(R)$. Under normal conditions these parts will be close to the classical turning point for central collisions on $V_*(R)$. Classically, the largest measured electron energy $\varepsilon_{\max}$ is then related to the value of the potential $V_+(R)$ at the turning point R_{tp} by

$$(\varepsilon_{\max} - \varepsilon_0) - E_k(\infty) \approx V_+(\infty) - V_+(R_{\mathrm{tp}})$$

(II.7)

If $V_+(R)$ is known, (II.7) yields a very direct estimate of the position of the turning point on $V_*(R)$; if $V_+(R)$ is not known, (II.7) yields information on the attraction in the ionized system.

3. Classical Relations

Denoting by l the angular momentum of the relative heavy-particle motion, the classical expression for the angular dependent transition probability at a certain distance R is given by

$$P_l(R) = \frac{2\Gamma(R)}{\hbar v_l(R)} \exp\left(-\int_{R_l}^{\infty} \frac{\Gamma(R)dR}{\hbar v_l(R)} \right) \cosh\left(\int_{R_l}^{R} \frac{\Gamma(R')dR'}{\hbar v_l(R)} \right)$$

(II.8)

where R_l is the l-dependent turning point on $V_*(R)$ and

$$V_l(R) = \sqrt{\frac{2}{\mu}E_k(\infty) + V_*(\infty) - V_*(R) - l(l+1)\hbar^2/2\mu R^2} \qquad \text{(II.9)}$$

μ is the reduced mass of the collision system, and R_l is determined as the outermost root of $v_l(R) = 0$. In expression (II.8) contributions from "on the way in" and "on the way out" are contained. Integration with respect to R leads to the transition probability along the whole trajectory with angular momentum l. We obtain the simple result

$$P_l = \int_{R_l}^{\infty} P_l(R)\,dR = 1 - \exp\left(-2\int_{R_l}^{\infty} \frac{\Gamma(R)\,dR}{\hbar v_l(R)}\right) \qquad \text{(II.10)}$$

and the total cross section is given by

$$\sigma_{\text{total}}(E_k(\infty)) = \frac{\pi}{k_0^2} \sum_{l=0}^{\infty} (2l+1)P_l \qquad \text{(II.11)}$$

From the size of the measured cross sections it can be concluded that the main contribution comes from large values of l. Therefore, the discrete angular momenta may be replaced as usual by the relation $bk_0 = (l + \frac{1}{2})$—with b the impact parameter and $\hbar k_0 = \sqrt{2\mu E_k(\infty)}$ —and by transforming the sum in (II.11) into an integral over b. We then obtain

$$\sigma_{\text{total}}(E_k(\infty)) = 2\pi \int_0^{\infty} P(b)b\,db \qquad \text{(II.12)}$$

where $P(b)$ is calculated from (II.9) and (II.10) with l replaced in the described way. The energy dependence of the total cross section is given through the energy-dependent local velocity (II.9) and through the energy-dependent turning points $R(b)$ appearing in (II.10). It should be noticed that the total cross section is determined by $V_*(R)$ and $\Gamma(R)$ and does not depend on $V_+(R)$. For known functions $V_*(R)$ and $\Gamma(R)$, it can easily be calculated numerically, using either (II.11) or (II.12). As quantum effects usually average out in the total cross section, these classical results should be rather accurate. For the special system $He(2^3S)$–Ar, this was actually found by comparison with quantum-mechanical calculations.[21] To explain general features of measured cross-section curves $\sigma_{\text{total}}[E_k(\infty)]$, we discuss in the paragraphs that follow the general predictions that can be made on the basis of (II.12) in the case of the two regions characterized by the conditions $[E_k(\infty)/D_*] \ll 1$ and $[E_k(\infty)/D_*] \gg 1$.

$[E_k(\infty)/D]\ll 1$: As b is varied from zero to larger values, the turning point $R(b)$ given by the outermost root of $v[R(b), E_k(\infty)]=0$ jumps at a certain value $b=b_c$ from small distances at the repulsive wall of $V_*(R)$ to large distances outside the rotational barrier of the effective potential. The term b_c is the impact parameter for which "orbiting" occurs. If $\Gamma(R)$ decreases steeply with R—as is usually the case—$P(b)$ drops at $b_c=b$ to almost zero, as can be seen from (II.10). Since for $b<b_c$ the trajectory is mainly determined by the attractive forces and almost independent of b and of $E_k(\infty)$, $P(b)$ will be nearly constant. Therefore, we may approximate

$$\sigma_{total}(E_k(\infty))=2\pi\int_0^\infty P(b)b\,db\approx 2\pi\int_0^{b_c} P(b)b\,db\approx \eta\pi b_c^2 \quad (II.13)$$

where $\eta=\overline{P(b)}$ is an average and nearly energy-independent transition probability per collision with impact parameter $b<b_c$. The energy dependence of b_c is determined by the long-range attractive forces of $V_*(R)$. If we approximate for large R as $V_*(R)\propto R^{-s}$, it follows that $b_c\propto E_k(\infty)^{-1/s}$, so that the total cross-section curve decreases with $E_k(\infty)$ as

$$\sigma_{total}[E_k(\infty)]\propto [E_k(\infty)]^{-2/s} \quad (II.14)$$

$[E_k(\infty)/D]\gg 1$: In this energy range the influence of the attractive part of $V_*(R)$ can be neglected and $V_*(R)$ approximated by a purely repulsive potential. For such a potential, the integration with respect to b can be carried out analytically if the assumption of a small transition probability per collision is made, so that we may approximate

$$P(b)\approx 2\int_{R(b)}^\infty \frac{\Gamma(R)dR}{\hbar v(b_iR)} \quad (II.15)$$

The cross section is then given by an integral over distance alone

$$\sigma_{total}[E_k(\infty)]\approx \int_{R(0)}^\infty \frac{4\pi R^2\Gamma(R)dR}{\hbar\sqrt{\dfrac{2}{\mu}[E_k(\infty)-V_*(R)]}} \quad (II.16)$$

Equation (II.16) is already a very useful formula; however, to determine a simple approximate energy dependence of the total cross section, we approximate further the repulsive potential by $V_*(R)=C\exp(-R/D)$ and the width by $\Gamma(R) = \hbar A \exp(-R/B)$. With the additional condition $R/B \gg 1$, we then obtain from (II.16)

$$\sigma_{\text{total}}\left[E_k(\infty) \right] \approx 2A \left(\frac{\mu}{2B} \right)^{1/2} (BD\pi)^{3/2} \left[\ln \frac{C}{E_k(\infty)} \right]^2 \left[\frac{E_k(\infty)}{C} \right]^{D/B-1/2}$$

(II.17)

In a log–log plot of the cross section versus energy, (II.17) yields approximately a straight line whose slope is given by

$$m \approx \frac{D}{B} - \tfrac{1}{2}$$

(II.18)

Measured cross sections usually show this behavior with $m > 0$. Deviations arise at low $E_k(\infty)$ when the attractive forces become important and at high $E_k(\infty)$ when the ionization probability per collision becomes of the order of unity so that condition (II.15) is violated. A measured slope within the region where (II.17) is a good approximation allows comparison of the "steepness" of $V_*(R)$, given by D^{-1}, with the steepness of $\Gamma(R)$, given by B^{-1}. A steep $\Gamma(R)$ as compared to $V_*(R)$ leads to a strong increase of the cross section with collision energy. It may be noted that condition (II.15) implies an isotope effect of the total cross section, $\sigma_{\text{total}} \propto \mu^{1/2}$, which shows up explicitly in (II.16) and (II.17).

From the classical relations (II.1) and (II.8) and by defining the electron distribution belonging to one angular momentum l of the heavy particle system, as

$$P_l(\varepsilon) d\varepsilon = \sum_l P_l(R_i) dR_1$$

(II.19)

we immediately find

$$P_l(\varepsilon) = \sum_i P_l(R_i) \left| \frac{d}{dR} \left[V_*(R) - V_+(R) \right] \right|^{-1}_{R_l = R_l(\varepsilon)}$$

(II.20)

In (II.20) contributions from transitions occurring at distances determined by (II.1) are summed. The total electron energy distribution is obtained by summing over l or integrating over impact parameter b.

$$P(\varepsilon) = \sum_l (2l+1) P_l(\varepsilon) \approx 2k_0^2 \int_0^{\infty} b P(\varepsilon_1 b) db$$

(II.21)

In the application of (II.21) a difficulty arises because $P_l(\varepsilon)$ is singular (1) at the ε value corresponding to the l-dependent turning point R_l—because here $v_l(R_l)=0$ and $P_l(R_l)\to\infty$—and (2) at the ε values for which the derivative of the difference potential vanishes. Both types of singularity are avoided if a finite energy resolution for the measurement of $P(\varepsilon)$ is included by introducing the measured distribution $\overline{P_l(\varepsilon)}$ by

$$\overline{P_l(\varepsilon)} = \frac{1}{2\Delta} \int_{\varepsilon-\Delta}^{\varepsilon+\Delta} P_l(\varepsilon^1)d\varepsilon^1 \qquad (\text{II.22})$$

Peaks arising from the $l=$ dependent turning-point singularity average out in the total distribution; however, the position of the singularity caused by the vanishing of the derivative of the difference potential is l independent and leads to a peak in the classically calculated distribution. Under conditions that we called "normal" and that are also assumed for the potential curve diagram in Fig. 1, the difference potential has only one minimum, at R_{min}^{diff}, and $P(\varepsilon)$ thus has a peak at $\varepsilon_* = V_*(R_{min}^{diff}) - V_+(R_{min}^{diff})$, as indicated in Fig. 1. Under these normal conditions the sum in (II.20) contains two terms for $\varepsilon < \varepsilon_0$ and one term for $\varepsilon > \varepsilon_0$.

So far the observable quantities were obtainable from the functions $V_*(R)$, $V_+(R)$, and $\Gamma(R)$. To calculate angular intensity distributions of Penning electrons measured in the laboratory with respect to the axis of relative motion of A* and B, it is necessary to allow for an angular dependent "internal" transition probability. Denoting by γ the angle with respect to the internuclear axis, we have to replace $W(R)=[\Gamma(R)/\hbar]$ by $W(R,\gamma)=[\Gamma(R,\gamma)/\hbar]$. A measured angular distribution $I(\theta)$ is nonisotropic only if both the following conditions are met: (1) the axis of the diatomic A–B must on the average be oriented in space during electron ejection, and (2) the internal distribution must be nonisotropic. Since no *ab initio* knowledge on internal distributions of PgI systems is available, it is, of course, desirable to extract such information from measured distributions $I(\theta)$. This can be done by model calculations in which a function $W(R,\gamma)$ —containing free parameters to be determined by comparison with the measured $I(\theta)$—is assumed. Within the classical treatment of the heavy-particle motion we proceed as follows.[23,24] Classical trajectories on $V_*(R)$ for the chosen kinetic energy and for certain impact parameter vectors **b** are calculated. At each distance—corresponding to a defined orientation of the collision system in space—the intensity at the detector is calculated for the assumed function $W(R,\gamma)$, and contributions from all distances along the trajectory are integrated. To obtain the total intensity at the detector, $I(\theta)$, we then integrate over contributions calculated in the same way for trajectories with different orientations of **b** and with all the relevant values $|b|$. The calculations are greatly simplified if the internal distribution does not change appreciably within the relevant region of R. It is then possible to approximate

$$W(R,\gamma) \approx W(R)P(\gamma) \qquad (II.23)$$

A possible nonisotropic internal distribution was not considered in the derivation of formulas for $P(\varepsilon)$. In general, since electron energy and the orientation of the autoionizing system are correlated, energy distributions will depend on the detection angle, $P(\varepsilon,\theta)$. Measured distributions have until present not shown this effect.

At high collision energies the calculation of angular distributions is further complicated by the fact that the very large values of the angular momentum of the heavy-particle motion cause the small angular-momentum exchange between electronic motion and heavy-particle motion on electron ejection to lead to a very appreciable energy exchange. This exchange depends on the ejection angle of the electron and leads to an angular-dependent shift of the measured distribution. Classically, this effect corresponds to the Doppler shift of the energy of an electron ejected from a source moving with respect to the detector. A somewhat more detailed discussion is given in connection with measurements of this effect.

4. Quantum-mechanical Relations

This section is based mainly on work published by Miller and Schafer[17] and Hickman and Morgner.[22] As in the preceding section, the intention is to show how the electronic quantities $V_*(R)$, $V_+(R)$, and $\Gamma(R)$ can be used to calculate—this time quantum mechanically—observable quantities.

Let $\varphi_d(\mathbf{R},\mathbf{r})$ be the initial electronic state corresponding to A*–B with $\mathbf{r}$ symbolizing the electronic coordinates. An important feature of $\varphi_d(\mathbf{R},\mathbf{r})$ is that it is not an eigenstate of the total electronic Hamiltonian, H, because spontaneous transitions to the final states $A + B^+ + e^-(\varepsilon,\lambda,\mu)$—with a free electron of energy ε and angular momentum quantum numbers (λ,μ)—are possible. Denoting the final-state wave functions as $\varphi_+(\mathbf{R},\mathbf{r})\varphi_{\varepsilon\lambda\mu}(\mathbf{R},\mathbf{r})$, the functions governing the PgI process are given by

$$V_*(R) = \langle \varphi_d(\mathbf{R},\mathbf{r}) | H | \varphi_d(\mathbf{R},\mathbf{r}) \rangle \qquad (II.24)$$

$$V_+(\mathbf{R}) = \langle \varphi_+(\mathbf{R},\mathbf{r}) | H | \varphi_+(\mathbf{R},\mathbf{r}) \rangle \qquad (II.25)$$

$$\Gamma(R) = 2\pi\rho \sum_\lambda |V_{\varepsilon\lambda}(R)|^2 \qquad (II.26)$$

$$V_{\varepsilon\lambda}(R) = \langle \varphi_d(\mathbf{R},\mathbf{r}) | H | \varphi_+(\mathbf{R},\mathbf{r})\varphi_{\varepsilon\lambda}(\mathbf{R},\mathbf{r}) \rangle \qquad (II.27)$$

where ρ is the density of continuum states at energy ε. In principle, in (II.27) transitions into different substates (λ, μ) should be distinguished. So far, however, the discussion in Hickman and Morgner[22] was restricted to the case $\mu = 0$. Equation (II.27) is calculated for fixed values of the parameter R, and thus the energy-conservation principle postulates that the energy of the ejected electron is given by the classical relation (II.1). If, however, the heavy-particle system is allowed to have finite relative kinetic energy, only the total energy—consisting of electronic and heavy-particle energy—must be conserved in the transition. Energy ε can deviate from the classical value, given by (II.1), by the same amount that the instantaneous kinetic energy of the heavy particles can change in the transition, so that total energy conservation, given by (II.3), is fulfilled. The possible change of the heavy-particle energy, on the other hand, is correctly taken into account if the corresponding change of state of the nuclear motion is treated quantum mechanically. It follows that, in the quantum-mechanical treatment of the nuclear motion, the probability $\Gamma(R)/\hbar$, which is calculated as a probability for a transition leading to a final state with defined energy ε of the ejected electron, has to be used as a probability for a transition to a final state with unspecified energy of the ejected electron.

As in the classical treatment, some of the observable quantities are determined by $V_*(R)$ and $\Gamma(R)$ alone, that is, by the scattering wave function for the complex potential $\tilde{V}_*(R) = V_*(R) - [i\Gamma(R)/2]$. Let $\xi_d^l(R)$ be the lth partial wave of the expansion of the solution $\xi_d(\mathbf{R})$ to the single-channel Schrödinger equation with potential $\tilde{V}_*(R)$. All the information depending on the initial channel alone is then contained in the complex phase shifts δ_l appearing in the asymptotic form of the partial waves

$$\lim_{R \to \infty} \xi_d^l(R) = k_0^{-1/2} \sin k_0 R - \frac{l\pi}{2} + \delta_l \qquad \text{(II.28)}$$

The complex phase shift can be obtained from exact numerical solution of the radial Schrödinger equation.[21] The following quantities can immediately be given in terms of δ_l.

The differential elastic cross section in the center-of-mass system

$$\sigma_{\text{el}}(\Theta) = \frac{1}{4k_0^2} \sum_l (2l+1) P_l(\cos\Theta)(e^{2i\delta_l} - 1) \qquad \text{(II.29)}$$

The total elastic cross section

$$\sigma_{\text{el}} = \frac{\pi}{k_0^2} \sum_l (2l+1)|e^{2i\delta_l} - 1|^2 \qquad \text{(II.30)}$$

The total ionization cross section

$$\sigma_{\text{total}} = \frac{\pi}{k_0^2} \sum_l (2l+1)\left(1 - |e^{2i\delta_l}|^2\right) \tag{II.31}$$

Observable quantities that depend on the final state of the Penning ionized system are more complicated to calculate. The most detailed amplitude that can be calculated is differential in the final direction of the heavy particles $(\mathbf{k}_f)$ and in energy and direction of the ejected electron $(\varepsilon, \hat{\varepsilon})$. This amplitude is given by

$$f(\hat{k}_f, \varepsilon, \hat{\varepsilon}) = -\left(\frac{1}{4\pi}\right)^2 \left(\frac{2\mu}{\hbar^2}\right)\left(\frac{k_f}{k_0}\right)^{1/2} \rho^{1/2} \langle \xi_e(\mathbf{R}) | V_{e\hat{\varepsilon}}(R) | \xi_d(\mathbf{R}) \rangle \tag{II.32}$$

where $\xi_e(R)$ is the solution of the Schrödinger equation with the real potential $V_+(R)$ and with asymptotic relative kinetic energy

$$E_k'(\infty) = \frac{\hbar^2 k^2}{2\mu} = E_0 - \varepsilon - V_+(\infty) \tag{II.33}$$

Cross sections are obtained from (II.32) by squaring (II.32) and integrating with respect to the variables that are not determined in the experiment. The electronic coupling $V_{e\hat{\varepsilon}}(\mathbf{R})$ of the initial state to a final state with specified direction of the ejected electron with respect to the internuclear axis is connected to the matrix elements $V_{e\lambda}(R)$, from which $\Gamma(R)$ is calculated [equations (II.26) and (II.27)], by the expansion

$$V_{e\hat{\varepsilon}}(\mathbf{R}) = \sum_\lambda y_{\lambda 0}(\hat{\varepsilon}, \hat{R}) i^{-\lambda} e^{i\sigma_\lambda} V_{e\lambda}(R) \tag{II.34}$$

where σ_λ represents the asymptotic phase shifts of the partial Coulomb waves of the expansion of the ejected electron wave function. The square of $V_{e\hat{\varepsilon}}(R)$ is the "internal" intensity distribution used in the classical description, with the angle $\gamma = (\hat{\varepsilon}, \hat{R})$.

To calculate angular distributions for an actual collision, $V_{e\hat{\varepsilon}}(R)$ must be written in terms of angles with respect to the external frame, that is, with respect to the relative heavy-particle velocity direction. This is done by applying the addition theorem of spherical harmonics

$$V_{e\hat{\varepsilon}}(\mathbf{R}) = \sum_{\lambda\mu} \left(\frac{4\pi}{2\lambda+1}\right)^{1/2} i^{-\lambda} e^{i\sigma_\lambda} y_{\lambda\mu}^*(\hat{\varepsilon}) Y_{\lambda\mu}(\hat{R}) V_{e\lambda}(R) \tag{II.35}$$

Further, the nuclear wave functions are expanded as

$$\xi_d(\mathbf{R}) = (4\pi)^{1/2} \sum_l (2l+1) i^l e^{i\delta_l} \frac{\xi_d^l(R)}{k_0^{1/2}R} y_{l0}(\hat{R}) \tag{II.36}$$

$$\xi_\varepsilon(\mathbf{R}) = 4\pi \sum_{l'm'} i^{-l'} e^{-i\eta_{l'}} \frac{\xi_\varepsilon^{l'}(R)}{k_f^{1/2}R} y_{l'm'}(\hat{R}) y_{l'm'}(\hat{k}_f) \tag{II.37}$$

With expansions (II.35) to (II.37) inserted into (II.32) we can, in principle, calculate all observable quantities.

The angular distribution of ejected electrons is

$$\sigma(\hat{\varepsilon}) = \int \int d\hat{k}_f d\varepsilon |f(\hat{k}_f, \hat{\varepsilon}, \varepsilon)|^2 \tag{II.38}$$

The angular distribution of Penning ions B^+ (associative ions are excluded) with energy corresponding to the relative kinetic energy $E_k'(\infty) > 0$, given by (II.33), is

$$\sigma(\hat{k}_f, \varepsilon) = \int d\hat{\varepsilon} |f(\hat{k}_f, \hat{\varepsilon}, \varepsilon)|^2 \tag{II.39}$$

The energy distribution of Penning electrons is

$$\sigma(\varepsilon) = \int \int d\hat{\varepsilon} d\hat{k}_f |f(\hat{k}_f, \hat{\varepsilon}, \varepsilon)|^2 \tag{II.40}$$

Relations (II.38) to (II.40), with the full expressions (II.35) to (II.37) in the amplitude $f(\hat{k}_f, \hat{\varepsilon}, \varepsilon)$, have not yet been used for actual calculations. For such calculations it would be necessary to know—or to assume—the "internal" electron angular distribution given by (II.34) instead of the total rate $\Gamma(R)/h$ given by (II.26).

For $\sigma(\hat{k}_f, \varepsilon)$ and $\sigma(\varepsilon)$, which do not depend on the internal electron distribution, approximate expressions have been derived.[22] The following approximations were made:

$$\int \xi_\varepsilon^l V_{\varepsilon\lambda} \xi_d^{l'} dR \approx \int \xi_\varepsilon^l V_{\varepsilon\lambda} \xi_d^l dR \tag{II.41}$$

$$V_{\varepsilon\lambda}(R) \approx \alpha_\lambda(R) \left(\frac{P(R)}{2\pi\rho} \right)^{1/2} \tag{II.42}$$

with $\alpha_\lambda(R)$ assumed to "vary slowly." We then obtain from (II.40)

$$\sigma(\varepsilon) = \frac{4\pi}{k_0^2}\left(\frac{2\mu}{\hbar^2}\right)^2 \frac{1}{2\pi} \sum_l (2l+1)\exp(-2Im\delta_l) \left| \int \xi_\varepsilon^l(R)(\Gamma(R))^{1/2}\xi_d^l(R)dR \right|^2$$

(II.43)

Equation (II.41) assumes that the change in angular momentum of the heavy-particle motion on ejection of the Penning electron does not change the weighted overlap integral of the initial and final states of nuclear radial motion appreciably. Since l is on the average large for real systems, this approximation is not severe. By stating that $\alpha_\lambda(R)$—the fraction of transitions at distance R belonging to electronic angular momentum λ—"varies slowly," it is meant that it varies slowly enough to be taken out of the integral in (II.32). Because of the oscillating phase factor contained in the integrand, usually only a limited R range contributes to a special amplitude. Hence the error introduced by taking $\alpha_\lambda(R)$ out of the integral—and approximating it by its value at the point of stationary phase—will also be small. For any R, it follows from the definition of $\alpha_\lambda(R)$ that $\Sigma|\alpha_\lambda(R)|^2 = 1$. Thus in the cross-section formula the summation over λ can be carried out without even knowing explicitly the values $\alpha_\lambda(R)$ at the points of stationary phase.

If, in addition to (II.41) and (II.42), it is assumed that

$$\int y_{l0} y_{l'm'} y_{\lambda\mu} d\hat{R} \approx \delta_{ll'}\delta_{m'\mu}$$

(II.44)

we also obtain for the detailed cross section $\sigma(\hat{k}_f,\varepsilon)$ a rather simple expression

$$\sigma(\hat{k}_f,\varepsilon) = \frac{4\pi}{k_0^2}\left| \sum_l (2l+1)S_\varepsilon^l P_l\left(\hat{k}_f,\hat{k}_0\right) \right|^2$$

(II.45)

with S_ε^l given by

$$S_\varepsilon^l = -2i\left(\frac{2\mu}{\hbar^2}\right)\exp(i\delta_l + i\eta_l^\varepsilon)\int \xi_\varepsilon^l(R)\left(\frac{\Gamma(R)}{2\pi}\right)^{1/2}\xi_d^l(R)dR$$

(II.46)

Since $(\hat{k}_f,\hat{k}_0) = \cos\theta$, with θ the center-of-mass scattering angle, the left-hand side of (II.45) may also be written as $\sigma(\theta,\varepsilon)$. By energy-conservation ε fixes the heavy-particle relative energy [see relation (II.3)].

So far only PgI into final states with $E_k'(\infty) > 0$ was discussed. For $\varepsilon > [\varepsilon_0 + E_k(\infty)]$—corresponding to $E_k'(\infty) < 0$—PgI occurs into a bound rotational vibrational state with a wave function that is an eigenfunction to the Hamiltonian with potential $V_+(R) + l(l+1)\hbar^2/2\mu R^2$ and a discrete negative energy $E_n = [\varepsilon_0 + E_k(\infty) - \varepsilon_n]$. In this range of electron energies ε_n the wave functions ξ_ε^l have to be replaced by the rotational vibrational wave functions ξ_n^l. The ξ_n^l can also be obtained from direct numerical integration of the Schrödinger equation. They are normalized by $\int dR |\xi_n^l(R)|^2 = 1$. The cross section for populating such a bound (n, l) state is given by

$$\sigma(u, l) = \frac{2\pi}{k_0^2} \frac{2\mu}{\hbar^2} (2l+1) \exp(-2 Im\delta_l) \left| \int \xi_n^l(R) [\Gamma(R)]^{1/2} \xi_d^l(R) dR \right|^2$$

(II.47)

where the same approximations as in the derivation of (II.43) are used.

5. Semiclassical Approximations

Starting from the expression for the electron energy distribution $\sigma(\varepsilon)$ [equation (II.43)], simple semiclassical relations can be derived under special conditions regarding the potentials $V(R)$ and $V_+(R)$.[17,25] This is done by replacing in the integral in (II.43) the wave functions by WKB-wavefunctions

$$\xi_d^l = [k_d^l(R)]^{-1/2} \sin \varphi^l(R)$$

(II.48)

$$\xi_\varepsilon^l = [k_\varepsilon^l(R)]^{-1/2} \sin \varphi_\varepsilon^l(R)$$

(II.49)

with the phases

$$\varphi^l(R) = \int_{R_l}^{R} k_d^l(R') dR' + \frac{\pi}{4} + \delta_l$$

(II.50)

$$\varphi_\varepsilon^l(R) = \int_{R_{l,\varepsilon}}^{R} k_\varepsilon^l(R') dR' + \frac{\pi}{4} + \eta_l^\varepsilon$$

(II.51)

and carrying out the integration with respect to R by applying the stationary phase approximation. The terms $\hbar k_d^l(R)$ and $\hbar k_\varepsilon^l(R)$ are the instantaneous radial momenta on $V_*(R)$ and $V_+(R)$. Denoting the relative phase as

$$F_l(R, \varepsilon) = \varphi^l(R) - \varphi_\varepsilon^l(R)$$

(II.52)

the points of stationary phase R_i—from the neighborhood of which the main contributions to the integral in (II.43) originate—are determined by the condition

$$\frac{dF_l(R,\varepsilon)}{dR} = 0 \tag{II.53}$$

This condition is identical to the classical conditions (II.1) and (II.2) for the distances R_i at which a transition can occur. For the case where the difference potential is monotonic, so that there exists only one point of stationary phase, R_0, we obtain in this semiclassical treatment immediately the classical result [cf. with equations (II.20) and (II.21)] for $\sigma(\varepsilon)$, except for an oscillatory factor:

$$\sigma(\varepsilon) = \frac{\pi}{k_0^2} \sum_l (2l+1) P_l(R_0) \left| \frac{d}{dR}\left(V_*(R) - V_+(R)\right) \right|^{-1} \left(1 - \sin 2F_l(R_0,\varepsilon)\right)$$

$$\tag{II.54}$$

Here $P_l(R_0)$ is given by relations (II.8). The oscillatory factor reflects interferences between the two contributions arising in the region $R \approx R_0$ from transitions "on the way in" and "on the way out." Since the phase of the oscillatory factor depends on l, the oscillations will average out in the summation and not be visible in the distribution.

A more interesting case, where the semiclassical approximation leads to very useful formulas, arises in the case of two neighboring points of the stationary phase—given by condition (II.53)—in the vicinity of an extremum of the difference potential, $\Delta V(R)$. As seen from Fig. 1, in the normal case the difference potential has a minimum, which leads to the edge in the electron distribution discussed earlier. In the region of this edge (i.e., for $\varepsilon \sim \varepsilon_*$), expansion of the phase function (II.52) about the position of the minimum $R_{\min}^{\text{diff}}$ leads to the following approximate expression:

$$\sigma(\varepsilon) = \frac{\pi}{k_0^2} \sum_l (2l+1) P_l(R_{\min}^{\text{diff}}) 2\pi \left(4/\Delta V''(R_{\min}^{\text{diff}})\hbar v_l(R_{\min}^{\text{diff}})\right)^{1/3} Ai^2(z_l)$$

$$z_l = \left(\tfrac{1}{2}\Delta V''(R^{\text{diff}}\text{min})\right)^{-1/3} \left(\hbar v_l(R_{\min}^{\text{diff}})\right)^{-2/3} (\varepsilon_* - \varepsilon) = C_l(\varepsilon_* - \varepsilon)$$

$$\tag{II.55}$$

An oscillatory term with an l-dependent phase as in (II.54) has been neglected in (II.55). The term $Ai^2(z_l)$ is the square of the Airy function, and $\Delta V''(R_{\min}^{\text{diff}})$ is the curvature of the difference potential at the minimum. The singularity that arose in the classical expression of $\sigma(\varepsilon)$ at $\varepsilon = \varepsilon_*$ is removed, and the nonclassical extension of $\sigma(\varepsilon)$ into the region $\varepsilon < \varepsilon_*$ is described.

The argument z_l of the Airy function depends on l only through the velocity $v_l(R_{\min}^{\text{diff}})$ at the minimum of the difference potential. If the well depth of $V_*(R)$ is large compared to the collision energy, $D_* \gg E_k(\infty)$, this velocity is nearly independent on l [see relation (II.9)], and we can write

$$\sigma(\varepsilon) \propto Ai^2(z); \qquad z = c(\varepsilon_* - \varepsilon) \tag{II.56}$$

The first maximum of $Ai^2(z)$ lies at $z = 1.018$, and at $z = 0$ it has decreased to 44% of its value at the first maximum. In a measured distribution, therefore, the 44% value at the low-energy side of the low-energy peak determines the position of ε_* from which by relation (II.6) the well depth D_* of $V_*(R)$ can be obtained. The measured energy difference between the peak position ε_p and ε_* allows determination of the constant c by

$$c = \frac{1.018}{(\varepsilon_p - \varepsilon_*)} \tag{II.57}$$

and an approximation to the curvature $\Delta V''(R_{\min}^{\text{diff}}) \approx V_*''(R_{\min})$ from (57) as

$$V_*''(R_{\min}) \approx \frac{\mu}{\hbar^2} D_* \left(\frac{\varepsilon_p - \varepsilon_*}{1.018} \right)^3 \tag{II.58}$$

Expressions (II.55) and (II.56) are only good approximations for describing the low-energy peak of $\sigma(\varepsilon)$ if the transition probability $\Gamma(R)\hbar$ varies slowly with R, so that its variation is negligible within the R-region contribution. Since the R-region contribution increases as the collision velocity increases, the expressions would presumably lose their validity at higher collision energies. This is actually found to be true at collision energies of several electron volts. An improved semiclassical expression, which approximately accounts for the variation of $P(R)$, is obtained by expanding $\Gamma(R)$ around $R_{\min}^{\text{diff}}$ and retaining the first derivative $\Gamma'(R_{\min}^{\text{diff}})$. Instead of (II.56), we then obtain

$$\sigma(\varepsilon) \propto Ai^2(z) + \frac{1}{4} \left[\frac{\Gamma'(R_{\min}^{\text{diff}})}{\Gamma(R_{\min}^{\text{diff}})} \right]^2 Ai'^2(z) \tag{II.59}$$

Recent measurements of the spectra for the system He^+–Ca in a large-collision energy range showed that (II.59) cannot satisfactorily reproduce the observed energy dependence of the low-energy peak of the spectra either.

For the evaluation of these data a still more advanced semiclassical formula was derived[26] that accounts for the functional dependence of $\Gamma(R)$ and hence is not restricted to the region $\varepsilon \sim \varepsilon_*$. The reader is referred to the original article for details.[26]

B. Experimental Results and Their Evaluation

In this section we do not intend to present a complete review of experiments and their results, but rather a description of our present knowledge on PgI as it is obtained from experimental studies. The theoretical background provided in Section II. A will enable a good understanding of the experimental results in terms of the physical quantities. The following discussion is subdivided according to the different observable quantities determined in the experiments. Since the entire field has rapidly expanded recently and as the more recent experiments tend to be more detailed for obvious reasons, mostly rather recent experiments are discussed.

1. Elastic Scattering

Elastic-scattering data on PgI systems are still rather scarce. After the first measurements of the velocity dependence of the total cross section for the system $He(2^3S)-Ar$ in the thermal-energy range in 1965,[27,28] the first differential cross-section measurements for the systems $He(2^1S)-Ne,Ar$ and $He(2^3S)-Ne,Ar$ were not reported until 1974.[29] More recently, the systems $He(2^3S)-Ar$ and $He(2^1S)-Ar$ have again been studied in great detail[30,31] by measuring differential elastic cross sections in a wide relative velocity range. Other systems are currently under investigation.[32]

We discuss these recent data[30,31] in some detail. The central part of the apparatus is shown in Fig. 2. Supersonic helium and argon beams are formed by expansion at high pressure through a small nozzle. The kinetic energy of the beam particles is $\frac{5}{2}kT$ and can be varied—by varying the source temperature between $80°K$ and $1600°K$—between 16.5 me and 350 meV. The spread of velocities is approximately $(\Delta v/v) = 1.5\%$. The helium beam traverses a region where part of the atoms is excited by electron impact at 150 to 200 eV. After a few μsec it contains, in addition to ground-state atoms, only atoms in the metastable states $He(2^1S)$ and $He(2^3S)$ and a small number of atoms in high Rydberg states. These are removed by field ionization in a "quench condensor" shown in Fig. 2. To obtain data separately for $He(2^3S)$ and $He(2^1S)$ metastables, the beam passes a region where light from a helium discharge contained in a glass spiral is irradiated. This light contains photons from transitions $He(2^1P) \rightarrow He(2^1S) + h\nu(2\ \mu m)$, which quench the $He(2^1S)$ metastables with high

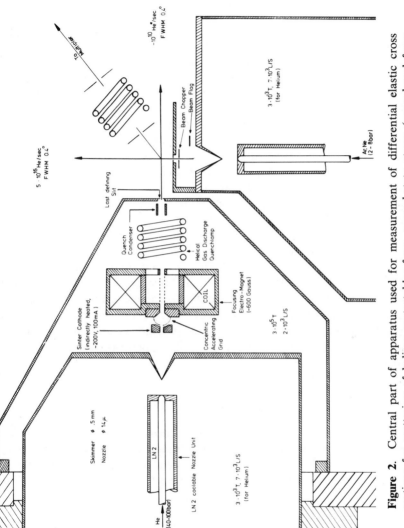

Figure 2. Central part of apparatus used for measurement of differential elastic cross sections for scattering of helium metastables from ground-state atoms (reproduced from Brutschy et al.[30])

efficiency through the process

$$He^*(2^1S) + h\nu(2\mu m) \rightarrow He^*(2^1P) \rightarrow He(1^1S) + h\nu \qquad (II.60)$$

In electron-spectroscopic measurements to be discussed later it was shown[12] that in this way the $He(2^1S)$ component can be removed to $\lesssim 1\%$ from thermal beams. The helium and the argon beams, which cross at $90°$ in the scattering center, are both collimated to $0.4°$. The helium intensity —about equal amounts of $He(2^3S)$ and $He(2^1S)$ with the quench lamp off —was typically $\sim$ about 10^{10} He/sec, and the argon intensity of approximately $\sim 5 \cdot 10^{13}$ ar/sec, in the scattering center. The scattered metastable atoms are detected by a multiplier of angular acceptance $0.4°$, which can be rotated about the scattering center in a plane containing the beams.

Laboratory angular distributions of the scattered $He(2^3S)$ intensity measured with the described apparatus at various center-of-mass kinetic energies are shown in Fig. 3, which is a reproduction of a figure in Brutschy et al.[30]

The goal of the evaluation of such data is, of course, the determination of the complex potential $\tilde{V}_*(R) = V_*(R) - [i\Gamma(R)/2]$. Other than in the case of scattering from a real potential, the potential cannot be obtained by direct inversion of the scattering data.[33] Instead, we can assume certain functions $V_*(R)$ and $\Gamma(R)$, which contain parameters, and determine these parameters by fitting results of model calculations to the experimental data. Depending on the number of free parameters used as compared to the amount of information contained in the experimental data, this fitting procedure leads to more or less reliable functions $V_*(R)$ and $\Gamma(R)$.

Brutschy et al.[30] assumed $\Gamma(R)$ to be of the exponential form

$$\Gamma(R) = A \exp(-\alpha R) \qquad (II.61)$$

The values of parameters A and α were already approximately known[22,34] and needed only be varied in a narrow range. As the real part $V_*(R)$ a model function was used, which consisted of three parts in different regions of distance (atomic units are used): (1) a long-range part of the van der Waals form

$$V_*(R) = - \frac{C}{R^6} \qquad (II.62)$$

for $R \geqslant 1.5 R_{min}$, (2) an intermediate range part of the Morse form

$$V_*(R) = D_* \left[\exp\beta(1 - R/R_{min}) \right] \left[\exp\beta(1 - R/R_{min}) - 2 \right] \qquad (II.63)$$

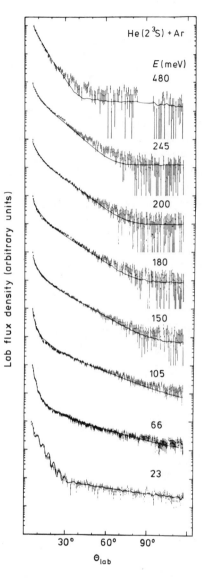

Figure 3. Differential elastic relative cross sections in laboratory system for He(2^3S) scattered from argon at various given collision energies. Solid lines through experimental data represent best fits of model calculations (see text; reproduced from Brutschy et al.[30])

for $5 \leqslant R \leqslant 1.2 R_{min}$, and (3) a short-range part of the shielded Coulomb form

$$V_*(R) = \frac{2}{R}\left(8\exp\frac{-R}{R_M} + 8\exp\frac{-R}{R_L} + 2\exp\frac{-R}{R_K}\right) \qquad \text{(II.64)}$$

for $R \leqslant 4.6$. In the regions $1.5 R_{min} - 1.2 R_{min}$ and 4.6 to 5, exponential spline interpolation functions were used to connect the three forms. The values $C = 6173, R_M = 0.93, R_L = 0.18, R_K = 0.057$ were taken from the literature. The only parameters in $V_*(R)$ that were varied are the well depth D_*, the minimum distance R_{min}, and the shape parameter β, in the Morse form (II.63). With the complex potential of the functional form given by (II.61) and (II.62) to (II.64) the exact complex phase shift δ_l is calculated by numerical solution of the radial Schrödinger equation, and then the differential center-of-mass cross section is computed using expression (II.29).[22] The center-of-mass cross section is then transformed into the laboratory system and averaged appropriately to simulate the experimental conditions. To find the best fit to the experimental data by variation of parameters, the iterative nonlinear least-squares method due to Marquard[35] was used. It was found that no satisfactory agreement could be achieved with the assumed form of $V_*(R)$. Therefore, the flexibility of the Morse potential was increased by allowing β to vary with distance in a certain way. In this way good agreement for all energies was achieved. The calculated distributions are shown in Fig. 3 as solid lines. The parameters determined by these goodness-of-fit calculations are the following:

$$A = 1360 \text{ eV}; \qquad \alpha = 2.05 \text{ au}; \qquad D = 0.0037 \text{ eV}; \qquad R_{min} = 10.7 \text{ au}$$

$$\beta = \begin{cases} 3.77\left(1.86 - 0.86\exp\left[-1.3(R - 5.37)^2\right]\right) & \text{for} \quad 5 \leqslant R \leqslant 5.37 \\ 3.77 & \text{for} \quad 5.37 \leqslant R \leqslant 1.2 R_{min} \end{cases}$$

$$\text{(II.65)}$$

The functions $V_*(R)$ and $\Gamma(R)$ corresponding to these parameters are shown in Fig. 4.

Functions $V_*(R)$ and $\Gamma(R)$ obtained in the same way for the system He(2^1S)–Ar have been reported by Haberland and Schmidt.[31] These functions are shown in Fig. 5. The peculiar double-well structure of $V_*(R)$ seems to be present also, although less pronounced, for the systems He(2^1S)–D_2,Kr,Xe[31]. The term $\Gamma(R)$ is composed of an exponential and a Gaussian. Recently the He(2^1S)–Ar data were reevaluated together with new data that were obtained with somewhat better statistics.[36] It turned

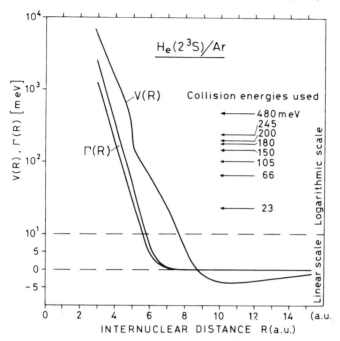

Figure 4. Values $V_*(R)$ and $\Gamma(R)$ for system He(2^3S)–Ar as determined from elastic-scattering data of Fig. 3. Collision energies at which scattering data were obtained are indicated (reproduced from Brutschy et al.[30]). Lower $\Gamma(R)$ curve from Illenberger and Niehaus.[34]

out that, although the double-well structure of $V_*(R)$ was again necessary to explain the data, $\Gamma(R)$ may also be represented by a simple exponential, as in the He(2^3S)–Ar case.

The unusual complex potential derived for He(2^1S)–Ar does not lead to contradictions with other experimental data, such as total ionization cross section and electron energy distribution, but rather explains some of the observed differences between the systems He(2^1S)–Ar and He(2^3S)–Ar.

For systems like He*–Ar, with real parts $V_*(R)$ having well depths D_* smaller or comparable to thermal energies, elastic differential scattering data obtained with thermal velocity selected beams yield, in connection with the rather well-developed theory, the most reliable information on $V_*(R)$ to date. The systems such as He*–H,Li with well depths D_* considerably larger than thermal energies cannot, however, be studied in this way. The most direct information on $V_*(R)$ in these cases is obtained from electron distributions, as is discussed later.

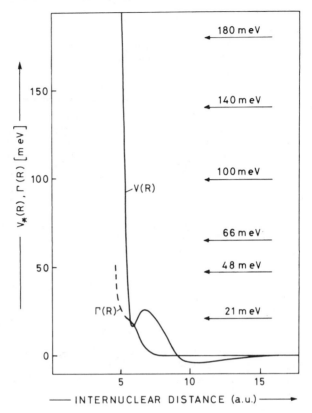

Figure 5. Values $V_*(R)$ and $\Gamma(R)$ for system He(2^1S)–Ar as determined from elastic-scattering data. Collision energies at which used data were obtained are indicated (reproduced from Haberland and Smith.[31])

2. Total Ionization Cross Sections

Measurements of absolute total cross sections for PgI are complicated by two facts: (1) it is not easy to reliably determine the metastable current used in a beam experiment; and (2) the cross sections usually vary considerably in the thermal-energy range, so that an exact knowledge of the energy distribution of the metastables used is necessary for estimation of the result of a measurement.

Recent measurements utilizing the crossed-beam technique have been performed as follows.[37] A metastable helium beam is formed by electron-impact excitation of a thermal helium beam effusing from a multichannel array. The optical quenching method[12] described earlier is applied to obtain results for He(2^1S) and He(2^3S) separately. The target gas beam is also formed by thermal effusion. The problem of absolutely determining the metastable flux is solved by using the "gas-cell method"[38] by which the secondary electron-emission coefficient of surfaces on impact of metastables can be measured *in situ* by essentially equating the number of Penning ions produced, to the number of metastables removed from the beam through Penning collisions, as measured by the secondary-electron detector. The velocity distribution of the metastables was determined in a separate experiment[39] by using the TOF method.

The other method recently applied for the measurement of absolute cross sections is the *flowing-afterglow* (FA) method.[40] In a flowing helium gas, at pressures around 1 torr, metastables are formed by electron impact and carried with the flowing ground-state atoms. Reactant gas is added to the stream, and downstream of the reactant-gas inlet the metastable density is monitored by optical methods as a function of the reactant flow rate. In this way absolute rates for the destruction of metastables can be obtained and by dividing these rates by the average thermal velocity, average destruction cross sections. The earlier FA measurements were performed at room temperature.[40,41] Comparison of these results with the results of beam measurements[37] showed discrepancies, especially for $He(2^3S)$ cross sections that were, in case of all simple systems, found to be about three times lower when determined by the FA method. This was later attributed to the strong velocity dependence of the PgI cross sections, which was evidenced by (1) variation of the absolute rate constants with temperature, as measured in more recent FA experiments[42] and (2), more directly, by beam measurements of PgI cross sections with velocity-selected metastables.[34,43]

These latter measurements led only to relative cross-section values. However, by comparison with absolute values of velocity-averaged cross sections, they can be put on an absolute scale. To do this, the absolute values obtained in FA measurements were used because here the velocity distribution is exactly known—a Maxwellian distribution $f(v, T)$ with the temperature of the buffer gas. Denoting the velocity-dependent relative total ionization cross section, obtained in the beam experiment, by $\sigma_{rel}(v)$ and the absolute total ionization rate constant obtained in the FA experiment by $R(T)$, then a normalization k may be determined by

$$k \int \sigma_{rel}(v) f(v, T) v \, dv = R(T) \tag{II.66}$$

With this factor the absolute velocity-dependent total ionization cross section is given as

$$\sigma_{total}(v) = k \sigma_{rel}(v) \tag{II.67}$$

Cross-section curves $\sigma_{total}(v)$ obtained in this way constitute probably the most reliable and most detailed experimental information on total ionization cross sections available at present. It is available for the PgI systems $He(2^1S)$–Ar,Kr,Xe,N_2 and $He(2^3S)$–Ar,Kr,Xe,N_2.[34] Unnormalized cross-section curves $\sigma_{rel}(v)$ are available for the systems $He(2^1S)$–Hg,Ba; $He(2^3S)$–Hg,Ba; $Ne(^3P_{2,0})$–Hg; and $Ar(^3P_{2,0})$–Hg.[34,44] Cross-section curves directly measured on an absolute scale, using a mechanical velocity selector, are available for the systems $Ne(^3P_{2/3})$–Ar,Kr,Xe.[45] As examples, we show cross-section curves for $He(2^1S)$, $He(2^3S)$–Ar, $He(2^1S)$, $He(2^3S)$–Hg, and Ne*–Kr in Figs. 6 to 8. In Fig. 8 results of the TOF method[34] are compared with results using mechanical velocity selection.[45] For the system $He(2^3S)$–Ar, it is shown in Fig. 9 that the temperature

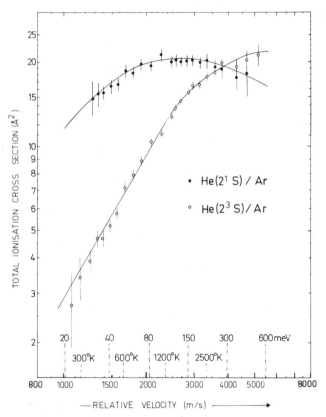

Figure 6. Velocity dependence of total ionization cross sections for systems $He(2^3S)$ and $He(2^1S)$–Ar. Absolute scale by comparison with absolute destruction rate constants (reproduced from Illenberger and Niehaus.[34])

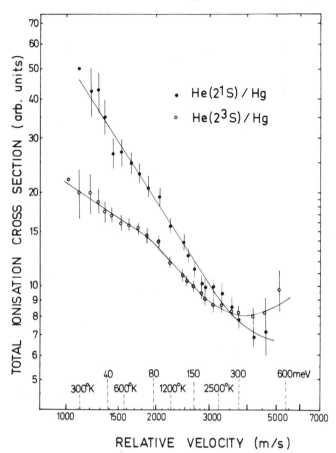

Figure 7. Velocity dependence of total ionization cross sections of systems
$He(2^3S)$ and $He(2^1S)$–Hg. Absolute scale is arbitrary but same for both systems
(reproduced from Illenberger and Niehaus.[34])

dependence of the rate constant, as calculated from $\sigma_{total}(v)$ by (II.66),
agrees very well with the experimental one, as determined by the FA
method.[42] Most of the qualitative features of the known cross-section
curves can be understood in terms outlined in Section II.A.3 The three
paragraphs that follow further illustrate these points.

First, systems $He(2^3S)$–Kr,Xe,N_2 are very similar to $He(2^3S)$–Ar shown
in Fig. 6. The corresponding cross-section curves show a steep—almost
linear—rise in a log–log scale. These are obviously systems for which in
the thermal range $E_k(\infty) \gg D_*$ and which are approximately described by

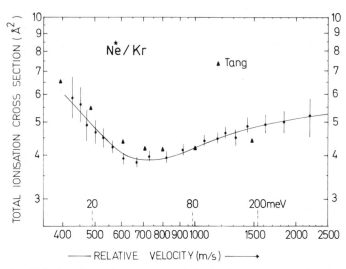

Figure 8. Velocity dependence of total ionization cross section for system Ne*-Kr. Cross-section scale is arbitrary Illenberger and Niehaus[34]; (Δ), From Tang et al.[45]; (φ), From Illenberger and Niehaus.[34]

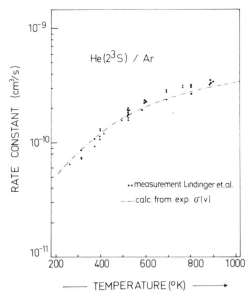

Figure 9. Temperature dependence of destruction rate constant of $He(2^3S)$ by argon. Data points from direct measurement in FA system[42] and dotted line calculated from cross-section curve of Fig. 6 Illenberger and Niehaus.[34]

formula (II.17). The ratios of the characteristic lengths, D/B, of the exponentials describing the real part (D) and the imaginary part (B) of $\tilde{V}_*(R)$, can be estimated from the slopes of the curves by relation (II.18). In this way the respective values $(D/B)\approx 2.2$ (N_2), ≈ 1.85 (Ar), ≈ 1.75 (Kr), and ≈ 1.60 (Xe) can be obtained. These numbers suggest that the following very simple picture is correct: the real part of the repulsive potential is determined by the exponentially decreasing density of the $2s$ electron on $He(2^3S)$, which has a characteristic length approximately proportional to $\{E_i[He(2^3S)]\}^{-1/2}$, with $E_i[He(2^3S)]=4.8$ eV the ionization energy of $He(2^3S)$, and the imaginary part is determined by the exponentially decreasing density of the outer electron shell of the target atom, which has a characteristic length approximately proportional to $[E_i(B)]^{-1/2}$, with $E_i(B)$ the ionization energy of the target atom. In this simple picture the ratio D/B is given by

$$\frac{D}{B} \approx \left\{ \frac{E_i(B)}{E_i[He(2^3S)]} \right\}^{1/2} \tag{II.68}$$

which leads to numbers that are in rather good agreement with the experimentally determined ones: $(D/B)\approx 1.9$ (N_2), ≈ 1.8 (Ar), ≈ 1.71 (Kr), and ≈ 1.6 (Xe). This picture implies that the actual ionization is caused by electron exchange whose probability is proportional to the (squared) overlap of the outer target electron orbitals with the unfilled inner orbital of the metastable.[11] This mechanism may be indicated by the following reaction scheme for PgI with metastables

$$A^*(1) + B(2) \rightarrow A(2) + B^+ + e^-(1)$$

$$\rightarrow A(2)B^+ + e^-(1) \tag{II.69}$$

where (1) and (2) denote the excited electron on the metastable and the target electron to be ionized, respectively.

Second, the cross section curves for $He(2^1S)$–N_2,Kr,Xe are very similar to the $He(2^1S)$–Ar curve shown in Fig. 6. Also, they lie in an energy region where $E_k(\infty)\gg D_*$—because no decrease as predicted for the region $E_k(\infty)\ll D_*$ [equation (II.14)] is observed at low collision energies— however, they are also not described by (II.17), which predicts a linear increase as observed for the triplet systems. Only at the lower velocities a nearly linear increase is observed. The deviation from the behavior predicted by (II.12) at higher velocities is probably caused by the fact that the ionization probability per collision in the relevant impact parameters range is already close to unity whereas $P(b)\ll 1$ was assumed in the derivation of

(II.17). The difference between PgI systems with $He(2^1S)$ and ones with $He(2^3S)$ can be explained by the different real parts of $\tilde{V}_*(R)$ observed in the elastic scattering data[30-32] and shown for argon in Figs. 4 and 5. Systems with $He(2^1S)$ can penetrate to smaller distances at low relative velocities, which leads to large cross sections and an earlier "saturation" caused by the condition $P(b) \sim 1$.

Third, similar to the curves for $He(2^1S)$, $He(2^3S)$–Hg shown in Fig. 7, the curves for Ne*,Ar*–Hg also decrease nearly linearly in the lower portions, which indicates that relation (II.14) can be used to approximately describe the energy dependence. From the slope of these low-energy portions the power (S) of the long-range attractive forces may be estimated. For the three systems $He(2^3S)$,Ar*,Ne*–Hg we obtain $S \approx 6$, indicating van der Waals attraction in the distance range of the rotational barrier at the lower thermal velocities. For $He(2^1S)$–Hg, we obtain $S = 2.66$, which indicates a chemical-type attractive interaction at the relevant distances. This is supported by the fact that the $He(2^1S)$–Hg potential has a well depth $D \approx 0.5$ eV, which is rather large as compared, for instance, to $D \approx 0.08$ for $He(2^3S)$–Hg.[21] A minimum in the cross-section curve as seen for $He(2^3S)$–Hg and for Ne*–Kr is also observed for Ar*,Ne*–Hg. Such a minimum arises in an energy range where $E_k(\infty) \sim D$ and where the behavior of the cross section gradually changes from the behavior predicted by (II.14) to the behavior predicted by (II.17).

Quantitative calculations of cross-section curves have been performed for the systems $He(2^1S)$–Ar and $He(2^3S)$–Ar. A comparison of theory and experiment in the case of $He(2^1S)$ is not so significant since the measured curve varies only little in the subtended energy range. Therefore, we compare in results for $He(2^3S)$–Ar Fig. 10:

1. The curve that essentially connects the data points is obtained by a classical calculation [relations (II.8) to (II.11)] using $V_*(R)$ as determined from elastic scattering data[30] [equations (II.62) to (II.64)] and $\Gamma(R) = A \exp(-R/B)$ with $A = 14$ au and $B = 0.526$, chosen to obtain the best fit to the experiment.
2. The curve lying slightly above but running parallel to the experimental points is the result of a quantum-mechanical calculation[30] [relation (II.31)] with exact phase shift δ_l using $\tilde{V}_*(R)$ as determined from elastic scattering data[30] (see Fig. 4).
3. The straight dotted lines in regions $D_* \gg E_k(\infty)$ and $D_* \ll E_k(\infty)$ are calculated from the approximate relations (II.14) and (II.17), using the same $\tilde{V}_*(R)$ as in item 2 (preceding). The position of these lines on the absolute scale is arbitrary.

The comparison in Fig. 10 shows that there is no significant disagreement

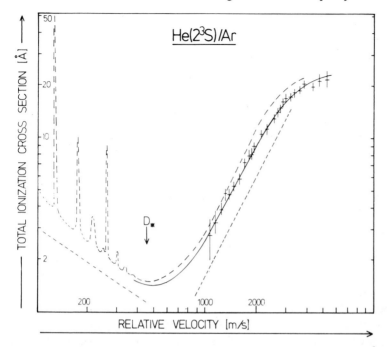

Figure 10. Velocity dependence of total ionization cross section of He(2^3S)–Ar system: $(++)$, measurement[34]; $(-)$, classical calculation[50];$(---)$, quantum-mechanical calculation[30]; straight dotted lines, approximate cross section formulae (see text).

between calculation and experiment. The small shift of the experimental points with respect to the quantum-mechanical result could be the result of either an error in the absolute normalization—the possible error of the rate constant used is $\pm 30\%$[42]—or a slightly excessive preexponential factor in the imaginary part of $\tilde{V}_*(R)$, to which the angular distributions are not very sensitive. At not too low velocities the quantum effects seem to average out so that the classical result coincides with the quantum result. Only at low velocities do orbiting resonances, which show up in an enhanced way in the ionization cross section, appear in the quantum calculation. The extent to which the dotted lines representing the approximate formulas (II.14) and (II.17) run parallel to the calculated cross sections indicates how well the physical quantities determining mainly the cross-section variation in the respective regions may be estimated from the slopes of these lines.

3. Electron Energy Distributions

The energy spectrum of electrons ejected in a PgI process contains two different kinds of information:

1. If more than one final state of the target ion can be reached, the measured electron spectrum reflects the population of these different states. In this respect PgI electron spectroscopy (PgIES) is analogous to the photoionization electron spectroscopy (PIES).

2. Since the PgI process occurs *during* the collision—with certain probabilities at certain distances between the collision partners—to each transition leading to the population of a certain final state, there belongs a *distribution* of electron energies that reflects the variation of the physical quantities during the collision.

In this section we deal with the second aspect of PgIES, and the first aspect is treated later, when the population of different final electronic and vibrational states is discussed.

The application of the PgIES method to the study of PgI was first introduced by Čermak[46] in 1966. Because of lack of energy resolution, however, only the information of type 1 could be obtained.[7, 8] The first studies with sufficiently high resolution and accurate energy calibration to observe characteristic broadenings, shapes, and shifts of individual distributions were reported in (1968).[10, 11] Because the measured distributions contain very direct information on the dynamics of the process itself, these first and the following high-resolution PgIES studies have contributed considerably to our present understanding of PgI[47] and to development of the theory.

Most of the PgIES studies have been performed with helium metastables. The reason is that helium combines several advantages: (1) metastables are easily produced; (2) the excitation energy of the metastables is high, so that the energy requirement for PgI is fulfilled for almost any target; (3) the excitation energies of the two metastable states $He(2^1S)$, and $He(2^3S)$ differ by the rather large amount of 0.8 eV, which in most cases is sufficient to lead to well-separated distributions in case of atomic targets; (4) in cases where the distributions from $He(2^1S)$ and $He(2^3S)$ overlap—which usually happens with molecular targets—it is still possible, by applying the quenching method,[12] to determine separate singlet and triplet distributions; and (5) the role played by the spin state can be investigated.

The principle of an experimental setup used for PgIES with helium metastables is shown in Fig. 11. The metastables are produced in a hot cathode discharge, running at approximately 10^{-2} torr and ~ 0.1 A in a differentially pumped chamber. The quenching lamp allows removal of

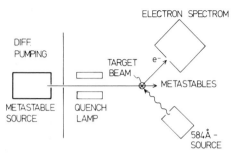

Figure 11. Schematic drawing of experimental setup used for Penning ionization electron spectroscopy using helium metastables.

He(2^1S) metastables from the thermal beam by reaction (II.59). Electrons emerging from the crossing region of the metastable beam with the target gas beam are detected at some angle by some electron spectrometer. Earlier measurements were performed with a retarding-field-type electrostatic spectrometer and later ones, with a 127° cylindrical condenser combined with a retarding field. The standard resolution of these spectrometers was 10 to 30 meV. The helium lamp—a small dc discharge running at ~5 torr and 20 mA that creates predominantly 584-Å photons in the UV region—serves to calibrate the analyzer with respect to resolution, absolute energy, and transmission, by measuring simultaneously with the Penning electrons the photoelectrons from the target gas (or gases). This technique allows an absolute energy calibration of a few milli-electron-volt accuracy.

As outlined in Section II.A, the width of a Penning electron distribution —or rather the quantity $\varepsilon_0 - \varepsilon_*$—is an approximate measure of the strength of the interaction between the metastable and the target particle, that is, a measure of the well depth D_* of $V_*(R)$ [see relation (II.6)]. If the observed $\varepsilon_0 - \varepsilon_*$ is large compared to the thermal energy, D_* is large and approximately equal to $\varepsilon_0 - \varepsilon_*$, and if the observed $\varepsilon_0 - \varepsilon_*$ is of the order of thermal energies, it can be concluded that D_* is also small. In this way we may immediately distinguish between cases of weak and strong interaction between metastable and target atom, merely from inspection of the measured spectra.

Figure 12 shows a selection of narrow distributions, indicating weak interaction.[21] As expected, weak interaction arises in cases where the target atom has a closed-shell structure. Wide distributions, indicating strong interaction, are shown in Fig. 13.[21] The distribution for the system He(2^3S)–Hg—which has an intermediate width, indicating a well depth of $D \approx 0.085$ eV—is shown in both figures for comparison.

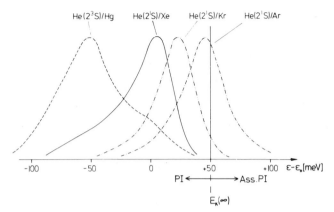

Figure 12. Penning electron energy distributions for systems with weak attraction between metastables and target.

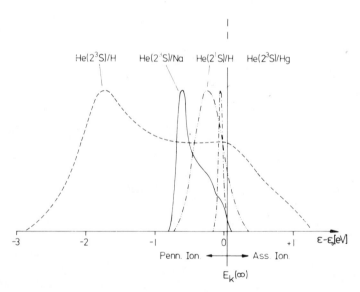

Figure 13. Penning electron energy distributions for systems with strong attraction between metastable and target. Distribution for system He(2^3S)–Hg is shown in Figs. 12 and 13 for comparison.

436

In principle, by comparing experimental distributions with model calculations, it is possible to derive rather quantitative information on the functions $V_*(R)$, $V_+(R)$, and $\Gamma(R)$. This is especially true for the strong interaction systems, as is demonstrated in the text that follows for the system $He(2^3S)$–H. In cases of weak interaction a complication arises because the measurements are carried out with a distribution of relative collision energies, and the shape of the distributions changes considerably within the energy range subtended. This dependence of the shape of a distribution on the collision energy has been studied for a few systems in the thermal collision energy region by velocity selecting the metastable beam.[47–49] The distributions measured for $He(2^3S)$–Ar at three different velocities (resolution $\Delta v/v \approx 18\%$) are shown in Fig. 14. In the case of $He(2^3S)$–Ar rather reliable functions $V_*(R)$, $\Gamma(R)$, and $V_+(R)$ are available from evaluation of elastic scattering data,[31] total ionization cross section curves,[34] and the observed velocity dependence of the associative PgI : PgI cross-section ratio[43] (see section II.B.5), respectively. Therefore, the measured distributions of Fig. 14 were calculated using the classical relations (II.1), (II.20), and (II.21) to test consistency.[50] The dotted lines in Fig. 14 show the results of these calculations, which take into account the finite width of the velocity distributions and energy analysis. The significant deviation of the theoretical result from the experimental one at the lowest average collision energy probably indicates that the well depth of $V_+(R)$ is somewhat larger than the assumed 19 meV. This is supported by the quantum-mechanical evaluation[22] of an associative PgI : PgI cross-section ratio curve that differs slightly from the one of Pesnelle et al.[43] and gives a well depth of 26.5 meV.

A recently measured distribution of the strong interaction system $He(2^3S)$–H[51] is shown in Fig. 15. The target beam was formed by dissociation in an RF source and consisted of a mixture of H atoms and H_2 molecules. The spectrum of Fig. 15 is obtained by subtracting the H_2 part from the measured Penning electron spectrum. The characteristic features of this distribution—which is very broad as compared to the width of the collision energy distribution—are determined by $V_*(R)$, $V_+(R)$, and $\Gamma(R)$ alone, and only the overall shape changes with collision energy within the thermal range. This was found by cooling the target gas down to liquid-nitrogen temperature. An analysis of the measured distribution by model calculations can thus yield more direct information on the model functions than was possible in the case of $He(2^3S)$–Ar. Since, in addition, $V_+(R)$ is known from accurate *ab initio* calculations[52] as well as from differential scattering experiments,[53] the number of parameters to be determined by comparison of calculated and measured spectrum is reduced, which leads to a more reliable determination of parameters in $V_*(R)$ and $\Gamma(R)$.

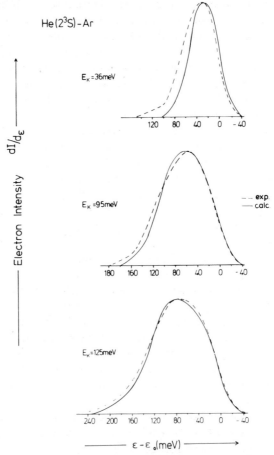

Figure 14. Penning electron energy distributions for He(2^3D)–Ar at three different selected collision energies. Dotted curves represent result of classical model calculation (see text; reproduced from Illenberger.[50])

Model quantum-mechanical calculations were carried out on the basis of formulas (II.43) and (II.47), and functions $V_*(R)$ and $\Gamma(R)$ were determined by comparison with the distribution shown in Fig. 15.[51] As can be seen from Fig. 15, the measured distribution extends considerably into the region of associative PgI characterized by condition (II.4a). In this region formula (II.47) was used, and in the PgI region, characterized by (II.4b), formula (II.43) was used, except for those special values of ε and where, for a certain collision energy, the final state of nuclear motion corresponds to a temporarily bound—or quasibound—HeH$^+$ system

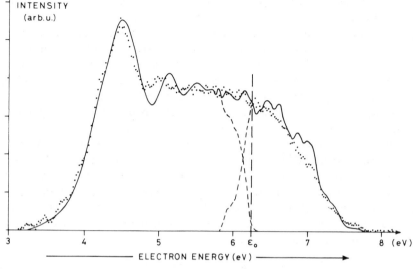

Figure 15. Penning electron energy distribution for $He(2^3S)$–H. Solid line represents result of quantal (model) calculations. Indicated part of calculated distribution at high electron energies belongs to associative (PgI) and formation of quasibound HeH^+ (see text).

These states arise only for small kinetic energies $E_k'(\infty)$. They have characteristic lifetimes and decay by "tunneling" through the rotational barrier of the effective potential. In the calculations the possibility of formation of these quasibound states was treated as in Hickman et al.[20] by determining approximate eigenstates of positive energy $E_k'(\infty)$ in the effective potentials and applying formula (II.47). Thus in the calculated spectrum, transitions leading to quasibound states result in finite values of the associative part of the electron spectrum at energies $\varepsilon \lesssim \varepsilon_0 + E_k(\infty)$.

The actual procedure of determining $V_*(R)$ and $\Gamma(R)$ in model calculations is complicated in the present case by the fact that the electron distribution varies rather significantly with the collision energy in the thermal range. Therefore, to achieve an accurate determination of parameters by comparison with the experiment, it is necessary to construct the theoretical spectrum as a weighted superposition of single spectra calculated at several discrete collision energies within the range of the collision energy distribution used in the experiment. The calculation of such an averaged distribution for a certain set of parameters in this way is rather time consuming, and hence it was not possible to determine the parameters by means of a normal fitting procedure. Instead, a "trial-and-error" procedure was used guided by the results of a semiclassical analysis of the experimental distribution that yields fairly accurate values of the well depth, the curvature at the minimum, and the zero crossing radius of $V_*(R)$. The choice of parameters was further restricted by the requirement that the measured absolute total cross section be reproduced, which, however, can always be achieved for any trial functions $V_*(R)$ and $\Gamma(R)$ by simply choosing the appropriate value of a prefactor contained in $\Gamma(R)$. Since the value of $V_*(R)$ at large R is reflected in a rather complicated way in the shape of the distribution, it was further required, to reduce the number of trial calculations, that the model function $V_*(R)$ smoothly joins the *ab initio* potential[19] at $R = 9a_0$. This procedure may be justified by the fact that theoretical potential curves tend to be more accurate at large R, together with the fact that the deviation of the *ab initio* potential from the finally determined $V_*(R)$ is rather small at any R, as can already be judged from the direct semiclassical analysis of the measured distribution.

The model functions used are

$$V(R) = \begin{cases} D_e\left(\exp 2\beta_1 \dfrac{R_e - R}{R_e} - 2\exp \beta_1 \dfrac{R_e - R}{R_e}\right); & R \leqslant R_e \\[2ex] D_e\left(\exp 2\beta_2 \dfrac{R_e - R}{R_e} - 2\exp \beta_2 \dfrac{R_e - R}{R_e}\right); & R_e \leqslant R \leqslant 4.5a_0 \\[2ex] \text{Two appropriate spline functions;} & 4.5a_0 \leqslant R \leqslant 9a_0 \\[1ex] G \cdot e^{\gamma R}; & R \geqslant 9a_0 \end{cases}$$

$$\text{(II.70)}$$

$$\Gamma(R) = \frac{1}{F} A \cdot e^{-\alpha_1 R}; \qquad F = \begin{cases} 1; & R \geqslant R_c \\ \cosh\left[\alpha_2(R_c - R)\right]; & R \leqslant R_c \end{cases} \qquad \text{(II.71)}$$

The parameter values determined by comparison of the appropriately thermally averaged theoretical spectrum with the experimental spectrum shown in Fig. 15 are

$$D_e = 2.26 \text{ eV} \qquad\qquad A = 42 \text{ au (atomic units)}$$

$$\left.\begin{array}{l} R_e = 3.3a_0 \\ \beta_1 = 2.0794 \\ \beta_2 = 1.800 \end{array}\right\} V(R) \qquad \left.\begin{array}{l} \alpha_1 = 3a_0^{-1} \\ \alpha_2 = 2.33a_0^{-1} \\ R_c = 3.2a_0 \end{array}\right\} \Gamma(R) \qquad \text{(II.72)}$$

$$G = -283 \text{ eV}$$

$$\gamma = -0.9935$$

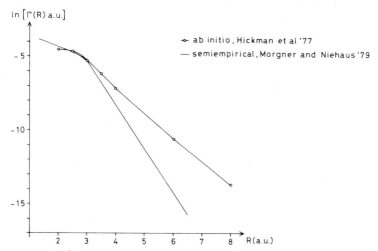

Figure 16. The $^2\Sigma$ interaction potential of He(2^3S)–H: ($-$), result of fit calculations;[51] (–o–), result of *ab initio* calculations.[19]

A graphical representation of the real and imaginary parts of the thus determined complex potential for He(2^3S)–H is given in Figs. 16 and 17.

With the determined complex potential given by (II.70) to (II.72) cross sections can be calculated as a function of collision energy. Calculated cross-section curves are shown in Fig. 18. The two almost coinciding upper curves show the total ionization cross section for He(2^3S)–H, calculated quantum mechanically as the total "absorption" cross section [relation (II.31)], and classically using relations (II.8) to (II.11). In these calculations only $V_*(R)$ and $\Gamma(R)$ enter. The other curves correspond, respectively, to the partial cross sections of Penning ionization [$\sigma(\text{PgI})$], associative ionization [$\sigma(\text{AI})$], and the formation of "quasibound" HeH$^+$ molecules [$\sigma(\text{QI})$], where these latter processes are characterized by the condition that the relative kinetic energy of the He–H$^+$ system is greater than zero as in Penning ionization, but that the system is temporarily bound inside a rotational barrier. In the calculation of the partial cross sections the potential $V_+(R)$ is, of course, also involved in addition to $V_*(R)$ and $\Gamma(R)$.

The calculated cross-section curves are compared with experimental data in Fig. 18. The fact that the total cross-section curves pass through the experimental point for an average collision energy of 45 meV was required in the fit calculations. As seen in Fig. 9, the value obtained by Howard et al.[54] for an average collision energy of 370 meV also lies on the calculated curve. However, for a given absolute calibration of $\Gamma(R)$, as the energy

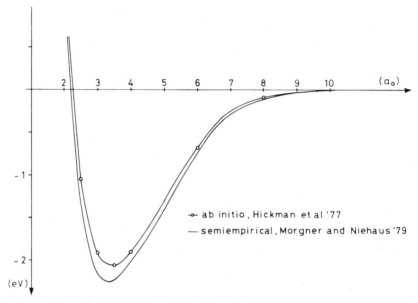

Figure 17.   Width function $\Gamma(R)$ of $He(2^3S)$–H: $(-)$, result of fit calculations;[51] $(-o-)$, result of *ab initio* calculations.[19]

dependence of the total cross section is rather independent of details of the complex potential at small R but almost entirely determined by $V_*(R)$ at large R, where the *ab initio* result[19] was assumed to be correct, this agreement only indicates the consistency of this assumption with the two mentioned experimental data points and thus essentially reproduces the conclusions drawn from cross-section calculations[55] using the *ab initio* functions.

Fort et al.[56] have reported partial Penning ionization and associative ionization cross sections, measured by the TOF method in the thermal-energy range. To compare their data with the calculated curves, it is necessary to know whether in their experiment the quasibound HeH^+ are detected as HeH^+ or as H^+. This is a question of the lifetime of these molecules with respect to autodissociation in comparison with the time span between formation and detection of the ions. Calculations show that only a small fraction, amounting to a cross section of about $1a_0^2$, survives long enough to be detected as a molecule. Therefore, the sum $\sigma(QI) + \sigma(PgI)$ of the calculated curves should be compared to the measured partial Penning ionization cross section of Fort et al.[56] In approximate agreement with the very slight decrease of the theoretical curve $\sigma(QI) + \sigma(PgI)$, they

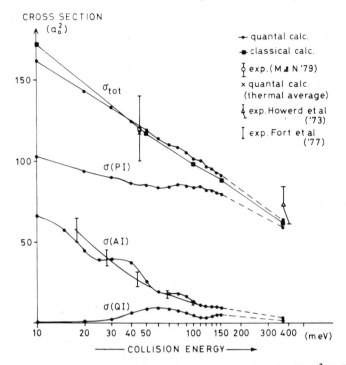

Figure 18.   Calculated cross sections and experimental data for He(2^3S)–H: σ_{total}, total ionization; σ(PgI), partial Penning ionization; σ(AI), partial associative ionization; QI, partial formation of quasibound heH$^+$; (–o–), quantal calculation;[51] (– –), classical calculation.[51] Experimental data: value obtained with thermal velocity distribution corresponding to average energy of 45 meV;[51] value obtained at average energy of 3 + 0 meV; I, renormalized data due to Howard et al.;[54] x, quantal calculation, thermal average for conditions under which point was obtained.[51]

find a nearly constant value in the range 10 to 150 meV. However, their absolute value, as well as their ratio σ(PgI)/σ(AI), is lower by a factor of approximately 0.7. This may be the result of discrimination against the rather fast H$^+$ ions (kinetic energy $\sim$2 eV) with respect to the thermal HeH$^+$ ions in their experiment. In the case of σ(AI), the agreement of the experimental curve with the calculated curve is satisfactory, except for the slight undulation of the theoretical curve, which is not observed in the experiment.

The discussion of the He(2^3S)–H system has shown the quantum-mechanical calculations to generally account quite well for the detailed experimental results. However, it should be pointed out that these calculations also indicate the limited validity of the underlying theoretical formulation, as is explained in the paragraphs that follow.

In the frame of the theoretical formulation, in which the Penning process is described by the local quantities $V_*(R)$, $\Gamma(R)$, and $V_+(R)$, the total cross section can be calculated as either (1) total "absorption" cross section σ_{total} from the complex phase shift for scattering by the complex potential $\tilde{V}(R) = V_*(R) - \frac{i}{2}\Gamma(R)$ or (2) as the sum of the partial cross sections $\sigma(PgI)$, $\sigma(AI)$, and $\sigma(QI)$, into whose calculation also $V_+(R)$ enters in the form of matrix elements involving nuclear wave functions in this potential.

If the results of calculations using these different methods deviate from each other, this indicates an internal inconsistency of the formulation, and the extent of the deviation in special cases should be a measure of the errors that can arise in its application to these cases. In Fig. 19 the relative deviation $[\sigma(PgI) + \sigma(AI) + \sigma(QI) - \sigma_{total}]/\sigma_{total}$ is plotted as a function of collision energy, as calculated for the present case of He(2^3S)–H. The relative deviation is seen to undulate with an amplitude of up to $\pm 5\%$ at small energies, as a result of the partial cross section for associative ionization. Although some additional minor approximations have been introduced in the practical calculations in the frame of the general formulation,[22] the deviation is much too large to be explained by these approximations and hence must be ascribed to a limitation of the general formulation, especially in case of the calculation of the partial cross section for associative ionization. Evidence of such a failure of the PgI formulation on the basis of a local complex potential has not yet been pointed out to our knowledge. However, Bardsley,[57] in his discussion of resonant scattering theory for electron molecule collisions, gives the formal conditions under which the nuclear motion of a molecule,—capable of decay by electron emission,—may be correctly described by a local complex potential. He concludes that these conditions are better fulfilled with a greater number of energetically accessible final states of nuclear motion. Applying this conclusion to PgI, we would thus expect that a failure of the present theory should—if at all—show up in describing the associative processes, where the nuclear motion in the final states—which are populated by electron ejection,—is constrained to a limited number of discrete vibrational–rotational states, as is actually evidenced in the calculations. It is an interesting result of these calculations that they furnish an estimate of the extent to which the present PgI theory fails through the use of a local complex potential. Further, since the relative deviation plotted in Fig. 19 approaches (approximately) zero as the partial cross section for AI approaches zero, we may conclude that the theory is quite accurate when the σ_{AI}/σ_i cross-section ratio is small.

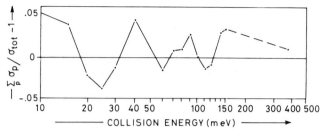

Figure 19. Deviation of sum of partial cross sections from total ionization cross section, normalized to total ionization cross section, is plotted against the collision energy. All cross sections are calculated on basis of description of Penning process by local complex potential. Deviation is measure of inconsistency of their desorption for case of system $He(2^3S)$-H.

4. Electron Angular Distributions

Angular distributions of PgI electrons with respect to the axis of relative velocity of the colliding particles have been measured for the systems $He(2^3S)$-Ar,Kr,Xe,N$_2$,CO,Hg and $He(2^1S)$-Ar,Kr,Xe,N$_2$,CO,Hg in the thermal-energy range[58,59] and in a range of 50 to 500 eV for He-Ar,Hg,[60,61] where "He" represents a mixture of $He(2^3S)$ and $He(2^1S)$. A schematic drawing of the essential parts of the apparatus used is shown in Fig. 20, where the thermal beam is produced as usual and the fast beam is formed by charge exchange with potassium. Electrons, produced by PgI in the scattering chamber containing the target gas, are detected by two analyzers of the 127° electrostatic condenser type, one positioned at 90° and one rotatable between 20° and 160° with respect to the primary beam direction. Electron spectra that are simultaneously measured by these two analyzers are stored in multichannel analyzers. The energy range subtended by the analyzers is chosen so that also photoelectrons, created by 584-Å radiation from the helium-discharge lamp, are measured in addition to PgI electrons. These photoelectrons are used to normalize the PgI electron angular distributions to the known photoelectron angular distributions.[62] In this way a possible influence of an angular dependent transmission on the PgI electron distribution is eliminated. From the following four electron intensities, obtained by integrating the corresponding peak areas in the simultaneously measured spectra,

$$I_{pg}(\vartheta) \quad \equiv \text{PgI electron intensity at } \vartheta$$

$$I_{ph}(\vartheta) \quad \equiv \text{photoelectron intensity at } \vartheta$$

$$I_{pg}(90) \quad \equiv \text{PgI electron intensity at } \vartheta = 90°$$

$$I_{ph}(90) \quad \equiv \text{photoelectron intensity at } \vartheta = 90°$$

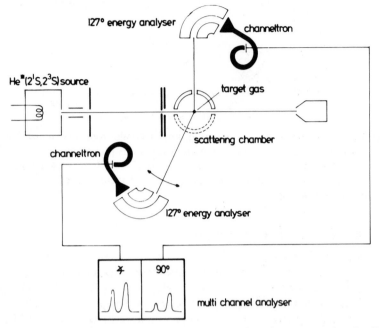

Figure 20.   Apparatus used for measurement of Penning electron angular distributions.[59]

the quantity

$$M(\vartheta) = \frac{I_{\text{Pg}}(\vartheta)/I_{\text{Pg}}(90)}{I_{\text{ph}}(\vartheta)/I_{\text{ph}}(90)} \tag{II.73}$$

is calculated, which then yields, with the known normalized photoelectron angular distribution $N_{\text{ph}}(\vartheta)$, the normalized PgI electron angular distribution

$$N_{\text{pg}}(\vartheta) = M(\vartheta) \cdot N_{\text{ph}}(\vartheta) \tag{II.74}$$

As examples distributions $N_{\text{pg}}(\vartheta)$ obtained in this way for He(2^3S), He(2^1S)–Ar, He(2^3S), and He(2^1S)–CO in the thermal-energy range are shown in Figs. 21 and 22. The angular distributions for nitrogen are almost identical to the ones for argon and the ones for krypton and xenon are very similar to the ones for carbon monoxide. In all cases the asymmetry is more pronounced for the triplet distributions. The distributions for mercury are practically isotropic in the thermal range. For He(2^3S)–Ar,

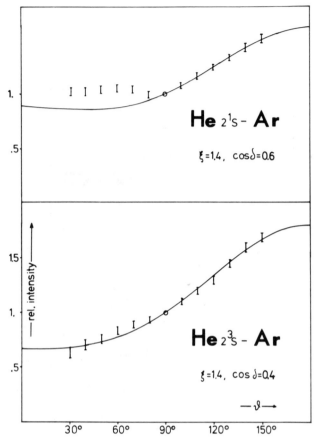

Figure 21. Normalized Penning electron angular distributions for He(2^1S)–Ar and He(2^3S)–Ar: (I) experimental points, showing statistical uncertainty (reproduced from Ebding and Niehaus.[59])

the variation of the angular distribution with collision energy is shown in Fig. 23. In the thermal range the different measurements at average energies 80 meV, 130 meV, 190 meV, and 350 meV were made by dividing the thermal distribution into four parts of similar intensity using the TOF technique.

As outlined in Section II.A, knowledge of $\Gamma(R)$, $V_*(R)$, and $V_+(R)$ is not sufficient to calculate angular electron distributions. The "internal" distribution with respect to the molecular axis must also be known. Model quantum calculations based on relation (II.38) with expressions (II.32) to (II.37) have not been performed until present. To explain the experimental

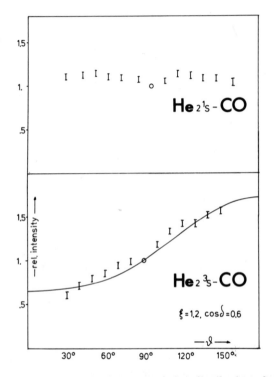

Figure 22. Normalized Penning electron angular distributions for $He(2^1S)-C)$, $CO^+X(v=0)$, and $He(2^3S)-CO$, $CO^+X(v=0)$: (I) experimental data, showing statistical uncertainty (reproduced from Ebding and Niehaus.[59])

results, classical calculations of the type described in Section II.A have been carried out for $He(2^3S)-Ar$.[61] For the internal distribution approximation (II.23) was assumed with the angular ejection probability $P(\gamma)$ given by

$$P(\gamma) = |P_0(\cos\gamma) + a_1 P_1(\cos\gamma)|^2 \qquad (II.75)$$

where P_0 and P_1 are the Legendre polynomials and a_1 is a complex number

$$a_1 = \xi e^{i\delta} \qquad (II.76)$$

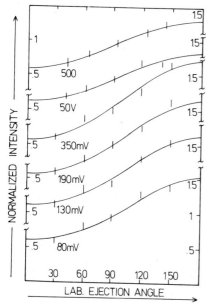

Figure 23. Variation of He(2^3S)–Ar Penning electron angular distribution with collision energy: (I) experimental data points, showing statistical uncertainty; solid lines represent results of classical calculations using a model "internal" angular distribution (see text).[60, 61]

The classical trajectories were calculated for the real potential $V_*(R)$ given by (II.62) to (II.65), and the transition probability at certain points of the trajectory was calculated with $\Gamma(R)$ as given by (II.61) with $A = 14$ au and $\alpha = 1.9$ au. These functions gave the excellent agreement between calculated and measured total ionization cross-section curves (see Fig. 10). Free parameters were only (ξ) and (δ) in (II.76). The fit obtained by adjusting (ξ) and (δ) at the different collision energies is shown by the solid lines in Fig. 23. The values of the parameters determined in this way are given in Table I, which also gives the average distances at which the ionization occurs at the different collision energies. A polar diagram of the internal distribution $P(\gamma)$ for the four average thermal energies is given in Fig. 24.

Recently the theory for evaluating Penning electron angular distributions in the frame of the quantum-mechanical PgI formulation has also been reported[63] and applied to the evaluation of the He*–Ar data.[64] In addition to the determination of an "internal" distribution similar to the one determined classically, this evaluation yields the result that the Penning process leads exclusively to He–Ar$^+$ states the spatial wave functions of which have Σ symmetry. This result can, in turn, be used to calculate the population of the Ar$^+(^2P_{3/2,1/2})$ fine-structure levels. The authors find excellent agreement with the available experimental data.

The experimental angular distribution of He*–Hg at high collision energies is very similar to that of He*–Ar. since almost every trajectory

Table I.

Values of Parameters ξ and $\cos\delta$ of "Internal" Angular Distribution of He(2^3S)–Ar System as Determined at different Collision Energies

COLLISION ENERGY (eV)	ξ	$\cos\delta$	$\bar{R}(a_0)^a$
0.08	0.62	0.60	6.4
0.13	0.64	0.62	6.0
0.19	0.68	0.66	5.6
0.35	0.73	0.70	5.2
50	0.69	0.64	3.1
100	0.71	0.67	2.9
500	0.74	0.71	2.5

^{a}Calculated average distance at which the ionization occurs.[59, 60]

that contributes to ionization is practically straight and depends very little on the potential at high collision energies, it can be concluded that the internal distribution of He*–Hg is similar to that of He*–Ar. The great difference between these distributions at thermal energies is a result of the different strengths of attractive interaction. The strong He*–Hg interactions lead to "spiraling" trajectories that do not give rise to an average orientation of the system during ionization.

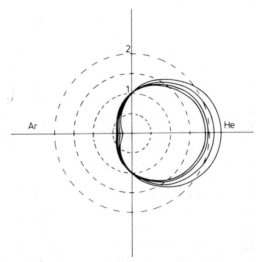

Figure 24. Polar diagram of "internal" angular distribution of He (2^3S)–Ar system as obtained from measurements at 80-mV, 130-mV, 190-mV, and 350-mV collision energies. Asymmetry increases from 80 to 350 mV.

At high collision energies it is observed[60,61] that the peak position of the electron energy distribution shifts appreciably with the observation angle. Classically, this may be attributed to the fact that the electron is ejected from a "moving source" that leads to a Doppler shift. Since the trajectories of the He* are essentially straight, the average velocity of the "moving source" may be approximated in a molecular model by $v_{He} \cdot \bar{s}/\bar{b}$, with v_{He} the helium velocity, $\bar{s}$ the average distance of the source from the center of mass at the distance of closest approach, and $\bar{b}$ the average impact parameter of trajectories leading to PgI. Neglecting small terms, the Doppler shift of electrons of velocity v_{el} ejected from the "moving source" is then given by (atomic units are used):

$$\Delta\varepsilon(\vartheta) \approx \frac{\bar{s}}{\bar{b}} v_{el} \cdot v_{He} \cdot \cos\vartheta \qquad (II.77)$$

The observed angular-dependent shifts at the various collision energies are well described by (II.77). From the observed shifts the physically interesting quantity $\bar{s}/\bar{b}$ can be determined by (II.77), which indicates the average location in the quasimolecule where the electron is ejected. It is found that $\bar{s}/\bar{b}$ is always close to unity, which means that the electron "comes from" the helium. This is very direct evidence of the exchange mechanism of PgI [see equation (II.69)].

In the case of He*–Ar the value $(\bar{s}/\bar{b}) = 1.4$ is found for all collision energies in the high-energy range. From the known functions $V_*(R)$ and $\Gamma(R)$ for this system, we can calculate $\bar{b}$ and obtain $\bar{b} \approx 2.2a_0$ for 500-eV collision energy, which then leads to $\bar{s} \approx 3.1a_0$. Obviously, this result indicates that the excited electron originally centered around helium is in the collision promoted. Indeed, this is what would be expected since the diabatic correlation diagram for He($1s2s$)–Ar predicts that the $2s$ helium orbital becomes a $4d$ orbital of the united calcium atom, with an average radius of $\sim 4a_0$.

The angular-dependent shift of the peak position of a PgI electron energy distribution at high collision energies can also be explained in terms of rotational transitions in the process of electron ejection, instead of using the classical concept of the Doppler shift. The average rotational angular momentum of the ionizing collision system is approximately (atomic units are used)

$$\bar{l} = \mu v_{He}\bar{b} \qquad (II.78)$$

The average change of angular momentum is similarly approximated by

$$\Delta\bar{l}(\vartheta) \approx v_{el}\bar{s}\cos\vartheta \qquad (II.79)$$

Since l is large, the energy change corresponding to the change of angular momentum is given by

$$\Delta\varepsilon(\vartheta)\approx\frac{\bar{l}\Delta\bar{l}(\vartheta)}{\mu\bar{b}^2}=\frac{\bar{s}}{\bar{b}}v_{el}v_{He}\cos\vartheta \qquad (II.80)$$

This is exactly the same result as derived by considering the Doppler effect. When solved for the averaged change of angular momentum, we obtain from the observed shift by (II.79) and (II.80)

$$\Delta\bar{l}(\vartheta)\approx\frac{\Delta\varepsilon(\vartheta)\bar{b}}{v_{He}}\approx1.7\cos\vartheta \qquad (II.81)$$

The angular-momentum change is, of course, carried away by the ejected electron. The value 1.7 for the maximum (average) angular momentum of the ejected electron fits very well with the prediction of the correlation diagram.

5. Associative (PgI): Normal PgI Ratio

Associative PgI is characterized by condition (II.4a). By inspection of the potential curve diagram (Fig. 1), it is immediately seen that this requirement—that the asymptotic energy $E'_k(\infty)$ is negative—will in normal cases only be met at collision energies that are lower than, or of the same order of, the well depth of $V_+(R)$. As the collision energy is increased, the impact parameter range for which (II.4a) is met will rapidly decrease, leading to a rapidly decreasing associative PgI: PgI ratio. In the limited collision energy range where an appreciable part of ionizations leads to molecule formation, the associative PgI: PgI ratio is thus a sensitive means to obtain information on the potentials $V_*(R)$ and $V_+(R)$. Early measurements of this ratio, using full thermal velocity distributions at different temperatures, can only yield qualitative information. Therefore, only some results of more recent studies applying some sort velocity selection are discussed.

Three experimental methods have been used and applied to a number of systems:

1. The TOF method was used to select the velocity of the metastables within the thermal energy range, and mass analysis was used to distinguish between associative PgI and PgI. The systems $He(2^1S)$ and $He(2^3S)$–Ar,D were studied,[43,65] as well as the systems Ne^*–Ar,Kr,Xe.[66]
2. In merging beam experiments, in which the metastable beam is formed by charge exchange with alkali atoms and the target beam by resonant

charge exchange, well-defined relative energies in the range 0.01 to $\sim$ 10 eV are realized in an interaction region of the two beams. Normal Penning and associative Penning ions formed in this interaction region are mass selectively detected. The method was applied to He* (mixed $2^3S, 2^1S$)-H,D[67,68] and Ne($^3P_{2,0}$)-Ar[69,70],Kr.

3. From Penning electron energy distributions measured with velocity selected metastables in the thermal range the ratio is directly obtained as the ratio of the areas of the distributions defined by conditions (II.4a) and (II.4b), respectively.

All investigated systems are similar in that the ratio of associative PgI: PgI ratio decreases strongly with the collision energy. As a typical example in Fig. 25, the results for the system Ne*–Kr, which has been studied by all three methods mentioned, are shown.

A quantitative evaluation of a measured energy dependence of the ratio has been made only for the system He(2^3S)-Ar for which $V_*(R)$ and $\Gamma(R)$ are known, so that the evaluation leads to a determination of parameters of $V_+(R)$. In classical model calculations,[43] using a semiempirically determined potential $V_*(R)$[71] that only slightly deviates from the one determined from elastic scattering[30] and $\Gamma(R) = 4000 \ \exp(-R/0.36)$ (au), which was determined by the requirements that the total ionization cross-section curve due to Pesnelle et al.[43] be reproduced with the chosen $V_*(R)$, for a Morse potential $V_+(R)$ the following parameter values were determined: well depth 16 meV, equilibrium distance $5.67a_0$, and shape

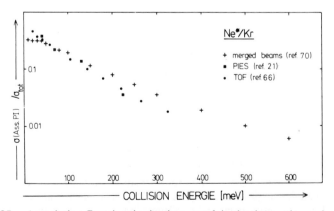

Figure 25. Associative Penning ionization: total ionization ratio as function of collision energy for Ne($^3P_{2,0}$)-Kr. Results obtained by three different experimental methods are compared.

parameter 4.92. Quantum-mechanical calculations on the basis of relations (II.43) and (II.47) and with functions $V_*(R)$ given by (II.61) to (II.64) and $\Gamma(R) = 14 \cdot \exp(-1.9R)$, have also been carried out,[22] and by comparison with the measured[43] energy dependence of the ratio, the parameters of a Morse potential $V_+(R)$ were determined as: well depth 19 meV, equilibrium distance $5.2a_0$, and shape parameter 4.92.

Because of the occurrence of quasibound states it was necessary to calculate the lifetimes of these states and to compute the associative PgI: PgI ratio for times relevant in the actual experiment. For a collision energy of 42 meV, the dependence of the computed "quasi associative" PgI cross section on the time (T) elapsed since electron ejection is shown in Fig. 26. Under typical experimental conditions the associative ion is detected at $T \sim 10^{-5}$ sec. It is seen that at these times the quasibound states contribute only little to the experimentally determined associative PgI cross section of $3a_0^2$ at 42 meV. For comparison with experiment, a computed cross section for $T = 13$ μsec was used.

In both classical and quantum-mechanical calculations, the measured velocity dependence of the associative PgI: PgI cross-section ratio is well reproduced with the assumed functions $V_*(R)$ and $\Gamma(R)$ and the determined function $V_+(R)$.

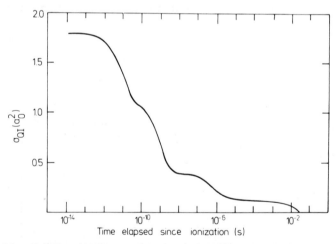

Figure 26. Calculated cross section for formation of "quasibound" diatomics $HeAr^+$ by Penning ionization of Ar by $He(2^3S)$ at 42-meV collision energy, as function of time elapsed between ionization and observation (reproduced from Hickman and Morgner.[22])

6. Comparison of Results for He(2^3S) and He(2^1S)

The helium metastables differ only in the spin state. Thus pronounced differences in the functions $V_*(R)$ and $\Gamma(R)$ for Penning systems with He(2^3S) and He(2^1S) should be attributable to an influence of the spin. A quantitative discussion of this influence lies in the regime of the electronic structure and has not been given as yet. Hence we are limited in this section to stressing some of the observed similarities and differences between the helium metastables regarding PgI and to try to give plausible qualitative explanations.

The cross-section curves of the He*–rare-gas systems showed rather strong differences for the two metastables. However, these differences are caused by rather small differences of the functions $V_*(R)$, which could already be explained by the slightly different electron-density distributions of the metastables and by the different admixture of higher excited states —because of the 0.8-eV higher energy of the singlet metastable—at small separation of the collision partners. A more striking feature of the results for the He*–rare-gas systems is that, in spite of the different spin states, the widths $\Gamma(R)$ are virtually identical at the same distances. This is explained by the exchange mechanism (II.69) of PgI: the magnitude of the exchange matrix element is proportional to the overlap of the spin orbitals of the target electrons to be ionized, with the unfilled $1s$ spin orbital at the excited helium atom. This overlap is identical for He(2^3S) and He(2^1S)—which differ only in the orientation of the $2s$ electron relative to the $1s$ electron—as long as the target atom has no total spin angular momentum, which would, for He(2^3S) and He(2^1S), lead to different orientations of the spins of the target electrons with respect to the spin of the $1s$ spin orbital.

It has been argued that the observed differences between singlet and triplet cross-section curves are evidence of a significant contribution of a radiative-type PgI mechanism in the case of He(2^1S), which—in contrast to the exchange mechanism (II.69)—may be indicated as

$$A^*(1) + B(2) \rightarrow A(1) + B^+ + e^-(2) \qquad (II.82)$$

Since (II.82) implies the transition $A^*(1) \rightarrow A(1)$, which is less strongly forbidden in the case of He(2^1S), a larger contribution to ionization in the He(2^1S) systems would be expected. However, since the main differences between singlet and triplet cross-section curves are already explained by the slightly different potentials $V_*(R)$ determined from elastic scattering data[30,31] and since the internal angular distribution of the ejected PgI electrons are very similar for He(2^3S) and He(2^1S) systems, it seems very unprobable that (II.82) contributes significantly to PgI with helium metastables.

Significant differences are observed for PgI systems with strong attractive interaction in the incoming channel $V_*(R)$. These are usually systems with target atoms having an open shell and correspondingly nonzero spin. As mentioned earlier, differences are also expected for such systems on the basis of the exchange mechanism. Examples are given in Table II, where The well depths of $V_*(R)$ for four atoms with $He(2^3S)$ and $He(2^1S)$ are given. The well depths are determined from analyses of PgI electron energy spectra. For comparison, the well depths of interaction potentials of lithium with the four atoms are also listed.[48] The symmetries of the respective potentials are given in brackets behind the numbers. For $He(2^3S)$ interacting with an atom in a ^2S state, the two potential curves of symmetry $^2\Sigma$ and $^4\Sigma$ arise. Because of the need to conserve total spin in the ionizing transition, PgI from the $^4\Sigma$-curve is forbidden if the final-state potential curve $V_+(R)$ has $^1\Sigma$ symmetry, as is the case for the hydrogen, sodium, and potassium atoms. Therefore, it is the $^2\Sigma$-curve that is determined by analyzing the PgI electron energy spectra. In Table II two facts are noted: (1) the well depths for $Li(^2S)$ and $He(2^3S)$ are very similar; and (2) the well depths for $He(2^1S)$ and $He(2^3S)$ are very different. In case of the ^2S-target atoms this is easily explained.

Regarding Fact 1 in the preceding paragraph, the bonding between lithium and the ^2S atoms in the $^2\Sigma$ molecular state is dominated by the effect of two electrons with antiparallel spin in the bonding orbital formed from the lithium $2s$ and the outer s electron of the target atom. Exactly the same is true for the bonding between $He(2^3S)$ and the ^2S atoms in the $^2\Sigma$-molecular state. This explains the similar well depths for the $^2\Sigma$ potential curves. Analogously, the $^3\Sigma$-potential curves, arising from $Li(^2S)$ and the ^2S atoms, should be rather similar to the $^4\Sigma$ curves of the $He(2^3S)$–^2S systems. Regarding Fact 2, $He(2^1S)$ forms only one $^2\Sigma$-potential curve with ^2S atoms, in contrast to both lithium and $He(2^3S)$, which explains the different well depths. The weaker attraction of the $He(2^1S)$–^2S systems is

Table II.

Potential Well Depths (in electron Volts)[a]

	$Li(^2S)$	$He(2^3S)$	$He(2^1S)$
$H(1^1S)$	$2.52(^2\Sigma)$	$2.26 \pm 0.02(^2\Sigma)$	$0.46 \pm 0.05(^2\Sigma)$
$Na(3^2S)$	$0.93(^1\Sigma)$	$0.73 \pm 0.05(^2\Sigma)$	$0.03 \pm 0.05(^2\Sigma)$
$K(4^2S)$	$0.77(^1\Sigma)$	$0.60 \pm 0.05(^2\Sigma)$	$0.23 \pm 0.05(^2\Sigma)$
$Hg(^1S)$	$0.108(^2\Sigma)$	$0.085 \pm 0.05 \pm 0.05(^3\Sigma)$	$0.49 \pm 0.05(^1\Sigma)$

[a]Values taken from Hotop [48], except for the new value for $He(2^3S)$–H.[51] Well depths of potentials with helium metastables are derived from Penning electron spectra.

probably because configurations corresponding to two paired outer electrons (bonding), as well as to two outer electrons with parallel spin (antibonding), can contribute to the $^2\Sigma$-potential.

For He*–Hg(^1S) systems, the situation is not so clear because for both He(2^3S) and He(2^1S), only one potential curve is formed. No explanation for the considerably deeper well in systems with He(2^1S) has been proposed.

Of course, predictable differences between He(2^1S) and He(2^3S) occur regarding the polarization of Penning ions and electrons caused by total spin conservation. The fact that a component of spin angular momentum is conserved in PgI has been demonstrated by the observation of the transfer of spin polarization from optically pumped He(2^3S) atoms to the Penning ions of cadmium, zinc, and strontium.[72] The polarization was detected by measuring the polarization of the light emitted from the excited $^2D_{5/2}$ ions in $^2D_{5/2} \rightarrow {}^2D_{3/2}$ transitions. If a component of spin angular momentum is conserved, we may write the PgI process as

$$^3S_1(\uparrow\uparrow) + {}^1S_0(\uparrow\downarrow) \rightarrow {}^1S_0(\uparrow\downarrow) + {}^2D_{5/2}(\uparrow) + e^-(\uparrow) \qquad (\text{II.83})$$

Assuming that the ion is formed with equal probability in states of different m_l, the population of the different m_j sublevels of $^2D_{5/2}$ is in the ratios 0:1:2:3:4:5, which gives rise to polarization of the observed light emitted in the $^2D_{5/2} \rightarrow {}^2D_{3/2}$ transition.[72] It may be noted that the exchange mechanism of PgI[11] implies the conservation of a component of spin angular momentum.

Conservation of a component of spin angular momentum in PgI was also demonstrated by showing that PgI electrons formed by ionization of argon are polarized by optically oriented He(2^3S) in a flowing afterglow.[73]

7. Final-state Population

If in a simple PgI system A*–B more than one potential curves $V_{+i}(R)$ lie energetically below the potential $V_*(R)$—corresponding to the condition that the excitation energy of A* is larger than the ionization energies $E_i(B)$ of the target—then the total transition probability function $\Gamma(R)/\hbar$ branches into the individual transition probabilities $\Gamma_i(R)/\hbar$. The electronic branching ratios, defined by

$$a_i(R) = \frac{\Gamma_i(R)}{\Gamma(R)} \; ; \Gamma(R) = \sum_i \Gamma_i(R) \qquad (\text{II.84})$$

will in general depend on R and hence are not directly observable in an experiment as the corresponding cross-section ratio $\sigma_i/\sigma_{\text{total}}$. Instead, the

$\Gamma_i(R)$ have to be extracted from experimental information on the individual transitions, in the same way as that described for a single transition in the previous sections. Such detailed studies have not been made. So far only cross section ratios have been discussed, and it has been tacitly assumed that these ratios are good approximations to the electronic branching ratios. From the classical formulas (II.10) and (II.11) it is evident that this assumption is correct under the following two conditions: (1) the true branching ratios $a_i(R)$ must be nearly constant within the relevant R range, so that we may write

$$\Gamma_i(R) \approx a_i \Gamma(R) \tag{II.85}$$

and (2) in the relevant range of l values the "opacity" P_l given by (II.10) must be small compared to one, so that we may approximate, using (II.85)

$$P_l^i \approx 2 \int_{R_l}^{\infty} \frac{\Gamma_i(R)\,dR}{\hbar v_l(R)} \approx 2a_i \int_{R_l}^{\infty} \frac{\Gamma(R)\,dR}{\hbar v_l(R)} = a_i P_l \tag{II.86}$$

If approximation (II.86) is inserted into cross-section formula (II.11), it follows immediately that

$$\frac{\sigma_i}{\sigma_{\text{total}}} \approx a_i \tag{II.87}$$

Although (II.87) may be used for qualitative discussions, it should be kept in mind that assumptions (1) and (2) are rather severe.

The electronic branching ratios are determined by the electronic matrix elements responsible for PgI. As outlined, two types of matrix elements—corresponding to the mechanisms (II.69) "exchange" and (II.82), "radiative"—could essentially contribute. Photoionization of atom B involves the same type of matrix elements for the B→B⁺ transition as does the "radiative" mechanism (II.82). Therefore, if the "radiative" mechanism were dominant for PgI, we would expect branching ratios very similar to the ones in the photoionization process, and this would be observable in the corresponding electron spectra that directly show cross section ratios and, with (II.87), approximately the branching ratios. Comparisons of electron spectra from helium metastables, with photoelectron spectra obtained with 584-Å ($\hat{=}$ 21.22-eV) radiation, have been made for a large number of target atoms and molecules.[7, 8, 12, 74] It is generally found that the cross-section ratios $\sigma_i/\sigma_{\text{total}}$ are rather different for PgI and photoionization, indicating that the different types of matrix elements, corresponding to (II.69) for PgI and (II.82) for photoionization, are dominant. For example, in Fig. 27 the He*–Hg electron spectrum is compared

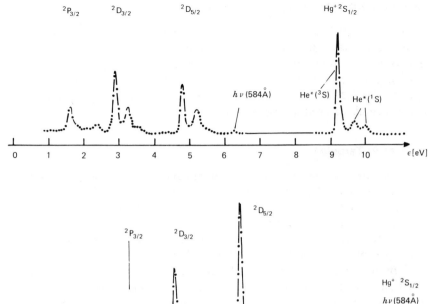

Figure 27. Population of electronic states of Hg^+ by Penning ionization with $He(2^1S)$ and $He(2^3S)$ (upper spectrum), as compared to photoionization with $h\nu$ (584 Å) (lower spectrum) (reproduced from Niehaus.[21])

to the 584-Å–Hg electron spectrum.[21] We recognize that the ionic ground state $He^+(^2S_{1/2})$ is strongly populated in PgI by both, $He(2^1S)$ and $He(2^3S)$, whereas the population by photoionization is rather small. Also, the relative population of the $^2D_{5/2}$ and $^2D_{3/2}$ states is different for PgI and photoionization. The greatest difference arises for the $Hg^+(5d^{10}6p)^2P_{3/2}$ state, whose population involves a two-electron transition.

The available data on branching ratios in PgI are in accord with the following simple rules: (1) only ionic states are populated that can be reached without violating the total spin conservation rule; (2) states that can be reached by a single electron transition are strongly preferred; and (3) a rough relative order of values of branching ratios into states that involve only a single electron rupture is given by the relative overlap (during the collision) of the orbital to be ionized, with the unfilled $1s$ orbital on helium, times the number of electrons in the respective target particle orbitals.[11,21]

III. PENNING IONIZATION—COMPLICATIONS

Many real PgI systems deviate from the somewhat ideal "simple systems" as we have defined them in Section II.A. As we had to introduce many restrictions in their definition, there are also many possibilities for real systems that deviate from this definition. We cannot deal with all these possibilities but will rather concentrate on some selected aspects for which experimental material has been reported.

A. Atomic Targets

If the entry channel adiabatic potential curve $V_*(R)$ is not well isolated from other curves $V'_*(R)$, transitions to these curves may arise as a result of the finite relative velocity of the colliding particles. The theory of PgI as described is then inapplicable without appropriate modification. Close separations between $V_*(R)$ and some other $V'(R)$ arise at "avoided crossings" of the adiabatic curves. Since $V_*(R)$ is just one special curve of an infinite number of curves $V'_*(R)$ of the diatomic system, it is only a question of collision energy whether the system penetrates to the critical distances. Good examples of the gradual change—from simple PgI systems to systems complicated by curve crossings—as the collision energy is increased are systems A*–B with A* = He* and B = Ar,Kr,Xe.[75,76] Let us denote by B** rare-gas atoms in low-lying autoionizing states of the type B** ($nsnp^6nl$) or B** ($ns^2np^4nl,n'l'$) and by $V'_*(\text{He–B**})$ potential curves dissociating into He + B**. Because of the rather small energy difference between the asymptotic energy of the entry channel curve $V_*(\text{He–B})$ and the curve $V'_*(\text{He–B**})$, curve crossings at which transitions $V_* \rightarrow V'_*$ can occur become important already at rather low collision energy. The ionization mechanism is then a mixture of PgI, and autoionization of the rare-gas atom (AAI):

$$\text{He*} + \text{B} \rightarrow (\text{He*} - \text{B}) \overset{(\text{PgI})}{\rightarrow} \text{He} + \text{B}^+ + e^- (P_{\text{PgI}}(\varepsilon))$$

$$(\text{He} - \text{B**}) \rightarrow \text{He} + \quad \text{B**}$$

$$\downarrow (\text{AAI})$$

$$\text{B}^+ + e^- (\varepsilon_i = E^i_{**} - E(\text{B}))$$

$$(\text{III.1})$$

where PgI leads to broad distributions $P_{\text{PgI}}(\varepsilon)$ that become even broader when the collision energy is increased, whereas AAI gives rise to narrow lines that become narrower with increased collision energy and approach the positions $\varepsilon_i = E^i_{**} - E(\text{B})$, with E^i_{**} the excitation energy of the

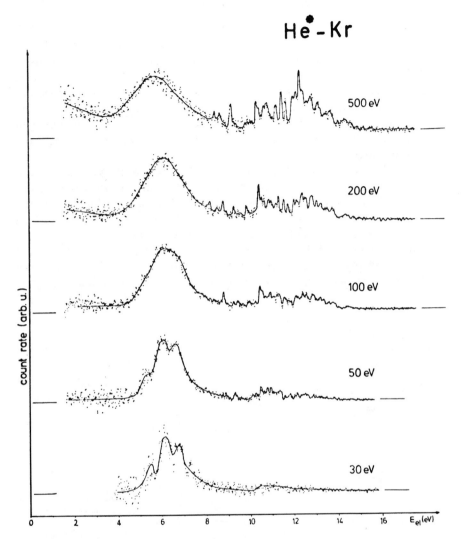

Figure 28. Electron spectrum for collision system He*–Kr at various collision energies. Broad distribution at low electron energies is a result of Penning ionization, and narrow peaks arise from atomic autoionization of krypton following excitation transfer from He* to Kr.[77]

461

individual electronic states of B** and E(B) the ionization energy of B. Therefore, in the electron spectrum the contributions from PgI and AAI can be distinguished, and the relative importance of these mechanisms estimated. As an example, spectra for the system He*–Kr at different collision energies are shown in Fig. 28. The measurements were carried out with a He* beam, formed by passage of He$^+$ through cesium vapor, which contained about 95% He(2^3S) metastables.[77] It is clearly seen that with increased collision energy the fraction of total ionization resulting from AAI increases.

The electrons are detected at $\vartheta = 90°$ with respect to the projectile beam. Neglecting the possibility of different average angular distributions for PgI and AAI, the ratio of integrated electron intensity belonging in the measured spectra to AAI, divided by the total integrated observed electron intensity, is equal to the fraction of ionization caused by diabatic $V_* \rightarrow V'_*$ transitions followed by AAI. This ratio is plotted in Fig. 29 against collision energy for the systems He*–Ar,Kr,Xe. The fact that the fraction of AAI increases from argon to xenon at a certain collision energy, is

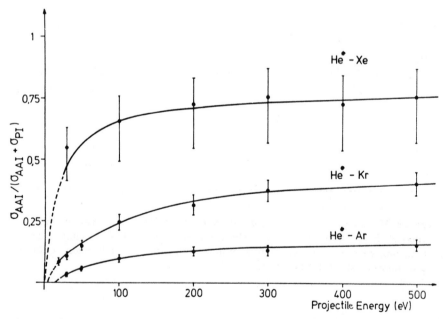

Figure 29. Ratio of cross section for excitation transfer followed by atomic autoionization (AAI), to total ionization cross section, as function of collision energy for systems He*–Ar, Kr, Xe.[77]

explained by the condition that the asymptotic energy gap between V_* and the curves V'_* [given by $E_{**} - E_*(\text{He*})$] decreases from argon to xenon. Usually a smaller energy gap leads to larger critical distances of the avoided crossings and this, in turn, to larger cross sections.

The fact that the AAI lines are narrow even at low collision energies proves that the AAI lifetimes are long compared to the collision time. This is quite in contrast to the average PgI lifetimes during collision, which are of the order of the collision time as was shown, for example, by analysis of the total ionization cross-section curves (Section II.B2.).

In principle, excitation transfer at curve crossings may also occur at thermal collision energies, but only under rather restricted conditions, because at low collision energies the system will usually follow the adiabatic curve $V_*(R)$. The following two exceptions may arise:

1. An avoided curve crossing exists, with an energy separation between V_* and V'_* that is only of the order of thermal energies and hence allows diabatic transitions with appreciable probability. Such a special crossing will usually arise only at large distances—because at small distances the $V_* - V'_*$ separations are usually larger—and thus an accidental very near resonance between E_* of the projectile and the excitation energy $E'_*(\text{B})$ of the target is necessary.

2. A strong chemical attraction of $V_*(R)$ causes a sufficient increase of instantaneous relative velocity, so that even at small separations a diabatic transition in the attractive part of $V_*(R)$ is possible in thermal collisions. A preliminary evaluation of PgI electron spectra of the system $\text{He}(2^3S)/\text{O}(^3P)$ at thermal energy[78] indicates that a considerable fraction of the total observed ionization is caused by the autoionization $\text{O}^{**}[2s2p^3(^2D)nl] \rightarrow \text{O}^+(2s^22p^2)^4S$, with the excited O^{**} atoms formed in a diabatic transition $V_*(\text{He*-O}) \rightarrow V'_*(\text{He-O}^{**})$ in the attractive part of $V_*(R)$. An interesting aspect of this proposed mechanism is that the large amount of energy ($\sim$3 eV) is transformed from electronic to relative kinetic energy in the exit channel $V'_*(R)$.

B. Molecular Targets

Obvious complications as compared to simple systems arise in the case of molecular targets. The potentials $V_*(R)$ and $V_+(R)$ and also the width $\Gamma(R)$ depend on the relative orientation of the molecule to the metastable–target direction. In addition, there is the possibility of vibrational and rotational transitions in the ionization process as well as the possibility of reaction before and after the ionization. No general extension of the PgI theory to the case of molecular targets has been formulated.

Studies of the PgI electron spectra have shown that, in spite of all the mentioned complications, many PgI systems with molecular targets can still be well described within the theory of simple PgI, if only the possibility of vibrational transitions is incorporated into the function $\Gamma(R)$. Within the Born–Oppenheimer approximation for both the projectile–target motion and the intramolecular motion, this is done in the following way. We denote by $\Gamma_i(R)$ the width belonging to a certain final electronic state, defined as in (II.85). Then $\Gamma_i(R)$ can, at any distance R, be decomposed as

$$\Gamma_{iv}(R) = b_v^i(R)\Gamma_i(R), \quad \Gamma_i(R) = \sum_v \Gamma_{iv}(R) \qquad \text{(III.2)}$$

The branching ratios b_v^i are proportional to the Franck–Condon factors for the molecular ionization transition, with the projectile at distance R from the molecule. The $b_v^i(R)$ depend parametrically on R. For the $b_v^i(R)$ to be well defined in an actual collision with velocity $\dot{R}$, it must be true that $\dot{R}$ is small compared to the vibrational intramolecular motion. This is a severe limitation, even at thermal velocities. Therefore, only when the interaction between the projectile and the molecule is weak, leading to a correspondingly weak variation of the $b_v^i(R)$ in the course of the collision, approximation (III.2) may be used. We may then further approximate by introducing an average branching ration defined by

$$\Gamma_{iv}(R) \approx b_i(v)\Gamma_i(R) \qquad \text{(III.3)}$$

With the classical relations (II.10) and (II.11) it follows then from (III.2) that

$$\frac{\sigma_i(v)}{\sigma_i} \approx b_i(v) \qquad \text{(III.4)}$$

In this approximation the average branching ratios—or the average Franck–Condon factors—can be taken directly from the observed electron spectra as the appropriately normalized areas below peaks $P_i^v(\varepsilon)$ belonging to individual vibrational states of a certain electronic state.

The strength of the interaction in the course of a collision leading to population of vibrational states of a certain electronic state can be judged from the widths of the individual vibrational lines in a measured spectrum. If the lines are narrow and well separated, the average interaction is weak, and we can expect that the approximations leading to (III.4) are valid. In all these cases, however, it is also expected that the $b_i(v)$ deviate very little from the Franck–Condon factors for the same ionizing transition in the unperturbed molecule caused by photoionization.

A great number of excited metastable–molecule systems have been studied by comparison of PgI electron spectra with photoelectron spectra using the resonance photons of the rare gas whose energy is nearly equal to the excitation energy of the metastables.[7,74,79,80] In this way the PgI electrons have energy very similar to that of the photoelectrons, so that the vibrational distributions for the two processes can be directly compared and deviations be attributed to the effect of perturbation of the molecule by the projectile particle. In the majority of cases the diatomic molecules form weak interaction systems with He metastables and show vibrational populations of the Penning ionized molecule that are very similar to the populations arising from photoionization, as predicted on the basis of the theoretical considerations. Figure 30 shows as an example the $He(2^3S)/N_2$

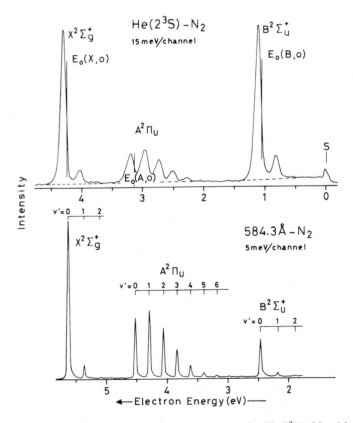

Figure 30. Comparison of Penning electron spectrum for $He(2^3S)$–N_2 with photo-electron spectrum for $h\upsilon$ (584 Å)–N_2.[81]

PgI and the 584-Å/N_2 photoionization electron spectra.[81] For the three different electronic final states, energetically accessible in the PgI transition from the single entry channel potential curve, rather narrow unshifted individual distributions and vibrational populations very similar to those for 584-Å photoionization are observed. In Fig. 31 the population factors —differently normalized $b_i(v)$—for PgI and photoionization are compared for some systems with well-resolved vibrational lines.[48, 74]

In the case of polyatomic molecules it is found that PgI electron spectra are usually unresolved and shifted.[79, 80] Population factors can be only crudely estimated. The observed shifts with respect to the nominal energy are usually towards lower electron energy, which indicates that the attraction in the entry channel is stronger than in the exit channel after ionization. It is further observed that the shifts increase as the dipole moment of the molecule increases and that they are considerably larger for $He(2^1S)$ than for $He(2^3S)$.[81] It seems that most of the observed features for polyatomic molecules can still be explained within the same picture valid for the weak interaction systems, by allowing for broader individual distributions caused by stronger interaction and for a correspondingly stronger average distortion of the molecule during the collision, which, in turn, leads to changed average $b_i(v)$.

Excitation transfer followed by autoionization of the target—as discussed in Section III.A for atomic targets—is, of course, also a possibility in PgI systems with molecules. Indeed, since molecular excited states usually are more densely spaced, the existence of crossings between $V_*(R)$ and some $V'_*(R)$ at large separation will arise more often, so that at thermal-collision energies the possibility of excitation transfer is expected to be given more often than in the atomic case. The occurrence of this process should be detectable in the electron energy spectrum by the following two features: (1) the vibrational population factors should deviate drastically from those for photoionization from the ground state; and (2) at the same time the individual vibrational lines should be narrow. As outlined earlier, deviations of the $b_i(v)$ from the corresponding ones for photoionization can arise in PgI only because of perturbation by the projectile, which is always accompanied by broad distributions. Only in one case does an observed PgI spectrum clearly show the combined features of (1) and (2), the $Ne(^3P_{2,0})-O_2$ spectrum[82] for population of the ground state of O_2^+. The spectrum is reproduced in Fig. 32. It consists of narrow lines with an envelope that differs drastically from that for photoionization from the ground state of O_2, and, in addition, the lines are unshifted with respect to the nominal energy. Thus the spectrum is attributed to a quasiresonant excitation transfer to an autoionizing neutral state of O_2, followed by autoionization into $O_2^+(X^2\Pi_g)$ after the collision when the O_2 is no longer perturbed by the neon atom.

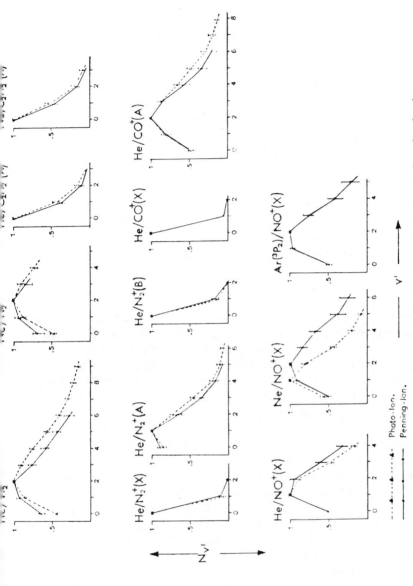

FIGURE 31. Comparison of population of vibrational levels in denoted molecular ion states as produced in Penning ionization by $He(2^3S)$, $Ne(^3P_{2,0})$, or $Ar(^3P_{2,0})$ and in photo-ionization by helium (584 Å) or neon (736 Å) resonance photons.[48]

467

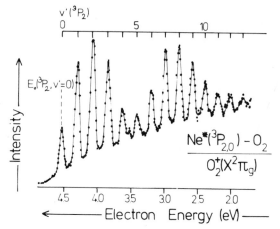

Figure 32. Electron spectra showing population of vibrational levels of $O_2^+(X^2\Pi_g)$ in Penning ionization by $Ne(^3P_{2,0})$ metastables.

Penning ionization electron spectra reflect the average instantaneous population of vibrational states of the molecular ion formed. Immediately after ionization, however, the vibrating molecule is still close to the projectile particle and can, as in a normal collision, react with it or can transfer vibrational energy to the energy of relative motion. Therefore, the vibrational population observed by PgI electron spectra is different generally from the population observed after the completed collision. The latter is observed by measuring light emission from molecular Penning ions. This method is known as PgI optical spectroscopy (PgIOS).[83] A good example of a case where, as a result of interaction in the "second half" of the Penning collision, very different vibrational populations are observed by PgIES and PgIOS is the system $He(2^3S)/HCl$, which has been studied by both methods.[84,85,86] In Fig. 33 the population factors observed for ionization into $HCl^+(A^2\Sigma)$ by the two methods are compared. Obviously, the molecular ion, initially in the PgI transition "Franck–Condon-like" vibrationally excited, loses a large part of its vibrational energy in the "second half" of the Penning collision.

Evidence of interactions in the "second half" of the PgI collision is also provided by the observation of reaction products containing the projectile atom. For example, consider the $He^*–H_2$ system. From Fig. 31 it is seen that the initial population of H_2^+ is "Franck–Condon-like." Hence it can be concluded that, on the average, at the instant of electron ejection, the

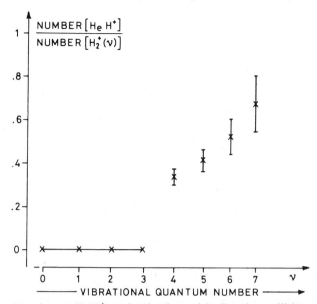

Figure 33. Number of HeH^+ molecules formed in Penning collision of $He(2^3S)$ with H_2, divided by number of H_2^+ molecules formed in certain vibrational states (ν) in initial ionization step. To make data comparable to results from collision experiments of $H_2^+(\nu)$ with helium, they are normalized to number of $H_2^+(\nu)$ ions for which He–$H_2^+(\nu)$ system is either bound (18% of all), or "on the way in" (see text, 41%).

H–H and the H–H$^+$ potential curves are only negligibly perturbed by He* and He, respectively. This means, of course, that no reactions in which reactions products containing helium are formed occur prior to the ionization. On the other hand, it is found in mass-spectrometric studies that, in addition to H_2^+, the products HeH_2^+ and HeH^+ are formed.[9,87,88] These products can only arise from reactions, in the second half of the (PI)-collision, between the vibrationally excited H_2^+ and He. We may indicate these conditions by the following scheme:

$$He(2^3S) + H_2 \rightarrow \left(He - H_2^{+\ddagger}\right) + e^- \left[\, P_v(\varepsilon) \,\right] \tag{III.5}$$

$$
\begin{vmatrix}
\rightarrow He + H_2^{+\ddagger} \\
\rightarrow HeH_2^+ \\
\rightarrow HeH^+ + H
\end{vmatrix}
$$

The second step may be viewed as the collision of the vibrationally excited hydrogen molecular ion $[H_2^{+\ddagger}(v)]$ with helium at collision energies determined by $P_v(\varepsilon)$ via energy conservation (II.5). The collision energy distribution $P[E_k'(\infty)]$ is independent of v. The measured electron energy distributions $P_v(\varepsilon)$ show that the individual peaks are shifted with respect to their nominal energy by about 70 mV toward higher energies.[74] Since this shift is of the order of the average collision energy, it is probable that a considerable fraction of the systems initially formed have negative energy $E_k'(\infty)$ and are thus initially bound. These systems can only dissociate by transforming vibrational energy of $H_2^{+\ddagger}$ into relative H_2^+–He kinetic energy. If no vibrational energy is available (i.e., for $v = 0$), stable HeH_2^+ is formed. Therefore, from the measured fraction of ionization processes leading to HeH_2^+ formation[89] and from the measured fraction[74] of ionization processes leading to initial population of the $v = 0$ states of H_2^+, we can calculate the fraction of PgI processes leading to initially bound systems.

With the fractions $f(HeH_2^+) = 2\%$, measured for thermal-energy distribution at approximately $380°K$, and $f(v = 0) = 13\%$, obtained from the PgI electron spectrum, we calculate that about 15% of the initially formed He–$H_2^{+\ddagger}$ systems are bound. Systems with H_2^+ formed initially in $v \geqslant 4$ states can either lead to HeH^+ or to H_2^+ formation. The branching into these channels will depend on v and on the initial state of relative motion, especially on whether the system is initially bound. An average branching ratio, which neglects these dependencies, can again be directly calculated from the measured fraction of ionization processes leading to HeH^+ and from the fraction of processes leading to vibrational states with $v \geqslant 4$. We then obtain $b(HeH^+) = 33\%$. For the vibrationally excited $H_2^{+\ddagger}$ to react with helium to give HeH^+, the He–$H_2^{+\ddagger}$ distance must become small in the second half of the PgI collision. For the 85% initially unbound systems this arises only in about 50% of the cases, when the ionization occurs "on the way in." Therefore, the branching ratio is zero for about 42.5% of the initially unbound systems.

Taking this into account, we obtain for the unbound system "on the way in" and for the bou. d systems an average branching ratio of approximately 60%. Since for the bound systems the branching ratio is probably higher, we may differentiate and state that, for the initially bound systems with ($v \geqslant 4$), the branching ratio can vary in the limits $60\% \leqslant b_b(HeH^+) \leqslant 100\%$ and for the unbound systems "on the way in" between the limits $60\% \geqslant b_u(HeH^+) \geqslant 40\%$. The combined information obtained from PgI electron spectroscopy and mass spectrometry on the $He(2^3S)$–H_2 system at thermal energies ($\sim 380°K$) may be summarized in the following detailed scheme:

$$He(2^3S) + H_2$$

	43% → $He-H_2^+(v=0)$	100% → HeH_2^+
$\dfrac{0.15 \cdot \sigma_{total}}{\text{(bound)}}$	54% → $He-H_2^{+\ddagger}(0<v<4)$	100% → $He + H_2^{+\ddagger}(v<3)$
	33% → $He-H_2^{+\ddagger}(v \geqslant 4)$	$\begin{cases} \text{40-0\%} \to He + H_2^{+\ddagger}(v) \\ \text{60-100\%} \to HeH^+ + H \end{cases}$
	13% → $He-H_2^+(v=0)$	100% → $He + H_2^+(v=0)$
$\dfrac{0.425 \cdot \sigma_{tot}}{\text{(unbound "in")}}$	54% → $He-H_2^{+\ddagger}(0<v<4)$	100% → $He + H_2^{+\ddagger}(0 \leqslant v<4)$
	33% → $He-H_2^{+\ddagger}(v \geqslant 4)$	$\begin{cases} \text{40-60\%} \to He + H_2^+(v) \\ \text{60-40\%} \to HeH^+ + H \end{cases}$
	13% → $He-H_2^+(v=0)$	100% → $He + H_2^+(v=0)$
$\dfrac{0.425 \cdot \sigma_{total}}{\text{(unbound "out")}}$	54% → $He-H_2^{+\ddagger}(0<v<4)$	100% → $He + H_2^{+\ddagger}(0<v<4)$
	33% → $He-H_2^{+\ddagger}(v \geqslant 4)$	100% → $He + H_2^{+\ddagger}(v \geqslant 4)$

$$(\text{III.6})$$

Recently the two-step mechanism (III.5) and scheme (III.6) have been proven correct in a direct way by coincidence measurements.[90] The fragment ions H_2^+, HeH^+ and HeH_2^+ were measured in coincidence with the Penning electrons. In this way separate electron spectra were obtained for the different fragment ions, from which branching ratios for the ion–molecule reaction as a function of the vibrational quantum number of the H_2^+ formed by PgI can be obtained. It is found that HeH_2^+ is, indeed, only formed from $H_2^+(v=0)$, as proposed in (III.6), and that the fraction of ionizations leading to initially bound $He-H_2^{+\ddagger}$ systems is 0.18, to be compared with 0.15 of (III.6). Further, it is found that HeH^+ is only formed from $H_2^{+\ddagger}$ $(v \geqslant 4)$ and that the average fraction of $H_2^{+\ddagger}$ $(v \geqslant 4)$ leading to HeH^+ formation is 0.37. For "unbound in" systems and bound systems, this leads to an average branching ratio of 0.63, in good agreement with (III.6). Most importantly, the variation of this branching ratio with v is also obtained (see Fig. 33).

IV. OTHER SPONTANEOUS IONIZATION MECHANISMS

A. True Associative Ionization

In contrast to the associative process arising in PgI systems when the ejected electron carries away so much energy that the collision partners cannot separate, we deal in this section with ionization processes that can only occur because energy is gained by formation of a bond between the collision partners after ionization. Such true associative ionization (AI) systems are characterized by the energy conditions

$$V_+(\infty) - D_+ - E_k(\infty) < V_*(\infty) < V_+(\infty) \tag{IV.1}$$

where D_+ is the well depth of the potential curve of the ionized system $V_+(R)$.

The first AI process ever observed was

$$Hg(^3P_1) + Hg(^3P_1) \rightarrow Hg_2^+ + e^- \tag{IV.2}$$

In mercury vapor irradiated by the 2537-Å mercury resonance line, ionization was observed[91] and later attributed[92] to the AI process (IV.2). Since then AI was reported for a large number of systems.[93] Especially the homonuclear AI, also called *Hornbeck–Molnar process*,[94]

$$A^* + A \rightarrow A_2^+ + e^- \tag{IV.3}$$

was studied intensively regarding the excited states of A.

True associative ionization may also occur if two ground-state atoms or molecules collide. The energy condition (IV.1) must then hold for the ground-state potential curve $V(R)$; that is, for low collision energies, the ionization energy of one of the particles must be smaller than the binding energy D_+ in the ionized system. Many such systems exist. Unique identification of such AI processes, which are usually termed *chemiionization reactions*, and whose importance in hydrocarbon–oxygen flames has been recognized for a long time, was only rather recently reported for several metal–oxygen systems.[95] Examples are

$$U + O \rightarrow UO^+ + e^- \qquad (IV.4)$$

$$U + O_2 \rightarrow UO_i^+ + e^- \qquad (IV.5)$$

which are exoergic by 2.0 and 4.1 eV, respectively.

Two different mechanisms have been considered to explain the occurrence of AI.[6,96,97,98] The potential curves corresponding to these mechanisms are shown in Fig. 34 for the case of atomic collision partners. The mechanism of Fig. 34a is characterized by a potential curve $V_*(R)$ that at all accessible distances R lies below $V_+(R)$; therefore, $V_*(R)$ is real and ionization can only occur by diabatic coupling. An electron ejected in such an ionization has to get its kinetic energy from the heavy-particle motion, which wants to retain its instantaneous state. Hence the ejected electron spectrum will have considerable intensity only close to zero energy in thermal collisions. The mechanism shown in Fig. 34b is characterized by a potential curve $V_*(R)$ that crosses into the continuum at some distance R_c. In this case $\tilde{V}_*(R)$ is for $R < R_c$ complex with an imaginary part, $i\Gamma(R)/2$, so that the system can decay into the continuum in an electronic transition of probability $\Gamma(R)/\hbar$, as in the case of PgI. For $R < R_c$, therefore, the PgI theory can be applied with the appropriate modifications.[97] In contrast to the mechanism shown in Fig. 34a, here the energy of the ejected electrons is not limited to small values. Hence the two mechanisms may generally be distinguished by their respective electron spectra.

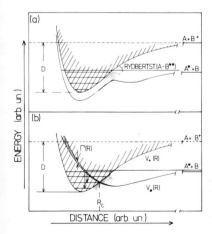

Figure 34. Two possible mechanisms for true associative ionization: (a) adiabatic path and ionization by dynamic coupling; (b) diabatic path and ionization by electronic coupling.

Until recently, experimental studies of AI were limited to the identification of the process and, in some cases, to the determinations of cross sections or rate constants. It was not possible to draw definite conclusions from this experimental information regarding the involved mechanisms. In recent studies of the AI systems R*–H, with R* = Ar($^3P_{2,0}$), Kr($^3P_{2,0}$), Xe($^3P_{2,0}$), it was verified by electron spectroscopy that the mechanism of Fig. 34b is dominant for these systems.[99–101] In all three cases the observed electron spectra extended to the rather high energies of ε_{max} = 1.45 eV (Ar), = 1.0 eV (Kr), and 1.2 eV (Xe) and showed structure resulting from population of different vibrational rotational states as expected for the mechanism of Fig. 34b. As an example, the AI electron spectrum for Ar($^3P_{2,0}$)–H is shown in Fig. 35.

By carrying out model calculations on the basis of the PgI theory and comparing the resulting calculated spectrum with the observed one, quantitative information on $V_*(R)$ and $V_+(R)$ for $R < R_c$ can be obtained, as in the case of PgI. The potential curves $V_+(R)$ are known for the systems R–H$^+$ from elastic scattering data,[53] so that rather direct information on $\tilde{V}_*(R)$ can be obtained for the R*–H systems. In a preliminary model calculation of the quantum-mechanical version[101] it was found that the measured Ar*–H spectrum can be well reproduced.

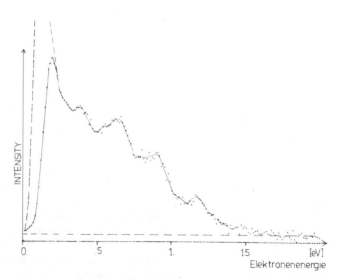

Figure 35. Electron spectrum for thermal collision system Ar($^3P_{2, 0}$)–H. Dashed line at low energy indicates estimated transmission corrected intensity.[99, 100]

The turning point for central collisions, R_0, at $V_*(R)$ can be directly determined from the observed maximum electron energy, ε_{max}. From the potential curve model of Fig. 34b it can be seen that

$$\varepsilon_{max} = E_k(\infty) - \Delta + V_+(\infty) - V_+(R_0) \tag{IV.6}$$

with $\Delta = V_+(\infty) - V_*(\infty)$. Since $V_+(R)$ is known, R_0 can be calculated from (IV.6). We then obtain $R_0 = 3.2a_0$ (Ar), $3.2a_0$ (Kr), and $3.7a_0$ (Xe), with a possible error of $\sim 0.1a_0$. The metastables are in triplet states that form doublet and quartet potential curves with hydrogen atoms. The information extracted from the electron spectra concerns only the doublet curves because ionization from the quartet curves to the singlet curves $V_+(R\text{–}H^+)$ is spin forbidden. The metastable beam formed in a discharge contained metastables in the states $^3P_2$ and $^3P_0$ at a ratio of approximately 5:1. Thus the electron spectra contain contributions at approximately this ratio from the doublet curves dissociating into $R(^3P_2) - H(^2S)$ and $R(^3P_0) - H(^2S)$. The observed Ar*–H spectrum shows indeed reproducibly a double structure consistent with the assumption that it is composed of two contributions that arise from vibrational population of the same final states with the same relative probabilities from two initial curves that run approximately parallel and are spaced by approximately the asymptotic energy difference between the $^3P_2$ and $^3P_0$ metastables. Thus relation (IV.6) yields the same turning-point values for the two potential curves. The same is true for krypton and xenon.

The main result of these first detailed experimental studies of the AI mechanism is that, even at thermal velocity, the system R*–H follows a potential curve $V_*(R)$ that diabatically crosses an infinite number of potential curves dissociating into $R + H^{**}$, H^{**} being hydrogen in a Rydberg state, and continues to be well defined in the ionization continuum. We might mention that results on AI are also interesting because AI is the reverse reaction to the important process of dissociative recombination of molecular ions.

B. Spontaneous Ionization by Electron Transfer

The adiabatic potential curve $\tilde{V}_*(R)$ for a collision of an ion in its ground state and an atom or molecule in its ground state can lie in the ionization continuum of one or more states of that system so that, similar to the case of PgI systems, spontaneous decay may occur into the continuum. Including the possibility of multiply charged ions as projectiles the process may be schematically described as

$$A^{n+} + B \rightarrow A^{(n-1)+} + B^{++} + e^- \tag{IV.7}$$

Since the energy needed for electron ejection is provided by transfer of an electron from the target to the projectile ion, the term *transfer ionization* (TI)[101] has been proposed for this type of process. The energy requirement for TI to be energetically possible without the use of relative collision energy is given by

$$RE > E_1 + E_2 \qquad \qquad \text{(IV.8)}$$

where RE is the highest recombination energy of A^{n+} and E_1 and E_2 are the energies needed to remove the first and the second electron, respectively, from the target B. There exists a great number of systems for which (IV.8) holds, even for the case of singly charged ions A^+. Rather little attention has been paid to TI in the past. The first evidence for its occurrence was found in measurements of total ionization cross sections at high collision energies (3 to 30 keV) for the systems $Ne^{n+}-Xe$ and $Xe^{n+}-Ne$.[103] The results were later analyzed and the possible spontaneous ionization mechanisms discussed in a theoretical paper.[104] Latush and Sem[105] found TI to be an important source of doubly ionized impurity atoms in gas discharges, and Arrathoon et al.[106] reported the first total ionization cross sections for thermal He^+-Ba,Mg,Ca collisions in a flowing afterglow. Kulander and Dahler[102] applied the classical theory of (PGI) to a theoretical study of the He^+-Mg system. Some indirect information on (TI) for the systems $He^{++}-Ne,Ar,Kr$ was obtained by Austin et al.[107] from differential scattering at collision energies of 8 to 60 eV. These experimental data have not been sufficiently detailed as to allow definite conclusions regarding the specific mechanisms of TI. As in the case of PgI and AI, such detailed data have recently been obtained also for TI by the method of electron spectroscopy. Results for the systems $He^{++}-Hg,Xe$[76,108,109] and He^+-Ba,Ca[100,110] and preliminary results on He^+-Mg are available.[111] Results for He^+-Ba,Ca and $He^{++}-Hg$ are discussed in the following paragraphs.

Electron spectra measured for the He^+-Ba system at various collision energies are shown in Fig. 36. A large number of narrow lines is observed, which do not broaden as the collision energy is increased. This can only be explained if the responsible ionization mechanism is autoionization of the barium particle at rather large distances. Therefore, the obvious mechanism in this case is as follows: (1) at distances where the imaginary part of $\tilde{V}_*(R)$ is still small, diabatic transitions $\tilde{V}_*(R) \rightarrow \tilde{V}'_*(R)$ occur at avoided crossings, where $V'_*(R)$ are potential curves that dissociate into $He + Ba^{+*}$ and whose imaginary part is small enough within the R range covered during the entire collision that the molecular autoionization probability per collision is small compared to one; and (2) at large distances between He and Ba^{+*} the Ba^{+*} autoionize. From the energy of the

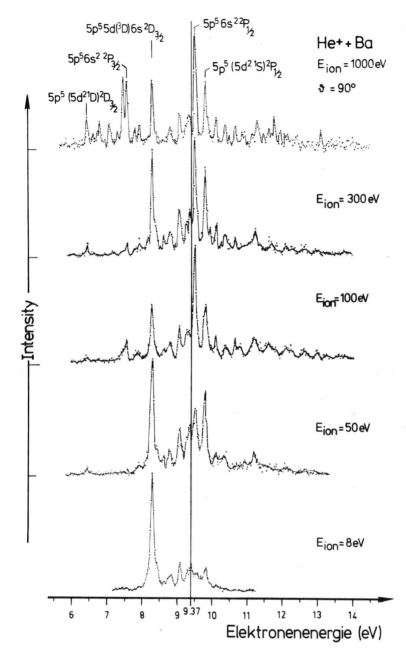

Figure 36. Electron spectrum for transfer ionization system He$^+$–Ba at various collision energies.[110] Indicated a tentative identification of some autoionizing states of Ba$^+$ formed by charge transfer to He$^+$.

observed peaks—given by the relation $\varepsilon_k = E_*^k - E_2$—we can conclude that the excited barium ions are formed by transfer of a $5p$ electron to He^+. Thus the whole TI process may be schematically described as

$$He^+ + Ba(5p^6 6s^2) \rightarrow He + Ba^{+*}(5p^5 nl.n'l')$$
$$\qquad\qquad\qquad\qquad\quad \downarrow$$
$$\qquad\qquad\qquad\qquad\quad \longrightarrow B^{++}(5p^6) + e^-(\varepsilon_k) \qquad\qquad \text{(IV.9)}$$

The nominal electron energy, defined as in the case of PgI, is given by $\varepsilon_0 = RE(He^+) - E_1(Ba) - E_2(Ba) = 9.37$ eV. The fact that the observed spectrum centers around this value at low collision energies reflects the fact that at low collision energies only curve crossings lying in energy close to $V_*(\infty)$ are accessible. The behavior of the peak at 8.28 eV indicates that the curve crossing responsible for the excitation of the corresponding state lies in the attractive part of $V_*(R)$. Since below the 8.28-eV peak no other peak is observed at low collision energies, this curve crossing is the lowest lying accessible within the well of $V_*(R)$. Assuming that the potential curve $V'_*(R)$, to which the transition in the bottom of the well of $V_*(R)$ occurs, has at the corresponding distance still approximately its asymptotic value, we can estimate the well depth of $V_*(R)$ by $D \approx (9.37 - 8.28)$ eV = 1.1 eV. Figure 37 shows the collision energy dependence of the 8.28-eV peak area. Also shown is the result of a Landau–Zener calculation for a curve crossing positioned 1.1 eV below $V_*(\infty)$. The agreement with the shape of the experimental fractional cross-section curve is evidence of the validity of the interpretation.

Electron spectra for the He^+–Ca system at various collision energies are presented in Fig. 38. These spectra consist of just one broad distribution whose width increases with increasing collision energy. This behavior proves that in this case the autoionization occurs in the quasimolecule

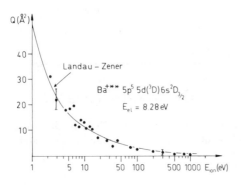

Figure 37. Variation with collision energy of cross section, leading by autoionization to main electron peak at 8.25 eV. Solid line represents result of Landau–Zener calculation (see text).

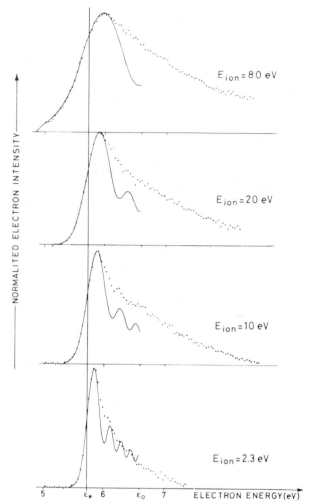

Figure 38. Electron spectra for transfer ionization system He^+-Ca at different collision energies. Nominal energy ε_0 and position ε_* are indicated.[110] Solid line: result of semiclassical calculations with model functions V_* and Γ.

during the collision, as in PgI. The distributions may thus be analyzed using the PgI theory. The semiclassical formulas mentioned in the theoretical section were applied; the simple expression (II.56) obviously fails to describe the low-energy peak of the spectra correctly in that it predicts that the intensity at ε_* should be 44% of the peak value independent of collision energy, whereas it is observed that $\sigma(\varepsilon_*)/\sigma(\varepsilon_p)$ varies. Expression (II.59) describes the variation of $\sigma(\varepsilon_*)/\sigma(\varepsilon_p)$ correctly, but the shape of the low-energy edge is not well reproduced for the 80-eV spectrum. The formula derived by Hultzsch et al.[26] yields the solid lines shown in Fig. 38. It is seen that the low-energy edge is well described in the whole collision energy range. In addition, the slightly indicated oscillations in the observed spectra are reproduced. The fit to the experimental spectra yields parameters of the used model functions $V_*(R)$ and $\Gamma(R)$. The striking difference between the He^+-Ba and the He^+-Ca systems comes from the fact that the transfer of a $4p$ electron of Ca to He^+ is energetically not possible.

Additional interesting information on the He^+-Ca TI mechanism is obtained from the angular dependence of the energy position of the electron distribution at high collision energies. As in the case of PgI, a rather large angular-dependent Doppler shift is observed, from which a ratio $\bar{s}/\bar{b} \approx 1.2$ is derived using relation (II.77). This clearly shows that, as in the case of PgI, the TI-mechanism for the He^+-Ca system is also characterized by ejection of the electron from the helium part of the collision system. This seems at the first sight surprising because it is the calcium that loses two electrons in the TI process. The following explanation is proposed. The adiabatic potential curve $V_*(R)$, corresponding at large R to the state He^+-Ca, changes character at an avoided crossing with the state $He^*(2^3S)-Ca^+(^2S)$, and electron ejection occurs from the curve with this character at smaller separations. Regarding the involved electronic matrix elements, this spontaneous ionization is thus a PgI of Ca^+ by the metastable $He(2^3S)$, of which we know that it occurs by electron exchange and implies ejection of the electron from the helium of the diatomic system. Thus the TI mechanism in the He^+-Ca system may schematically described as

$$He^+ + Ca(1,2) \frac{(\text{adiabatic})}{(\text{rearrangement})} He(1)2^3S + Ca^+(2)$$

$$\times \frac{(\text{PgI})}{} He(2) + Ca^{++} + e^-(1) \qquad (IV.10)$$

The arabics denote the two outer s electrons of calcium, and the first arrow indicates the adiabatic rearrangement at the avoided crossing. The potential curve diagram resulting from the described analysis is shown in Fig. 39.

The TI mechanism for the system $He^{++}-Hg$ is again entirely different from the two mechanisms found for He^+-Ba, and He^+-Ca. The electron spectra measured for this system at various collision energies are shown in Fig. 40. The broadening of the electron distributions with increasing collision energy reveals the ionization mechanism to be a molecular one.

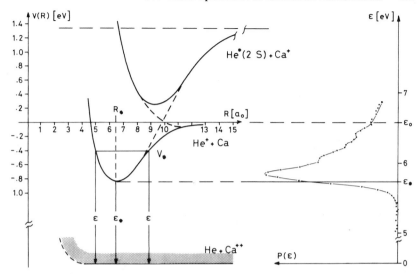

Figure 39. Potential curve diagram for system He^+–Ca as derived from semi-classical analysis of electron spectra. Relation of $V(R)$ and the spectrum $P(\varepsilon)$ at 2.3-eV collision energy is also shown. Oscillations in $P(\varepsilon)$ in range $\varepsilon < \varepsilon < \varepsilon_0$ are caused by interference of the two contributions to $P(\varepsilon)$ at a certain ε.

At low collision energies the spectrum consists of only three peaks whose intensity ratio is nearly constant. This suggests that here the ionization occurs from one initial potential curve into different final state curves. Under the assumption that the observed energy separations between the peaks are similar to the separation of the atomic states into which these final state curves dissociate, we identify as final states of the atoms: He^+ in its ground state and Hg^{++} in its five lowest states $5d^{10}\,^1S_0$, $5d^9(^2D_{3/2})6s$ $^3D_2$, $^1D_2$, and $5d^9(^2D_{5/2})6s\,^3D_{3,2}$. The variation of the observed spectra with collision energy suggests that the initial state curve $V_*(R)$ branches at curve crossings into three curves, each of which populates by ionization the five final states mentioned. For instance, the prominent peak group between 16 and 19 eV corresponds to transitions from these three initial curves into the final curve that dissociates into $He^+ + Hg^{++}(^1S_0)$. The solid lines drawn into the measured spectra show individual contributions corresponding to defined initial and final potential curve. They are obtained by composing the measured spectrum on the basis of the interpretation described.

The three curves into which the entry channel $V_*(R)$ branches as the distance decreases dissociate into $He^{+*}(n=2) + Hg^+(5d^{10}6s)$. Depending on the distance between the charge at Hg^+ and $He^{+*}(n=2)$, the nearly degenerate helium states are split by the linear Stark effect, which leads to

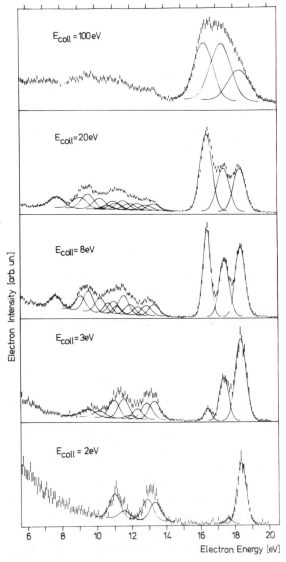

Figure 40.   Electron spectrum for transfer ionization system He^{++}–Hg at various indicated collision energies.[109]

splitting into three potential curves, $V_*^+(\Sigma)$, $V_*(\Pi)$; and $V_*^-(\Sigma)$. Down to not too small R values these three curves may be approximated by Coulomb curves modified by the respective R-dependent Stark effect. Since the branching into these three curves, as well as the ionization from these curves occur at rather large distances and since we are mainly interested in the determination of the TI mechanism, rather than in a

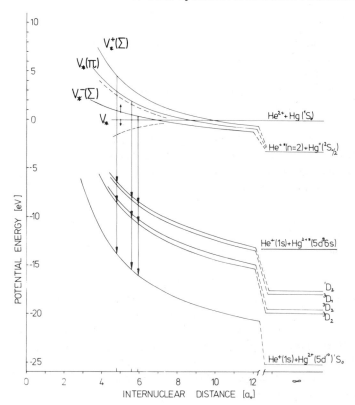

Figure 41.   Potential curve diagram for He^{++}–Hg transfer ionization system as derived from analysis of electron spectra. Ionization occurs after electron rearrangement at $R \sim 5\ a_0$, as indicated. (see text).[109]

quantitative discussion, we neglect corrections to the potentials that could be made at small R. Similarly, we approximate $V_*(R)$ by a constant. With these approximations the crossings of $V_*(R)$ with the curves $V_*^+(\Sigma)$, $V_*(\Pi)$, and $V_*^-(\Sigma)$ arise at $R(^+\Sigma) = 9.8 a_0$, $R(\Pi) = 8.7 a_0$, and $R(^-\Sigma) = 6.9 a_0$, respectively. Assuming now the final-state curves to be Coulomb curves $(Z = 2)$, the observed electron-peak positions determine within the potential curve model the distances at which the ionization occurs from the three initial curves. The complete potential curve model developed on the basis of the electron spectra is shown in Fig. 41. Energy analysis of the superelastically formed He^+ ions leads to complete agreement with the predictions of the model. Calculations carried out on the basis of the model,[113] which imply a combination of Landau–Zener theory to describe the branching and of the classical theory of PgI, enable us to reproduce the measured energy dependence of fractional cross sections belonging to ionization from the different initial curves.

484 Spontaneous Ionization in Slow Collisions

REFERENCES

1. F. M. Penning, *Naturwissenschaften*, **15**, 818 (1927).
2. F. M. Penning, *Physica*, **1**, 1028 (1934).
3. A. A. Kruithof and M. J. Druivestein, *Physika*, **4**, 450 (1937).
4. T. G. Schut and J. A. Smit, *Physica*, **10**, 440 (1943).
5. H. W. Webb, *Phys. Rev.*, **24**, 113 (1924).
6. V. Čermak, *Collect. Czech. Chem. Commun.*, **31**, 649 (1966).
7. V. Čermak, *Collect. Czech. Chem. Commun.*, **33**, 2739 (1968).
8. V. Čermak and Z. Herman, *Chem. Phys. Lett.*, **2**, 359 (1968).
9. H. Hotop and A. Niehaus, *Z. Phys.*, **215**, 395 (1968).
10. V. Fuchs and A. Niehaus, *Phys. Rev. Lett.*, **21**, 1136 (1968).
11. H. Hotop and A. Niehaus, *Z. Phys.*, **228**, 68 (1969).
12. H. Hotop, A. Niehaus, and A. L. Schmeltekopf, *Z. Phys.*, **229**, 1 (1969).
13. H. Hotop and A. Niehaus, *Chem. Phys. Lett.*, **3**, 687 (1969).
14. H. Hotop and A. Niehaus, *Z. Phys.*, **238**, 452 (1970).
15. H. Nakamura, *J. Phys. Soc. Jap.*, **26**, 1473 (1969).
16. W. H. Miller, *J. Chem. Phys.*, **52**, 3563 (1970).
17. W. H. Miller and H. F. Schäfer, III, *J. Chem. Phys.*, **53**, 1421 (1970).
18. W. H. Miller, C. A. Slocomb, and H. F. Schäfer, III, *J. Chem. Phys.*, **56**, 1347 (1972).
19. A. P. Hickman, A. D. Isaacson, and W. H. Miller, *J. Chem. Phys.*, **66**, 1483 (1977).
20. A. P. Hickman, A. D. Isaacson, and W. H. Miller, *J. Chem. Phys.*, **66**, 1492 (1977).
21. A. Niehaus, *Ber. Bunsenges. Phys. Chemie*, **77**, 632 (1973).
22. A. P. Hickman and H. Morgner, *J. Phys.*, **B9**, 1765 (1976).
23. T. Ebding and A. Niehaus, *Z. Phys.*, **270**, 43 (1974).
24. A. Niehaus and T. Ebding, "Electronic and Atomic Collisions," in *Abstracts, IXth ICPEAC*, Seattle, 1975, p. 1075.
25. H. Morgner, Ph.D. thesis, University of Freiburg, 1976.
26. W. Hultzsch, W. Koonast, A. Niehaus, and M. W. Ruf, *J. Phys.* (in press).
27. E. W. Rothe and R. H. Neynaber, *J. Chem. Phys.*, **42**, 3306 (1965).
28. E. W. Rothe, R. H. Neynaber, and S. M. Trujillo, *J. Chem. Phys.*, **42**, 3310 (1965).
29. C. H. Chen, H. Haberland, and Y. T. Lee, *J. Chem. Phys.*, **61**, 3095 (1974).
30. B. Brutschy, H. Haberland, and K. Schmidt, *J. Phys.*, **B9**, 2693 (1976).
31. H. Haberland and K. Schmidt, *J. Phys.*, **B10**, 695 (1977).
32. H. Haberland, private communication.
33. U. Buck, in *Advances in Chemical Physics*, Vol. 30, I. Prigogine and S. A. Rice, Eds., Wiley, London, 1975, p. 313.
34. E. Illenberger and A. Niehaus, *Z. Phys.*, **B20**, 33 (1975).
35. P. R. Bevington, *Data Reduction and Error Analysis for the Physical Sciences*, McGraw-Hill, New York, 1969, p. 235.
36. H. Haberland, private communication.
37. J. P. Riola, J. S. Howard, R. D. Rundel, and R. F. Stebbings, *J. Phys.*, **B7**, 376 (1974).
38. R. D. Rundel, F. B. Dunning, J. S. Howard, J. P. Riola, and R. F. Stebbings, *Rev. Sci. Instrum.*, **44**, 60 (1973).
39. R. D. Rundel, F. B. Dunning, and R. F. Stebbings, *Rev. Sci. Instrum.*, **44**, 60 (1973).

40. A. L. Schmeltekopf and F. C. Fehsenfeld, *J. Chem. Phys.*, **53**, 3173 (1970).

41. R. C. Bolden, R. S. Hemsworth, M. F. Shaw, and N. D. Twiddy, *J. Phys.*, **B3**, 61 (1971).

42. W. Lindinger, A. L. Schmeltekopf, and F. C. Fehsenfeld, *J. Chem. Phys.*, **61**, 2890 (1974).

43. A. Pesnelle, G. Watel, and C. Manus, *J. Chem. Phys.*, **62**, 3590 (1975).

44. K. Gérard and H. Hotop, *Chem. Phys. Lett.* (in press).

45. S. Y. Tang, A. B. Marcus, and E. E. Muschlitz, *J. Chem. Phys.*, **56**, 566 (1971).

46. V. Čermak, *J. Chem. Phys.*, **44**, 3781 (1966).

47. A. Niehaus, in B. C. Cobic and M. V. Curepa, Eds., *The Physics of Electronic and Atomic Collisions*, invited lectures and progress reports, VIIth ICPEAC, Belgrade, 1973, p. 649.

48. H. Hotop, *Radiat. Res.* **59**, 379 (1974).

49. H. Hotop, E. Illenberger, H. Morgner, and A. Niehaus, *Abstracts, VIIth ICPEAC*, Amsterdam, 1971, p. 1101.

50. E. Illenberger, Ph.D. thesis, University of Freiburg, 1976.

51. H. Morgner and A. Niehaus, *J. Phys.* (in press).

52. W. Kolos and J. M. Peek, *Chem. Phys.*, **12**, 381 (1976).

53. H. U. Mittmann, H. P. Weise, A. Ding, and A. Henglein, *Zeitschr. Naturforsch.* **26a**, 1112 (1971).

54. J. S. Howard, J. P. Riola, R. D. Rundel, and R. F. Stebbings, *J. Phys.*, **B6**, L109 (1973).

55. A. P. Hickman and H. Morgner, *J. Chem. Phys.*, **67**, 5484 (1977).

56. J. Fort, J. J. Laucagne, A. Pesnelle, and G. Watel, *Phys. Rev.*, **A18**, 2063 (1978).

57. J. N. Bardsley, *J. Phys.*, **B1**, 349 (1968).

58. H. Hotop and A. Niehaus, *Chem. Phys. Lett.*, **8**, 497 (1971).

59. T. Ebding and A. Niehaus, *Z. Phys.*, **270**, 43 (1974).

60. A. Niehaus and T. Ebding, "Electronic and Atomic Collisions," in *Abstracts IXth ICPEAC*, Seattle, 1975 p. 1075.

61. T. Ebding and A. Niehaus, unpublished results.

62. A. Niehaus and M. W. Ruf, *Z. Phys.*, **252**, 84 (1972).

63. H. Morgner, *J. Phys.*, **B11**, 269 (1978).

64. V. Hoffmann and H. Morgner, *J. Phys*, (in press).

65. J. Fort, J. J. Laucagne, A. Pesnelle, and G. Watel, *Chem. Phys. Lett.*, **37**, 60 (1976).

66. L. Kaufhold (1975), in Ref. 68.

67. G. D. Magnuson and R. H. Neynaber, *J. Chem. Phys.*, **60**, 3385 (1974).

68. R. H. Neynaber and G. D. Magnuson, *J. Chem. Phys.*, **62**, 4953 (1975).

69. R. H. Neynaber, G. D. Magnuson, *Phys. Rev.*, **A11**, 865 (1974).

70. R. H. Neynaber, G. D. Magnuson, *Phys. Rev.*, **A14**, 961 (1976).

71. R. E. Olson, *Phys. Rev.*, **A6**, 1031 (1972).

72. L. D. Shearer, *Phys. Rev. Lett.*, **22**, 629 (1969).

73. L. D. Shearer, *Phys. Lett.*, **31A**, 457 (1970).

74. H. Hotop and A. Niehaus, *Internat. J. Mass Spectrom. Ion Phys.*, **5**, 415 (1970).

75. G. Lantschner and A. Niehaus, *Chem. Phys. Lett.*, **23**, 223 (1973).

76. A. Niehaus, *Radiation Research, Biomedical, Chemical, and Physical Perspectives*, Academic, New York, 1975, p. 227.

77. G. Lantschner, Ph.D. thesis, University of Freiburg, 1975.

78. E. Illenberger, Ph.D. thesis, University of Freiburg, 1976.

79. D. S. C. Yee and C. E. Brion, *J. Electron. Spectr.* **8**, 377 (1976).

80. V. Čermak, *J. Electron Spectr.* **9**, 419 (1976).

81. H. Hotop and G. Hübler, *J. Electron. Spectr.* **11**, 101 (1977).

82. H. Hotop and A. Zastrow, Abstracts, *Xth ICPEAC*, Paris, 1977, p. 306.

83. W. C. Richardson and D. W. Setser, *J. Chem. Phys.*, **58**, 1809 (1973).

84. W. C. Richardson, D. W. Setser, D. L. Albritten, and A. L. Schmeltekopf, *Chem. Phys. Lett.*, **12**, 349 (1971).

85. H. Hotop, G. Hübler, and L. Kaufhold, *Internat. J. Mass Spectrom. Ion Phys.*, **17**, 163 (1975).

86. V. Čermak, *J. Electron Spectrosc.*, **8**, 325 (1976).

87. L. T. Specht, K. D. Foster, and E. E. Muschlitz, *J. Chem. Phys.*, **63**, 1582 (1975).

88. R. H. Neynaber, G. D. Magnuson, and J. K. Layton, *J. Chem. Phys.*, **57**, 5128 (1972).

89. B. Brutschy, Diplom-Thesis, University of Freiburg, 1973.

90. A. Münzer and A. Niehaus, manuscript in preparation.

91. W. Steubing, *Physik Z.*, **10**, 787 (1909).

92. F. G. Houtermans, *Z. Physik*, **41**, 619 (1927).

93. (a) F. W. Lampe, in J. L. Franklin, Ed., *Ion–Molecule Reactions*, Vol. 2, Plenum, New York, 1972, Chapter 13; (b) A. Fontijn, *Prog. React. Kinet.*, **6**, 75 (1972); A. Fontijn, *Pure Appl. Chem.*, **39**, 287 (1974).

94. J. A. Hornbeck, and J. P. Molnar, *Phys. Rev.*, **84**, 621 (1951).

95. W. L. Fite and P. Irving, *J. Chem. Phys.*, **56**, 4227 (1972).

96. R. S. Berry, in C. Schlier, Ed., *Proceedings of the International School of Physics "Enrico Fermi," Course 44*, Academic, New York, 1970, p. 193.

97. H. Nakamura, *J. Phys. Soc. Jap.*, **31**, 574 (1971).

98. F. Kolke and H. Nakamura, *J. Phys. Soc. Jap.*, **33**, 1426 (1972).

99. H. Morgner, and A. Niehaus, "Electronic and Atomic Collisions," in J. S. Risley, and R. Geballe, Eds., *Abstracts, IXth ICPEAC*, Seattle, 1975, p. 1073.

100. A. Niehaus, in B. Navinsek and J. Stefan, Eds., *Physics of Ionized Gases*, Ljubljana Institute, 1976, p. 143.

101. H. Morgner, Ph.D. thesis, University of Freiburg, 1976.

102. K. C. Kulander and J. S. Dahler, *J. Phys.*, **B8**, 460, 2679 (1975).

103. I. P. Flaks, G. N. Ogurtsov, and N. V. Fedorenko, *Sov. Phys. JETP*, **14**, 1027 (1962).

104. L. M. Kishinevski and E. S. Parilis, *Sov. Phys. JETP*, **28**, 1020 (1968).

105. E. L. Latush and M. F. Sem, *Zh. Eksp. Teor. Fis. Pis. Red.*, **15**, 645 (1972).

106. R. Arrathoon, I. M. Littlewood, and C. E. Webb, *Phys, Rev. Lett.*, **31**, 1168 (1973).

107. T. M. Austin, J. M. Mullen, and T. L. Bayley, *Chem. Phys.*, **10**, 117 (1975).

108. A. Niehaus and M. W. Ruf, *Abstract, IVth International Conference on Atomic Physics*, Heidelberg, 1974, p. 501.

109. A. Niehaus and M. W. Ruf, *J. Phys.*, **B9**, 1401 (1976).

110. W. Hultzsch, W. Kronast, A. Niehaus, and M. W. Ruf, *J. Phys.* (in press).

111. G. R. Branton and A. Niehaus, manuscript in preparation.

112. G. Gerber and A. Niehaus, *J. Phys.*, **B9**, 123 (1976).

113. M. W. Ruf, and A. Niehaus, *Abstracts, Xth ICPEAC*, Paris, 1977, p. 1040.

CHAPTER SIX

SCATTERING OF NOBLE-GAS METASTABLE ATOMS IN MOLECULAR BEAMS

H. Haberland

Fakultät für Physik der Universität Freiburg, Freiburg im Breisgau, Germany

Y. T. Lee

Department of Chemistry and Lawrence Radiation Laboratory, University of California, Berkeley California 94720

P. E. Siska

Department of Chemistry, University of Pittsburgh, Pittsburgh, Pennsylvania 15260

Contents

I INTRODUCTION   489

II BRIEF SUMMARY OF POTENTIAL SCATTERING THEORY   496
  A Elastic Differential Scattering in Nonreactive Systems   497
  B Elastic Scattering in Inelastic Systems   499
  C Scattering in Symmetric Systems with One Atom Excited   505
  D Inelastic Events   506

III EXPERIMENTAL DIFFERENTIAL CROSS-SECTION MEASUREMENTS AND INTERPRETATION   510
  A Special Techniques for Excited-state Scattering Studies   510
  B Pure Elastic Scattering   521
  C Symmetric Noble-gas Systems   523
  D Asymmetric Noble-gas Systems   545

IV FUTURE DIRECTIONS   579

I. INTRODUCTION

Although the important role played by metastable noble-gas atoms in the ionization process was recognized as long as half a century ago, investigations of microscopic processes involving metastable noble-gas atoms have only recently become one of the more active research areas.

In 1927 Frans Michel Penning (1894–1953) at Philips Laboratories discovered that breakdown voltages and the ionization coefficients of neon and argon were markedly influenced by the addition of minute impurities and concluded from a series of experiments that the impurities were ionized by energy transfer from metastable neon or argon atoms that were present in discharges.[1] In addition to Penning ionization (PgI), one other type of ionization process involving electronically excited atoms was also observed in the 1920s and 1930s. It was found that the vapors of cesium and rubidium could be photoionized by light in the discrete absorption region of the atomic spectra with energy less than the atomic ionization energy.[2-5] The proper explanation of this ionization process was given by Franck[6] in 1928 as associative ionization to form diatomic molecule ions in two-body collisions between an excited atom and a ground-state atom; this was verified by Mohler and Boeckner in their quantitative study of the formation of Cs_2^+ in 1930.[7]

It was 20 years before the important microscopic processes involved in early findings were "rediscovered" in several studies. In 1951 Hornbeck and Molnar[8] carried out the first thorough examination of the formation of molecule ions from electronically excited noble gases in a mass spectrometer. The presence of molecule ions was shown to be the result of associative ionization, and this process is now often referred to as the *Hornbeck–Molnar process*, in recognition of their work. In 1949 Jesse discovered in his studies of high-energy ionizing radiations, such as α and β particles, that the absorbed energy per ion pair produced in helium gas was extremely sensitive to the presence of minute amounts of almost any impurity. This sensitivity is called the *Jesse effect* in the field of radiation research. Although the Penning and Jesse effects describe different macroscopic phenomena, the microscopic processes that produce these effects, mainly Penning and associative ionization, are undoubtedly the same.

Substantial advances in the understanding of microscopic processes of electronically excited atoms began only in the 1960s, when improvements in experimental methods and available technology allowed scientists to obtain quantitative microscopic information under well-defined conditions.

This is especially true for metastable noble-gas atoms. Because the lifetimes of metastable noble-gas atoms are much longer than the transit time of atoms in a typical high-vacuum apparatus ($\sim 10^{-3}$ sec), it became possible to perform experiments under the single-collision conditions of the molecular beam. The measurements of ionization cross sections via total ion collection by Sholette and Muschlitz[9], the velocity dependence of the total scattering cross sections by Rothe, et al.,[10] and the studies of energy distributions of electrons from PgI by Čermak[11] mark the beginning of a rapid growth in the amount of information related to the dynamics of microscopic processes involving metastable noble-gas atoms.

Many of the chemical and dynamical properties of metastable noble-gas atoms can be easily appreciated by noting that their excitation energies are higher than the ionization potentials of many diatomic and polyatomic molecules, and their polarizabilities and ionization potentials are quite close to those of alkali atoms. In addition, the long lifetime, which prevents radiative decay from competing with the efficient conversion of stored energy in collisional processes, makes it possible to use energy transfer from metastable noble-gas atoms as pumping processes for gas lasers. Table I lists some important properties of metastable noble-gas atoms, and Fig. 1 illustrates them.

The interactions between metastable noble-gas atoms and ground-state noble-gas atoms are relatively simple and have been investigated quite extensively. If the excitation energy is lower than the ionization potential of the collision partner, the only important inelastic process is the transfer of excitation energy.[12] The excitation transfer is usually very efficient when the process is near resonant. The process that is responsible for the operation of the He–Ne laser,[13]

$$\text{Excitation transfer:} \qquad \text{He*}(2^1S) + \text{Ne} \rightarrow \text{He} + \text{Ne*}(3s)$$

is one such example. The other type of excitation transfer is the formation of excimers. For example, in the process

$$\text{Excimer formation:} \qquad \text{Xe*} + \text{Xe} \rightarrow \text{Xe}_2^* \rightarrow 2\text{Xe} + h\nu$$

the delocalization of excitation energy from one atom to two causes a drastic decrease in the radiative lifetime. The excitation energy is then transformed into a photon and the kinetic energy between two ground-state atoms. The excimer laser is based on this process.[14–16] If the ground-state atom has a lower ionization potential than the excitation energy of the metastable atom, energy-transfer collisions will mainly take the form of

Table I.
Characteristics of Some Metastable Species

ATOM	STATE	EXCITATION[A] ENERGY (eV)	IONIZATION[A] POTENTIAL (eV)	LIFETIME (sec)	POLARIZABILITY ($Å^3$)[b]
Helium	2^3S_1	19.820	4.768	$4.2 \times 10^{3\,c}$	46.86^e
				$6.2 \times 10^{5\,d}$	(24.3^f)
	$^1S_0$	20.616	3.972	$3.8 \times 10^{-2\,g}$	118.9^e
				$2.0 \times 10^{-2\,h}$	
				$1.95 \times 10^{-2\,i}$	
Neon	$^3P_2$	16.619	4.946	24.4^k	27.8^f
				$>0.8^i$	
	$^3P_0$	16.716	4.849	430^k	(23.6^f)
Argon	$^3P_2$	11.548	4.211	55.9^k	47.9^f
				$>1.3^i$	
	$^3P_0$	11.723	4.036	44.9^k	(43.4^f)
Krypton	$^3P_0$	9.915	4.084	85.1^k	50.7^f
				$>1^i$	
	$^3P_0$	10.563	3.437	0.488^k	(47.3^f)
Xenon	$^3P_0$	8.315	3.815	149.5^k	63.6^f
	$^3P_0$	9.447	2.683	0.078^k	(59.6^f)
Hydrogen 2	$^3S_{1/2}$	10.199	3.400	$1/7^l$	

C. E. Moore, *Atomic Energy Levels*, Vols. I to III, U.S. Government Printing Office, Washington, D. C., 1949.
The values in parenthesis are those for the corresponding alkali atoms.
H. W. Moss and R. J. Woodworth, *Phys. Rev. Lett.*, **30**, 775 (1973).
H. R. Grim, *Astrophys. J.*, **156**, L103 (1969).
G. A. Victor, A. Dalgarno, and A. J. Taylor, *J. Phys.*, **B1**, 13 (1968).
R. A. Molof, H. L. Schwartz, T. M. Miller, and B. Bederson, *Phys. Rev.*, **A10**, 1131 1974).
A. S. Pearl. *Phys. Rev. Lett.*, **24**, 703 (1970).
R. S. Van Dyck, Jr., C. E. Johnson, and H. A. Shugart, *Phys. Rev. Lett.*, **25**, 1403 1970).
G. W. F. Drake, G. A. Victor, and A. Dalgarno, *Phys. Rev.*, **180**, 25 (1969).
Lifetimes measured for the mixture of $^3P_0$ and $^3P_2$ states; see R. S. Van Dyck, Jr., C. E. Johnson, and H. W. Shugart, *Phys. Rev.*, **A5**, 991 (1972).
N. E. Small-Warren and L. Y. Chin, *Phys. Rev.*, **A11**, 1777 (1975).
G. Breit and E. Teller, *Astrophys. J.*, **91**, 215 (1940).

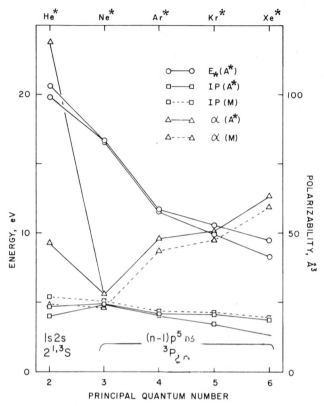

Figure 1.   Graphic display of atomic properties of metastable noble gases. Solid lines correspond to metastables, dashed to analogous alkali atom (He* corresponds to Li, etc.); –o–, excitation energies; –□–, ionization potentials; –Δ–, polarizabilities, with angular momentum substates of metastables shown separately (see Table I for values).

ionization processes.[17, 18] For example, the reaction

Penning ionization: $He^*(2^1S, 2^3S) + Ar \rightarrow He + Ar^+ + e$

Associative ionization: $\rightarrow HeAr^+ + e$

produces both atomic and molecular ions through Penning and associative ionization processes. There is no mechanistic difference between Penning and associative ionization with regard to electronic energy transfer. Whether a specific two-body collision will produce an atomic or molecular ion depends entirely on whether the electron ejected will remove sufficient energy to leave He–Ar$^+$ in a bound negative energy state. Since the electronic motion is only weakly coupled to the motion of the nuclei, the kinetic energy of the electron ejected is approximately the difference between the potential energies of the ionic state and the excited state at an internuclear distance where the transition takes place. Consequently, if the gain in potential energy during the transition is larger than the kinetic energy, a molecular ion will be formed. Of course, the transition probability as a function of internuclear distance depends not only on the coupling between the excited state and the ionic state, but also on the kinematic conditions such as relative kinetic energy and orbital angular momentum.

In the interactions between metastable atoms and diatomic and polyatomic molecules, excitation transfer and ionization processes may proceed initially with mechanisms similar to those with ground-state noble-gas atoms, but the outcome of the process is often more complicated.[12] After excitation transfer the newly formed electronically excited molecules sometimes dissociate before the excitation energy is removed by radiative decay, whereas after ionization, in addition to possible dissociation of excited molecule ions, ion–molecule rearrangements may also occur. However, there is a type of interaction of metastable noble-gas atoms with molecules that is fundamentally different from the interaction with noble-gas atoms. Since the ionization potentials of metastable noble-gas atoms are comparable to those of the corresponding alkali atoms, the interactions with molecules with appreciable electron affinities, such as diatomic halogen molecules or certain halogen-containing polyatomic molecules, may proceed initially with electron transfer from the metastable atom to the molecule. This is followed by the formation of an electronically excited rare-gas halide molecule in a reaction similar to the formation of an alkali halide by the "harpooning" mechanism.[19] For example,[20]

$$Xe^* + F_2 \rightarrow [Xe^+ F_2^-] \rightarrow XeF^* + F$$

This alkalilike behavior of metastable noble-gas atoms effectively transforms the excitation energy of the metastable noble-gas atom into electronic energy of a rare-gas halide molecule with large reaction cross section. Because the electronically excited noble-gas halides have short radiative lifetimes and the ground-state noble-gas halides are not strongly bound, the process of formation of electronically excited noble-gas halides from metastable noble-gas atoms has been shown to be ideal for the operation of the electronic transition laser and has been successfully used in high-efficiency rare-gas halide lasers in recent years.[21–23]

Accompanying the recent improvements in the microscopic experiments exploring the dynamics of elementary processes involving metastable noble-gas atoms, there have also been substantial advances in the theory,[24-25] especially for the ionization processes involving metastable noble-gas and ground-state atoms or simple diatomic molecules. For simple systems such as $He^* + H$ and $He^* + H_2$, it has become possible to calculate the interaction potentials in the excited state and the ionic state and the coupling strength between the resonance and continuum states from first principles.[26, 27] The collision process can now be described by either semiclassical or quantum-mechanical treatments in terms of interaction potentials of excited states and ionic states and the coupling strength between them.[27-29] Indeed, if these quantities are known, direct calculations of such observables as Penning electron energy spectra, velocity dependence of the ionization cross section, the branching ratio between Penning and associative ionization, and total and differential elastic scattering cross sections are reasonably straightforward. On the other hand, accurate results of microscopic experiments will enable us to understand the details of reaction dynamics by finding, in terms of an optical potential model,[30, 31] the real and imaginary parts of the interaction potentials between metastable noble-gas atoms and ground-state atoms and the interaction potentials of ionic states. The advancement in this direction is best exemplified in the pioneering experiments of Niehaus and co-workers in Penning electron and ion spectroscopy,[17, 18, 32] and differential elastic scattering cross section measurements which will be discussed in this chapter. But the derivation of these quantities from microscopic experiments is by no means simple because three quantities must be determined as a function of internuclear distance, and the observations might only reflect certain relations between these quantities. For example, Penning electron spectra are sensitive to the difference between the potentials of the excited state and the ionic state and its relation to the imaginary part of the potential. Consequently, only if the real part of the potential of the excited state or the potential of the ionic state are accurately known will it become possible to derive the imaginary part of the potential and the other interaction potential by measuring Penning electron spectra, preferably at more than one collision energy. Total ionization cross sections as a function of collision energy provide a range of combinations of real and imaginary parts of the potential that gives a satisfactory description of experimental results; the determination of these quantities from such measurements alone is not considered unique.[33] For a given real and imaginary part of the excited potential, it is possible to derive the potential of the ionic state by combining measurements of the branching ratio between Penning and associative ionization and of the total ionization cross section as a function of collision energy. It is clear that the precise determination of these potentials requires more than one type of precise microscopic experiment, and it is also important to include the type of microscopic experiment that is particularly sensitive to one or two of the potential functions.[33] Consequently, the measurements of elastic differential cross sections of the metastable noble-gas atoms not only supplement Penning electron spectra and total ionization cross sections as a function of collision energy, but also contribute significantly to the understanding of the dynamics of the collisions of metastable noble-gas atoms.

It is well known that one of the most direct and reliable methods of obtaining detailed information on interaction potentials of atomic systems is the measurement of elastic-scattering cross sections.[34] In the past decade precise interaction potentials for alkali–mercury, alkali–noble gas, and noble-gas–noble-gas systems have been derived mainly from high-resolution crossed-molecular-beam measurements of scattering intensities as a function of scattering angle at several collision energies. Well-resolved structures in the differential elastic scattering cross sections, such as rainbows, supernumerary rainbows, and fast oscillations, as well as measurements covering a wide angular range, are necessary for the precise determination of interaction potentials. It has also been shown in studies of reactive systems, such as alkali atoms with diatomic halogen molecules or halogen-containing polyatomic molecules, that it is possible to derive some information on the reaction probability as a function of the impact parameter from depletion of the elastic-scattering cross section induced by chemical reaction.[35]

Crossed-molecular-beam studies of differential scattering of metastable noble-gas atoms with ground-state noble-gas atoms or simple molecules is the major topic of this chapter. These studies have been carried out recently at several laboratories with the common goal of finding both real and imaginary parts of the interaction potentials to further our understanding of the dynamics of collision processes involving metastable noble-gas atoms.

Experimentally, the scattering studies of metastable noble-gas atoms enjoy an advantage similar to that of alkali-atom experiments in the ease of detection of metastable atoms. The high specificity of the detection of metastable noble-gas atoms by Auger electron emission,[36] (See Note on page 513.) similar in some sense to the detection of alkali atoms by a hot-wire detector, is immune to the presence of background gas, whose collision rate with the detector surface is usually many orders of magnitude higher than that of the scattered metastable atoms. This partly explains why so much information has been obtained with relatively modest experimental arrangements in a rather short period of time. Of course, the development of the high intensity, monoenergetic free-jet molecular-beam source[37,38] in the past decade has also contributed significantly to the possibility of carrying out differential scattering cross-section measurements efficiently and with reasonably high resolution, so as to provide reliable information on interaction potentials. The first precise differential scattering cross-section measurements of metastable rare-gas by Chen, et al.[39] and by Winicur and Fraites,[40] as well as more extensive studies of Haberland and Siska and their co-workers, are all carried out with a free-jet beam source.

In this chapter we summarize the current status of the low-energy scattering of noble-gas metastable atoms in molecular beams. A brief summary of potential scattering theory that is relevant to the understanding of collision dynamics, as well as a description of the experimental method, precedes the presentation of experimental findings. The experimental results presented are mainly from the authors' laboratories.

As is seen here and in other related chapters in this volume, the field of collision dynamics of metastable noble-gas atoms has made some significant advances in recent years. These advances undoubtedly will give some insight into many macroscopic phenomena involving metastable noble-gas atoms. The recent development of several potentially useful lasers involving noble-gas atoms makes the understanding of the collision dynamics of elementary processes involving metastable noble-gas atoms even more important. We hope this chapter will serve as a useful guide to the kind and quality of information obtainable on metastable noble-gas atom interactions from beam-scattering methods.

II. BRIEF SUMMARY OF POTENTIAL SCATTERING THEORY

Two-body collision theory for a central potential can be applied with a rigor limited mainly by the Born–Oppenheimer approximation to a large and interesting class of metastable noble-gas collisions, namely those with other atomic species. In the atom–atom cases, even when electronically inelastic channels are open, the optical model may often be applied to describe the effect on the elastic scattering. If only a few channels are important, a close-coupling treatment is feasible because of the absence of rotational and vibrational degrees of freedom and the consequent reduction in computational effort. Penning ionization is an excellent case for the application of the optical model, since the rapid departure of the Penning electron renders the process irreversible. Excitation transfer in systems such as $Ar^* + Kr$ or $He^* + Ne$ promises to be analyzable in a close-coupling framework with only a few states included, and analysis of the elastic scattering using a simple optical model may be adequate if the coupling is weak. Analysis of product angular distributions, of course, requires a complete close-coupling description.

Collisions with diatomic and polyatomic molecules are, as usual, difficult to describe because of the internal degrees of freedom, as well as the anisotropy in the potential function. For many target molecules involved in PgI, however, nearly vertical ionization occurs as indicated by Franck–Condon vibrational populations, thus inviting treatment neglecting the internal degrees of freedom. Particularly at the level of Penning ion product angular distributions, small molecules often give results very similar to the atomic cases. Effects of vibrational and rotational inelasticity and potential anisotropy prevent the extraction of truly quantitative information from nonreactive scattering with molecules, but analysis of these data using two-body mechanics can still be of value.

Many books and reviews give exemplary treatments of two-body collision theory, especially for our purposes those treating molecular collisions.[41–45] Therefore, we confine ourselves to brief discussion and statement of results essential to understanding the differential scattering, giving both quantum and classical mechanical formulas where possible.

A. Elastic Differential Scattering in Nonreactive Systems

In the absence of inelastic processes, and assuming a single potential curve, the outcome of a collision—specified by the relative energy E, collisional reduced mass μ, and impact parameter b or angular momentum quantum number l—is described classically by the deflection angle $\chi(b, E)$ or quantally by the phase shift $\eta_l(E)$, both asymptotic quantities. In the semiclassical (large l) limit, we may identify $l + \frac{1}{2} = kb$, k equal to the wave number $(2\mu E)^{1/2}/\hbar$, and $\chi = 2\partial\eta_l/\partial l$. In terms of these, the differential cross section is given by

$$\sigma_C(\theta, E) = \frac{\sum_i b_i}{\left(\sin\theta \,|\partial\chi(b, E)/\partial b|_{b_i}\right)} \, ; \qquad |\chi(b_i, E)| \underset{\text{mod } \pi}{=} \theta \qquad (\text{II.1})$$

classically, and

$$\sigma_Q(\theta, E) = |f(\theta, E)|^2 \qquad (\text{II.2})$$

quantally, with the scattering amplitude

$$f(\theta, E) = (2ik)^{-1} \sum_l (2l + 1)\left[e^{2i\eta_l(E)} - 1\right] P_l(\cos\theta) \qquad (\text{II.3})$$

The value χ may be obtained from the potential $V(r)$ by a simple quadrature,

$$\chi(b,E) = \pi - 2b \int_{r_0(b,E)}^{\infty} - \frac{dr}{r^2 \left[1 - V(r)/E - b^2/r^2 \right]^{1/2}} \qquad (\text{II.4})$$

where $r_0(b,E)$ is the outermost turning point, whereas η results from integration of the radial Schrödinger equation,

$$\frac{d^2 G_l(r)}{dr^2} + k^2 \left[1 - \frac{V(r)}{E} - \frac{l(l+1)}{k^2 r^2} \right] G_l(r) = 0 \qquad (\text{II.5})$$

into the asymptotic region, where the solution becomes

$$G_l(r) \sim kr \left[\sin \eta_l j_l(kr) - \cos \eta_l n_l(kr) \right] \qquad (\text{II.6})$$

where j_l and n_l are the spherical Bessel and Neumann functions respectively. Semiclassically, $\chi = (2 \partial \eta_l / \partial l)$ may be employed to give the JWKB expression,

$$\eta_l = \frac{\pi}{2}\left(l + \tfrac{1}{2}\right) - kr_0 + k \int_{r_0}^{\infty} \left\{ \left[1 - \frac{V(r)}{E} - \frac{\left(l + \tfrac{1}{2}\right)^2}{k^2 r^2} \right]^{1/2} - 1 \right\} dr, \qquad (\text{II.7})$$

which yields very accurate molecular cross sections when substituted in equation (II.3).

We note that, formally, the scattering angle θ bears no necessary relation to the angular momentum l in quantum mechanics [equation (II.3)], whereas classically a given b leads uniquely to a deflection angle χ through a classical trajectory. In the limit of large l, however, stationary-phase regions of the summand in (II.3) lead to different groups of l values, each contributing mainly to a small range of angles. Thus for molecular collisions, where the sum in (II.3) generally extends to $l \sim 10^2$ to 10^4, a close connection persists between the results of quantum calculations and the classical picture. Interference features in the differential cross section, such as rainbow structure and the rapid oscillatory pattern often observed superimposed on it, can usually be interpreted as a superposition of two or more interfering classical trajectories.

The noble-gas metastable atoms Ne* through Xe* are $^3P_{2,0}$ atoms, so that more than one potential-energy curve governs the scattering, even with S-state collision partners. Because spin–orbit interactions are large whereas weak van der Waals interactions imply only weak coupling to the internuclear axis, Hund's case (c) is probably the most appropriate coupling to use, with the projection Ω of the total electronic angular momentum on the internuclear axis being conserved. Thus $^3P_2$ metastables scattering from $^1S_0$ partners gives $\Omega = 2, 1, 0^-$, whereas $^3P_0$ yields $\Omega = 0^-$ only. At the detector, $^3P_2$ atoms scattered on the various curves are indistinguishable because Ω no longer has meaning. The differential cross section is thus

$$\sigma_Q(\theta) = \left| \sqrt{\tfrac{2}{5}} \, f_2(\theta) + \sqrt{\tfrac{2}{5}} \, f_1(\theta) + \sqrt{\tfrac{1}{5}} \, f_0 - (\theta) \right|^2 \qquad \text{(II.8)}$$

for $^3P_2$ scattering, whereas only one amplitude contributes for $^3P_0$ scattering. Metastable helium $(2^1S, 2^3S)$ scatters from other S-state atoms on a single potential curve. For non-S-state collision partners, formula (II.8) is easily generalized to a sum over allowed values of the "good" electronic quantum number. If the coupling between the electronic angular momentum and nuclear rotation is strong, [e.g., Hund's case (d)], formulas such as (II.8) cannot be used, as the individual angular momenta need not be conserved. A theory that accounts for intermultiplet transitions must then be invoked. On the other hand, it is possible that the Ω quantum number does not affect the electronic energy in the case of van der Waals interactions, so that only the possible values of J, the total electronic angular momentum, need be considered. Then $^3P_2 + {}^1S_0$ results in only one state or potential-energy curve labeled by $J = 2$, and the problem for collisions with non-S-state partners is correspondingly simplified. The superposition of amplitudes in (II.8) may produce an interference pattern in the angular distribution, but thus far only the Ar*–Kr system has shown any wide-angle structure (see Section II.B), and this is likely the result of other effects.

B. Elastic Scattering in Inelastic Systems

Molecular nonreactive scattering angular distributions in systems with nonelastic channels open have been traditionally analyzed in terms of the *optical model*, a tool borrowed from nuclear scattering. Micha[31] has recently reviewed the use of the optical model in molecular collisions. In the quantal version of the model the inelastic processes are viewed as removing or "absorbing" flux from each elastic partial wave (a view that is quite rigorous). Since the actual: collisionless outgoing partial wave-amplitude ratio is given by the elastic S matrix

$$S_l = e^{2i\eta_l} \qquad \text{(II.9)}$$

attenuation of an outgoing wave can be represented by allowing the phase shift η_l to take on complex values

$$\eta_l = \delta_l + i\xi_l \tag{II.10}$$

with the condition $\xi_l > 0$ assuring that the square modulus of S_l

$$|S_l|^2 = e^{-4\xi_l} \tag{II.11}$$

is no greater than unity. It may be shown rigorously, and is easily seen by inspecting the Born approximation for η_l, that a complex potential

$$V(r) = V_0(r) - \frac{i}{2}\Gamma(r) \tag{II.12}$$

where $\Gamma(r) > 0$ gives rise to a complex phase shift, [equation II.(10)]. The term $\Gamma/\hbar$ may be interpreted as the absorption rate at internuclear distance r. The quantal differential cross-section formulas, [equations (II.2) and (II.3)] remain unchanged.

Ideally, η_l is obtained from integration of the radial equation for a complex potential. This resolves (II.5) into a pair of coupled differential equations. For

$$G_l(r) = H_l(r) + iI_l(r) \tag{II.13}$$

we obtain

$$\frac{d^2 H_l(r)}{dr^2} + k^2 \left[1 - \frac{V_0(r)}{E} - \frac{l(l+1)}{k^2 r^2} \right] H_l(r) = \frac{k^2}{2} \frac{\Gamma(r)}{E} I_l(R)$$

$$\frac{d^2 I_l(r)}{dr^2} + k^2 \left[1 - \frac{V_0(r)}{E} - \frac{l(l+1)}{k^2 r^2} \right] I_l(r) = -\frac{k^2}{2} \frac{\Gamma(r)}{E} H_l(r) \tag{II.14}$$

Asymptotic forms of $H_l(r)$ and $I_l(r)$ (both real functions) may be obtained by finding the real and imaginary parts of (II.6), from which the complex phase shift may be extracted. The value η_l in the JWKB approximation[25] may be obtained from (II.7) with (II.12) substituted for $V(r)$ (the turning point becomes complex); an expansion for $\Gamma(r) \ll V(r)$ yields separate formulas for δ_l and ξ_l. In this limit, δ_l is identical to η_l of (II.7) evaluated for the potential $V_0(r)$ only, and ξ_l is given by

$$\xi_l \simeq \frac{1}{2\hbar v} \int_{r_0(b,E)}^{\infty} \frac{\Gamma(r)\,dr}{\left[1 - V_0(r)/E - \left(l + \frac{1}{2}\right)^2 / k^2 r^2 \right]^{1/2}} \tag{II.15}$$

This perturbation result somewhat surprisingly corresponds exactly to the classical formulation (see text that follows) through identification of the classical *opacity function* $P(b)$ with its quantal form

$$P(l) = 1|S_l|^2 = 1 - e^{-4\xi_l} \tag{II.16}$$

Equation (II.16) gives the fractional absorption of each partial wave, and in terms of it the total inelastic cross section is

$$\sigma_Q^{\text{inel}} = \frac{\pi}{k^2} \sum_l (2l+1)P(l) \tag{II.17}$$

The validity of the perturbation theory requires that the influence of $\Gamma(r)$ on the trajectory be small, implying $P(l) \gtrsim 0.3$ for accurate calculations using it. This condition is seldom met in the PgI systems thus far studied.

In the classical version of the optical model, as developed mainly by Ross and co-workers,[35,46,47] each impact parameter b manifests a certain reaction probability $P(b)$, the classical opacity. Although a complex potential as in (II.12) is probably meaningless in a purely classical context, if we use the (rigorous) rate interpretation of $\Gamma(r)$, we may derive[24]

$$P(b) = 1 - \exp\left\{ -\int_{-\infty}^{\infty} \frac{\Gamma[r(t)]}{\hbar} dt \right\} \tag{II.18}$$

where $r(t)$ is obtained from the classical trajectory in the real potential $V_0(r)$. This result is equivalent to the JWKB perturbation formula [equation (II.15)] for ξ_l substituted in (II.16) and is obtained under the assumption of irreversibility of the inelastic process. The classical differential cross section is then

$$\sigma_C(\theta, E) = \sum_i \frac{b_i[1 - P(b_i)]}{\sin\theta |\partial\chi(b,E)/\partial b|_{b_i}} \tag{II.19}$$

with $\chi(b, E)$ given in (II.4) for the potential $V_0(r)$. Equation (II.19) is the basis for the straightforward data analysis proposed by Ross and co-workers[35,46,47] to obtain $P(b)$ directly. Typically, $P(b)$ differs from zero only at small b and thus affects the scattering only at wide angles θ. In this case the sum over branches of the deflection function in (II.19) reduces to a single term, and

$$P(b, E) = \frac{\sigma^{(0)}(\theta) - \sigma(\theta)}{\sigma^{(0)}(\theta)} ; \qquad \theta = \chi(b, E) \tag{II.20}$$

where $\sigma^{(0)}(\theta)$ is calculated from a $V_0(r)$ determined from the scattering at small angles and $\sigma(\theta)$ is experimentally measured. The total inelastic cross section is then

$$\sigma_C^{\text{inel}} = 2\pi \int_0^\infty P(b)b\,db \qquad (\text{II.21})$$

a formula analogous to (II.17). A semiclassical extension of the classical result (II.19) to include attenuation of the rainbow structure has been given by Harris and Wilson.[48]

The classical theory makes especially clear the inherent ambiguity of data analysis with the optical model, and this ambiguity carries over into the quantum model. If we wish to use experimental differential cross sections to gain information about $V_0(r)$ and $P(b)$ or $\Gamma(r)$, we must assume a reasonable parametric form for $V_0(r)$ that determines the shape of the cross section "in the absence of reaction." The value $P(b)$ is then determined [or $\Gamma(r)$ chosen] by what is essentially an extrapolation of this parametric form. In the classical picture a $V_0(r)$ with a less steep repulsive wall yields a lower reaction probability from the same experimental cross-section data. The pair of functions $V_0(r), P(b)$ or $V_0(r), \Gamma(r)$ is thus underdetermined. The ambiguity may be relieved somewhat (to what extent is not yet known) by fitting several sets of data at different collision energies and, especially, by fitting other types of data such as total elastic and/or reactive cross sections simultaneously.

1. Optical Model in Penning Systems

As mentioned in Section II.A, the PgI process is ideal for the application of the optical model. This is clear in the classical and semiclassical PgI theory,[24,25] for which opacity and cross-section formulas are completely equivalent to those given earlier in this chapter. The quantal optical model is also rigorously related to the elastic component of the quantal PgI theory. Miller[49] has shown that $\Gamma(r)$, identified in PgI as the autoionization width of the excited electronic state, may be accurately obtained by a standard Born–Oppenheimer electronic structure calculation as

$$\Gamma(r) = 2\pi\rho |<\phi_\epsilon |H - E|\phi_0>|^2 \qquad (\text{II.22})$$

where $|\phi_0>$ is the initial (discrete) electronic state, $|\phi_\epsilon>$ is the final (continuum) electronic state degenerate with it, H and E are the electronic Hamiltonian and energy, respectively, and ρ is the density of final continuum states. The He*(2^3S)–H(1^2S) PgI width and potential curves were the first to be calculated with some accuracy.[27] The meeting ground between *ab initio* PgI theory and differential scattering experiments is currently in the He* $-$ H$_2$ system (see Section III).

2. Optical Model in Excitation-transfer Systems

In the projection operator formalism, which leads to a rigorous basis for the optical potential, the absorptive imaginary part is associated with transitions out of the elastic channel from which no return occurs. Whereas PgI transitions are in this category, excitation transfer (ET) transitions are not, since return ("virtual excitation") can occur during the ET collision. In the event that a localized avoided curve crossing with one other state dominates the inelastic process (expected for many endoergic transfers), the total absorption probability (opacity) can still be defined:

$$P_{01}(b) = 2p_{01}(1 - p_{01}) \qquad \text{(II.23)}$$

where p_{01} is the probability of making a diabatic transition (hopping between adiabatic potential curves), and the crossing point is outside and not too near the turning point. Probability p_{01} may be calculated by, Landau–Zener–Stückelberg (LZS) theory, for example.[50-52] The probability of purely elastic scattering on an adiabatic incoming potential $V_0(r)$ is then

$$P_0(b) = (1 - p_{01})^2 \qquad \text{(II.24)}$$

and the probability of making two diabatic transitions (return) is

$$P_{010}(b) = p_{01}^2 \qquad \text{(II.25)}$$

Naturally, $P_0 + P_{01} + P_{010} = 1$. To apply a *conventional* optical model analysis to elastic differential scattering in an ET system, we must have $P_{010} \ll 1$; in other words, the probability of return must be small. If not, the "recrossing" trajectories will at least alter the relative intensities of small- and wide-angle scattering and perhaps produce new interference patterns in the angular distribution. Such effects cannot be accounted for in fitting a local absorptive optical potential to experiment, since $\Gamma(r)$ accounts only for P_{01} (i.e., absorption), and $V_0(r)$ cannot reproduce the recrossing dynamics. If V_0 is chosen as a diabatic curve, the same argument ensues with P_0 and P_{010} interchanged.

The LZS form for p_{01} is

$$p_{01} = \exp\left(-\frac{v_{01}}{v_b}\right); \qquad r_0 < r_x \qquad \text{(II.26)}$$

$$= 0; \qquad r_0 > r_x$$

with

$$v_{01} = \frac{2\pi V_{01}^2(r_x)}{\hbar|(dV_0/dr) - (dV_1/dr)|_{r_x}} \tag{II.27}$$

where $V_0(r)$ and $V_1(r)$ are diabatic potential curves intersecting at r_x, $V_{01}(r)$ is the nonadiabatic coupling potential, and the velocity

$$v_b = v\left[1 - \frac{V_0(r_x)}{E} - \frac{b^2}{r_x^2}\right]^{1/2} \tag{II.28}$$

The overall LZS opacity [equation (II.23)] may be improved for the region $r_0 \sim r_x$ by properly considering the interference between incoming and outgoing transitions.[52] This produces oscillations in $P_{01}(b)$ that may appear in the differential scattering. For n open inelastic channels accessible by pairwise curve crossing from state 0, equation (II.23) can be generalized to give $P_{0j}(b)$ for the $0 \to j$ transition, $j < n$.[53] The total opacity for the elastic channel is then

$$P(b) = \sum_{j=1}^{n} P_{0j}(b) \tag{II.29}$$

The sum over states is likely to damp out any oscillations in $P(b)$.

For exoergic channels, there is often no accessible avoided crossing, in which case the trajectory assumptions underlying the LZS theory are violated. The nonadiabatic coupling region may extend over a considerable range of internuclear distance, and semiclassical methods using exact classical trajectories represent the minimal necessary improvement over LZS.

3. Close-coupling Treatment in Excitation-transfer Systems

Assumptions limiting the usefulness of LZS theory and its extensions may be removed only at the expense of losing the closed-form solution. The coupling of LZS to an optical model analysis introduces further limitations. The optical model may be applied without recourse to the LZS interpretation of $P(b)$, but when done rigorously it is as difficult as a two-state close-coupled calculation. Since excellent computational algorithms now exist for solving coupled sets of Schrödinger equations for multichannel scattering on today's high-speed computers, this approach has become a feasible alternative, especially if only a few channels are involved. In addition, the case of no near approach of the channel potentials can be treated on an equal footing with the familiar diabatic curve-crossing case. For two-body nonadiabatic collisions, the coupled equations are

$$\frac{d^2 G_l^j(r)}{dr^2} + k_j^2 \left[1 - \frac{V_j(r)}{E_j} - \frac{l(l+1)}{k_j^2 r^2} \right] G_l^j(r) = k_j^2 \sum_{i \neq j}^n \frac{V_{ij}}{E_j}(r) G_l^i(r)$$

$$j = 0, 1, \ldots, n \qquad (\text{II.30})$$

where V is a potential matrix, chosen in most cases to be in the diabatic representation, and

$$E_j = E_0 - V_j(\infty) \left[V_0(\infty) \equiv 0 \right]$$

$$= \frac{\hbar^2 k_j^2}{2\mu} \qquad (\text{II.31})$$

is the final kinetic energy for channel j. The equations uncouple asymptotically, and the S matrix is obtained formally as the ratio of amplitudes of the outgoing spherical partial wave in the jth channel to the collisionless outgoing wave. Various numerical algorithms are available for evaluating the S matrix from close-coupled solution of equations (II.30).[54] The similarity between the resonant two-state equations and optical model equations (II.14) is notable.

C. Scattering in Symmetric Systems with One Atom Excited

When the collision partner is the ground state of the same atom as the excited one, special resonance effects come into play. Classically, we may speak of "direct" and "resonant exchange" collisions, but quantally these are inextricably intertwined in principle, because of the indistinguishability of $A^* + A$ from $A + A^*$. Proper symmetrization of the electronic wave function with respect to these two arrangements leads to a *gerade* and an *ungerade* electronic state for each value of the electronic angular momentum projection Ω. The identity of the atoms allows a strong "chemical" (as opposed to only van der Waals) exchange interaction with bond energies of at least 1 eV for the favorable (for the noble gases, u) overlap of the atomic wave functions.

The cross section may be computed from the scattering amplitudes for the g and u potentials using

$$\sigma(\theta) = \tfrac{1}{4} \left| f_g(\theta) + f_g(\pi - \theta) + f_u(\theta) - f_u(\pi - \theta) \right|^2 \qquad (\text{II.32})$$

for spinless nuclei (^{4}He, ^{20}Ne, ^{40}Ar, ^{84}Kr, ^{132}Xe). The isotopes with nonzero spin require the admixture of a term similar to the right-hand side (RHS) of (II.32) with the signs of the $\pi - \theta$ amplitudes reversed. When various projections Ω are possible, each g,u amplitude must be calculated as a weighted sum of projection amplitudes, as in (II.8). The expected coherent interference between $f(\theta)$ and $f(\pi - \theta)$ is similar in nature to that for symmetric ground-state scattering, as in He–He. This interference may be discussed in terms of deflection functions for the g,u potentials. A detailed discussion specialized to the case of He*$(2^1S, 2^3S)$ + He is given in Section III.C.

D. Inelastic Events

1. Penning Ionization Heavy-particle Angular Distributions

In a classical picture of Penning ionization,[24] the molecules approach along a trajectory on the initial A* + B (real) potential $V_0(r)$. Ionization occurs at a specified (but random) value of the internuclear distance, r_i, and the products then complete their trajectories on an ion–molecule potential $V_+(r)$ for A + B$^+$. Neglecting the momentum of the ejected electron, deflection functions can be computed according to whether the ionization occurs on the *in*coming or *out*going part of the $V_0(r)$ trajectory. These are

$$\chi_{in}(b, E, E') = \chi_+(b', E') - \Delta\chi_+(b', E') + \Delta\chi_0(b, E)$$

$$\chi_{out}(b, E, E') = \chi_0(b, E) - \Delta\chi_0(b, E) + \Delta\chi_+(b', E') \tag{II.33}$$

with

$$\chi_0(b, E) = \pi - 2b \int_{r_0}^{\infty} \frac{dr}{r^2 \left[1 - V_0(r)/(E - b^2/r^2) \right]^{1/2}}$$

$$\Delta\chi_0(b, E) = -b \int_{r_i}^{\infty} \frac{dr}{r^2 \left[1 - V_0(r)/(E - b^2/r^2) \right]^{1/2}} \tag{II.34}$$

where E' is the recoil energy of the products and r_0 is the turning point on V_0. The values χ_1 and $\Delta\chi_1$ are given by expressions identical to equations (II.34) with $b \to b'$, $E \to E'$, $r_0 \to r_+$ and $V_0 \to V_+$. Conservation of orbital angular momentum requires $b' = b(E/E')^{1/2}$. A vertical (Franck principle) transition from V_0 to V_+ is assumed, wherein the local kinetic energy of the nuclei at $r = r_i$ is conserved. This determines E' through

$$E' = E - V_0(r_i) + V_+(r_i) \tag{II.35}$$

with the potentials both referred to zero energy at $r = \infty$. The energy of the ionized electron is then

$$\varepsilon = E - E' + \varepsilon_0$$

$$= \varepsilon_0 + V_0(r_i) - V_+(r_i) \qquad (II.36)$$

where ε_0 is the ionization exoergicity at $r = \infty$, given by the difference between the excitation energy of A* and the ionization potential of B and second equality obtained from (II.35). The impact parameter b must be small enough to make $r_i > \max(r_0, r_+)$. The classical differential cross section is then

$$\sigma_R(\theta, E, E') = \sum_{j = \text{in, out}} \sum_{\nu} \frac{P_j(b_\nu, E, r_i) b_\nu}{|\sin \chi_j \partial \chi_j / \partial b_\nu|} ; \quad |\chi_j(b_\nu)|_{\text{mod} \pi} = \theta \qquad (II.37)$$

The sum over ν is analogous to the sum over i in (II.1); the values of b contributing will depend of j. The term $P_j(b, E, r_i)$ is the probability of ionization at r_i, given by

$$P_{\text{in}}(b, E, r_i) = \frac{\Gamma(r_i)}{\hbar v_b(r_i)} \exp\left[-2\Delta \xi(b, E) \right]$$

$$\qquad (II.38)$$

$$P_{\text{out}}(b, E, r_i) = \frac{\Gamma(r_i)}{\hbar v_b(r_i)} \exp\left[-4\xi(b, E) + 2\Delta \xi(b, E) \right]$$

where

$$\Delta \xi(b, E) = \tfrac{1}{2} \int_{r_i}^{\infty} \frac{\Gamma(r) \, dr}{\hbar v_b(r)}$$

$$\qquad (II.39)$$

$$\xi(b, E) = \tfrac{1}{2} \int_{r_0}^{\infty} \frac{\Gamma(r) \, dr}{\hbar v_b(r)}$$

and

$$v_b(r) = v \left[\frac{1 - V_0(r)}{E - b^2 / r^2} \right]^{1/2} \qquad (II.40)$$

The term $\xi(b, E)$ may be identified as the classical limit expression for the imaginary part of the phase shift [equation (II.15)], and $v_b(r)$ is the local velocity. The total reaction cross section is

$$\sigma_R(E) = 2\pi \int_0^\infty b\, db \int_{r_0}^\infty dr_i \big[P_{in} + P_{out} \big] \tag{II.41}$$

which reduces to the optical model expression, [equation (II.21)] on integrating over r_i analytically.

Miller[24] gives the quantum-mechanical S matrix as

$$S_l(E, E') = -2i\left(\frac{2\mu}{\hbar^2}\right)^{1/2} (EE')^{-1/4} \exp\big[i(\eta_l + \eta_l^+) \big]$$

$$x \int_0^\infty dr\big[G_l(r) \big]^* \mathfrak{v}(E' - E; r) G_l^+(r) \tag{II.42}$$

where η_l and η_l^+ are phase shifts in $V_0(r) - \frac{i}{2}\Gamma(r)$ at E and $V_+(r)$ at E', respectively, and G_l and G_l^+ are the corresponding radial wave functions, normalized to unit amplitude sine functions asymptotically. The term $\mathfrak{v}(E' - E; r)$ is an electronic discrete–continuum matrix element given by

$$\mathfrak{v}(E' - E; r) = \langle \phi_\epsilon | H - E | \phi_0 \rangle \tag{II.43}$$

as in (II.22), with ϵ, the energy of the ionized electron, given by $\epsilon = \epsilon_0 - E' + E$. It is noted that

$$\Gamma(r) = 2\pi\rho|\mathfrak{v}(E' - E; r)|^2; \qquad E' - E = V_0(r) - V_+(r) \tag{II.44}$$

by comparing (II.43) and (II.22). The scattering amplitude for the heavy particles is then

$$f_R(\theta, E, E') = (2ik)^{-1} \sum_l (2l+1) S_l(E, E') P_l(\cos\theta) \tag{II.45}$$

Miller shows[24] that stationary-phase evaluation of the integral in (II.42) using WKB wave functions leads to a semiclassical theory, in which the classical Franck condition [equation (II.35)] holds and $S_l(E, E')$ is given in the classical limit by

$$S_l(E, E') = \sum_i \left[\frac{P_{in}(l, E, r_i)}{(d/dr)\big[V_0(r) - V_+(r) \big]_{r=r_i}} \right]^{1/2} \exp i\phi_{in}(r_i)$$

$$+ \left[\frac{P_{out}(l, E, r_i)}{(d/dr)\big[V_0(r) - V_+(r) \big]_{r=r_i}} \right]^{1/2} \exp i\phi_{out}(r_i) \tag{II.46}$$

where the sum is over r_i satisfying (II.35), the P values are given in (II.38), and the phases are

$$\phi_{in}(r_i) = 2\eta_l^+ - \Delta\eta_l^+ + \Delta\eta_l^0$$

$$\phi_{out}(r_i) = 2\eta_l^0 - \Delta\eta_l^0 + \Delta\eta_l^+ \qquad (II.47)$$

with η_l^0, η_l^+ JWKB phase shifts in $V_0(r)$ at E and $V_+(r)$ at E' as in (II.7), and

$$\Delta\eta_l^0 = k \int_{r_i}^{\infty} \left[\frac{1 - V_0(r)}{E - \left(l + \frac{1}{2}\right)^2 / k^2 r^2} \right]^{1/2} dr \qquad (II.48)$$

with transcriptions for $\Delta\eta_l^+$ as earlier for χ, with $k \to k'$. Thus in the classical limit a real trajectory is specified on V_0 and V_+ by imposing the classical Franck condition. The probability moduli in (II.46) will possess singularities at extrema in the potential difference as well as when $r_i = r_0$. These can only be removed by redoing the stationary-phase integration in a "uniform" manner. Hickman and Morgner[30] have used a model that removes these singularities while still employing a local $\Gamma(r)$. They approximate the $\mathfrak{v}$ matrix by

$$\mathfrak{v}(E' - E; r) \simeq \left(\frac{\Gamma(r)}{2\pi\rho} \right)^{1/2}, \qquad (II.49)$$

while performing the radial integral in (II.42) numerically. This essentially replaces the classical Franck (stationary-phase) approximation by a Franck–Condon-like approximation and treats all the singularities and interferences uniformly. The theory was used to calculate the PgI angular distribution for $He^*(2^3S) + Ar$ (see Section III.D.1.b).

2. Excitation-transfer Differential Cross Sections

Just as calculation of the Penning ion angular distributions requires V_0, V_+ and Γ, the angular distribution of products of electronic energy transfer in a two-state approximation demands V_0 and V_1, where V_1 is the product potential, along with the transition probability, contained in the off-diagonal coupling V_{01}. It is usually convenient (see preceding sections) to work in the diabatic representation, and we assume that V_0 and V_1 are diabatic potentials that may cross at some real value of r. The classical formulation [equations (II.33), (II.34), and (II.36)] ensues nearly exactly as given earlier if we specify a transition radius r_t in place of the ionization radius r_i. In this case the product kinetic energy E' is determined by asymptotic energies only:

$$E' = E + \varepsilon(A^*) - \varepsilon(B^*) \qquad (II.50)$$

where the ε values are bound-state electronic energies, taken relative to the ground state of each atom or molecule. Thus the classical Franck condition (II.35) holds only if the transition occurs at a crossing point, r_x, of V_0 and V_1. For $r_t = r_x$, evaluation of the transition probabilities appearing in (II.36) can be done using the LZS theory (which employs classical trajectories). The total transfer cross section in the LZS approximation may be evaluated as a one-dimensional quadrature if straight-line trajectories are used.[55]

Quantally, this problem is in the usual close-coupled equations category. For analysis of product angular distributions, it seem reasonable that a two-state analysis might suffice, provided that the different product channels are not strongly coupled themselves and that the absorption into channels other than the one considered is weak.

III. EXPERIMENTAL DIFFERENTIAL CROSS-SECTION MEASUREMENTS AND INTERPRETATION

A. Special Techniques for Excited-state Scattering Studies

Scattering studies with metastable atoms are in many cases easier (and less expensive) than experiments with ground-state atoms. The discussion that follows is mainly concerned with helium, as most of the information is available for this atom. Figure 2 shows a skeletal setup of the experiment. A helium beam from a supersonic nozzle source is excited by electron impact to its two metastable states. The singlet state can be quenched by the 2μ radiation from a helium-gas discharge lamp:

$$He(2^1S) + h\nu(2\mu) \rightarrow He(2^1P) \rightarrow He(1^1S) + h\nu(584\ \text{Å})$$

$$\rightarrow He(2^1S) + h\nu(2\mu)$$

The branching ratio is $1 : 1000$ in favor of the ground state. The beam is scattered from a second supersonic beam, and the electronically excited atoms are detected. As excitation transfer can occur during the collision (e.g., $He^* + Ne \rightarrow He + Ne^*$), a second quench lamp is sometimes installed in front of the detector, to enable study of the energy-transfer process separately.

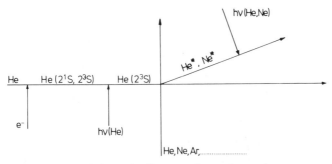

Figure 2.   Schematic diagram of experimental setup.

1. Detector

The detector used is very simple. It consists of an open multiplier or channeltron, usually operated in the pulse-counting mode. The high electronic excitation of the particle causes electron emission on the first dynode or entrance cone. The probability for electron emission has been measured absolutely by Stebbings et al.[56] and Borst.[57] It depends on the material of the detector and its gas coverage and past history. It is universally assumed that the emission probability is independent of the kinetic energy of the excited atom. But only recently Brutschy[58] has shown that this is true for kinetic energies in the range 17 to 86 meV, at least for He* on a dirty (10^{-6}-torr) surface. The differential cross section for scattering of He(2^1S) from helium was measured twice at the same center-of-mass energy ($E = 42$ meV), but under different laboratory conditions. One of the two beams was alternately cooled to liquid-nitrogen temperature and the other one left at room temperature. Figure 3 shows the two Newton diagrams for the two runs. The velocity of the He* impinging on a surface changes by a factor of 2.2. The largest energy ($E = 86$ meV) is obtained for $\theta = 45°$. Figure 4 shows the experimental results. They differ by up to a factor of 4 at some angles because of the Jacobian transformation factor. After converting the two differential cross sections to the center of mass system, they should be identical, save for the velocity dependence of the detection probability. Figure 5 shows the center-of-mass distributions. They are identical within experimental error. This implies that the emission probability is independent of the kinetic energy in the range 17 to 86 meV. The group at Saclay (Manus, Watel, and

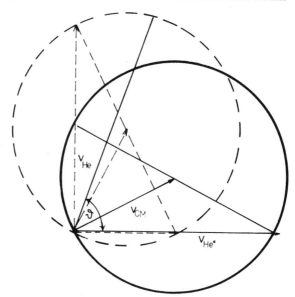

Figure 3. Newton diagrams for the two sets of measurements shown in Figs. 4 and 5. On left, He* beam source is cooled to liquid-nitrogen temperature and ground-state helium beam is at room temperature, whereas beam temperatures are interchanged on right. This gives same center-of-mass kinetic energy but different laboratory energies for scattered atoms.

co-workers) is planning to measure the velocity dependence of the emission coefficient absolutely with a laser technique.[59]*

[NOTE: The electron emission process has recently been studied in more detail (1). A He beam, which is very similar to that shown on the left-hand side of Fig. 6, is scattered from a well defined Pd (111)-surface. Either an atomically clean or a surface covered with half a monolayer of CO was studied. The energy spectra of the emitted electrons were measured for different electron emission angles. For a gas covered surface one observes that:

1. The electron emission is very sensitive to the surface. Half a monolayer of CO is sufficient to completely suppress the emission from the Pd-crystal.
2. The electron spectra show clearly that the electron is mainly emitted by the Penning process, and not by the two-step mechanism assumed earlier (2-4).

Further work on the problem of electron emission from well defined surfaces by rare gas metastables is in progress.

(1) H. Conrad, G. Ertl, S. W. Wang, K. Gérard, and H. Haberland, *Phys. Rev. Lett.* **42**, 1082 (1979).
(2) References 36, 56, and 57 of this chapter.
(3) P. D. Johnson and T. A. Delchar, *Surface Sci.* **77**, 400 (1978).
(4) J. Roussel and C. Boizeau, *J. de Phys.* **38**, 757 (1977).]

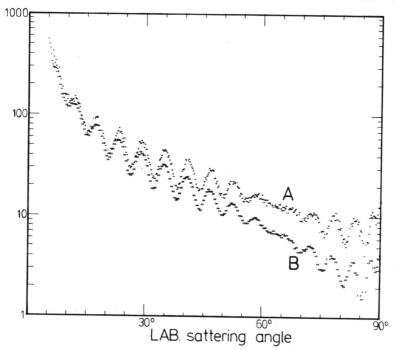

Figure 4. He*(2^3S) + He laboratory angular distributions at same energy for (A) cold excited beam and (B) cold ground-center-of-mass state beam. Relative intensities differ because of velocity dependence of intensity transformation Jacobian.

It is generally not necessary to differentially pump the detector. An electric field perpendicular to the detector axis is used to prevent any charged particles from reaching the detector. The background count rate, with the He* beam turned off and the quench lamp(s) burning, should not exceed 5 to 10 counts per second. The large-angle falloff of the He* beam profile is also much better than in ground-state scattering. For sufficiently collimated beams the large angle ($\theta > 30°$) beam profile is determined mainly by scattering of the He* beam from the background gas. For a 10^{-6}-torr vacuum, a value of 10^{-9} is typically attained between the intensities at $\theta = 0°$ and $\theta = 90°$. This low background is two to three orders of magnitude lower than that for the best detectors for ground-state particles, which use electron-impact ionization and UHV-techniques. Because of the extreme simplicity of the detector, a stationary monitor detector can be installed easily, whose count rate can be used to compensate for shifts and fluctuations in the beams and quench lamps, although not for gain shifts in the two detectors.

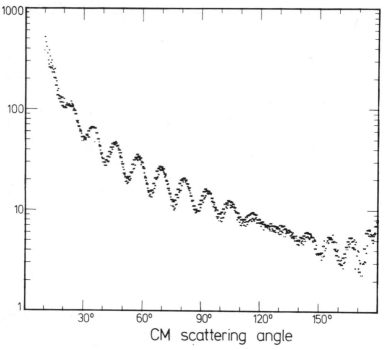

CM scattering angle

Figure 5.   Center-of-mass angular distributions derived from data of Fig. 4. Results are identical within experimental error, showing velocity independence of detection efficiency in energy range 17 to 86 meV.

The intensity at small scattering angles is very high, which can lead to count-rate-dependent gain variations and a rapid deterioration of the detector. Copper–beryllium multipliers can be rejuvenated easily by heating them first with a hot air gun to 100 to 120°C and then rinsing the hot multiplier in cold methanol.

2. Quench Lamp

Gas-discharge lamps are used to optically pump the metastable helium atoms into a higher excited electronic state, which has a dipole-allowed transition to the ground state. Only He*(2^1S) can be pumped selectively, thereby producing pure He(2^3S) beams. For the heavier rare gases, both metastable states are equally pumped by gas-discharge lamps. The use of cutoff filters to selectively pump one state is not adequate because of the temperature dependence of the filter transmission and the low f numbers of the pumping transition. Metastable neon can be selectively pumped by a continuous wave (cw) dye laser,[60] whereas Ar*, Kr*, and Xe* have so far only been selectively pumped by pulsed dye lasers.[61]

Table II.
Helium Quench-lamp Characteristics

Operating current	50 to 75 mA
Operating voltage	2.5 to 3.0 kV DC
Startup voltage	4.5 kV
Helium pressure	1 to 5 torr (flowing)

A detailed investigation of the 2μ output of capillary helium gas discharges has shown the features collected in Table II to be optimal.[58] Four to five turns of a 3-mm inner diameter Pyrex tube is wound helically around a 35-mm mandrel. The anode is a small (1 mm diameter) tungsten pin. The hollow cathode is made from aluminum of 20 to 30 mm diameter and 50 mm height. A slow, continuous stream of gas flows from anode to cathode. The lamp has to be cooled effectively for smooth operation. The following procedure has been proven very satisfactory. All outer parts of the lamp are coated by a thin (1 to 2 mm) layer of a silicone rubber[62] and put into a splittable aluminum housing. The space between the housing and the lamp is then filled with the same silicone rubber, which gives good thermal contact and electrical isolation. The lamp housing is clamped on a precision-machined water-cooled rod. The total length of the lamp and housing is 40 mm. For a He* beam above a kinetic energy of about 100 meV, a larger lamp is needed for adequate ($>99\%$) quenching power. It is more convenient to work with two short lamps instead of a longer one. For many experiments, the lamps have to be switched on and off periodically. The maximal usable frequency is around 10 kHz; above this limit the pulse shape deteriorates because of the long-lived afterglow.[58]

3. Description of a Scattering Apparatus

Figure 6 shows that central part of the Freiburg scattering apparatus for which the best resolution has been obtained so far. The helium gas is expanded from a very high pressure (20 to 100 bar) through a small nozzle hole (12 to 100 μm) into the first differential pumping chamber. The temperature of the gas before the expansion can be varied between $T=80°$K and 1600°K. Figure 6 shows beam sources that are used at and below 300°K.* The final kinetic energy of an atomic nozzle beam is given by $E=\frac{5}{2}kT$ ($\frac{3}{2}kT$ from the random motion, $1\,kT$ from the work gained in the expansion), so that the beam energy can be varied between 16.5 meV and approximately 350 meV. The central part of the beam passes through

*[NOTE: The high-pressure, high-temperature supersonic beam source has been described in detail by B. Brutschy and H. Haberland, *J. Phy.* **E13**, 150 (1980).]

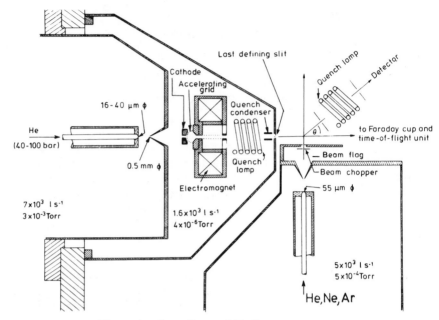

Figure 6.   Central part of Freiburg apparatus.

a Campargue-type skimmer[63] into the excitation chamber. It traverses a hole in the indirectly heated cathode, which has a spherical electron-emitting surface. The electrons are accelerated by a concentric grid to typically 150 to 200 eV of kinetic energy. The two beams interact over a 4-cm distance in the electromagnet, whose field compensates the diverging effect of the space charge of the electrons. After the excitation region the singlet metastables can be deexcited by the quench lamp. Charged particles and helium atoms in very high Rydberg states are removed from the beam by the quench condenser. The last defining slit collimates the beam to 0.4°. A second quench lamp in front of the detector is used for kinetic energies above 100 meV, when one lamp is not sufficient for an adequate (>99%) quenching efficiency, or to quench metastable neon atoms, which have been produced by excitation transfer (see Section III.D.2). A more detailed description has been given elsewhere.[58, 64]

4. Velocity Distributions

The excitation of an atomic beam by electron impact is surprisingly intricate, although little more than momentum and energy conservation is needed for a basic discussion. The kinematics of the excitation process has

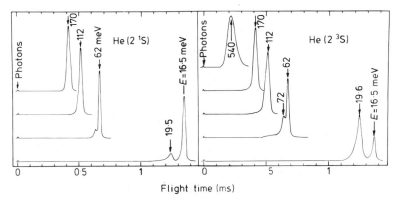

Figure 7. Time-of-flight distributions of metastable helium beams. Singlet distributions are generally narrower than triplet ones. The 540-meV distribution was obtained with plasma-jet source (see Section III.A.6).

been discussed in detail and is not repeated here.[64] It turns out that the narrow velocity distribution of the ground-state beam is best preserved by directing the electron beam parallel or antiparallel to the gas beam. The detailed shape of the velocity distribution depends on a number of parameters, such as electron energy and current, compensating magnetic field, angle of divergence of the electron beam, and background and gas pressure. In the apparatus shown in Fig. 6 the velocity distributions of the He* beam can be observed during the measurement of an angular distribution by the time-of-flight (TOF) method. The beam is mechanically chopped and the TOF spectra recorded by a fast multiscaler. Figure 7 shows some TOF distributions. The overall experimental resolution is $t = 10$ μsec for a flight path of 121 cm. A detailed analysis of the surprising bimodal structure clearly visible in Fig. 6 for flight times greater than 0.5 msec has been given elsewhere.[64] Only the results are quoted here. The peak at smaller flight times is attributable to He* atoms from the primary excitation process:

$$He + e^- \rightarrow He^* + e^-$$

The second peak, at larger flight times, which has exactly the velocity of the unexcited helium beam, is caused by a subsequent resonant energy transfer:

$$He^* + He \rightarrow He + He^*$$

The difference in flight time between the two peaks is caused by the momentum transfer of the electron to the helium atom during the excitation process. The cross section for the resonant energy-transfer process is to a first approximation proportional to the $g-u$ splitting of the two corresponding excited-state potential curves of the He_2 molecule.[65] The van der Waals constants for the singlet state differ by 15%, whereas they are equal for the triplet state.[66] This leads to a larger splitting and thus an increased rate of energy transfer and a better velocity resolution of the singlet He* beams. The exchange peaks at longer flight times are in general larger than the direct excitation peak. This means that most He* atoms will have experienced more than one excitation transfer collision in the beam. A simple calculation shows that $>98\%$ of the excitation transfers occur before the atoms reach the scattering center, so that the measured velocity distributions are the appropriate ones. At higher stagnation temperatures the velocity distribution from nozzle beams deteriorates. The relative momentum transfer from the electron also becomes smaller for increased He* velocities so that the two peaks finally coalesce.

The TOF distribution for $E = 540$ meV has been obtained in a completely different way. The helium plasma of a high-current arc discharge has been expanded through a small nozzle hole, giving directly a supersonic $He(2^3S)$ beam of higher kinetic energy (see Section III.A.6).

5. Intensities

The intensity of the He* beam of the apparatus shown in Fig. 5 collimated to 0.4° full width at half maximum (FWHM) is typically 10^{10} singlets/sec (or $3 \cdot 10^{14}$ atoms/sec·sr) and $1.5 \cdot 10^9$ triplets/sec for beam energies of 66 to 350 meV and about 20% of this value at 16.5 meV (liquid-nitrogen-cooled nozzle). The triplet intensity can be increased at the expense of a poorer velocity resolution by a lower electron-acceleration voltage.

There is a difference of roughly seven orders of magnitude in the intensity of the ground-state helium beam compared to the metastable helium beam. This large ratio may appear unfavorable, but it cannot be improved easily without unfavorable effects on the velocity and angular resolution. The electron beam excites $\approx 10^{-3}$ of the ground-state beam to the metastable state, a fraction that cannot be increased very easily under beam conditions. But only $\approx 10^{-4}$ of the excited He* atoms remain in the final beam. All others have too large scattering angles and hit some collimating diaphragms and hence are therefore removed from the He* beam before it can enter the collision chamber. A comparison of beam intensities for different designs has been given.[64]

6. Other He* Sources

The highest kinetic energies available from a nozzle source are limited by the melting point and tensile strength of the nozzle materials to energies below about 400 meV. Beams energies above approximately 10 eV can be obtained by charge exchange in cesium vapor.[93,94] To obtain He* beams between these two extremes, two different gas-discharge devices have been tried.

Searcy[68] operated a low-current, high-voltage (0.1-mA, 8 kV) discharge between a positively charged needle through a nozzle hole onto a skimmer. The He* atom beam has an energy of 4.6 eV and an intensity of 10^{10} atoms/sr·sec. One scattering experiment has been reported using this source.[69]

Schmidt[70] has tried a low-voltage, high-current arc discharge (30 V, 40 A). The hot helium plasma expands through a relatively large nozzle (~0.1 to 0.3 mm). A sufficient number of He* atoms, more than 98% in the triplet state, are produced directly by this plasma jet. The beam intensity is roughly a factor of 5 less intense, compared to electron-impact excitation, but nearly four orders of magnitude more intense than that from Searcy's design. By varying the electrical power, gas pressure, and nozzle diameter, the kinetic energy can be varied up to 0.8 eV for He* and 1.6 eV for Ar*. The beam intensity and energy is (sometimes) sufficiently stable for several hours for angular distributions to be measured (see Figs. 18 and 30). Higher energies could not be obtained because of the very low-momentum transfer cross section between electrons and helium atoms (Ramsauer minimum) and cooling of the beam at the water cooled nozzle hole. Because of the large power density in the discharge region (2 to 5 kW/cm^3), efficient water cooling was necessary. A modified version of Maecker's cascade plate design[70,71] was used for confining the arc.

For ground-state helium beams, Knuth's group[72,73] has obtained energies up to 5 eV with a different arc source. Large nozzle holes (~2 mm) and very high pumping speeds were used in the first chamber. It would be surprising if this design could not also be made to yield higher-energy metastables. The larger nozzle hole might drastically reduce the cooling of the hot helium plasma.

7. Influence of Photons on Differential Cross Sections

Two cases are known where the experimental differential cross sections are strongly perturbed over a narrow angular range by UV photons. Both cases are discussed in detail to facilitate the identification of similar effects in future work.

For the scattering of $He(2^3S) + He$, Haberland et al.[74] observed a very narrow spike at $\theta_{lab} = 90°$ in the angular distribution as shown in Fig. 8.

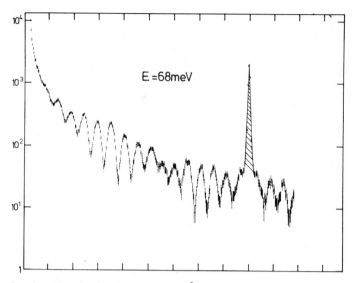

Figure 8. Angular distribution for He*(2^3S)+He showing sharp 90° peak partially caused by photoexcitation of crossed beam.

The ratio of the intensity at $\theta = 90°$ to that at some other angle was independent of nearly every experimental parameter (beam intensity, electron current and voltage, magnet current, and background and secondary-beam pressure); consequently, it was thought that the 90°-spike was not an experimental artifact. Only after it was impossible to produce a similar spike in a theoretical computation did a second detailed investigation reveal the following cause. There are always some photons from the helium resonance transition ($2^1P \rightarrow 1^1S$, $E = 21.21$ eV) in the beam. Because of the extremely large cross sections for resonance absorption, they can be effectively trapped in a strong beam. Some of those photons are absorbed by helium atoms from the secondary beam in the scattering center. Most of them decay back to the ground state, but a fraction of 10^{-3} make a transition to the metastable 2^1S state, and these atoms can be detected on the multiplier. The shape of the 90° peak is given exactly by the convolution of angular profiles of the detector and secondary beam, as the momentum transfer by the photon is negligible in this case.

A similar peak at $\theta_{lab}^\circ = 90°$ was observed by Martin et al.[75] and by Haberland et al.[76] in scattering studies of He*+Ne. An interpretation similar to that for He*-He was given first by the latter authors but was proven incorrect.[75] The following two-step process is presently thought to be the most likely cause.[77] An excitation transfer occurs from He*(2^1S) to

Ne*($3s_2$) in the scattering center. The $3s_2$ state has a lifetime of only about 10^{-8} sec, so that most of the atoms radiate when the two atoms have separated but are still in the scattering center. More than 50% of the excited neon atoms make a transition to the ground state. The emitted UV photon can be reabsorbed by any neon atom nearby. The strongest absorption occurs from atoms in the intense (10^{-4} T) secondary beam. If these decay to the metastable states, they are detected at $\theta_{lab} = 90°$.

In summary, any sharp peak at $\theta_{lab} = 90°$ is very valuable for testing the resolution of the apparatus but should be regarded with suspicion.

B. Pure Elastic Scattering

As outlined in Section II.B, optical model analysis of elastic scattering, even in the Penning systems, may be ambiguous because of the interplay between the real and imaginary parts of the optical potential. Therefore, it was desirable to examine systems for which the absorption or inelasticity would be vanishingly small. Although this is difficult for electronically excited atoms, evidence from static afterglow experiments[78-80] attests to fairly slow quenching of metastables in pure noble-gas discharges (although very rapid resonant energy exchange). For neon through xenon, quenching by the bath gas occurs mainly by intermultiplet mixing,[2,3] for example,

$$\text{Ne*}\left(^3P_{2,0}\right) + \text{Ne} \rightarrow \text{Ne*}\left(^3P_1, {}^1P_1\right) + \text{Ne} \rightarrow \text{Ne} + \text{Ne} + h\nu$$

and the measured rates correspond to thermal cross sections of approximately 5×10^{-3} Å^2 or less. If the ground-state noble-gas partner is lighter [e.g., Ne*($^3P_{2,0}$) + He], intermultiplet mixing is the only possible inelastic process at thermal energy. Such systems are thus likely to behave nearly elastically, and one can safely use a real two-body potential to analyze the angular distributions.

Experimentally, it is not feasible to detect the scattered ground-state atom because of the enormous background of ground-state atoms in the excited beam (see Section III.A). Although detection of the heavier scattered atom is kinematically unfavorable because its recoil velocity is smaller than the center-of-mass velocity, analysis of differential scattering can still be carried out, since the center-of-mass→laboratory transformation is unambiguous. In addition, it becomes possible to observe backward scattering in the normal range of laboratory angles. In the fitting procedure, a parametric potential function is assumed, a center-of-mass angular distribution calculated [usually by numerical integration of equation (II.5)] and transformed to the laboratory system with appropriate resolution averaging for comparison with experiment. Parameter values are adjusted by the Marquardt nonlinear least-squares method[81] to give a best fit.

Figure 9 shows scattering data for Ne* + He and Ar* + He[82] from the Pittsburgh laboratory. The wide-angle maxima are kinematic artifacts ("Jacobian rainbows") because of the infinite density of center-of-mass scattering angles contributing at the edge of the recoil velocity shell for the heavier atom. If the beams were monochromatic and angular resolution perfect, no scattering would be observed outside the Jacobian rainbow angle. The observed scattering intensity thus drops off rapidly and is a measure mainly of the velocity resolution. The calculated curves are fits from assumed potential functions. The absence of interference features attests to the repulsive character of the interactions as well as to the lack of significant splitting of the potential curves for different Ω. A sufficiently fine-grained oscillatory pattern would not be resolved, however. Because the Jacobian rainbows are very sensitive to apparatus resolution, their angular region was excluded in the fitting procedure.

Potential functions used[82] were an exp-6 function with two free repulsive parameters and a modified exp-6 with three repulsive parameters. The modification was introduced as a screened ion-induced dipole attraction, which softens the repulsion at small r. The fitted potentials are shown in Fig. 10. The dashed curves in Fig. 9 show the fits obtained with the simple exponential repulsion and the solid curves with the modified repulsion.

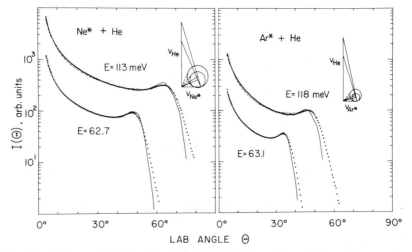

Figure 9.   Laboratory angular distributions for Ne* and Ar* + He. Solid points are experimental; solid and dashed curves are calculated from potentials of Fig. 10.

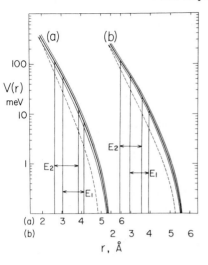

Figure 10. Interatomic potentials for (a) Ne* + He and (b) Ar* + He derived from fitting the data of Fig. 8. Vertical lines show the range of internuclear distance probed at each energy E_i, with $E_1 < E_2$ as given in Fig. 9.

The bend in the potentials at $r \simeq 3$ Å is clearly required for a good fit to the higher energy. The part of the repulsion probed at each energy is indicated in Fig. 10. Fitting of He* scattering has also required modification of the low-energy repulsion (see Section III.C). Further experimental work on these and other systems of the heavy-metastable–light-atom variety is in progress.

C. Symmetric Noble-gas Systems

1. Symmetry Properties of Potentials and Scattering Amplitudes

We begin by specializing the discussion of Section II.C to the He*–He case. If a metastable helium atom collides with a ground-state helium atom, the interaction to two identical atoms in different electronic states results. This simple fact has a strong influence on the scattering patterns.

Asymptotically, one beam can be prepared to carry electronically excited atoms (A) and the other ground-state atoms (B), and thus the wave function can be written as

$$\phi_1(A) \cdot \phi_0(B)$$

where $\phi_1(A)$ means that atom A is in the excited state and $\phi_0(B)$, that B is in the ground state. As the Hamiltonian is invariant with respect to interchange of atom A and B

$$\phi_1(B) \cdot \phi_0(A)$$

is also a possible wave function. A linear combination will then give the correct asymptotic eigenfunction

$$\chi^{g,u}(r\to\infty)\sim\phi_1(A)\phi_0(B)\pm\phi_1(B)\phi_0(A)$$

where g (*gerade*) corresponds to the plus sign and u (*ungerade*) to the minus sign. The asymptotically prepared state is not an eigenstate of the Hamiltonian when the atoms are close together, leading to a rapid exchange of the excitation energy between the two atoms. The electronic wave functions χ^u and χ^g are orthogonal for all internuclear distances r; in addition, all coupling terms vanish for identical isotopes. Within the Born–Oppenheimer approximation the total interaction is averaged over the electronic wave functions for fixed r. As we average over different electronic wave functions, we obtain different interaction potentials V_g and V_u. These are shown for the lowest electronic states of He$_2$ in Fig. 11 and are discussed in the text that follows. Within the Born–Oppenheimer

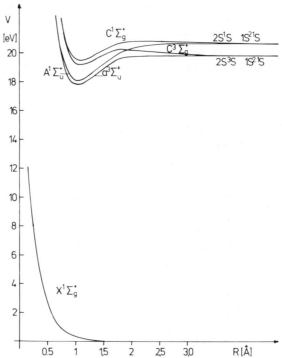

Figure 11. Potentials for He$_2$. Figures 14 and 17 show long-range parts of excited-state curves in greater detail.

approximation the potential does not depend on the mass of the heavy particles; therefore, the potentials for $^4He^4He$, $^4He^3He$, and $^3He^3He$ are identical.

The scattering amplitudes, on the other hand, depend strongly on the combination of isotopes. The total scattering amplitude for distinguishable particles (e.g., $^4He + ^3He$) becomes

$$f(\theta) = \tfrac{1}{2}\left[f_g(\theta) + f_u(\theta) \right];$$

For indistinguishable particles, this has to by symmetrized appropriately. The ^4He nucleus is a boson, and hence the total wave function does not change sign when interchanging the nuclei. One obtains for the properly symmetrized scattering amplitude

$$f(\theta) = \tfrac{1}{2}\left[f_g(\theta) + f_g(\pi - \theta) + f_u(\theta) - f_u(\pi - \theta) \right]$$

The amplitude f_u is antisymmetric with respect to interchange of the nuclei, which is a direct reflection of the symmetry property of the corresponding electronic wave function. This implies that the cross section need not be symmetric about $\theta_{CM} = 90°$. We can define a scattering amplitude $f_d(\theta)$ for direct scattering

$$f_d(\theta) = \frac{1}{2}\left[f_g(\theta) + f_u(\theta) \right]$$

and for exchange scattering,

$$f_{ex}(\theta) = \frac{1}{2}\left[f_g(\pi - \theta) - f_u(\pi - \theta) \right].$$

The total scattering amplitude becomes

$$f(\theta) = f_d(\theta) + f_{ex}(\theta)$$

The corresponding scattering process can be visualized as shown in Fig. 12. The direct and exchanged excited particles come from different beams. But the detector, in principle, cannot distinguish between them; thus the amplitudes add coherently. If the overlap between direct and exchange amplitudes is small, the exchange contribution can be isolated.[65] For noble-gas exchange scattering at thermal energies, however, the overlap is substantial (see Section III.C.5).

Other exchange processes may be described similarly, such as spin exchange, charge exchange, or—at much higher kinetic energies—neutron or π^- exchange.[83]

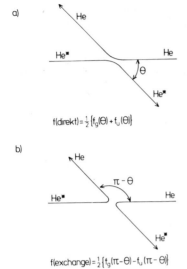

Figure 12. Classical picture of direct and exchange scattering.

2. General Features of Noble-gas Excimer States

The potential-energy curves of the noble-gas diatomic molecules are rather unusual.[65,84,85] The ground state of the He_2 molecule is purely repulsive save for a weak van der Waals minimum (well depth ~ 1 meV), which might not support a bound state. The next four excited states correlate asymptotically with $He(2^1S)$ and $He(2^3S)$, respectively. As can be seen from Fig. 11, these states have deep chemical wells at about 1 Å and intermediate maxima at 2 to 3 Å.

The $He_2 A^1\Sigma_u^+$ state has nearly the same dissociation energy as the $He_2^+ (^2\Sigma_u)$ ion. This supports the idea that the excited He_2 configurations can be described at small interatomic distances as an inner He_2^+ core with an outer Rydberg orbital. This description is less quantitative for the heavier rare-gas pairs. The unusual maxima result either from curve crossing (e.g., $C^1\Sigma_g^+$) or as for the $A^1\Sigma_u^+$ state by a changeover in the dominant exchange energy. At large r the electron clouds overlap only weakly, and the usual repulsion from the Pauli principle results. At smaller r the attractive He_2^+ ionic core is formed. An extensive discussion of the He_2 potentials has been given by Guberman and Goddard.[84]

Transitions from the $A^1\Sigma_u^+$ potential become optically allowed for small internuclear distances and give rise to the well-known Hopfield continuum and the 600-Å emission and absorption bands.[86] The inner attractive parts of the potentials have been determined quite accurately from analysis of the optical spectrum. Information on the long-range parts, however, has been only semiquantitative at best.

For the heavier noble gases, the core multiplicity of the metastable states give rise to eight potential curves (six for $^3P_2$ and two for $^2P_0$). The Ne$_2$ potentials have been calculated by Schneider and Cohen,[87] who have also performed scattering calculations for this system. Theoretical and experimental data exist for Ar$_2$,[88] whereas only qualitative estimates of the potentials are available for Kr$_2$ and Xe$_2$. These excited states play a prominent role in the rare-gas excimer lasers.[89,90]*

3. Experimental Results and Potentials

Figures 13, 16, 18, and 19 show the experimental results for He*–He scattering in the laboratory system. The intensity in arbitrary units is plotted against the laboratory scattering angle. For two particles of equal mass—as is the case here—the center-of-mass scattering angle is obtained by multiplying the laboratory angle by a factor of 2. For the thermal-energy results, the potential parameters were determined by trial and error. Piecewise analytical functions, coupled by spline interpolation polynomials, were used to represent the shape of the potential. The phase shifts were calculated numerically by the Numerov procedure. The calculated cross section was then transformed into the laboratory system and averaged over experimental resolution. The broadening of the data as a result of the limited velocity resolution was taken into account by increasing the breadth of the angular resolution appropriately so as to conserve computer time. The free potential parameters are determined by the Marquardt nonlinear least-squares routine.[81] The analytical form of the potential is rather complicated, but not too much effort was made to keep the number of free parameters small because of the complicated shapes of the potentials. Because of the extensive interference patterns, the potential parameters had to be close to their final values to allow convergence of the Marquardt routine; thus a rather large amount of manual adjustment of the potential parameters was necessary.

a. He*(2^1S) + He

The angular distributions for He*(2^1S) + He are shown in Fig. 13 for six different kinetic energies. The part of the potential that can be derived from the data is given in Fig. 14. (Note the change in energy scale above -50 meV.) The horizontal arrows give the collision energies used in the experiments. The potential is given in Table III (energies are in electron volts and distances in angstroms). A Morse-type potential V_1 has been obtained by Sando[86] for the inner part ($r \leqslant 1.7$ Å) or the $A\,^1\Sigma_u^+$ potential

---

*[NOTE: References to recent experimental and theoretical work on the excited Xe$_2$ states are given in U. Buck, L. Mattera, D. Pust, and D. Haaks, *Chem. Phys. Lett.* **62**, 562 (1979).]

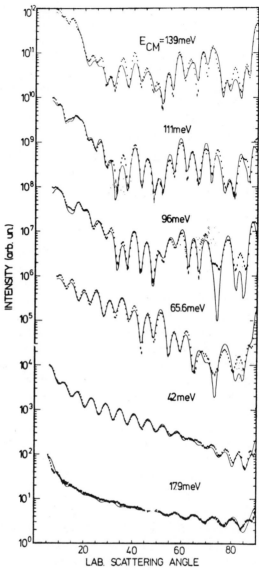

Figure 13. Laboratory differential cross sections for He*(2^1S) + He at six kinetic energies.

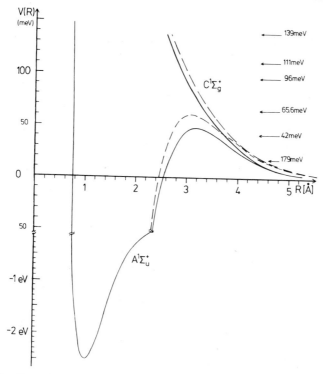

Figure 14. Potentials for He*(2^1S) + He derived from data of Fig. 13. Apparent discontinuity at 50 meV results from change of ordinate scale. Dashed lines are GVB *ab initio* results due to Guberman and Goddard.[84] Collision energies are given at right. Potentials are tabulated in Table III.

from the analysis of optical data. He also gives some numerical values for larger r, which were fitted by subtracting a sine function from $V_1(r)$, giving $V_2(r)$. The small r side of the potential maximum was represented by a parabola (V_4) that was smoothly joined at both ends by cubic spline interpolation (V_3, V_5). The long-range part was found to be well represented by a modified exponential function (V_6). The long-range part of the $C^1\Sigma_g^+$ curve could also be represented by a modified exponential (V_g). Only the parameters of V_4, V_6, and V_g were varied by the Marquardt routine. Values for V_1 and V_2 have been determined by Sando, and V_3, and V_4 are spline interpolations.

The potential minima of the van der Waals attraction[58,91] are at $r \geqslant 6$ Å, where they have only a negligible influence on the differential cross section. Therefore, they were neglected in the calculation. At $r = 6$ Å the

Table III.

Numerical Values for Singlet Potentials $V_g = C^1\Sigma_g^+$ and $V_u = A^1\Sigma_u^+$

R (Å)	V_g (meV)	V_u (meV)
0.5		9867.0
0.75		−501.0
1.00		−2472.0
Minimum (1.05)		−2492.0
1.25		−2122.0
1.50		−1359.0
1.75		−704.0
2.00		−303.0
2.25		−101.0
2.50	163.0	−82.0
2.80	109.0	32.6
3.00	90.1	44.9
Maximum (3.15)	77.2	47.4
3.20	73.4	47.2
3.40	59.8	44.1
3.60	48.0	38.4
3.80	37.5	31.5
4.00	28.2	24.7
4.20	20.3	18.5
4.40	14.2	13.3
4.60	9.6	9.2
4.80	6.4	6.3
5.00	4.1	4.2
5.25	2.4	2.6

well depth would be smaller than 1 meV. The 15% difference in the van der Waals constants for the two potentials thus affects the differential cross sections only indirectly.[109] But this difference has a significant influence on the form of the velocity distribution of the He* beam,[64] where the relative kinetic energies are much lower (10^{-2} to 10^{-5} eV). The fit to the data is quite good, especially at lower energies, but could still be improved.

For one particular energy, a much better fit could usually be obtained, but then the fits at the other energies deteriorate rapidly. The χ^2 values for the fits are 8.4, 13.6, 108, 63, 57, and 131 from the lowest to the highest energy. These values are relatively large because of the very small error bars of the experimental results. A more flexible potential and proper

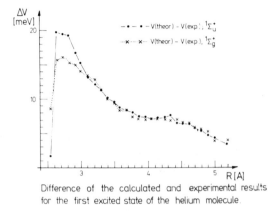

Difference of the calculated and experimental results
for the first excited state of the helium molecule

Figure 15. Difference between calculated and experimental potentials for first excited $^1\Sigma$ states of He$_2$ molecule. Within experimental error the two curves are identical for $r > 3$ Å.

treatment of the velocity averaging could probably yield much smaller χ^2 values.

Buckingham and Dalgarno[92] were the first to calculate the He*(2^1S)–He interaction. Recently Guberman and Goddard[84] performed a generalized valence-bond (GVB) calculation for many excited He$_2$ states. Their results are given by the dashed lines in Fig. 14. The calculated barrier height is 60.7 meV at 3.09 Å, compared to $47 \pm^2_1$ meV at 3.14 ± 0.05 Å from the analysis of the differential cross section. A GVB calculation always gives an upper limit to the exact result;[84] in fact, the theoretical results are 5 to 20 meV higher everywhere.

Figure 15 shows the difference between the experimental and calculated potentials. Because the splitting is very small for $R > 3$ Å, this difference is nearly independent of the g–u symmetry, reflecting differences mainly in the mean potential. Guberman and Goddard propose a reduction of their results by 10 to 20%, to account for the neglected part of the correlation energy, but this reduction is sufficient only between 3.2 and 4.0 Å. Assuming the accuracy of the potentials derived from differential scattering, Fig. 15 represents, for large r, the remaining correlation energy.

Electronic transitions are optically forbidden only for large internuclear distances r. For finite r, dipole transitions to the $X\,^1\Sigma_g^+$ ground state are possible. They give rise to the well-known Hopfield continuum and 600-Å emission and absorption bands.[86] Only those collision partners that surmount the barrier of the *ungerade* potential are likely to radiate. The cross section for light emission[65] is typically 10^{-4} Å^2, which is much too small to have a noticeable influence on the differential cross sections and hence was therefore neglected.

b. He(2^3S) + He

The thermal-energy results are shown in Fig. 16. The data are noisier as the triplet intensity is a factor 5 to 7 smaller than the singlet intensity. The overall structure of the data is similar to that for singlet scattering. The solid line gives again the differential cross section from the potential shown in Table IV and Fig. 17. The maximum in the interaction potential is at a smaller distance and roughly 10 meV higher than in the singlet case.

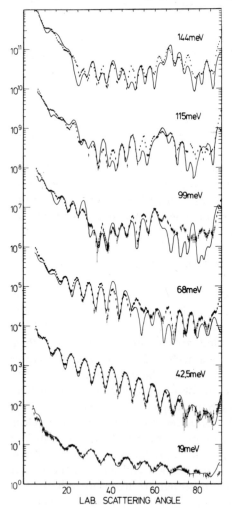

Figure 16. Laboratory differential cross sections for He*(2^3S) + He at six kinetic energies.

Table IV.

Numerical Values for Triplet Potentials $V_g = c^3\Sigma_g^+$ and $V_u = a^3\Sigma_u^+$

R (Å)	V_G (meV)	V_U (meV)
0.5		10990.0
0.75		98.0
1.00		−1920.0
Minimum (1.05)		−1947.0
1.25		−1615.0
1.50		−842.0
1.75		−232.0
2.00		−36.1
2.25		37.6
2.40		52.0
2.60	144.7	56.8
Maximum (2.75)	127.6	57.0
3.00	97.9	53.5
3.20	74.8	46.5
3.40	53.8	37.3
3.60	36.3	27.6
3.80	23.6	18.9
4.00	15.1	12.6
4.20	9.6	8.3
4.40	6.1	5.4
4.60	3.8	3.5
4.80	2.4	2.2
5.00	1.5	1.4
5.25	0.82	0.78

The potential maximum could not be fitted by a parabola as in the singlet case; a r^4 functional dependence was found to be more adequate. The deep chemical well at 1.045 Å was represented by a Morse function, which reproduced Ginter's spectroscopic results.[160]

The long-range part of the potential has been calculated by Das[161] in a multiconfiguration self-consistent-field (SCF) computation. He obtains a van der Waals minimum at about 7 Å with a well depth of 0.16 meV, which is consistent with our results. The dotted lines have been determined by Hickman and Lane[162] from thermal diffusion and exchange measurements. The agreement is satisfactory below 40 meV. The data they analyzed were limited to this energy range. Earlier attempts to obtain the triplet potentials from bulb experiments have been reviewed by Fugol.[163]

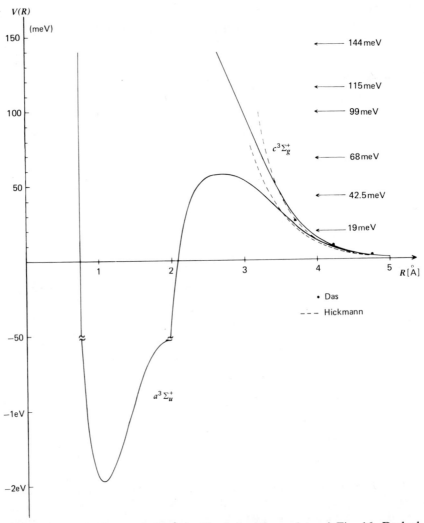

Figure 17. Potentials for He*(2^3S) + He derived from data of Fig. 16. Dashed lines are derived from analysis of spin-exchange experiments, whereas points are taken from an *ab initio* calculation. Potentials are tabulated in Table IV.

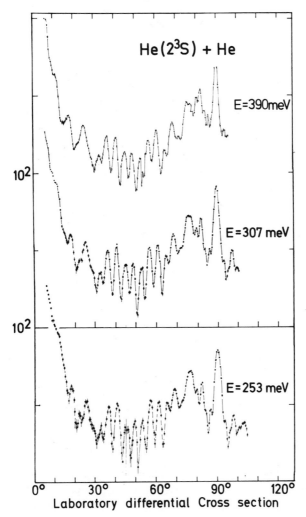

Figure 18. Hyperthermal energy differential cross sections for He*(2^3S) + He, obtained with plasma He* source.

Figure 18 shows differential cross sections measured using the helium plasma jet described in Section III.A.6. It is surprising that so much structure is still resolved, even though the velocity resolution of the beam is only 30%. No fit has so far been attempted for these data.

It is also possible to produce He(2^3S) atoms by the charge exchange of He$^+$ ions in cesium-vapor.[93] This technique has been used extensively by the Stanford Research Institute (SRI) group to measure differential cross sections at higher energies (5 to 10 eV). Some of the data[94] in reduced units

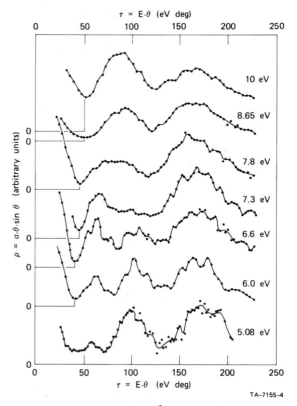

Figure 19. Reduced-variable plot of He*(2^3S) + He differential cross sections, obtained with charge-exchange He* source.

are shown in Fig. 19. The kinetic energy is high enough that endoergic inelastic processes substantially affect the scattering. Analysis of these data requires a multichannel treatment.

4. Heavier Noble-gas Symmetric Systems

Scattering has been measured for Ne* + Ne, Ar* + Ar, and Dr* + Kr by different groups at thermal energies, but no data and potentials have been published so far, as the analysis is quite involved. First, it is impossible to quench one of the two metastable states without an expensive laser, so that an investigator is generally forced to work with mixtures; and second, six potentials contribute coherently for the dominant $^3P_2$ species. The cross sections for fine-structure changing collisions are small[78-80] and hence can be neglected. At higher kinetic energies the Ne* + Ne and Ar* + Ar scattering has been studied by the SRI group.[93]*

5. Total Cross Sections

The total cross sections calculated from the potentials in Table III and IV are shown in Figs. 20 and 21. The solid line is for identical particles, whereas the dotted line has been calculated assuming distinguishable particles. For $He(2^3S)$ + He the experimental data of Trujillo[95] are included in Fig. 21. The total cross section for $He^*(2^1S)$ is roughly 40 Å^2 larger than that for $He^*(2^3S)$. For energies below the barrier heights, the cross sections for distinguishable particles can be well approximated by

$$Q = A \cdot v^{-B} [\overset{\circ}{A}{}^2]$$

with $A = 503$, $B = 0.14$ for $He^*(2^1S)$ and $A = 529$, $B = 0.18$ for $He^*(2^3S)$, and v in m/sec. The oscillations below approximately 50 meV must be symmetry osscillations as they vanish for distinguishable particles. They result from the interference of collisions with large impact parameters, which are nearly forward scattered ($\theta \approx 0$), with energy-transfer collisions at small impact parameters and $\theta \approx \pi$. The relative difference between the two cross sections is given by curve I (arbitrary scale). The position of the different maxima is nearly entirely given by the energy dependence of the S-wave phase shift, as discussed elsewhere.[58]

The sharp structures above 50 meV are the result of orbiting resonances from the deep attractive well of V_u. All particles can tunnel through the maximum in V_u. The amplitude for finding a particle inside the maximum will be resonantly enhanced if the kinetic energy matches the energy of a quasibound state of V_u. For energies much below the barrier height, the

---

*[NOTE: Scattering studies of metastable neon ($^3P_0$ and $^3P_2$) have been performed at Freiburg. A single mode dye-laser was used to selectively quench one of the two fine-structure states. The natural isotope composition of neon is 90.5% ^{20}Ne and 9.2% ^{22}Ne. As the isotope shift (1.7 GH) of the pumping transition is much larger than its experimental linewidth (.2 GH, mainly residual Doppler width and power broadening) the two isotopes and the two fine-structure states can be labelled separately by the laser. The natural isotope composition of the beam is therefore an extra bonus and not a disadvantage as usually. By switching the laser on and off before and after the scattering center angular distributions for excitation, transfer processes like

$$^{20}Ne^* + {}^{22}Ne \rightarrow {}^{20}Ne + {}^{22}Ne^*$$

can be measured in addition to the elastic processes. Scattering studies for a series of atoms and molecules are in progress (1979). In some cases large differences in the differential cross sections for the two fine-structure components are found; see W. Beyer, H. Haberland, and D. Hausamann, unpublished data.]

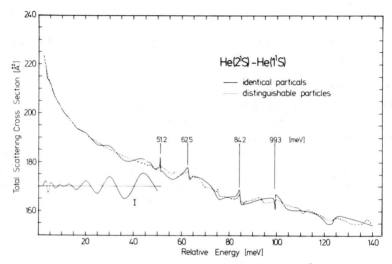

Figure 20. Energy dependence of total cross section for He*(2^1S)+He calculated from potentials of Fig. 14 and Table III. Oscillations at low energies attributable to nuclear-symmetry. Glory effect is amplified in curve I, in which difference between cross sections for identical and distinguishable particles is plotted on an expanded scale.

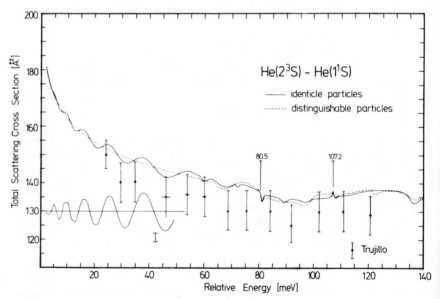

Figure 21. Energy dependence of total cross section for He*(2^3S)+He calculated from potentials of Fig. 17 and Table IV. Data points represent measurements of Trujillo.[95] Format as in Fig. 20.

tunneling probability and hence also the width of the resonance will be small and could, therefore, only accidentally be detected with the grid used in calculating the curves for the total cross section. If the kinetic energy is only a bit smaller than the barrier height, the width of the resonance will become larger. These orbiting or shape resonances play a large role in the calculation of the 600-Å band emitted by $He(2^1S)$ particles crossing the barrier. This spectrum and the resonances have been calculated by Sando[86] using his potential (V_1 of Table III). The resonances can be classified according to their vibrational (v) and rotational (J) quantum number, which can be deduced from inspection of the calculated wave functions.

6. Excitation-transfer Cross Section

In principle, there is no way to directly measure the exchange process for $^4He^* + {}^4He$ scattering, as the particles are indistinguishable. But the cross section for metastability exchange can of course be calculated from the determined potentials assuming distinguishable particles.[65] The expression for the total excitation transfer cross section is

$$\sigma_{\text{trans}} = \frac{\pi}{k^2} \sum_{l=0}^{\infty} (2l+1)\sin^2(\eta_l^g - \eta_l^u)$$

where $\eta_l^{g,\,u}$ are the lth phase shift calculated from $V_{g,\,u}$. The transfer cross section is determined mainly by the difference potential. The calculated excitation-transfer cross section for $^3He(2^3S) + {}^3He$ is shown in Fig. 22 as a function of the kinetic energy. Because of the increasing splitting of the two potentials for smaller r, the cross section rises with kinetic energy. It begins to oscillate when the energy becomes larger than the barrier in the *ungerade* potential. The rate of excitation transfer [σ_{trans}·relative velocity, averaged over a Maxwellian distribution] is compared to experimental results in Fig. 23. These rates have been measured in two remarkable optical pumping experiments in ^3He.[96, 97]

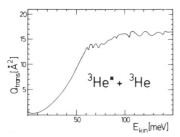

Total Cross Section for Excitation Transfer

Figure 22. Energy dependence of excitation transfer cross section for ^3He-$*(2^3S) + {}^3He$, calculated from potentials of Fig. 17 and Table IV.

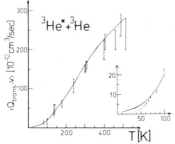

Rate Constant for Excitation Transfer

Figure 23. Temperature dependence of rate constant for excitation transfer in ^{3}He*(2^3S) +^{3}He, calculated from potentials of Fig. 17 and Table IV. Data are derived from analysis of optical pumping experiments.

The ^{3}He nucleus has spin $\frac{1}{2}$, so that the hyperfine state can be different before and after a collision. This leads to a loss of coherence and a broader line width in the optical pumping experiments. The linewidth is measured as a function of temperature, and the rate of excitation transfer is obtained after an involved analysis.[97, 98] The agreement with the higer-temperature data of Colegrove et al.[96] is very good if the correction factor of $\frac{9}{4}$ is applied to their data, as shown by Dupont-Roc et al.[98] The agreement is not as good with the lower-temperature results due to Rosner and Pipkin shown in the insert. This may be due to the neglect of the van der Waals attraction, which has a negligible influence on the differential cross sections.

7. Discussion of Interference Structure

The interference structure in the differential cross sections is quite complicated—first, because two potentials of unusual shape contribute coherently and second, because of the identical nuclei. The effect of the latter is easily studied by calculating the differential cross section assuming distinguishable particles. This is shown in Fig. 24 for He(2^1S)+He using for the total scattering amplitude $f(\theta) = \frac{1}{2}[f_g(\theta)+f_u(\theta)]$ as discussed earlier. The regular oscillations at lower energies are completely absent; therefore, they must be attributable to nuclear symmetry. For higher energies the peak at 90° is much smaller but still present and the intensity at large angles is markedly decreased because of the loss of the exchange contribution. This is indicated by the hatched area of the 139-meV curve in Fig. 24. At low kinetic energies only the long-range part of the potentials is probed, where the splitting of the potentials is rather small (i.e., $V_g \approx V_u$). Therefore, the scattering amplitudes will also be similar ($f_g \approx f_u$). Inserting this into the properly symmetrized scattering amplitude for identical particles, we obtain the result that the symmetrized and unsymmetrized scattering

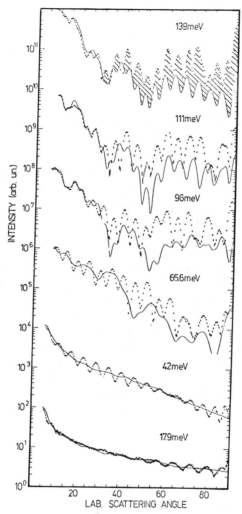

Figure 24. Calculations for He*(2^1S)+He assuming distinguishable particles compared to experiment.

amplitudes are identical for $V_g = V_u$. As this is a very good approximation for He(2^1S)+He at large r, the symmetry oscillations are washed out at the two lowest energies (see also Fig. 24).

The discussion of interferences in ground-state atom–atom scattering relies heavily on the classical deflection function, which can be calculated from equation (II.4) if the potential is known. For He*–He scattering, phase shifts can be calculated quantally only for integer l. It has proved very convenient to define a "quantal deflection function"[99] by analogy with semiclassical equivalence as

$$\chi\left(l+\tfrac{1}{2}\right)=2(\eta_{l+1}-\eta_l)$$

The phase shifts are calculated only modulo 2π, but the variation from one l to the next is rarely larger than 2π, so that deflection functions are easily constructed. Figure 25 shows the (quantal) deflection function for the 42-meV measurement. It shows the behavior expected for the scattering from a purely repulsive wall. The classical deflection function would give $\chi(l=0)=\pi$, and it is surprising to realize how closely this value is attained. Figure 26 shows the deflection functions for the higher kinetic energies. For the *gerade*, potentials we still get the same monotonic behavior as at lower energies, but dramatic differences can be seen for the deflection functions for the *ungerade* potential χ^u. The very sharp minima result from orbiting in the deep inner well, whereas the structure on the rainbow maxima is attributable to orbiting resonances (see earlier). The classical differential cross section can be calculated from the deflection function using (II.1). If $\sin\theta$ vanishes ($\chi=n\pi, n=0,1,2\ldots$), glory scattering results; if $\partial\chi/\partial b$ is zero, rainbow scattering occurs. The rainbow peaks are very small in this case. They are indicated by vertical arrows in Fig. 27, which compares the differential cross section calculated from the *ungerade* potential only $[f(\theta)=f_u(\theta)]$ with the experimental result.

The other oscillations can be understood with the help of Fig. 28, which shows schematically a typical deflection function including the effect of

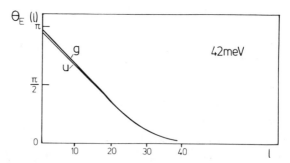

Figure 25. Quantal deflection functions for He*(2^1S)+He at 42 meV. Small splitting at large l causes damping of symmetry oscillations.

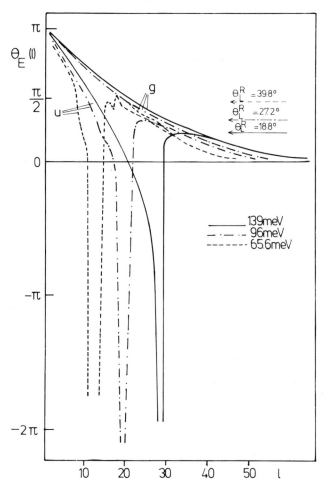

Figure 26. Quantal deflection functions at higher energies for He $*(2^1S)$+He. Orbiting spikes result from trajectories that spiral into inner minimum of *ungerade* potential.

nuclear symmetry. The dashed lines correspond to the exchange contributions. The large angle oscillations of Fig. 15 result from interference of l_1 with l_3, the $g-u$ oscillation from interference of l_1, and l_3 with l_2. This qualitative discussion can be made quantitative, as shown elsewhere.[58] The wavelengths of the oscillations $[\Delta\Theta = 2\pi/|l_i \pm l_j|]$ read off from the deflection function agree with a remarkable precision with the experimental results.

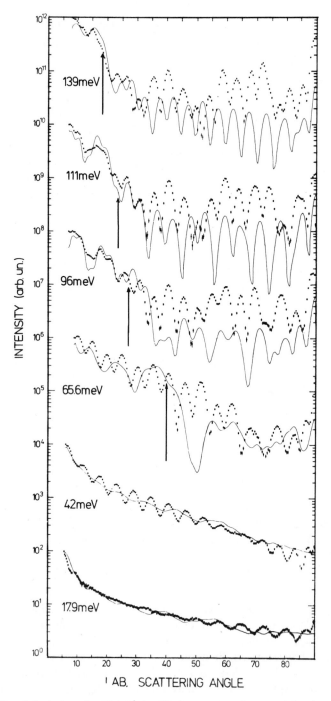

Figure 27. Calculations for He*(2^1S) + He from *ungerade* potential only compared to experiment. Classical rainbow angles are indicated by vertical arrows.

544

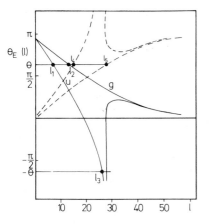

Figure 28. Typical deflection functions for energies above *ungerade* barrier. Dashed lines give contributions from exchange scattering.

D. Asymmetric Noble-gas Systems

1. Scattering in Penning Systems

a. Elastic Scattering

As outlined in Section III.A, recent advances in experimental technique have made possible a new series of measurements on the He* Penning systems with sufficient resolution to expose quantum interference structure at small angles in the differential cross section. This allows a much better determination of the long-range part of the potential, where $\Gamma(r)$ is small and the scattering is determined by $V_0(r)$ only. These experiments have been carried on mainly in Freiburg by Haberland and collaborators[100-104] with the Penning target species Ar, Kr, D_2, N_2, and CO over a wide energy range. With one exception, earlier published measurements by Grosser and Haberland[105], Lee et al.,[33, 39, 106] and Winicur and Fraites[40] on He* Penning scattering showed, at most, hints of the elusive undulatory structure. Somewhat better-resolved structure was reported by Bentley et al.[107] on He* + Kr, but He*(2^1S) and He*(2^3S) were not separated by optical quenching, and the measurements were restricted to small scattering angles. Jordan et al.[108, 109] have recently reported He*(2^1S) + Ar, Kr, and Xe scattering data of resolution comparable to those of Haberland et al.; Ne*($^3P_{2,0}$) also Penning ionizes the heavier noble gases and all known molecules. At present there is only a limited amount of scattering data available,[40, 109, 110] which are reviewed later in this section. Much of the work presented in this section is only recently published, in press, or in preparation, making this description as much a progress report as a review.

The well-studied systems $He^*(2^1S, 2^3S) + Ar$ comprise a natural and interesting prototype for excited-state intermolecular forces and Penning ionization; in several ways these systems are unique, however, as is seen later. We are struck at the outset by the experimental observation of quite different angular distributions for singlet and triplet; this is illustrated in Fig. 29[100-102] for a collision energy of 66 meV. The singlet scattering shows a pronounced maximum at $\theta_{lab} = 30°$, whereas the triplet curve is monotonic with only subtle changes in slope. The quantum structure at small angles is also much better resolved for the singlet, whereas the wide angle intensity is much lower relative to small angles than for triplet. The 30° maximum in $He^*(2^1S) + Ar$, which shifts with collision energy in much the same way as a rainbow maximum (see Fig. 30),[101, 102] has been the subject of some interesting qualitative speculation and quantitative interpretation. First observed as a shoulder by Chen et al.,[39] it was analyzed as a quantum reflection from a steeply rising opacity function. Later Bentley et al.[107] speculated that the hump might be electronically excited argon formed by direct excitation transfer from He^*. More recently, Haberland and Schmidt[102] presented a complete optical analysis of their singlet data in which they interpreted the hump, now a well-resolved maximum, as a rainbow arising from a local maximum embedded in the low-energy

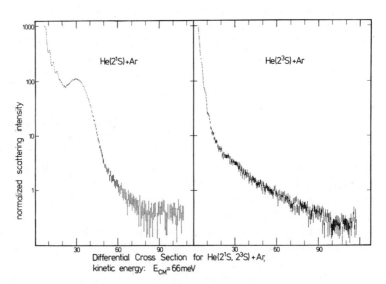

Figure 29. Laboratory angular distributions for $He^*(2^1S, 2^3S) + Ar$ at 66 meV. Relative intensities of singlet and triplet are as measured.

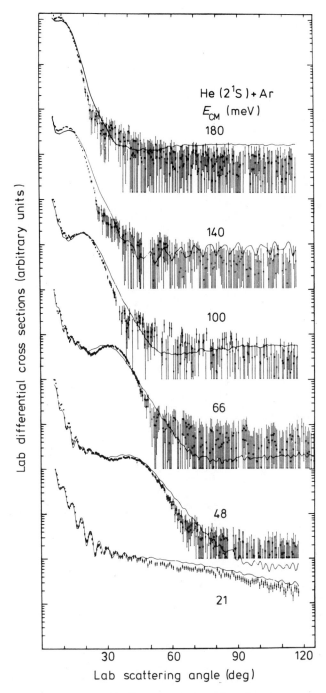

Figure 30. Laboratory angular distributions for He*(2^1S)+Ar at six collision energies from Freiburg laboratory. Solid curves are calculated from optical potential due to Haberland and Schmidt.[102] (see Figs. 32 and 35).

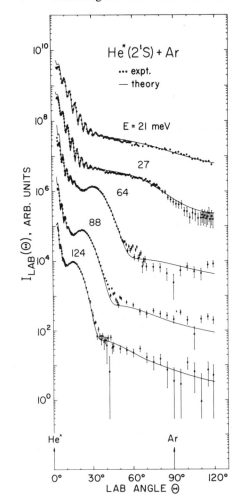

Figure 31. Laboratory angular distributions for He*(2^1S) + Ar at five collision energies from Pittsburgh laboratory. Solid curves are calculated from optical potential due to Martin et al.[109] (see Figs. 32 and 35).

repulsion of the real part of the optical potential. Figure 30 shows their experimental results over a range of energy and the fit they obtained. The existence of a barrier in the singlet potential had been postulated earlier[111, 112] on the basis of Penning ion angular distribution data (see Section III.D.1.b). Jordan et al.[108] have demonstrated by TOF measurements that the hump is almost certainly purely elastic, ruling out excitation transfer as a possibility. They also showed that the rainbow maximum in the cross section can be reproduced by a real part of the optical potential having a *slope* maximum in the repulsive part, without an actual barrier. Figure 31 shows the data and fit obtained at Pittsburgh.[109] The real parts of the potentials observed by Haberland and Schmidt[102] and Jordan et al.[108, 109]

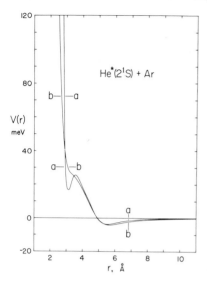

Figure 32. Comparison of potentials for He*(2^1S) + Ar (real part) derived by the Freiburg (*a*) and Pittsburgh (*b*) groups.

are compared in Fig. 32. Since the data from the two groups compare very well at 21 meV and 66(63) meV, the potentials are obtained on the same basis. The optimum potential has not yet been chosen; this may require further refinements and perhaps further experiments. The results for large *r* are in excellent agreement, whereas those at small *r*, at and inside the repulsive structure, disagree mainly because of the different $\Gamma(r)$ functions used (see following paragraphs). It may be fairly concluded that the potential is not well determined at small *r* in detail, although its gross features (e.g., the repulsive structure) are nearly beyond question.*

*[NOTE: Since this paragraph has been written, the He(2^1S) + Ar differential cross sections have been remeasured with much bettter statistics (1). Above a kinetic energy of roughly 100 eV a new structure is observed on the dark side of the rainbow, which at lower energies is only apparent as a change in slope. This can be seen at the left-hand side of Fig. 29, where at $\theta \approx 50°$ the near exponential fall-off of the rainbow changes slope. A fit to the data was performed using the analytical form of the potential proposed by Jordan et al. (references 108 and 109 of this chapter) for the real part of the potential, and a single exponential for its width. The potential proposed by Jordan et al., which is shown in Fig. 32, curve b, gives an excellent fit to the small angle part of the new data up to the rainbow region. The large angle part—including this new structure—can be fitted better with a potential having a maximum instead of only a slope maximum.

A quantum mechanical calculation of the Penning electron energy spectra for the two different potentials would be quite interesting. A comparison with recent high-resolution results (2) would give an independent experimental check on the existence or nonexistence of the maximum in the potential.

(1) R. Altpeter, B. Brutschy, H. Haberland, and F. Werner, to be published.
(2) H. Hotop, XI ICPEAC, Kyoto 1979, Book of Invited Talks.]

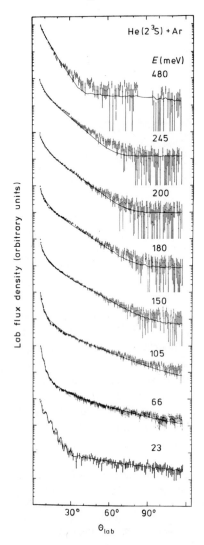

Figure 33. Laboratory angular distributions for He*(2^3S) + Ar at eight collision energies. Solid curves are calculated from optical potential due to Brutschy et al.[100] (see Fig. 34).

In contrast, the lack of an intensity maximum in He*(2^3S) + Ar scattering shown in Fig. 33, again over a wide energy range,[100] augurs against such structure in the triplet potential. The fit derived by Brutschy et al.,[100] also shown in Fig. 33, produced a much smoother, although still structured potential function given in Fig. 34, with two unusual bends in the repulsive part. (These features become apparent only on a semilogarithmic plot.) The bend at lower energy, where the slope of the potential markedly decreases, is needed to describe the "flattening out" of the angular dis-

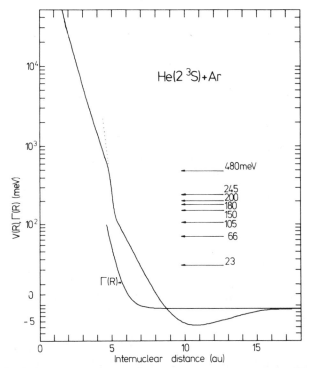

Figure 34. Optical potential for He*(2^3S) + Ar, due to Brutschy et al.[100] (see Note 115 in References Section).

tributions at wide angles for the higher collision energies, whereas the high-energy bend results from joining the fitted potential to the He$^+$–Ar ion–atom repulsion of Smith et al.[113, 114] and is not determined by experiment. The unusual nature of the repulsion in both singlet and triplet systems is discussed in Section III.D.1.d.

The widths derived are also of interest, since it has long been believed that they should be at least approximately exponential. Haberland and Schmidt[102] adopted an exponential-plus-floating Gaussian form for He*(2^1S) + Ar, obtaining a width with a shoulder (arising from the Gaussian) approximately at the position of the minimum inside their barrier maximum.[115] This enhancement of the width, which enabled a good fit to the data (Fig. 30), preventing some of the interference structure that would have been caused by the barrier from appearing in the calculated cross section. Jordan et al.[108, 109] on the other hand, used a simple exponential width in deriving their potential, since the slope maximum does not produce an extensive interference pattern.[109] The He*(2^3S) + Ar scatter-

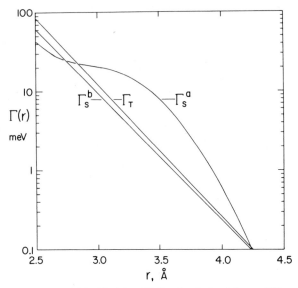

Figure 35.   Resonance widths $\Gamma(r)$ for He*+Ar derived from differential scattering: Γ_S^a, singlet;[102] Γ_S^b, singlet;[109] Γ_T, triplet.[100]

ing[100] was fit satisfactorily with an exponential Γ. The widths are compared in Fig. 35.*

Current work in the Freiburg[103, 104] and Pittsburgh[108, 109] laboratories gives evidence that the type of repulsive structure inferred for the He*+Ar potentials is probably a general phenomenon for noble-gas partners as well as D_2. In a recent communication Altpeter et al.[103] compared the He*+Ar potentials observed by Brutschy et al.[100] and Haberland and Schmidt[102] to He*+Kr, D_2 potentials obtained from scattering data to be published.[104] Jordan et al.[108, 109] have also given data and potentials for He*(2^1S)+Kr, Xe. These potentials are presented in Figs. 36 and 37. Although the potentials from the two groups differ in certain details and arise from two

---

*[NOTE: Differential cross section for He(2^1S)+Na and He(2^1S)+Hg have recently been measured (1). As the well depth ($\varepsilon = 300$ and 500 meV respectively) is larger than the collision energy, orbiting occurs for all energies. For He(2^1S)+Na, a backward glory is observed. All other oscillations are heavily damped by the residual velocity spread of the supersonic metal atom beams. The potentials derived from a fit to the data have a very soft long-range part and are in agreement with results from Penning electron energy spectroscopy (see the article by A. Niehaus in this book).

(1) H. Haberland, O. Schmidt, and W. Weber, to be published.]

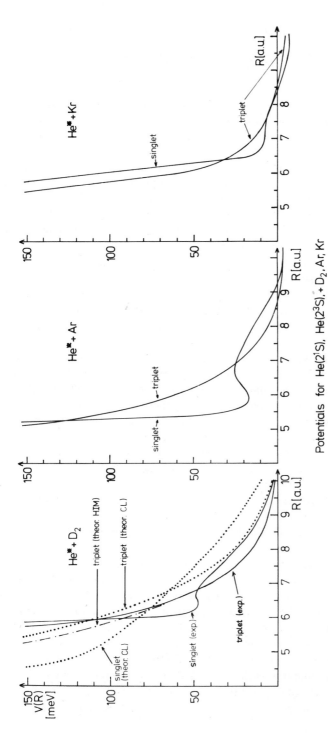

Figure 36. Potentials for He*+D₂, Ar, Kr from Haberland et al.[104] Dotted and chain curves for He*+D₂ are *ab initio* results.

quite different parametric potential functions, several conclusions may be drawn:

1. The van der Waals well depth ε increases monotonically as the polarizability of the ground-state partner increase, whereas the location of the well, r_m, remains roughly constant at about 6Å. This is expected by analogy with the well-known alkali–noble-gas potentials,[116-119] although the r_m values are larger than for the alkali case, as shown in Table V.
2. The potential energy V_s at which the repulsive structure for He*(2^1S) occurs decreases monotonically as the polarizability of the ground-state partner increases.
3. The position, r_s, of the repulsive structure increases for heavier partners. This point is less certain since, as usual for rainbow scattering, the experimental maxima correlate strongly with the energy of the structure in the potential, only weakly with its range.
4. At internuclear distances outside the repulsive structure, the triplet repulsive energy lies lower than the singlet, for argon, krypton, and deuterium partners,[103] as well as for helium (see Section III.C).

The He* + D$_2$ system is of particular interest, since *ab initio* calculations by two groups[26, 120, 121] have now appeared. Cohen and Lane's calculation[120] showed a substantial difference between singlet and triplet potentials but very similar, nearly exponential, widths. Although both interactions were found to be only weakly anisotropic, the spherically symmetric part of the interaction showed a shallow slope maximum in the repulsion for the singlet but smoother behavior for the triplet. These potentials are compared with experimental ones in Fig. 36. The experimental potentials were derived assuming zero anisotropy. On the other hand, Hickman and colleagues,[26, 121] who also obtained markedly different singlet and triplet interactions, found a highly anisotropic singlet potential surface, with a pronounced shoulder in the repulsion for C_{2v} geometry and a relatively smooth curve for $C_{\infty v}$. There is generally good agreement between both theoretical potentials and the spherically symmetric experimental potential for triplet as shown in Fig. 36, but the three diverge substantially for singlet. If the large singlet anisotropy found by Isaacson et al.[121] proves to be valid, the scattering analysis becomes much more complicated because of rotationally inelastic collisions, and the experimental potential is probably incorrect. Preston and Cohen[122] have initiated classical trajectory-surface-leaking (TSL) dynamics calculations on this system but on potential surfaces[120] with only weak anisotropy.

Predictions of total ionization cross-sections and quenching rate constants from these potentials are compared with experiment in Section

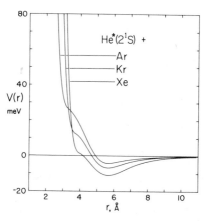

Figure 37. Potentials for He*(2^1S)+Ar, Kr, Xe (real part), from Martin et al.[109]

Table V.
Van der Waals Parameters for He*(2^1S, 2^3S)-Li Interactions

	He*(2^1S)		He*(2^3S)		LITHIUM	
PARTNER	E (meV)	r_m (Å)	E (meV)	r_m (Å)	E (meV)	r_m (Å)
Neon	$0.4^{a,b}$ $0.6^{b,c}$	$7.14^{a,b}$ $6.2^{b,c}$	$0.4^{a,b}$	$7.1^{a,b}$	$0.1^{b,d}$	$6.20^{b,d}$
Argon	$3.8^{b,e}$ $4.2^{b,f}$	$5.55^{b,e}$ $5.67^{b,f}$	$3.6^{b,g}$ $5.1^{b,f}$	$5.66^{b,g}$ $5.17^{b,f}$	5.6^h	4.86^h
Krypton	6.5^i 6.8^f	5.6^i 5.69^f	5.5^i	5.4^i	8.4^d	4.95^d
Xenon	11.0^f	5.69^f			12.8^d	4.95^d

[a] Haberland et al.[77]
[b] These parameters may not be reliable because of the insensitivity of the scattering data to very weak van der Waals attraction.
[c] Siska and Fukuyama.[145]
[d] Dehmer and Wharton.[119]
[e] Haberland and Schmidt.[102]
[f] Jordan et al.[108, 109] and unpublished work.
[g] Brutschy et al.[100]
[h] Ury and Wharton.[118]
[i] Haberland et al.[103, 104]

555

III.D.1.c, and the structure in the potentials is discussed qualitatively in Section III.D.1.d.

Published work on Ne* + Kr scattering[40] (see NOTE on page 552). From the Notre Dame laboratory has been interpreted through the use of the potential function derived by Buck and Pauly[116] for the alkali–rare-gas systems. This is a two-piece Lennard–Jones (LJ) potential, with the region $r < r_m$ described by a LJ (11,4) function and $r > r_m$ by another LJ (14,6). It was tacitly assumed that the four potential curves resulting from the various electronic angular momentum states are identical. The measurements give no evidence to the contrary. The general conclusion is that the van der Waals well depth and position are nearly identical to those of Na + Kr.[116] The data, however, did not extend to wide scattering angles (maximum angle reported was 22° laboratory), and the cross section from fitted potential actually fell below the data at the widest angles, making an optical model analysis impossible. Figure 38 shows center-of-mass cross sections extending to wide scattering angles, from unpublished work at

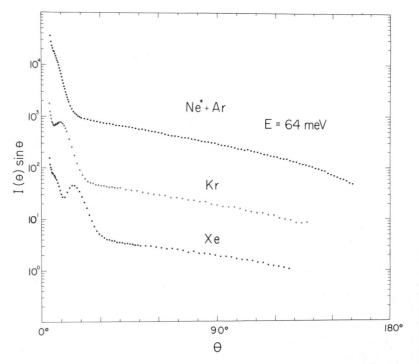

Figure 38. Center-of-mass angular distributions for scattering of Ne*($^3P_{2,0}$) by argon, krypton, and xenon.

Pittsburgh, for Ne* + Ar,Kr,Xe. The absence of structure in the wide-angle intensity is in marked contrast to the He*(2^1S) case, with optical model analysis requiring recourse to other types of data, such as total ionization cross sections (see Section III.C.1.c). An immediate conclusion is that the van der Waals repulsion in these systems does not bear the same relation to the attraction as for the alkalis, although the difference is not as great as for He* interactions versus those of lithium.

b. Product Angular Distributions

Leu and Siska[111, 112] have communicated measurements of Penning ion angular distributions for He* + Ar,H_2,N_2,CO,O_2 at several collision energies. Unpublished work from the Pittsburgh laboratory includes He* + CO_2,CH_4,C_2H_6,C_2H_4 and Ne* + Ar, with measurements on all observable fragment ions in the polyatomic systems. These experiments are performed with beam sources similar to those described in Section III.A (although with somewhat lower Mach numbers; ~20 for each beam). The open-electron multiplier detector is replaced by a quadrupole mass filter and scintillation ion counter of the type described by Lee et al.[123] Care was taken to eliminate stray electric fields near the collision volume; this was accomplished by enclosing the volume in a stainless-steel plate-and-mesh cage whose inner surfaces were coated thinly with aquadag. After 4 cm of free flight, the ions were accelerated and focused into the quadrupole filter. A small ionizer and retarding-field energy analyzer placed in front of the ion lens system allowed calibration of the mass filter and energy analysis of ionic collision products. Since the scintillation counter was an off-axis type, elastically scattered metastables and product photons were not detected, and the background was attributable only to the dark-counting rate of the ion counter, always less than 5 cps. Signal-counting rates typically ranged 5 to 1000 cps. Product ions with energies as low as 100 meV were successfully detected. Many of the experiments were run with the crossed beam seeded in 85 to 99% H_2 or He. This assured efficient collection of ions because of the resulting high centroid velocity and laboratory energy of the product ions. With unseeded beams, a likely wide spread in laboratory energy of products enhances undesirable discrimination against the lower energy ions. However, the kinematics and dynamics of the He* systems often favor a narrow laboratory distribution, enabling reasonably low-energy measurements. A singlet quenching lamp (Section III.A) had not yet been installed for these experiments. At the 250-eV electron energy used, the He* beam consists of approximately 85% singlets, so that the measurements are representative of the singlet ionization process.

Figure 39 shows angular distributions of Ar^+ and $HeAr^+$ for He*(2^1S) + Ar over a range of collision energies. The recoil momentum of the

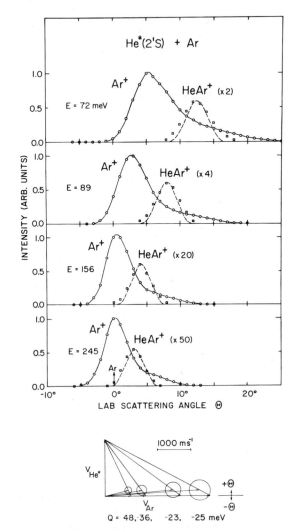

Figure 39.   Penning ion angular distributions for He*(2^1S)+Ar.

Penning electron may be neglected for all but the smallest heavy-particle center-of-mass recoil energies. The HeAr$^+$ ion is then constrained to travel with the center of mass. The measured HeAr$^+$ angular distributions agree well with calculated centroid distributions averaged over the acceptance angle (3°) of the detector; this indicates that stray fields near the collision volume or flight path do not exceed a few millivolts per centimeter. The Ar$^+$ ions are pitched sharply forward at each energy, with the sharpness of the peaks limited mainly by the angular resolution. The narrow angular range implies recoil energies comparable to the initial energy, as expected from Penning electron spectra.[18] An approximate transformation to the center-of-mass system using the fixed recoil velocity approximation[124] yields $Q = E' - E$ for each energy, as given in Fig. 39. The reaction is translationally endoergic at all energies studied, with Q approximately constant at -25 ± 5 meV at higher energies. This feature is nicely explained by the repulsive structure in the $He^*(2^1S) + Ar$ potential found from the elastic scattering (see earlier). For impact parameters allowing passage over the structure, the local velocity near the turning point is reduced, and the ionization probability, $\Gamma(r)/\hbar v_b(r)$, is thereby enhanced. In addition, the difference function $V_0(r) - V_+(r)$ may have extremum there, which produces a strong peak (classically infinite) in the energy distribution. If $V_+(r)$ is very weak, as expected for the $Ar^+ - He$ interaction, the peak translational endoergicity Q should be very nearly equal to the potential energy V_s at which the structure in $V_0(r)$ occurs. Reasonably close agreement between Q and V_s is evident by inspection of Fig. 32.

The derived center-of-mass angular distributions (see Fig. 40) are not quantitative because of the approximations in the laboratory center-of-mass transformation, but their form is highly suggestive of rainbow scattering, nonforward peaking at lower energies shifting to forward at higher energies. This also can be plausibly attributed to the repulsive structure, following reasoning similar to that used for the elastic angular distribution. Quantitative calculations of doubly differential cross sections $\sigma_{PgI}(\theta, E, E')$ using theory outlined in Section II.D.1 are currently in progress at Pittsburgh. Experiments employing ion energy analysis have been carried out for $He^* + Ar$ at $E = 154$ meV that support the kinematic analysis results

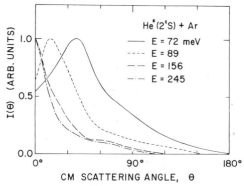

Figure 40. Approximate center-of-mass Penning ion angular distributions for $He^*(2^1S) + Ar$.

but are of insufficient resolution to allow construction of a contour map of the cross section. Time-of-flight experiments should improve this situation.

Hickman and Morgner[30] have used a quantum-mechanical Franck–Condon model (see Section II.D) to calculate a center-of-mass angular distribution for $He^*(2^3S) + Ar$. Although the singlet and triplet systems are governed by different excited-state potentials, the calculation shows strong forward scattering, in qualitative agreement with the experiments, which pertain to $He^*(2^1S)$. The cross section is actually differential in the recoil energy E' also, so the calculation (for which E' was averaged over) cannot be compared in a more quantitative way with experiment.

For canonical examples of diatomic molecules, we select H_2 and O_2. Figure 41 shows measured angular distributions[112] for Penning ionization of these molecules by $He^*(2^1S)$. The strongly forward scattering with $E' < E$ for H_2 is quite similar to the situation for argon. Again, the Q value is very close to V_s as found in the nonreactive scattering analysis (see Fig. 36). Similar results (not shown) are obtained for N_2 and CO targets;[112] the nonreactive results for these systems[104] are in preparation for publication.

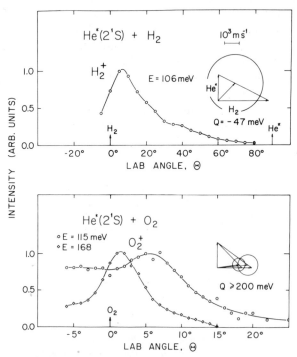

Figure 41. Penning ion angular distributions for $He^*(2^1S) + H_2$ and O_2.

Reliable distributions for production of HeH^+ through rearrangement ionization have not yet been obtained, although abundant HeH^+ product has been observed in a total ion collection mode. Merged-beam results of Neynaber et al.[125] strongly support forward scattering of HeH^+ at higher relative energies; these data pertain to a singlet–triplet mixture ($\sim 12:1\,^3S$ to ^1S). Further discussion of the merged beam experiments is given in the following paragraphs and in Section III.D.1.c. The observation of a nearly Franck–Condon vibrational population in H_2^{+} [18] is consistent with the relatively large internuclear distances for ionization required by the potential surface, as well as with the similarity between the H_2 and Ar angular distributions.

Penning ionization of O_2 apparently proceeds through interactions that are qualitatively different from those of closed-shell atoms and molecules. The angular distributions are much broader, with considerable energy released into translation, as indicated in Fig. 41. It is plausible that the lowest $^3A''$ potential surface (C_s symmetry) is highly attractive because of covalent–ionic-avoided crossing with another $^3A''$ state (crossing radius ~ 4 Å); ionizing transitions occurring over a deep potential well will give enhanced product translation in a two-body approximation. The attractive surface also is likely to cross lower-lying repulsive surfaces correlating with dissociating O_2 states, thus giving rise to competition between PgI and dissociative excitation. Observation of atomic oxygen emission in flowing afterglow spectroscopy[126] by collision of O_2 with triplet He* has been similarly interpreted. The form of the angular distributions, with a bump occurring near the usual closed-shell peak position and overlying a broader, flatter curve, suggests that both the usual weakly repulsive (in this case diabatic) and the attractive (adiabatic) mechanisms are at play. Product energy analysis will be highly informative on this point. Other support for the attractive mechanism comes from the observation of "sticky collision bumps" in the angular distribution of $K+O_2$.[127] The non-Franck–Condon vibrational distributions in O_2^+ states formed from reaction with ^3S He[126] is also consistent with an attractive surface, since harder collisions at smaller distances would strongly perturb O_2 during the ionizing transition. Reaction to form HeO* on the ionic surface is not energetically possible for O_2 but can be expected for more weakly bound oxygen atoms on the basis of similar reactions seen with Ar*.[128]

Measurements on He*(2^1S) PgI of closed-shell polyatomic molecules[129] give results very similar to argon and the closed-shell diatomic systems. Figure 42 displays CO_2^+ and CH_4^+ angular distributions from ionization of the parent molecules. Whereas $Q < 0$ for CH_4, $Q \simeq 0$ for CO_2, suggesting a somewhat less repulsive interaction in this case. As in the diatomic systems, relatively large amounts of electronic and vibrational energy (usually

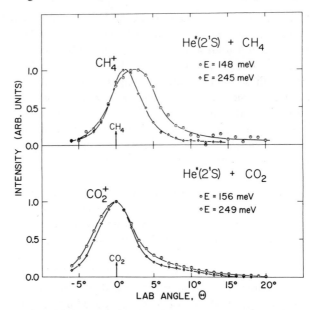

Figure 42.   Penning ion angular distributions for He*(2^1S)+CH$_4$ and CO$_2$.

several electron volts) are generally deposited in the Penning molecular ion, and especially in the polyatomic systems this produces fragmentation of the parent ion similar to that found in mass spectrometry. Measurements of the fragment ion angular distributions[129] show peaking at the same laboratory angle as the parent, with broadening resulting from recoil imparted by the neutral fragment. Figure 43 gives the results for CH$_3^+$

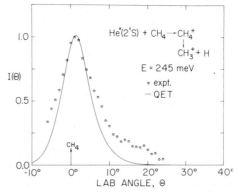

Figure 43.   Schematic diagram of CH$_3^+$ fragment ion angular distribution from He*(2^1S)+CH$_4$ compared with quasiequilibrium theory (QET).

from $He^*(2^1S) + CH_4$, compared to a prediction from the quasiequilibrium theory (QET) of mass spectra.[130] The energy deposition function for CH_4^+ is taken from the He I photoelectron spectrum, and the analysis assumes a two-step sequence of PgI followed by fragmentation.

Penning ionization in the $Ne^* + Ar$ system also yields sharp forward scattering, as shown in Fig. 44. Here the product distributions may be much more important to the determination of the incoming potential, since the elastic scattering (Fig. 38) is structureless at wide angles, and the kinematics of the product angular distributions are more favorable than for He^*. The Ne^* experimental work is still in progress at Pittsburgh.

Neynaber and Magnuson[125, 131–135] have carried out merged-beam experiments on a number of the Penning systems, including $He^* + H_2,H,D$ and $Ne^* + Ar,Kr$. Although the contributions of singlet and triplet helium cannot be separated experimentally, statistical arguments were used to deduce that the composition of the He^* beam, formed by charge exchange with cesium vapor, is $2^3S/2^1S = 12$. Although the merged-beams technique cannot yield angular distributions of the product ions, displacement of the measured product energy distributions with respect to the center of mass indicates the preferred hemisphere (forward or backward) for product scattering. Penning ionization of H_2 was not reported; however, as mentioned earlier, HeH^+ was found predominantly in the forward hemisphere.[125] Penning ionization of deuterium,[134] although beset by experimental difficulties, gave sharp forward scattering at high energies ($E > 1$ eV) and a distribution symmetric about the center of mass at 100 meV. The

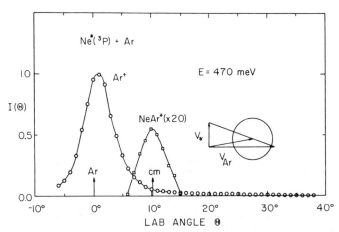

Figure 44.   Penning ion angular distribution for $Ne^* + Ar$.

symmetric distribution suggests that most of the D^+ is formed by tunneling through the centrifugal barrier in the HeD^+ effective potential. This phenomenon may also be responsible for the small but significant amount of backward scattering in the $He^*(2^1S) + Ar$ distributions. The Ar^+ and Kr^+ energy distributions from Ne^* PgI[133, 135] also indicated mainly forward scattering, in agreement with the crossed-beam results cited earlier.

c. Total and Ionization Cross Sections

The total cross sections can easily be calculated once the potential has been determined. As the magnitude of the total cross section is mainly determined by the long-range van der Waals attraction, the width of the potential has only a negligible influence. The velocity dependence of the total elastic cross section calculated from the potential for $He(2^3S) + Ar$ (see Fig. 34) is shown in Fig. 45. For energies above 10 meV, the cross section shows the well-known glory oscillations. At lower energies the effect of the orbiting resonances is clearly noticeable. The total cross section, that is, the sum of the elastic and inelastic cross sections, has been measured by Rothe et al.[10] and by Trujillo.[136] As the inelastic cross section

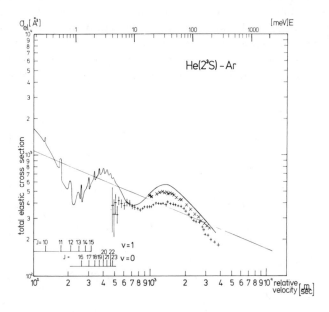

Figure 45. Velocity dependence of total cross section for $He^*(2^3S) + Ar$. Solid curve is calculated from potential due to Brutschy et al.;[100] points are measurements due to Trujillo.[136] Orbiting resonances are indicated by (v, J) quantum numbers.

is negligible compared with the elastic one, the data are directly compared with the calculation in Figure 45. The average cross section calculated from the Schiff–Landau–Lifshitz (SLL) approximation[42] is given by the straight line. The absolute size of the total cross section averaged over the glory oscillations is determined only by the van der Waals constant and should be given quite accurately by the SLL formula. Both experimental results lie below the predicted curve. Trujillo states an absolute uncertainty of 5% for his recent data, which are about 40% below our calculation at the glory maximum, 1.4 km sec. The van der Waals constant, C, has an estimated error of less than 15%,[137] which introduces a 6% error in the calculated cross section σ since $\sigma \propto C^{2/5}$. Therefore, the magnitudes of the experimental and theoretical curves are outside their stated uncertainties, although the positions of the glory maxima agree well.

The calculated total elastic cross section for $He(2^1S) + Ar$ is given elsewhere.[102] There are currently no data for comparison. Trujillo has also measured cross sections for $He(2^3S) + Ne, Kr$.[136] Using essentially the same apparatus Harper and Smith[138] have extended the cross section measurements to $He(2^3S) + H_2, CO, O_2, N_2$ and $Ne^* + He, Ne, Ar, Kr, H_2, CO, N_2, O_2$.

Assuming that Penning and associative ionization represent the only quenching channel, total quenching cross sections calculated from equation (II.17) using an optical potential fitted to the differential scattering data may be compared directly to measured total ionization cross sections. Figure 46 shows a comparison between the total ionization cross section energy dependence predicted from the potential of Fig. 34 and experimental results. Experimental data from three different groups are included. Illenberger and Niehaus[139] and Pesnelle et al.[140] have measured the velocity dependence of the total ionization cross section. Their relative data have been normalized to the absolute flowing afterglow rate constants given by Lindinger et al.[141] Within the quoted accuracy of less than 30% for the absolute value, all the data are in very good agreement. On a relative scale the data of Illenberger and Niehaus coincide with the calculation within experimental error, whereas the data of Pesnelle and co-workers disagree somewhat at higher collision energies. The one data point from Riola et al.[142] is an absolute determination in a crossed-beam experiment. The relative velocity was not measured at the same time as the total cross section, but was determined in a later experiment. As the He^* beam source was modified between the two experiments, it is believed that the velocity was changed by this modification.

The point at which the kinetic energy equals the well depth is indicated by ε in Fig. 46. At lower energies the expected increase in the ionization cross section is observed. The sharp peaks superimposed on the gradual rise are the result of orbiting or shape resonances. They are caused by

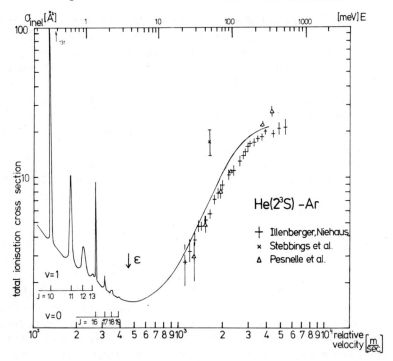

Figure 46.   Velocity dependence of total ionization cross section for He*(2^3S)+ Ar. Solid curve is calculated from optical potential due to Brutschy et al.[100] Symbols represent measurements by other authors.[139, 140, 142] (see text). Spikes result from orbiting resonances labeled by vibrational (v) and rotational (J) quantum number.

partial waves with high orbital angular momentum quantum numbers J, as indicated in Fig. 46, which can no longer overcome the centrifugal barrier but must tunnel through it. If the kinetic energy coincides with that of a quasibound, predissociating state, the probability for finding the particle inside the barrier becomes large. As these resonances are very narrow, the particle stays for a long time in a region where the width $\Gamma(r)$ is nonnegligible and is strongly absorbed.

The vibrational quantum number v for a resonance was obtained by counting the nodes in the wave function; the J value can be obtained directly by inspecting the calculated opacities. For $v=0$, there should also be resonances below $J = 16$, but they are so narrow that they have not been found, although the cross section has been calculated at 75 energies below ε. For the $v=1, J=10$ resonance, the cross section reaches a value of 131

$\mathring{A}^2$, which is off the scale in Fig. 46. For the higher partial waves the resonances, become broader and quickly blend into the smooth background. These resonances are very sensitive to the exact form of the attractive part of the potential. So far, they have not been observed experimentally. Their observation will be very difficult because they are so narrow and their energy lies below 1 meV.

For $He(2^1S)+Ar$ the shape of the total ionization cross section is substantially different (see Fig. 47). The solid curve has been calculated from the potential determined in Freiburg, and the dashed curve gives the Pittsburgh result. They both agree reasonably well with the experimental data, which have been measured by Illenberger and Niehaus[139] and Pesnelle et al.[140] Their relative data have again been normalized to the absolute rate constants measured by Lindinger et al.[141] On a relative scale the agreement is very good, except at low velocities where the resolution broadening of the experiment is largest. The absolute value of the rate constant is known to within 30%; thus the agreement is also satisfactory on an absolute scale. The total cross section has also been measured by Riola et al.[142] in a beam experiment. Their value is given by the one data point at 60 meV. Good agreement is obtained within experimental error.

The temperature dependence of the quenching rate constant $k(T)$ as measured by Lindinger et al.[141] can be obtained by a thermal average over the velocity dependence of the cross section. Good agreement is obtained for both $He(2^1S)$ and $He(2^3S)+Ar$ (see Fig. 48).*

d. Qualitative Interpretation of Structure in He* Potentials

In seeking to rationalize the repulsive structure found in the $He^*(2^1S)+$ Ar,Kr,Xe,D_2 interactions, we examine the model for noble-gas partners proposed by Siska et al.[108, 109, 112] and the rationale presented by Isaacson et al.[121] for $He^*(1^1S)+D_2$. In the model potential function of Siska et al.[108, 109] the low-energy repulsion is represented by a switchover from alkalilike—closed-shell repulsive behavior to ion core (He^+)—closed-shell Rydberg-like behavior with decreasing internuclear distance; in other words,

$$V_0(r) = \left[1 - f(r) \right] V_+(r) + f(r) V_*(r) \qquad (III.1)$$

*[NOTE: Recently the velocity dependence of total and the ionization cross section has been measured to very low kinetic energies (1 meV) for He^*+Ar (1). The predicted increase in the ionization cross section (see Figs. 46 and 47) has been seen for both the singlet and the triplet system. These data contain information mainly on the very long-range part of the interaction potential, which has not been probed otherwise.

(1) R. Feltgen, H. Pauly, Göttingen, private communication, 1979.]

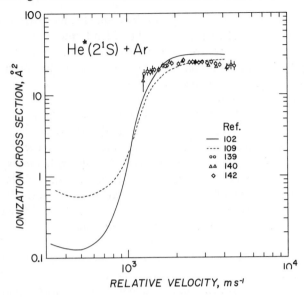

Figure 47.   Velocity dependence of total ionization cross section for He*(2^1S)+ Ar. Curves are calculated from optical potentials due to Haberland and Schmidt[102] and Jordan et al.;[108] symbols represent measurements by other authors.[139, 140, 142]

where $V_*(r)$ is the alkalilike potential, representing the dispersion attraction and overlap repulsion involving the $2s$ electron on He*; $V_+(r)$ is the ion–molecule interaction (He$^+$–X), including the attractive part; and $f(r)$ is a switchover function, $f\rightarrow(0, 1)$ as $r\rightarrow(0, \infty)$. Jordan et al.[108, 109] were able to fit He*(2^1S) scattering from argon, krypton, and xenon varying only $V_*(r)$ and $f(r)$, whereas leaving $V_+(r)$ fixed as determined by the ion–atom scattering experiments due to Smith et al.[113, 114] and Wiese and Mittmann.[144] This potential model appears to favor the substantial anisotropy found by Isaacson et al.[121] for He*(2^1S)+H$_2$, since the core He$^+$–H$_2$ interaction V_+ is likely to strongly resemble the Li$^+$–H$_2$ interaction, which is known from *ab initio* calculations and from experiment to be highly anisotropic. The model does not provide a way of rationalizing the marked dissimilarity between singlet and triplet interactions, but the *ab initio* calculations offer a reasonable explanation. Examination of the electronic wave function reveals that in the region of the repulsive structure for He*(2^1S)+H$_2$, the $2s$ orbital has acquired appreciable $2p$ character, that is, is hybridized. This enables the outer electron to remove itself largely to the far side of helium, since its exchange interaction with the partner's closed shell is repulsive. The He$^+$ core is thus partially bared to the

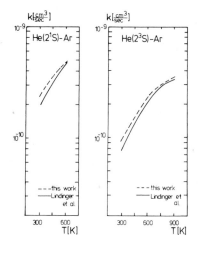

Figure 48. Temperature dependence of total quenching rate constant for He* + Ar. Solid curves represent measurements by Lindinger et al.;[141] dashed curves are calculated from optical potentials due to other authors.[100, 102]

partner, and the ion–atom interaction becomes important. The hybridization is enhanced by the small energy gap (0.602 eV) between the $1s2s2^1S$ and $1s2p2^1P$ states of helium. For triplet He*, the 2^3S-2^3P gap is nearly twice as large (1.144 eV) and hybridization becomes less favorable energetically; hence the ion–atom core interaction is more effectively shielded by the $2s$ electron. The triplet repulsion is thus expected to increase more smoothly in the low-energy regime, nicely illuminating the experimental finding of a much subtler structure in the repulsion for He*(2^3S) + Ar.[100]

This description appears to bear some relation to the avoided crossing between the $2s\sigma_g$ and $2p\sigma_g$ states in the kindred He*(2^1S) + He system.[84] This avoided crossing produces a $C\,^1\Sigma_g^+$ potential curve with a barrier maximum (217 meV, 2.06 Å) and a deep inner minimum. Although the corresponding states in He*(2^1S) + Ar are not expected to approach each other closely, some interaction between them seems plausible. Such an interaction would be weaker for the triplet system, again because of the larger asymptotic splitting.

2. Scattering in Excitation-transfer Systems

a. Elastic Scattering

As a general rule, when PgI is energetically possible, it is overwhelmingly preferred to other electronically inelastic channels. Thus He*

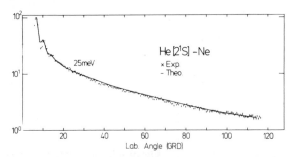

Figure 49.   Low-energy laboratory angular distribution for He*(2^1S)+Ne. Solid curve is a fit for an assumed single-channel real potential.

efficiently Penning ionizes every collision partner except helium and neon at thermal energies. Therefore, He*$(2^1S, 2^3S)$+Ne is a prototype system for electronic excitation transfer at thermal energies, both because of its relatively unique accessibility to *ab initio* as well as experimental methods and because it lends itself to model calculations using a spherically symmetric potential matrix.

Elastic-scattering measurements for He*+Ne have now been reported by a number of authors.[39, 75, 77, 145] The differential cross sections are complicated by the contributions of Ne* metastables, formed by excitation transfer and subsequent radiative cascade, and of product UV photons to the scattered intensity. The Ne* contribution was isolated in the double-quenching-lamp experiments due to Haberland et al.[76, 77] by use of a neon quenching lamp in the detector; however, because the neon lamp was less than 100% efficient ($\sim$70 to 85%), quantitative elastic-scattering data could not be extracted except at very low energy (25 meV), where the Ne* signal was negligibly small. These data are shown in Fig. 49, along with a fit for an assumed single-channel potential. Both the Pittsburgh and Freiburg laboratories have used TOF measurements to separate the elastic scattering from product Ne* and photons; these results are discussed in Section III.D.2.c.

In the 64-meV angular distribution for He*(2^1S)+Ne[145] shown in Fig. 50 along with a single-channel fit to small angles, a shoulder appears at $\theta_{lab}\sim$30 to 35°. This feature is similar in form to that for He*(2^1S)+D$_2$ and was interpreted again as a repulsive rainbow. However, from work in progress at Pittsburgh on the higher energies, a second hump appears at wider angles. This effect may only be interpretable in a multichannel

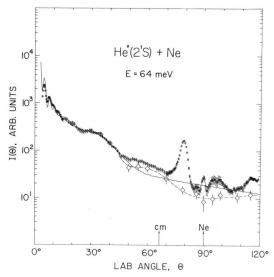

Figure 50. Laboratory angular distribution for He*(2^1S)+Ne. Circles are total angular distribution; diamonds are time-of-flight corrected for Ne* contribution. Solid curve is calculated by fitting a single-channel potential to small angles.

framework, where the effective single-channel potential is strongly perturbed by avoided crossings. Optical model and two-state close-coupling analysis have been performed for the 64-meV distribution.[145]*

Other work in atom–atom systems includes Ar*$(^3P_{2,0})$+Kr and Xe[75] and Kr*+Xe.[110] The Ar*+Kr angular distribution at 63 meV (see Fig. 51)

[NOTE: Differential elastic and excitation transfer cross sections have been measured for He(2^1S)+Ne and for He(2^3S)+Ne for energies between 25 and 370 meV (1). Some of the data are shown in Fig. 52. It was possible to measure the differential excitation cross sections for the triplet system, too. A semiclassical two-state calculation was performed for the pumping transition of the red line of the HeNe-laser He(2^1S)+Ne→He+Ne($5S$, $^1P_1$), which is the dominant transition for not too high energies (2). A satisfactory fit is obtained to the elastic and inelastic differential cross sections simultaneously, as well as to the known rate constant for excitation transfer. The He(2^1S)+Ne potential curve shows some mild structure, much less pronounced than those shown in Fig. 36. The excitation transfer for the triplet system goes almost certainly over two separate curve crossings. This explains easily the 80 meV threshold for this exothermic process as well as its small cross section, which is only 10% of that of the triplet system.

The relative cross sections for excitation transfer have also been measured by Krenos (3).

(1) P. Oesterlin, Ph.D. dissertation, Freiburg, 1979; and H. Haberland and P. Oesterlin, to be published.
(2) H. Haberland and P. Oesterlin, XI ICPEAC, Kyoto 1979, Book of Abstracts, p. 458.
(3) J. Krenos, VII International Symposium on Molecular Beams, Riva 1979, p. 75.]

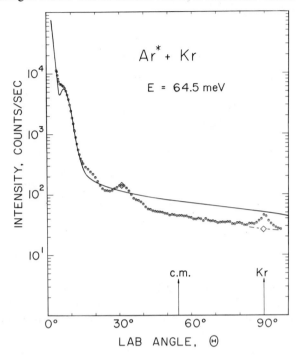

Figure 51. Laboratory angular distribution for Ar* + Kr. Diamonds are TOF corrected for Kr* contributions. Solid curve is calculated from a K + Kr potential.

shows, in addition to a normal rainbow expected by analogy with K + Kr, an additional array of bumps at wider angles centered around = 31°. Earlier interpreted as Kr* arising from the excitation transfer,[146] this structure has now been shown by TOF measurements[75] to be elastic scattering. The bump at 90° is established as a Kr* peak. The origin of the secondary features is likely related to curve crossing with the product channels, speculated to occur in the attractive part of the interaction for the major Kr $5p[\frac{3}{2}]_2$ channel.[147] Model calculations are under way at Pittsburgh to assess this question. The Ar* + Kr van der Waals parameters derived by Winicur et al.[146] establish a very close similarity with K + Kr. Results for Ar* + Xe, however, indicate an appreciably smaller well depth[75] than for K + Xe. It also seems quite likely that the low-energy repulsion is not as similar to that of the alkalis as is the attractive part of the potential. Winicur et al.[146] have carried out an approximate optical model analysis, but this suffers from the uncertainty in the repulsion for the input channel as well as from the basic limitations discussed in Section II.

Winicur et al.[149-151] have also published studies of $Ar^* + N_2^{150}$ and $HBr^{149, 151}$ scattering. Whereas the potential surfaces of the input channels may be highly anisotropic, optical model analysis in these cases is less likely to suffer from "recrossing" contributions because part of the electronic energy must be rapidly and nearly irreversibly converted to vibrational energy in the newly excited molecule. The $Ar^* + HBr$ cross section and van der Waals potential closely resemble those of $K + HBr$;[151] this suggests indirectly that covalent–ionic interaction may govern the quenching in this system. Reaction to form $ArBr^*$ is also energetically possible and is considered probable. Nonetheless, the total quenching cross section derived is nearly an order of magnitude smaller than that obtained from flowing-afterglow experiments.

b. Excitation-transfer Cross Sections

There is very little information available on magnitudes or energy dependences of electronic energy-transfer cross sections in atom–atom systems, aside from that inferred from classical kinetics. Some information on metastable noble-gas–molecule quenching cross sections from observation of molecular emission in crossed beams is now becoming available and is discussed in the text that follows. The usual optical model analysis of elastic scattering employs a local, absorptive potential that may not be very useful in the atom–atoms systems, and the cross sections therefrom may not be reliable. For $He^*(2^1S) + Ne$, the earlier results due to Chen, et al.[40] from optical model analysis of the elastic scattering at 63 meV are likely to be less reliable because the data were not corrected for Ne^* contributions. Fukuyama and Siska[145] have used both optical model and close-coupling (two-channel) methods at 64 meV on corrected data to obtain cross sections in the range 2 to 3.5 Å^2. These compare well with the flowing-afterglow work of Schmeltekopf and Fehsenfeld,[151] who find a thermal cross section of 4.7 Å^2. The crossed-beam experiments at Freiburg[76, 77, 152] and Pittsburgh[75, 145] show, from the appearance of the Ne^* contribution, that the cross section rises rapidly with collision energy. Each channel cross section appears to increase quickly near threshold and then level out, producing an increase in the total quenching as each new channel opens. Integration of transformed Ne^* angular distributions (see Section III.D.2.c) promises to provide the first reliable energy dependencies for the various channels. Flow-tube results of Lindinger et al.[141] on $He^*(2^3S) + Ne$ show a strong positive temperature dependence from $300°K$ to $900°K$, which correlates with a marked rise in scattered Ne^* intensity with collision energy observed by Oesterlin, et al.[152]

Cross-section energy dependences for specific product channels in the $Ar^* + N_2$ system have been directly investigated by Lee and Martin,[153-156] through observation of N_2^* emission from beam–gas cell or crossed-beam arrangements. Total quenching cross sections for Ar* and Kr* by O_2 have been reported by Gersh and Muschlitz[157] using the beam–gas method, in which the $^3P_0$ and $^3P_2$ states were resolved through the use of an inhomogeneous magnetic deflecting field. Whereas the $Ar^* + N_2(X) \rightarrow Ar + N_2^*(C)$ cross section shows a strong, line-of-centers energy dependence, the $Ar^*, Kr^* + O_2$ quenching cross sections are at most weakly energy dependent. This suggests that the N_2 system is repulsive in the incoming channel and the O_2, attractive. An anticipated correlation is evident then between these systems and the corresponding He* cases.

c. Product Angular Distributions

As introduced in Section III.D.2.a, angular distribution measurements in noble-gas excitation-transfer (non-Penning) systems have been shown to contain an unexpected bonus: angular distributions for formation of electronically excited products by direct electronic–electronic energy transfer. Although the new states freshly formed at the collision center are generally not metastable, the metastable states can be reached by one or a sequence of radiative transitions and will eventually (after $\sim 10^{-7}$ sec or less) be populated according to the relative overall emission rates for going to the ground and metastable states. Estimation of this branching ratio is difficult for states high in the spectrum of the newly excited atom, but for $Ar^*(^3P_{2,0}) + Kr$ the ratio has been measured by observation of Kr* emission.[158] The metastables are the only atomic remains of the inelastic collision surviving long enough to reach the detector surface. Far-UV photons produced by transitions to the ground state, also detectable by secondary emission, contribute appreciably to the observed intensity only when the total inelastic cross section becomes quite large or when the transitions are highly energetic, as for neon. This contribution is isotropic.

Two methods have been used for isolating the transfer angular distribution in He* + Ne. By installing a quench lamp operated with neon in the detector, Haberland et al.[76, 77, 152] have used a double-subtraction method to derive the total Ne* angular distributions; data for $He^*(2^1S) + Ne$ are shown in Fig. 52. State assignments were made by comparing the observed peak positions with a Newton diagram drawn for the most probable beam velocities and the known states of neon; an example is shown in Fig. 53. Although the assignments can be ambiguous, the distributions as a function of energy as shown in Fig. 52 often help to relieve the ambiguity. For $He^*(2^1S) + Ne$, strong forward scattering of the various product states enables the $3s_2$, $3s$, and $3s_5$ states of neon to be identified from the Newton diagram. As mentioned earlier, the neon quenching is not 100% efficient

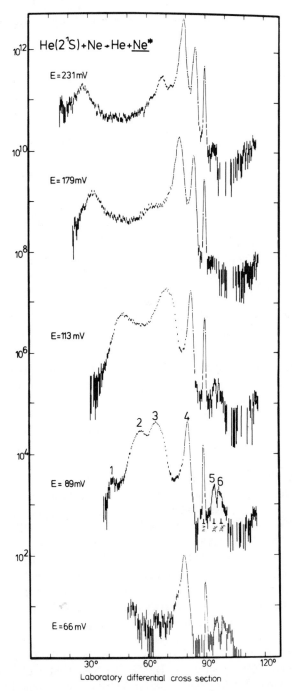

Figure 52. Neon* laboratory angular distributions from He*(2^1S)+Ne at five collision energies.

575

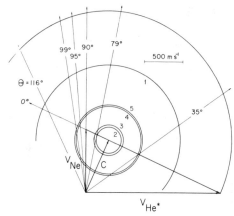

Figure 53.   Newton diagram for He*(2^1S)+Ne at 66 meV. Largest partial circle is locus of He* velocities from elastic collisions; smaller numbered ones represent inelastic production of Ne* in various final states. Numbers n correspond to subscripts $3s_n$ for states of neon (Paschen notation). Angular rays correspond to positions of maxima or shoulders in angular distribution of Fig. 50.

and hence is velocity dependent; Ne* atoms with higher laboratory velocities are less effectively quenched, and subtraction yields lower intensity for these atoms.

The second method, introduced by Martin et al.,[75] employs TOF measurements in regions of the angular distribution where excited atom products appear to be present. Both the Pittsburgh and Freiburg groups are now using the TOF method on the He* + Ne system. Rather than the usual pulsing of the scattered products with a mechanical chopper in front of the detector, the technique employs an electrostatically pulsed electron beam to excite the metastable beam. The mode (TOF or angular distribution) and TOF resolution are then selectable without modification of the experimental hardware. Although obtaining a complete product angular distribution entails many TOF spectra (each requiring 10^2 to 10^4 sec to collect), unambiguous state assignments can be made and the elastic and photon components of the intensity recovered at each angle. It has been found that the elastic scattering is reasonably smooth at wide angles for He* + Ne, and thus the various product angular distributions can be recovered with the original high-datum-point density by interpolating the underlying elastic distribution extracted from the TOF spectra. Figure 54 shows assigned TOF spectra for He*(2^1S) + Ne at 64 meV, and Fig. 55 presents the corresponding Ne* angular distribution obtained by the TOF method at the same energy (cf. Fig. 52, 66 meV). In unpublished work at

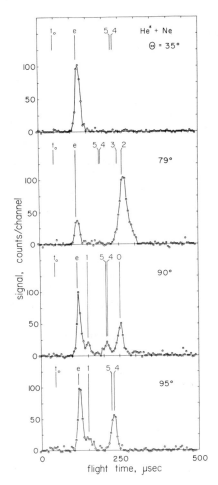

Figure 54. Sampling of TOF spectra for He* + Ne. Time t_0 is flight time from beam excitation region to collision center; e, expected elastic flight time derived from Newton diagram of Fig. 53, and numbered times those for Ne* in various final states (notation as in Fig. 53). Number zero corresponds to beam neon photoexcited by far-UV photons produced as result of energy transfer (see Section III.A.7).

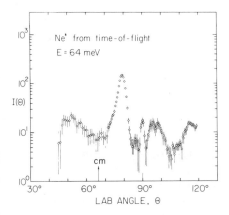

Figure 55. Neon* angular distribution from He*(2^1S) + Ne, derived from data of Fig. 50.

Freiburg and Pittsburgh the prominent peaks near the center of mass for He*(2^1S)+Ne at $E=89$ and 113 meV in Fig. 52 are shown to originate from a $2p^54d$ state and the wide-angle bump in the TOF angular distribution of Fig. 55, from the $2p^53d,3s_1$ manifold.

The singlet angular distributions are dominated by strong forward scattering with some backward intensity as well. Thus it is easy to assign the states of neon by reference to an appropriate Newton diagram. The width of the forward scattering peak for the endoergic states ($3s_2,2p^54d$) grows smaller with increasing collision energy, in accordance with the increasing number of orbital angular momenta that can contribute to each inelastic channel. For the exoergic states ($3s_4,3s_5,3s_1$) the widths are nearly invariant with energy. No triplet angular distributions have yet been measured. The triplet beam intensity is roughly a factor of 7 lower and the cross section a factor of 20 lower than in the singlet case. Therefore, the distributions published[152] as a conference abstract are heavily contaminated by singlets.

A two-state close-coupling analysis has been reported[145] at 64 meV for He*(2^1S)+Ne using $3s_2$ as the product channel and assuming the (diabatic) product potential curve crosses the incoming curve on its repulsive wall. The elastic scattering was fitted to yield a state-to-state cross section of 2.1 Å². The Ne* angular distribution calculated from the resulting

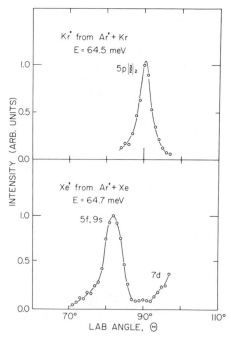

Figure 56. Angular distributions of Kr* and Xe* for excitation transfer from Ar*. State assignments are derived from TOF measurements.

potential matrix agreed fairly well with that part of the total Ne* scattering assigned to $3s_2$. The inclusion of a third, exoergic state does not substantially perturb the two-state results for cross sections into the exoergic channel as large as 0.5 Å^2. The opening of the highly probable $2p^54d$ channel at higher energies will complicate the analysis further.

Martin et al.[75] have also reported Kr* and Xe* peaks in Ar* + Kr,Xe scattering deduced from TOF spectra; the transfer angular distributions are shown in Fig. 56. These two systems are of contrasting degrees of complexity in that the spectrum of states is limited to two for Ar*($^3P_2$) + KR: $5p[\frac{3}{2}]_2$ and $5p[\frac{3}{2}]_1$,[158] but at least 28 states of Xe* are known to be populated from flowing-afterglow measurements.[159] The angular and TOF data again demonstrate sharp forward scattering for both exoergic and endoergic states and good agreement with major states found in the emission spectroscopy work. The energy dependence of the Xe* angular distribution is expected to be spectacular because of the myriad new channels opening up at higher energies. Measurements are currently in progress at Pittsburgh.

IV. FUTURE DIRECTIONS

The new levels of experimental resolution, data quantity and quality, and sophistication of interpretation of the differential scattering results presented here, combined with concomitant blossoming of the *ab initio* theory, have produced an exciting situation with few precedents in the field of molecular collision dynamics. Continued mutual feedback between theoretical and experimental efforts should rapidly increase and deepen our knowledge of excited-state collision dynamics. This interaction will become even more important when we move from atom–atom systems to atom–molecule systems in the future and will enable the tackling of even more challenging systems in which the alkalilike chemical nature of the noble-gas metastables competes in luminescent chemical reactions with the excitation transfer and ionization channels. Experimental studies of elastic and inelastic scattering of metastables by halogen-containing molecules are already under way, as well as measurements of optical emission in reactions under beam conditions. With recently improved potential curves for both ground and excited states of the diatomic noble-gas halides, chemiluminescence in metastable atom–halogen reactions can now provide information on energy partitioning in the initial products, which is largely unavailable for the corresponding alkali reactions. In these investigations we can draw on our extensive knowledge of alkali–halogen collision dynamics derived from beam studies of the past two decades to elucidate subtle differences between noble-gas halide excimer formation and the corresponding alkali reaction, as well as to clarify the role of competing processes.

Differential scattering experiments with Ne* and other beams state selected with a tuneable dye laser are near realization. Differences in the potential-energy curves and reaction probabilities for the $^3P_2$ and $^3P_0$ states will provide valuable insight into the role of the core ion on the collision dynamics and electronic structure as well as clarify the relative importance of the two states in macroscopic processes. Experiments using a metal-atom crossed beam, also currently in progress at Freiburg, promise a revealing contrast to the weak van der Waals interactions thus far studied.

Our hope is that, through continued experimental scattering work and theoretical studies on these and other systems, excited-state interactions will become as well characterized as those for the ground state are now. With important laser applications already discovered for systems containing metastable noble-gas atoms, unusually fast "transfer" of knowledge gained on microscopic collision dynamics in these studies to applied areas may be expected. However, because of the complexity of many of these highly energetic systems, future progress will likely entail careful investigations of all the competing processes by the combination of various experimental and theoretical methods.

ACKNOWLEDGMENTS

P. E. Siska wishes to acknowledge support from the National Science Foundation, the Petroleum Research Fund of the American Chemical Society, and the Alfred P. Sloan Foundation. H. Haberland wishes to acknowledge support from the Deutsche Forschungsgemeinschaft. Y. T. Lee wishes to acknowledge support from the Division of Chemical Sciences, Office of Basic Energy Sciences, U. S. Department of Energy under contract No. W-7405-Eng-48.

REFERENCES

1. F. M. Penning, *Naturwissenschaften*, **15**, 818 (1927); *Z. Phys.*, **46**, 335 (1928); *Z. Phys.*, **57**, 723 (1929); *Proc. Roy. Acad. Sci. (Amsterdam)*, **32**, 341 (1929); *Z. Phys.*, **72**, 338 (1931).

2. F. Mohler, P. Foote, and R. Chenault, *Phys. Rev.*, **27**, 37 (1926).

3. E. O. Lawrence and N. Edlefsen, *Phys. Rev.*, **34**, 233 (1929).

4. F. Mohler and C. Boeckner, *Nat. Bur. Stand. J. Res.*, **5**, 51 (1930).

5. K. Freudenberg, *Z. Phys.*, **67**, 417 (1931).

6. B. Gudden, *Lichtelektrische Erscheinungen*, Springer-Verlag, Berlin, 1928, p. 226.

7. F. Mohler and C. Boeckner, *Nat. Bur. Stand. J. Res.*, **5**, 399 (1930).

8. J. A. Hornbeck and J. P. Molnar, *Phys. Rev.*, **84**, 62 (1951).

9. W. P. Sholette and E. E. Muschlitz, Jr., *J. Chem. Phys.*, **36**, 3368 (1962).

10. E. W. Rothe, R. H. Neynaber, and S. M. Trujillo, *J. Chem. Phys.*, **42**, 3310 (1965).

11. V. Čermak, *J. Chem. Phys.*, **44**, 3781 (1966).

12. D. L. King and D. W. Setser, *Annu. Rev. Phys. Chem.*, **27**, 407 (1976).

13. R. J. Pressley, *Handbook of Lasers with Selected Data on Optical Technology*, Chemical Rubber Company, Cleveland, Ohio, 1971.

14. W. H. Hughes, J. Shannon, A. Kolb, E. Ault, and E. Bhaumik, *Appl. Phys. Lett.*, **23**, 385 (1973).

15. P. W. Hoff, J. C. Swingle, and C. K. Rhodes, *Appl. Phys. Lett.*, **23**, 245 (1973).

16. W. M. Hughes, J. Shannon, and R. Hunter, *Appl. Phys. Lett.*, **24**, 488 (1974).

17. A. Niehaus, *Ber. Bunsenges. Phys. Chem.*, **74**, 632 (1973).

18. H. Hotop, *Radiat. Res.*, **59**, 379 (1974).

19. D. R. Herschbach, *Adv. Chem. Phys.*, **10**, 319 (1966).

20. C. A. Brau and J. J. Ewing, *J. Chem. Phys.*, **63**, 4640 (1975).

21. J. E. Velazco and D. W. Setser, *J. Chem. Phys.*, **62**, 1990 (1975); L. A. Gundel, D. W. Setser, M. A. A. Clyne, J. A. Coxon, and W. Nip, *J. Chem. Phys.*, **64**, 4390 (1976) and references cited therein.

22. M. L. Bhaumik, R. S. Bradford, Jr., and E. R. Ault, *Appl. Phys. Lett.*, **28**, 23 (1976).

23. C. A. Brau and J. J. Ewing, *J. Chem. Phys.*, **63**, 4640 (1975).

24. W. H. Miller, *J. Chem. Phys.*, **52**, 3563 (1970).

25. H. Nakamura, *J. Chem. Soc. Jap.*, **26**, 1973 (1969); **31**, 574 (1971).

26. A. P. Hickman, A. D. Isaacson, and W. H. Miller, *J. Chem. Phys.*, **66**, 1483 (1977).

27. W. H. Miller, C. A. Slocomb, and H. F. Schaefer, III, *J. Chem. Phys.*, **56**, 1347 (1972).

28. A. P. Hickman, A. D. Isaacson, and W. H. Miller, *J. Chem. Phys.*, **66**, 1492 (1977).

29. J. C. Bellum and D. A. Micha, *Chem. Phys.*, **20**, 121 (1977).

30. A. P. Hickman and H. Morgner, *J. Phys.*, **B9**, 1765 (1976).

31. D. A. Micha, in W. H. Miller, Ed., *Modern Theoretical Chemistry*, Vol. II, Plenum, New York, 1976, p. 81.

32. T. Ebding and A. Niehaus, *Z. Physik*, **270**, 43 (1974).

33. Z. F. Wang, A. P. Hickman, K. Shobatake, and Y. T. Lee, *J. Chem. Phys.*, **65**, 1250 (1976).

34. See, for example, U. Buck, *Adv. Chem. Phys.*, **30**, 313 (1975); *Rev. Mod. Phys.*, **46**, 369 (1974).

35. E. F. Greene, A. L. Moursund, and J. Ross, *Adv. Chem. Phys.*, **10**, 135 (1966).

36. (a) H. D. Hagstrum, *Phys. Rev.*, **104**, 1516 (1956); (b) D. A. MacLeanner, *Phys. Rev.*, **148**, 218 (1966); (c) P. J. MacVicar-Whelan and W. L. Borst, *Phys. Rev.*, **A1**, 3141 (1970).

37. (a) J. B. Anderson, R. P. Andres, and J. B. Fenn, *Adv. Chem. Phys.*, **10**, 275 (1966); (b) J. B. Anderson, in P. P. Wegener, Ed., *Molecular Beams and Low Density Gas Dynamics*, Dekker, New York, 1974, Chapter 1.

38. G. Brunsdeylins, H.-D. Meyer, J. P. Toennies, and K. Winkelmann, "Rarefied Gas Dynamics," *Prog. Astron. Aeron.*, **51**, 1047 (1976).

39. C. H. Chen, H. Haberland, and Y. T. Lee, *J. Chem. Phys.*, **61**, 3095 (1974).

40. D. H. Winicur and J. L. Fraites, *J. Chem. Phys.*, **62**, 63 (1975).

41. N. F. Mott and H. S. W. Massey, *The Theory of Atomic Collisions*, 3rd ed., Oxford U.P., 1965.

42. R. B. Bernstein, *Adv. Chem. Phys.*, **10**, 75 (1966).

43. E. E. Nikitin, *Theory of Elementary Atomic and Molecular Processes in Gases*, Oxford U.P., 1974.

44. M. S. Child, *Molecular Collision Theory*, Academic, New York, 1974.

45. W. H. Miller, Ed., *Dynamics of Molecular Collisions*, Parts A and B, Vols. 1 and 2 of *Modern Theoretical Chemistry*, Plenum, New York, 1976.

46. D. Beck, E. F. Greene, and J. Ross, *J. Chem. Phys.*, **37**, 2895 (1962).

47. E. F. Greene and J. Ross, in C. Schlier, Ed., *Molecular Beams and Reaction Kinetics*, Academic, 1970, pp. 86–106.

48. R. M. Harris and J. F. Wilson, *J. Chem. Phys.*, **54**, 2088 (1971).

49. W. H. Miller, *Chem. Phys.* Lett., **4**, 627 (1970).

50. L. D. Landau, *Z. Phys. Sov. Union*, **2**, 46 (1932).

51. C. Zener, *Proc. Roy. Soc. (Lond.), Ser. A*, **137**, 696 (1932).

52. E. C. G. Stückelberg, *Helv. Phys. Acta*, **5**, 370 (1932).

53. J. S. Cohen, *Phys. Rev.*, **A13**, 99 (1976).

54. See, for example *Meth. Comp. Phys.*, **10** (1971).

55. R. E. Olson, F. T. Smith, and E. Bauer, *Appl. Opt.*, **10**, 1848 (1971).

56. R. D. Rundel, F. B. Dunning, J. P. Howard, J. P. Riola, and R. F. Stebbings, *Rev. Sci. Instrum.*, **44**, 60 (1973).

57. W. L. Borst, *Rev. Sci. Instrum.*, **42**, 1543 (1971).

58. B. Brutschy, Ph.D. thesis, Freiburg, 1977; B. Brutschy and H. Haberland, *Phys. Rev.*, **19**, 2232 (1979).

59. G. Watel, private communication.

60. R. J. Champeau and J. C. Keller, *J. Physique*, **36**, L161 (1975); H. Hotop and A. Zastrow, in G. Watel, Ed., *Abstracts, Xth ICPEAC*, Paris 1977, p. 306.

61. R. F. Stebbings, F. B. Dunning, and R. D. Rundel, *Proceedings of Fourth International Conference on Atomic Physics*, Heidelberg, 1974, book of Invited papers, p. 713. (Plenun 1975), Eds., G. Zu Putlitz, E. W. Weber, and A. Winnacher.

62. RTV 630, General Electric Company.

63. R. Campargue, and A. Lebehot, in M. Becker and M. Fiebig, Eds., *Ninth International Symposium on Rarefied Gas Dynamics*, Göttingen, 1974, p. C11.

64. B. Brutschy and H. Haberland, *J. Phys.*, **E10**, 90 (1977).

65. H. S. W. Massey, *Electronic and Ionic Impact Phenomena*, Vol. 3, *Slow Collisions of Heavy Particles*, Clarendon, Oxford, 1973.

66. G. A. Victor and K. Sando, *J. Chem. Phys.*, **55**, 5421 (1971).

67. R. E. Olson and F. T. Smith, *Phys. Rev.*, **A7**, 1529 (1973).

68. J. Q. Searcy, *Rev. Sci. Instrum.*, **45**, 589 (1974).

69. E. L. Leasure and C. R. Mueller, *J. Appl. Phys.*, **47**, 1062 (1976).

70. K. Schmidt, Ph. D. thesis, University of Freiburg, 1976.

71. H. Maecker and S. Steinberger, *Z. Angew. Physik*, **23**, 456 (1967).

72. W. S. Young, W. E. Rodgers, and E. L. Knuth, *Rev. Sci. Instrum.*, **40**, 1346 (1969).

73. W. J. Hays, W. E. Rodgers, and E. L. Knuth, *J. Chem. Phys.*, **56**, 1652 (1972).

74. H. Haberland, B. Brutschy, and K. Schmidt, in J. S. Risley and R. Geballe, Eds., *Abstracts, IXth ICPEAC*, Washington U.P., Seattle, 1975, p. 443.

75. D. W. Martin, T. Fukuyama, R. W. Gregor, R. M. Jordan, and P. E. Siska, *J. Chem. Phys.*, **65**, 3720 (1976).

76. H. Haberland, P. Oesterlin, and K. Schmidt, *J. Chem. Phys.*, **65**, 3374 (1976).

77. H. Haberland, P. Oesterlin, and K. Schmidt, *Sixth International Symposium on Molecular Beams*, Amsterdam, 1977.

78. A. V. Phelps and J. P. Molnar, *Phys. Rev.*, **89**, 1202 (1953).

79. A. H. Futch and F. A. Grant, *Phys. Rev.*, **104**, 356 (1956).

80. A. V. Phelps, *Phys. Rev.*, **114**, 1011 (1959).

81. D. W. Marquardt, *J. Soc. Ind. Appl. Math.*, **11**, 431 (1963).

82. T. Fukuyama and P. E. Siska, *Chem. Phys. Lett.*, **39**, 418 (1976).

83. W. V. Oertzen and H. G. Bohlen, *Phys. Rep.*, **C19**, 1 (1975).

84. S. L. Guberman and W. A. Goddard, *Phys. Rev.*, **A12**, 1203 (1975) and references cited therein.

85. G. Peach, *J. Phys.*, **B11**, 2107 (1978).

86. K. M. Sando, *Molec. Phys.*, **21**, 439; **23**, 413 (1971).

87. B. Schneider and J. Cohen, *J. Chem. Phys.*, **61**, 3240 (1974).

88. R. P. Saxon and B. Liu, *J. Chem. Phys.*, **64**, 3291 (1976).

89. D. C. Lorents, *Physica*, **82**, 19 (1975).

90. D. C. Lorents, in G. Watel, Ed., *Invited Lecture, Xth ICPEAC*, Paris, 1977.

91. B. Brutschy and H. Haberland, *Phys. Rev. Lett.*, **38**, 686 (1977).

92. R. A. Buckingham and A. Dalgarno, *Proc. Roy. Soc. (Lond.) Ser. A*, **213**, 506 (1952).

93. K. Gillen, in G. Watel, Ed., *Progress Report, Xth ICPEAC*, Paris, 1977.

94. R. Morgenstern, D. C. Lorents, J. R. Peterson, and R. E. Olson, *Phys. Rev.*, **A8**, 2372 (1973).

95. S. M. Trujillo, in J. S. Risley, R. Geballe, Eds., *Abstracts, IXth ICPEAC*, Seattle, 1975, p. 437.

96. F. D. Colegrove, L. K. Shearer, and G. K. Walters, *Phys. Rev.*, **A135**, 353 (1964).

97. S. K. Rosner, and F. M. Pipkin, *Phys. Rev.*, **A5**, 1909 (1972).

98. J. Dupont-Roc, M. Leduc, and F. Laloë, *Phys. Rev. Lett.*, **27**, 467 (1971).

99. N. K. Glendenning, *Rev. Mod. Phys.*, **47**, 659 (1975).

100. B. Brutschy, H. Haberland, and K. Schmidt, *J. Phys.*, **B9**, 2693 (1976).

101. B. Brutschy, H. Haberland, H. Morgner, and K. Schmidt, *Phys. Rev. Lett.*, **36**, 1299 (1976).

102. H. Haberland and K. Schmidt, *J. Phys.*, **B10**, 695 (1977).

103. P. Altpeter, H. Haberland, W. Konz, P. Oesterlin, and K. Schmidt, *J. Chem. Phys.*, **67**, 836 (1977).

104. H. Haberland et al., private communication.

105. J. Grosser and H. Haberland, *Phys. Lett.*, **27A**, 634 (1968).

106. H. Haberland, C. H. Chen, and Y. T. Lee, in S. J. Smith and G. K. Walters, Eds., *Atomic Physics*, Vol. 3, Plenum, New York, 1973, p. 339.

107. J. Bentley, J. L. Fraites, and D. H. Winicur, *J. Chem. Phys.*, **64**, 1757 (1976).

108. R. M. Jordan, D. W. Martin, and P. E. Siska, *J. Chem. Phys.*, **67**, 3392 (1977).

109. D. W. Martin, R. W. Gregor, R. M. Jordan, and P. E. Siska, *J. Chem. Phys.*, **69**, 2833 (1978); R. M. Jordan and P. E. Siska, *J. Chem. Phys.*, **69**, 4634 (1978).

110. C. H. Chen and Y. T. Lee, private communication.

111. M. T. Leu and P. E. Siska, *J. Chem. Phys.*, **60**, 2179 (1974).

112. M. T. Leu and P. E. Siska, *J. Chem. Phys.*, **60**, 4082 (1974).

113. F. T. Smith, R. P. Marchi, W. Aberth, D. C. Lorents, and O. Heinz, *Phys. Rev.*, **161**, 31 (1967).

114. F. T. Smith, H. H. Fleischmann, and R. A. Young, *Phys. Rev.*, **A2**, 379 (1970).

115. Because of a typographical error, the stated value of the C coefficient of the width function of Ref. 102 is a factor of 10 too large. It should read $C = 0.0326$ eV. The entire width function of Refs. 100 and 102 is too large by exactly a factor or two; the fits to the data and the resulting real part of the potential are not affected.

116. U. Buck and H. Pauly, *Z. Phys.*, **208**, 390 (1968).

117. R. Düren, C. P. Raabe, and C. Schlier, *Z. Phys.*, **214**, 410 (1968).

118. G. B. Ury and L. Wharton, *J. Chem. Phys.*, **56**, 5832 (1972).

119. P. M. Dehmer and L. Wharton, *J. Chem. Phys.*, **57**, 4821 (1972).

120. J. S. Cohen and N. F. Lane, *J. Chem. Phys.*, **66**, 586 (1977).

121. A. D. Isaacson, A. P. Hickman, and W. H. Miller, *J. Chem. Phys.*, **67**, 370 (1977).

122. R. K. Preston and J. S. Cohen, *J. Chem. Phys.*, **65**, 1589 (1976).

123. Y. T. Lee, J. D. McDonald, P. R. LeBreton, and D. R. Herschbach, *Rev. Sci. Instrum.*, **40**, 1402 (1969).

124. E. A. Entemann, Ph.D. thesis, Harvard University, Cambridge, Mass., 1967.

125. R. H. Neynaber, G. K. Magnuson, and J. K. Layton, *J. Chem. Phys.*, **57**, 5128 (1972).

126. W. C. Richardson and D. W. Setser, *J. Chem. Phys.*, **58**, 1809 (1973).

127. D. O. Ham and J. L. Kinsey, *J. Chem. Phys.*, **53**, 285 (1970).

128. M. G. Golde and B. A. Thrush, *Chem. Phys. Lett.*, **29**, 486 (1974).

129. R. W. Gregor, M. T. Leu, and P. E. Siska, manuscript in preparation.

130. See, for example, the review by A. L. Wahrhaftig in A. Maccoll, Ed., *Mass Spectrometry*, in *Physical Chemistry, Series 1*, Vol. 5 of *MTP International Review of Science*, Butterworths, London, 1972, pp. 1–24.

131. G. D. Magnuson and R. H. Neynaber, *J. Chem. Phys.*, **60**, 3385 (1974).

132. R. H. Neynaber and G. D. Magnuson, *J. Chem. Phys.*, **61**, 749 (1974).

133. R. H. Neynaber and G. D. Magnuson, *Phys. Rev.*, **A11**, 865 (1975).

134. R. H. Neynaber and G. D. Magnuson, *J. Chem. Phys.*, **62**, 4953 (1975).

135. R. H. Neynaber and G. D. Magnuson, *Phys. Rev. A*, **14**, 961 (1976).

136. S. M. Trujillo, in J. S. Risley and R. Geballe, Eds., *Abstracts, IXth ICPEAC*, Washington U.P., Seattle, 1975, p. 437.

137. K. L. Bell, A. Dalgarno, and A. E. Kingston, *J. Phys.*, **B1**, 18 (1968).

138. A. C. H. Smith, private communication. S. F. W. Harper and R. C. H. Smith, in G. Watel, Ed., *Abstracts, Xth ICPEAC*, Paris, 1977, p. 82.

139. E. Illenberger and A. Niehaus, *Z. Phys.*, **B20**, 33 (1975).

140. A. Pesnelle, G. Watel, and C. Manus, *J. Chem. Phys.*, **62**, 3590 (1975).

141. W. Lindinger, A. L. Schmeltekopf, and F. C. Fehsenfeld, *J. Chem. Phys.*, **61**, 2890 (1974).

142. J. P. Riola, J. S. Howard, R. D. Rundel, and R. F. Stebbings, *J. Phys.*, **B7**, 376 (1974).

143. T. F. O'Malley and H. S. Taylor, *Phys. Rev.*, **176**, 207 (1968).

144. H.-P. Weise and H.-U. Mittmann, *Z. Naturforsch., Teil A*, **28**, 714 (1973).

145. P. E. Siska and T. Fukuyama, in G. Watel, Ed., *Abstracts, Xth ICPEAC*, Paris, 1977, p. 552.

146. D. H. Winicur, J. L. Fraites, and J. Bentley, *J. Chem. Phys.*, **64**, 1724 (1976).

147. L. G. Piper, D. W. Setser, and M. A. A. Clyne, *J. Chem. Phys.*, **63**, 5018 (1975).

148. D. H. Winicur, J. L. Fraites, and F. A. Stackhouse, *Chem. Phys. Lett.*, **23**, 123 (1973).

149. D. H. Winicur and J. L. Fraites, *J. Chem. Phys.*, **61**, 1548 (1974).

150. D. H. Winicur and J. L. Fraites, *J. Chem. Phys.*, **64**, 89 (1976).

151. A. L. Schmeltekopf and F. C. Fehsenfeld, *J. Chem. Phys.*, **53**, 3173 (1970).

152. P. Oesterlin, H. Haberland, and K. Schmidt, in G. Watel, Ed., *Abstracts, Xth ICPEAC*, Paris, 1977, p. 558. Data presented are contaminated by contributions from He*(2^1S). Further work is in progress.

153. W. Lee and R. M. Martin, *J. Chem. Phys.*, **63**, 962 (1975).

154. A. N. Schweid, M. A. D. Fluendy, and E. E. Muschlitz, Jr., *Chem. Phys. Lett.*, **42**, 103 (1976).

155. R. A. Sanders, A. N. Schweid, M. Weiss, and E. E. Muschlitz, Jr., *J. Chem. Phys.*, **65**, 2700 (1976).

156. J. Krenos and J. Bel Bruno, *J. Chem. Phys.*, **65**, 5019 (1976); erratum, ibid, **66**, 5832 (1977).

157. M. E. Gersh and E. E. Muschlitz, Jr., *J. Chem. Phys.*, **59**, 3538 (1973).

158. L. G. Piper, D. W. Setser, and M. A. A. Clyne, *J. Chem. Phys.*, **63**, 5018 (1975).

159. D. L. King, L. G. Piper, and D. W. Setser, *J. Chem. Soc. Faraday Transact. II*, **73**, 177 (1977).

160. M. L. Ginter, *J. Chem. Phys.*, **42**, 561 (1965); M. L. Ginter and C. M. Brown, ibid., **56**, 672 (1972).

161. G. Das, *Phys. Rev.*, **A11**, 732 (1975).

162. A. P. Hickman and N. F. Lane, *Phys. Rev.*, **A10**, 444 (1974).

163. I. Ya. Fugol', *Sov. Phys. Uspekhi*, **12**, 182 (1969).

AUTHOR INDEX

Abbey, L. E., 87, 92, 111, 141, 208
Aberth, W., 152, 216, 551, 568, 584
Adam, M. Y., 41, 80
Adams, J. T., 198, 222
Adams, N. G., 181, 190, 222
Ajello, J. M., 87, 88, 96, 105, 107, 208, 209, 211
Alberti, F., 288, 339
Albritton, D. L., 87, 88, 108, 109, 110, 114, 115, 122, 123, 129, 130, 137, 145, 146, 162, 163, 168, 209, 212, 214, 215, 218, 220, 318, 340, 468, 486
Alderson, R. J., 146, 215
Alkemade, C. T. J., 347, 395, 396
Allen, L. C., 228, 338
Altpeter, P., 545, 549, 552, 554, 555, 583
Amme, R. C., 88, 94, 95, 96, 117, 164, 207, 208, 211, 218
Amusia, M. Ya., 24, 42, 43, 44, 76
Andersen, T., 152, 167, 217
Anderson, A., 142, 215, 268, 318, 339, 340
Anderson, J. B., 128, 214, 495, 581
Anderson, R. W., 212, 345, 395
Andlauer, B., 105, 212
Andreev, E. A., 357, 394, 397
Andres, R. P., 495, 581
Anicich, V. G., 117, 127, 148, 150, 151, 211, 213, 214
Ankudinov, V. A., 152, 216
Anlauf, K. G., 128, 214
Appell, J., 93, 176, 195, 203, 209, 222
Armstrong, B. H., 227, 337
Armstrong, E. B., 227, 337
Armstrong, L., 24, 76
Arrathoon, R., 476, 486
Asbrink, L., 61, 68, 79
Ast, T., 93, 141, 209
Asundi, R. K., 87, 106, 131, 208
Ault, E. R., 490, 493, 581
Ausloos, P., 83, 99, 115, 207, 211
Austin, T. M., 476, 486
Awan, A. M., 301, 339

Backx, C., 5, 7, 16, 19, 29, 30, 34, 36, 38,

40, 46, 49, 50, 51, 52, 53, 54, 57, 68, 69, 71, 72, 75, 77, 79, 80
Baer, M., 198, 206, 222
Baer, T., 96, 98, 104, 122, 123, 142, 143, 144, 210
Bagus, P. S., 230, 338
Bahr, J. L., 63, 73, 80
Bailey, T. L., 136, 164, 214, 218
Baker, A. D., 37, 77
Baker, D. J., 47, 52, 79
Bardsley, J. N., 357, 397, 444, 485
Barker, J. R., 344, 347, 349, 357, 395, 397
Bästlein, C., 347, 352, 396
Bates, D. R., 301, 339
Battacharya, A. K., 127, 214
Bauer, E., 355, 373, 396, 512, 582
Baumgarner, G., 347, 352, 396
Bauschlicher, C. W., Jr., 223
Bayes, K. D., 170, 171, 172, 184, 221
Bayley, T. L., 476, 486
Bearden, J. A., 16, 41, 78
Bearman, G. H., 155, 156, 157, 168, 169, 170, 174, 175, 187, 190, 191, 192, 193, 217, 220, 221, 222
Beaty, E. C., 31, 32, 76, 345, 377, 387, 389, 395
Beauchamp, J. L., 89, 93, 99, 115, 116, 117, 124, 126, 127, 200, 210, 211, 213
Beck, D., 393, 398, 501, 582
Becker, P. M., 87, 106, 107, 208
Beckey, H. D., 141, 215
Bederson, B., 491
Bedo, D. E., 47, 52, 79
Beebe, N., 268, 339
Bel Bruno, J., 573, 585
Bell, K. L., 22, 50, 75, 79, 565, 584
Bell, W. E., 182, 219
Bellum, J. C., 494, 581
Bender, C. F., 24, 80, 203, 223, 265, 339
Benes, E., 361, 397
Bennett, W. R., Jr., 166, 174, 182, 183, 219, 221
Benstein, R. B., 200, 223
Bente, P. F., III, 141, 215

Bentley, J., 545, 546, 572, 583, 585
Bergman, R. S., 119, 213
Bergmann, K., 361, 397
Berkowitz, J., 50, 52, 79, 94, 98, 102, 210, 211
Bernstein, R. B., 128, 163, 214, 218, 373, 374, 375, 397, 497, 565, 582
Berry, R. S., 23, 76, 103, 167, 168, 210, 219, 220, 473, 486
Bersohn, R., 396
Berta, M. A., 111, 212
Bertoncini, P. J., 227, 229, 230, 237, 337, 338
Bethe, H., 4, 74
Bevington, P. R., 424, 484
Beyer, W., 537
Beynon, J. H., 93, 141, 142, 155, 209, 215, 217
Bhattacharjee, R. C., 151, 216
Bhattacharya, A. K., 89, 210
Bhaumik, M. L., 490, 493, 581
Billingsley, F. P., 289, 339
Biondi, A., 174, 190, 221
Biondi, M. A., 119, 162, 166, 181, 182, 213, 218, 219
Birely, J. H., 171, 172, 221
Birkinshaw, K., 222
Bjerre, A., 352, 354, 355, 396
Blaauw, H. J., 32, 77
Black, J. H., 203
Blais, N. C., 353, 372, 398
Blake, A. J., 63, 73, 79, 80
Blakeley, C. R., 97, 111, 135, 148, 163, 211, 212
Blaney, B. L., 168, 220
Bloom, A. L., 182, 219
Blum, P., 361, 397
Bobashev, S. V., 152, 216
Bobrowicz, F. W., 356, 396
Boeckner, C., 489, 580
Boerboom, A. J. H., 157, 217
Bohlen, H. G., 525, 583
Bohme, D. K., 122, 150, 181, 190, 214, 216, 222
Boizeau, C., 512
Bolden, R. C., 122, 214, 427, 485
Boness, M. J. W., 301, 339
Bonham, R. A., 8, 29, 41, 75, 76, 78
Borst, W. L., 511, 581, 582
Bosse, G., 88, 111, 209

Botschwina, P., 389, 398
Bottcher, C., 356, 358, 391, 396, 397
Botz, F., 20, 151, 216
Bowers, M. T., 19, 87, 88, 96, 97, 106, 117, 122, 127, 131, 132, 135, 150, 151, 165, 184, 200, 208, 209, 211, 213, 214, 223
Brackmann, R. P., 119, 152, 213
Bradford, R. S., Jr., 493, 581
Brandt, D., 86, 119, 155, 169, 170, 176, 185, 193, 195, 203, 207, 222
Branton, G. R., 7, 17, 18, 20, 21, 29, 31, 32, 33, 36, 37, 41, 65, 66, 72, 73, 77, 476, 486
Brau, C. A., 493, 581
Braun, W., 351, 396
Bregman-Reisler, H., 119, 155, 171, 172, 173, 174, 213, 221
Brehm, B., 37, 77
Breit, G., 491
Brieger, M., 392, 398
Brion, C. E., 4, 5, 6, 7, 17, 18, 19, 20, 21, 23, 25, 26, 27, 28, 29, 30, 31, 32, 33, 34, 36, 37, 38, 39, 40, 41, 43, 46, 50, 54, 55, 57, 58, 59, 60, 61, 62, 63, 64, 65, 66, 67, 68, 69, 70, 71, 72, 73, 75, 76, 77, 79, 80, 465, 466, 486
Brion, H., 287, 339
Brocklehurst, B., 146, 215
Broida, H. P., 155, 169, 170, 171, 172, 173, 174, 184, 217, 221, 351, 396
Bromberg, J. P., 38, 77
Brooks, P. R., 128, 214
Brosa, B., 347, 352, 396
Brown, C. M., 533, 585
Brown, P. J., 197, 199, 222
Browning, R., 50, 79
Bruckmuller, R., 361, 397
Brunsdeylins, G., 495, 581
Brunt, J. N. H., 26, 76
Brus, L. E., 396
Brutschy, B., 420, 421, 422, 423, 425, 432, 433, 453, 455, 470, 484, 486, 511, 515, 516, 517, 518, 519, 529, 530, 537, 543, 545, 546, 549, 550, 551, 552, 555, 564, 566, 569, 582, 583
Buck, U., 358, 393, 397, 398, 422, 484, 495, 554, 556, 581, 584
Buckingham, A. D., 22, 23, 75
Buckingham, R. A., 531, 583
Bunker, D. L., 224

Bunker, R. J., 265, 339
Burke, P. G., 23, 29, 30, 42, 76, 78
Burnett, G. M., 145, 215
Burrow, P. D., 360, 397
Burt, J. A., 108, 112, 212
Bush, Y. A., 110, 114, 162, 212
Buttrill, S. E., Jr., 98, 99, 127, 200, 211, 223

Caddick, J., 152, 167, 168, 217, 220
Cairns, R. B., 39, 40, 42, 50, 77, 78
Callear, A. B., 146, 215
Campargue, R., 516, 582
Carleton, N. P., 155, 167, 168, 169, 170, 173, 217, 220
Carlson, R. W., 39, 40, 55, 57, 66, 68, 69, 78, 79, 80
Carlson, T. A., 4, 34, 77
Carney, G. D., 135, 203, 214, 223
Carter, G. M., 345, 395
Cartwright, D. C., 140, 215, 265, 339
Carver, J. H., 63, 73, 79, 80
Cellotta, R. J., 40, 78
Cermák, V., 83, 87, 88, 89, 106, 107, 131, 171, 207, 208, 209, 403, 404, 434, 458, 465, 466, 468, 474, 484, 485, 486, 490, 581
Chamberlain, G. E., 38, 77
Champeau, R. J., 514, 582
Champion, R. L., 19, 136, 164, 177, 214, 218, 219
Chandra, N., 23, 76
Chang, T. N., 46, 47, 77, 80
Chanin, L. M., 119, 174, 181, 193, 213
Chantry, P. J., 87, 106, 131, 208
Chapman, F. M., Jr., 198, 222
Chau, M., 88, 131, 209
Chen, C. H., 420, 484, 496, 545, 546, 570, 571, 581, 583, 584
Chenault, R., 489, 521, 580
Cheng, M. H., 89, 111, 136, 137, 157, 158, 209, 217
Cherepkov, N. A., 24, 42, 43, 44, 76
Chesnavich, J., 200, 223
Chiang, M. H., 88, 89, 111, 136, 137, 157, 158, 209, 217
Child, M. S., 497, 582
Chin, L. Y., 491
Chiu, Y. N., 146, 215
Chong, A. Y. S., 356, 396
Chow, G. K., 351, 396

Chupka, W. A., 50, 52, 79, 87, 88, 94, 96, 98, 100, 102, 103, 105, 115, 123, 126, 127, 133, 147, 207, 208, 209, 210, 211
Churchill, D. R., 227, 337
Clapp, T., 164, 218
Clark, K. C., 168, 169, 220
Clarke, E. M., 94, 211, 212
Clemens, E., 392, 398
Clow, R. P., 97, 127, 136, 142, 144, 211, 214
Clyne, M. A. A., 493, 572, 574, 579, 581, 585
Codling, K., 33, 41, 42, 43, 44, 46, 76, 77, 78
Coffey, D., Jr., 152, 216
Cohen, J. S., 504, 527, 554, 582, 583, 584
Cohen, R. B., 108, 109, 111, 162, 212
Colegrove, F. D., 539, 540, 583
Collin, J., 89, 209
Collins, C. B., 119, 172, 173, 174, 175, 181, 190, 193, 207, 213, 221
Collins, G. J., 119, 166, 167, 174, 181, 182, 183, 213, 219
Comer, J., 26, 76
Comes, F. J., 50, 79, 146, 215
Compton, D. M. J., 84, 85, 86, 88, 90, 91, 92, 207
Compton, R. N., 2, 100
Conrad, H., 512
Conrads, R. J., 116, 213
Cook, G. R., 40, 41, 50, 52, 78, 79
Cook, J. P. D., 73, 80
Cooks, R. G., 93, 141, 142, 155, 209, 215, 217
Cooper, J. W., 4, 11, 20, 22, 74, 75
Cooper, W. G., 198, 222
Coplan, M. A., 119, 170, 171, 172, 213, 220
Copley, G., 347, 349, 351, 396
Corderman, R. R., 99, 211
Cosby, P. C., 93, 95, 96, 107, 111, 116, 122, 157, 158, 209, 210, 211, 212, 214, 217
Cotter, R. J., 85, 86, 89, 123, 124, 130, 131, 132, 146, 207, 208, 209
Cottin, M., 87, 209
Cox, P. A., 22, 75
Coxon, J. A., 493, 581
Cress, M. C., 87, 106, 107, 208

Cross, R. J., Jr., 164, 218
Crowley, M. G., 170, 176, 195, 203, 221, 222
Csanak, G., 24, 80
Csizmadia, I. G., 223
Cunningham, A. J., 174, 190, 193, 207, 221
Curepa, M. V., 434, 437, 485
Curran, R. K., 87, 131, 208
Current, J. H., 150, 215
Curry, S. M., 190, 221
Cvejanovic, S., 15, 75
Cventanovic, R. S., 343, 394
Czajakowski, M., 351, 396

Dahler, J. S., 486
Dalgarno, A., 83, 203, 207, 227, 337, 491, 531, 565, 583, 584
Dana, L., 156, 217
Das, G., 203, 223, 227, 229, 230, 237, 303, 337, 338, 345, 533, 585
Datz, S., 34, 77, 157, 158, 217
Davenport, J. W., 23, 24, 67, 76
David, R., 161, 218
Davidovits, P., 360, 397
Davidson, E. R., 237, 338
Davydov, A. S., 9, 75
Debies, T. P., 22, 75
de Heer, F. J., 32, 61, 68, 77, 79, 107, 119, 136, 137, 152, 166, 168, 169, 179, 180, 184, 212, 213, 215, 216, 219, 221
Dehmer, J. L., 23, 24, 55, 57, 76, 79
Dehmer, P. M., 87, 96, 102, 105, 209, 210, 554, 584
Delchar, T. A., 512
Delos, J. B., 206, 224
Demtröder, W., 347, 349, 361, 396, 397
Derblom, H., 346, 395
Derrick, P. J., 88, 89, 106, 126, 131, 210
Dewar, M. J. S., 69, 71, 80
Dey, S., 72, 80
Dhez, P., 41, 46, 78, 80
Dibeler, V. H., 50, 79, 102, 210
Dill, D., 23, 24, 55, 57, 76, 79
Dillon, M. A., 19, 37, 39, 75, 77
D'Incan, J., 132, 173, 214
Dimpfl, W. L., 157, 217
Ding, A., 88, 111, 209, 437, 474, 485
Dittner, P. F., 157, 158, 217
Dixon, A. J., 72, 80

Dodd, J. N., 396
Doering, J. P., 119, 151, 155, 158, 159, 169, 170, 171, 172, 173, 174, 184, 185, 186, 188, 189, 213, 216, 217, 221, 222
Donnally, B. L., 164, 218
Donovan, R. J., 128, 214, 345, 395
Doolittle, J., 197, 222
Doverspike, L. D., 19, 136, 164, 177, 214, 218, 219
Downing, F. A., 146, 215
Drake, G. W. F., 491
Dressler, K., 244, 288, 338, 339
Druivestein, M. J., 402, 484
Dubrin, J., 109, 212
Ducas, T. W., 345, 395
Dunbar, R. C., 89, 93, 210
Dunkin, D. B., 150, 162, 181, 190, 216, 218, 222
Dunn, G. H., 54, 79, 168, 215, 220
Dunning, F. B., 427, 484, 511, 514, 582
Dunning, T. H., Jr., 140, 215, 356, 396
Dupont-Roc, J., 540, 583
Duren, R., 345, 395, 554, 584
Durup, J., 83, 93, 136, 137, 207, 209, 210, 215
Dworetsky, S. H., 119, 152, 166, 167, 179, 213, 216
Dyson, D. J., 181, 182, 219

Earl, B. L., 347, 396
Earl, J. D., 155, 156, 157, 168, 169, 170, 174, 175, 190, 191, 192, 193, 217, 220, 221
Eastman, D. E., 7, 23, 33, 55, 62, 67, 68, 73, 75, 80
Ebding, T., 411, 445, 446, 447, 448, 449, 450, 451, 484, 485, 494, 581
Ederer, D. L., 29, 30, 43, 76, 78
Edlefsen, N., 489, 494, 521, 580
Edmiston, C., 197, 222
Ehrhardt, H., 34, 77, 152, 166, 216, 356, 396
El-Sherbini, T., 38, 43, 46, 50, 55, 56, 58, 59, 60, 61, 77, 78, 79
Eland, J. H. D., 37, 77, 104, 210
Elleman, D. D., 122, 214
Elsenaar, R. J., 347, 396
England, W. B., 203, 223
Entemann, E. A., 559, 584
Erko, V. F., 168, 169, 170, 184, 220, 221

Ertl, G., 512
Evans, S., 24, 75
Evers, N. S., 198, 222
Ewing, J. J., 493, 581
Eyring, H., 222

Fair, J. A., 201, 202, 223
Faisal, F. H. M., 23, 76
Fan, C. Y., 169, 220
Fano, U., 11, 75, 367, 381, 397
Farrar, J. M., 86, 123, 124, 125, 201, 208
Farrow, L. A., 152, 166, 167, 179, 216
Faubel, M., 161, 218
Federenko, N. V., 476, 486
Fee, D. C., 166, 179, 180, 219
Fehsenfeld, F. C., 67, 88, 93, 108, 109, 110, 113, 114, 115, 120, 129, 130, 145, 146, 150, 162, 163, 166, 168, 169, 209, 212, 213, 216, 218, 220, 427, 429, 430, 433, 485, 565, 567, 569, 573, 585
Feltgen, R., 567
Fenistein, S., 119, 168, 169, 213
Fenn, J. B., 495, 581
Ferguson, E. E., 83, 87, 88, 108, 109, 110, 113, 114, 115, 120, 129, 130, 145, 146, 150, 162, 163, 166, 168, 169, 181, 190, 207, 209, 212, 213, 216, 218, 220, 222, 303, 339
Ferguson, H. I. S., 169, 170, 187, 220
Fiaux, A., 97, 106, 135, 148, 149, 211
Field, F. H., 87, 88, 127, 131, 208, 214
Field, R., 339
Fink, M., 8, 75
Fink, R. W., 54, 79
Fishburne, E. S., 170, 187, 220
Fisher, E. R., 355, 356, 357, 373, 396
Fite, W. L., 119, 121, 152, 213, 393, 398, 473, 486
Flaks, I. P., 476, 486
Flannery, M. R., 95, 96, 107, 116, 122, 123, 161, 163, 209, 211, 214, 218
Fleischmann, H. H., 177, 179, 219, 551, 568, 584
Fluendy, M. A. D., 573, 585
Flynn, G. W., 345, 395
Fogel, Ya. M., 168, 169, 170, 184, 220, 221
Foley, H. M., 152, 153, 216
Fontijn, A., 472, 486

Foote, P., 489, 494, 521, 580
Forst, W., 151, 216
Fort, J., 442, 452, 485
Fortune, P. J., 203, 223
Foster, K. D., 469, 486
Fournier, J., 168, 220
Fournier, P., 93, 136, 137, 209, 215
Fournier, P. G., 168, 220
Fowles, G. R., 166, 167, 182, 207, 219
Fraites, J. L., 496, 545, 546, 556, 572, 573, 581, 583, 585
Francis, W. E., 120, 185, 213, 221
Franck, J., 4, 74
Franklin, J. L., 62, 79, 83, 87, 88, 131, 191, 207, 208, 209
Freeman, C. G., 146, 215
Freeman, J. P., 151, 216
Freiser, B. S., 93, 117, 210, 213
Freudenberg, K., 489, 580
Fricke, J., 393, 398
Fridh, C., 61, 68, 79
Friedman, L., 83, 87, 88, 94, 95, 96, 97, 101, 106, 127, 131, 133, 135, 142, 144, 158, 207, 208, 209, 211, 214
Fryar, J., 50, 79
Fu, E., 93, 210
Fuchs, R., 87, 131, 208
Fuchs, V., 37, 77, 403, 404, 434, 484
Fugol', I. Ya., 533, 585
Fukuyama, T., 520, 570, 571, 572, 573, 576, 579, 583, 585
Fullerton, D. C., 188, 189, 221
Futch, A. H., 521, 536, 583
Futrell, J. H., 83, 87, 89, 97, 106, 111, 117, 118, 126, 135, 144, 148, 149, 151, 163, 207, 208, 210, 211, 212, 213, 215, 216

Gaily, T. D., 93, 210, 361, 397
Galileo Electro-Optics Corp., 34, 77
Gallagher, A., 396
Gallagher, J. W., 345, 377, 387, 389, 395
Galli, A., 87, 208
Garcia Santibanez, F., 34, 77
Gardner, A. B., 33, 41, 43, 78, 40
Gardner, J. L., 23, 30, 31, 33, 41, 42, 43, 52, 54, 55, 61, 62, 63, 65, 67, 68, 73, 76, 77, 78, 79, 80
Gaydon, A. G., 338, 348, 396
Geballe, R., 19, 152, 166, 168, 216, 220, 564, 565, 584

592   Author Index

Geiger, J., 8, 25, 38, 75, 77
Gelius, U., 34, 77
Gellender, M. E., 37, 77
Gentry, W. R., 159, 160, 203, 217, 223
Gérard, K., 428, 485, 512
Gerard, M., 119, 168, 169, 172, 213, 220, 221
Gerber, G., 486
Gerch, M. E., 574, 585
Gersing, E., 392, 398
Gianturco, F., 23, 76
Giardini-Guidoni, A., 87, 127, 208, 214
Gibson, D. K., 136, 215
Giese, C. F., 159, 160, 161, 203, 217, 223
Gilbert, T. L., 227, 229, 230, 237, 337, 338
Gilbody, H. B., 86, 208
Gill, P. S., 150, 216
Gillen, K., 519, 535, 537, 583
Gilman, G. I., 109, 162, 212
Gilmore, F. R., 140, 215, 228, 243, 265, 288, 318, 338, 343, 355, 373, 395, 396
Ginter, M. L., 533, 585
Gislason, E. A., 88, 89, 111, 123, 136, 137, 157, 158, 164, 209, 214, 217, 218
Givens, W., 231, 338
Glendenning, N. K., 542, 583
Glennon, B. M., 106, 212
Goddard, T. P., 345, 395
Goddard, W. A., 526, 531, 569, 583
Golde, M. G., 561, 584
Gorden, R., 127, 214
Gottscho, L., 339
Gould, R. F., 83, 207
Govers, T. R., 61, 68, 79, 110, 114, 119, 136, 137, 162, 168, 169, 172, 184, 212, 213, 215, 220, 221
Graham, E., IV., 119, 166, 174, 182, 190, 213, 221
Graham, I. G., 116, 212
Grajower, R., 100, 211
Grant, F. A., 521, 536, 583
Green, J. M., 166, 167, 219
Green, S., 289, 339
Green, T. A., 137, 142, 215
Greene, E. F., 495, 501, 581, 582
Gregor, R. M., 520, 570, 571, 572, 576, 579, 583
Gregor, R. W., 530, 545, 548, 551, 552, 555, 561, 562, 567, 568, 584

Grim, H. R., 491
Gross, M. L., 99, 211
Grosser, J., 545, 583
Grossheim, T. R., 106, 142, 143, 144, 211
Guberman, S. L., 526, 531, 569, 583
Gudat, W., 7, 23, 33, 55, 62, 67, 68, 73, 75, 80
Gudden, B., 489, 580
Gundel, L. A., 119, 168, 169, 174, 175, 184, 190, 193, 213, 493, 581
Gusev, V. A., 168, 169, 170, 220
Gustafsson, T., 7, 23, 33, 43, 55, 62, 67, 68, 73, 75, 80

Haberland, H., 420, 421, 422, 423, 424, 425, 426, 432, 433, 437, 453, 455, 484, 496, 511, 512, 515, 516, 517, 518, 519, 520, 529, 530, 537, 543, 545, 546, 547, 548, 549, 550, 551, 552, 554, 555, 560, 564, 565, 566, 568, 569, 570, 571, 573, 574, 575, 578, 581, 582, 583, 585
Habitz, P., 389, 398
Haddad, G. N., 52, 54, 55, 73, 79, 80
Hagstrum, H. D., 581
Ham, D. O., 561, 584
Hamill, W. H., 89, 126, 210
Hamlet, P., 127, 214
Hammer, D., 361, 397
Hamnett, A., 4, 7, 17, 18, 20, 21, 22, 23, 29, 31, 32, 33, 34, 36, 37, 54, 55, 60, 61, 62, 63, 64, 65, 66, 67, 68, 69, 70, 71, 72, 73, 74, 75, 77, 79, 80
Hansen, S. G., 86, 123, 124, 125, 201, 208
Hanson, H. D., 211
Harang, O., 227, 337
Hariri, A., 344, 347, 351, 359, 395
Harland, P. W., 87, 116, 122, 208
Harper, S. F. W., 565, 584
Harris, F. E., 227, 230, 231, 238, 240, 265, 337, 338, 339
Harris, H. H., 87, 106, 119, 142, 143, 144, 155, 156, 157, 164, 168, 169, 170, 174, 175, 176, 187, 190, 191, 192, 193, 194, 195, 203, 208, 211, 217, 218, 220, 221, 222
Harris, R. M., 502, 582
Harrison, A. G., 89, 126, 150, 210, 216
Harting, E., 26, 29, 30, 76
Hasted, J. B., 86, 120, 122, 208, 213, 214, 301, 339, 356, 396

Hatfield, L. L., 168, 220
Haugh, M. J., 119, 170, 171, 172, 173, 184, 185, 187, 189, 213, 221
Haugsjaa, P. O., 96, 164, 208, 211, 218
Havemann, U., 222
Hay, P. J., 356, 396
Hayden, H. C., 94, 95, 117, 211
Hayes, E. F., 197, 198, 199, 222
Hays, W. J., 519, 582
Heckel, E. F., 19, 151, 216
Heddle, D. W. O., 26, 76
Hedrick, A. F., 87, 88, 89, 92, 122, 136, 141, 208, 214
Hefter, U., 361, 397
Heidner, R. F., III, 128, 214
Heinz, O., 551, 568, 584
Hellner, L., 99, 115, 211
Helm, H., 87, 208
Hemsworth, R. S., 122, 214, 427, 485
Henchman, M., 112, 166, 212
Henderson, W. R., 121, 213
Henglein, A., 87, 88, 89, 111, 164, 209, 218, 437, 474, 485
Herbst, E., 203, 206, 224
Hering, P., 361, 397
Herm, R. R., 347, 396
Herman, Z., 87, 88, 89, 106, 107, 131, 164, 171, 208, 209, 218, 222, 223, 403, 404, 434, 458, 484
Hermann, H. W., 367, 381, 383, 384, 385, 397
Hermann, V., 160, 161, 218
Herod, A. A., 89, 126, 210
Herrero, F. A., 151, 159, 216, 217
Herron, J. T., 87, 209
Herschbach, D. R., 223, 493, 557, 581, 584
Hertel, G. R., 111, 212
Hertel, I. V., 39, 77, 295, 344, 345, 353, 360, 361, 365, 367, 368, 372, 373, 380, 381, 383, 384, 385, 387, 389, 394, 395, 397, 398
Hertz, G., 4, 74
Herzberg, G., 338
Hesser, J. E., 3, 74
Hesterman, V. W., 152, 216
Heydtmann, H., 344, 347, 350, 359, 395
Hickman, A. P., 403, 404, 406, 412, 413, 415, 422, 424, 437, 439, 440, 442, 444, 454, 484, 485, 494, 495, 509, 533, 545,

554, 560, 567, 568, 581, 584, 585
Hierl, M., 62, 79
Hierl, P., 164, 218
Hierl, P. M., 205, 223
Hirota, F., 23, 76
Hitchcock, A., 73, 80
Hoffmann, V., 449, 485
Hofmann, H., 295, 345, 360, 361, 368, 373, 385, 387, 394, 397
Holbrook, K. A., 199, 222
Holland, R. F., 140, 168, 169, 184, 187, 208, 220
Hollstein, M., 119, 168, 213, 220
Holt, R. A., 93, 210, 361, 397
Honma, K., 96, 210
Hood, S. T., 15, 34, 35, 72, 75, 77
Hooymayers, H. P., 347, 377, 379, 395, 396, 397
Hoppe, H. O., 345, 395
Hopper, D. G., 136, 163, 168, 203, 214, 218, 220, 223
Hornbeck, J. A., 472, 486, 489, 581
Hornstein, J. V., 122, 209
Horwitz, H., 396
Hotop, H., 30, 76, 403, 404, 422, 426, 428, 431, 434, 437, 445, 456, 457, 458, 459, 465, 466, 467, 468, 469, 470, 484, 485, 486, 492, 494, 549, 559, 561, 581
Hott, P. W., 490, 581
Houlgate, R. G., 33, 41, 42, 43, 44, 46, 76, 77, 78
Houriet, R., 151, 216
Houtermans, F. G., 472, 486
Howard, C. J., 120, 213
Howard, J. P., 511, 582
Howard, J. S., 426, 427, 441, 443, 484, 485, 565, 566, 567, 568, 585
Howe, I., 141, 215
Hsu, D. S. Y., 344, 347, 350, 352, 359, 395
Hübler, G., 465, 466, 468, 486
Hudson, R. D., 40, 78
Huebner, R. H., 40, 78
Hughes, A. L., 25, 76
Hughes, B. J., 119, 166, 168, 179, 180, 213
Hughes, B. M., 85, 86, 88, 90, 92, 97, 111, 136, 142, 144, 166, 168, 171, 179, 180, 207, 211, 219, 220, 221
Hughes, R. H., 168, 220
Hughes, W. M., 490, 581

Hulpe, E., 347, 396
Hultzsch, W., 420, 477, 479, 480, 484, 486
Hunten, D. M., 346, 395
Hunter, R., 490, 581
Huntress, W. T., Jr., 87, 88, 95, 96, 97, 98, 99, 103, 106, 115, 116, 122, 124, 133, 135, 147, 148, 184, 200, 208, 209, 210, 211
Hurle, I. R., 351, 396
Hurt, W. B., 174, 221
Husain, D., 128, 214
Husinsky, W., 361, 397
Hussain, M., 87, 208

Ikelaar, P., 36, 37, 75, 77
Illenberger, E., 422, 425, 427, 428, 429, 430, 433, 437, 438, 463, 484, 485, 565, 584
Imhof, R. E., 26, 76
Inel, Y., 150, 216
Ingrham, M. G., 37, 71, 77, 80
Inn, E. C. Y., 168, 184, 220
Inokuti, M., 4, 5, 7, 15, 20, 74
Irving, P., 473, 486
Isaacson, A. D., 403, 404, 439, 440, 442, 484, 494, 554, 567, 568, 581, 584
Isler, R. C., 119, 156, 166, 167, 168, 177, 178, 179, 180, 213, 217, 219, 220
Itikawa, Y., 23, 75
Itoh, T., 265, 339

Jaecks, D., 166, 179, 180, 219
Jaffe, S., 88, 208
Jaicks, D., 168, 220
James, P. B., 155, 156, 157, 217, 222
Jansen, R. H. J., 32, 77
Jarmain, W. R., 96, 265, 338
Jenkins, D. R., 347, 396
Jennings, D. A., 351, 396
Jennings, K. R., 89, 126, 141, 210, 215
Jensen, R. C., 166, 167, 174, 181, 182, 207, 219
Johnsen, R., 119, 162, 166, 174, 181, 182, 190, 213, 218, 219, 221
Johnson, B. M., 190, 221
Johnson, C. E., 491
Johnson, P. D., 512
Johnson, R. R., 4, 74, 227, 337
Johnson Laboratories Inc., 34, 77

Jones, A. E., 51, 52, 79
Jones, C. A., 86, 111, 163, 203, 207, 212, 218
Jones, E. G., 89, 119, 127, 166, 168, 171, 179, 180, 210, 213, 214, 219, 220, 221
Jopson, R. C., 55, 79
Jordan, R. M., 520, 530, 545, 548, 551, 552, 555, 567, 568, 570, 571, 572, 576, 579, 583, 584
Judge, D. L., 39, 40, 55, 57, 66, 68, 69, 78, 79, 80
Juduyama, T., 522, 583
Julienne, P. S., 265, 339
Jungen, C., 288, 339
Junk, G., 87, 131, 208

Kalff, P. J., 393, 398
Kallne, E., 46, 78
Kamin, M., 119, 174, 181, 193, 213
Kano, H., 166, 167, 181, 182, 183, 219
Kaplan, M., 345, 395
Kari, R. E., 223
Karl, G., 344, 347, 350, 359, 395
Karmohapatrd, S. B., 208
Karplus, M., 205, 224
Kasper, J. V. V., 128, 214
Kassal, T. T., 170, 187, 220
Kaufman, J. J., 86, 111, 162, 203, 207, 212, 218, 223, 362
Kaufold, L., 452, 568, 485, 486
Kaul, W., 87, 131, 208
Kay, R. B., 57, 80
Kebarle, P., 37, 77, 150, 216
Keiffer, L. J., 40, 54, 78, 79
Keller, J. C., 514, 582
Kelley, J. D., 187, 221
Kelly, H. P., 24, 76
Kelly, P. S., 227, 337
Kemper, P. R., 87, 88, 131, 132, 208, 209
Kempter, V., 392, 398
Kennedy, D. J., 42, 43, 78
Kermin, L., 94, 137, 142, 212
Kern, C. W., 265, 339
Kerstetter, J., 164, 218
Kerwin, L., 85, 87, 88, 92, 95, 101, 136, 137, 138, 140, 141, 208, 211
Kessel, Q. C., 34, 77
Kessler, K. G., 38, 77
Kevan, L., 122, 214

Kibble, B. P., 347, 349, 351, 396
Kim, J. -K., 98, 147, 148, 211
Kim, K. C., 141, 215
Kim, Y. K., 4, 29, 74
King, D. L., 490, 493, 579, 581, 585
Kingston, A. E., 22, 50, 75, 79, 565, 584
Kinsey, J. L., 561, 584
Kisninevski, L. M., 476, 486
Kistemaker, J., 152, 166, 216
Klein, F. S., 88, 94, 95, 115, 133, 208, 211, 212
Klein, M. B., 167, 219
Klein, O., 243, 338
Klemperer, W., 203
Klewer, M., 7, 68, 80
Klots, C. E., 200, 223
Knewstubb, P. F., 200, 223
Knuth, E. L., 519, 582
Kolb, A., 490, 581
Kolke, F., 473, 486
Kolos, W., 437, 485
Kompa, K. L., 343, 394
Konz, W., 545, 552, 554, 555, 583
Koonast, W., 420, 480, 484
Kormornicki, A., 69, 71, 80
Kornfeld, R., 141, 215
Koski, W. S., 85, 86, 89, 111, 123, 124, 130, 131, 132, 146, 162, 163, 203, 207, 208, 209, 212, 218
Kouri, D. J., 198, 206, 222
Koyano, I., 96, 210
Kramer, J. M., 89, 210
Kraus, J., 152, 153, 154, 167, 179, 217
Krause, H. F., 393, 398
Krause, L., 345, 347, 348, 349, 351, 395, 396
Krause, M. O., 43, 56, 78, 79
Krauss, M., 52, 79, 203, 222, 223, 227, 302, 303, 337, 339
Krenos, J. R., 205, 223, 571, 573, 585
Kritskii, V. A., 152, 216
Kronast, W., 477, 479, 486
Kruitof, A. A., 402, 484
Krupenie, P. H., 212, 242, 265, 338
Kruus, P., 344, 347, 350, 359, 395
Kulander, K. C., 486
Kumar, V., 63, 73, 80
Kummler, R. H., 343, 395
Kuntz, P. J., 198, 205, 206, 222, 223, 224

Kunzemuller, H., 46, 78
Kuprianov, S. E., 87, 209
Kurzweg, L., 119, 152, 213
Kusonoki, I., 176, 196, 222
Kuyatt, C. E., 26, 38, 40, 76, 77, 78
Kwei, G. H., 353, 372, 398

Lagerquist, A., 288, 339
Laidler, K. J., 352, 396
Laloe, F., 540, 583
Lampe, F. W., 87, 106, 107, 208, 472, 486
Landau, L. D., 503, 582
Lane, A. L., 88, 96, 209
Lane, N. F., 533, 554, 584, 585
Lang, J., 40, 78
Langhoff, P. W., 24, 80
Lantschner, G., 460, 461, 462, 485
Lao, R. C. C., 86, 124, 207
Larkin, I. W., 301, 339
Lassettre, E. N., 4, 5, 8, 16, 19, 26, 29, 37, 38, 39, 46, 47, 49, 51, 52, 73, 74, 75, 77, 79, 80
Latush, E. L., 486
Laucagne, J. J., 452, 485
Laudenslager, J. B., 19, 87, 88, 96, 98, 122, 131, 132, 165, 184, 190, 207, 208, 211
Laures, P., 156, 217
Lavroskaza, G. K., 87, 208
Lawrence, E. O., 230, 231, 265, 394, 489, 494, 521, 580
Lawrence, T. R., 168, 220
Layton, J. K., 89, 111, 112, 210, 212, 469, 486, 561, 563, 584
Leasure, E. L., 519, 582
Lebehot, A., 516, 582
Leblanc, F. J., 170, 205, 221
Lebreton, P. R., 88, 89, 96, 99, 115, 116, 124, 200, 209, 210, 211, 557, 584
Leduc, M., 540, 583
Lee, A. R., 120, 213, 223
Lee, F. W., 119, 181, 190, 213
Lee. L. C., 39, 40, 55, 57, 66, 68, 69, 78, 79, 80
Lee, W., 573, 585
Lee, Y. T., 163, 218, 420, 484, 494, 495, 496, 545, 546, 557, 570, 571, 581, 583, 584
Lefebvre-Brion, H., 318, 340
Lehmann, K. K., 205, 223
Lemont, S., 345, 395

596   Author Index

Leone, S. R., 343, 344, 347, 351, 359, 394, 395
Leoni, M., 244, 338
Leroy, R. L., 223
Lessmann, W., 50, 79
Lessner, E., 393, 398
Lester, W. A., Jr., 203, 223
Letokhov, V. S., 343, 394
Leu, M. T., 181, 219, 548, 557, 560, 561, 562, 567, 584
Leventhal, J. J., 87, 88, 96, 97, 106, 119, 131, 135, 142, 143, 144, 155, 156, 157, 164, 168, 169, 170, 174, 175, 176, 187, 190, 191, 192, 193, 194, 195, 203, 208, 211, 217, 218, 220, 221, 222
Levine, R. D., 137, 163, 197, 200, 215, 218, 223, 373, 374, 375, 376, 397
Levsen, K., 141, 215
Lias, S. G., 83, 122, 207, 214
Libby, W. F., 127, 214
Lichten, W., 19, 152, 153, 154, 167, 179, 216, 217
Lifshitz, C., 100, 200, 211, 222, 223
Light, J. C., 197, 222
Lijnse, P. L., 345, 347, 377, 379, 395, 396, 397
Lin, C. D., 43, 80
Lin, J., 197, 222
Lin, K. C., 85, 86, 123, 124, 130, 131, 132, 146, 207, 208
Lin, M. C., 344, 347, 350, 352, 359, 395
Lin, S., 168, 220
Lin, S. M., 343, 347, 395, 396
Lin, Y. N., 150, 215
Lindemann, E., 85, 89, 209
Linder, F., 160, 161, 218
Lindholm, E., 86, 87, 88, 89, 105, 106, 126, 131, 207, 208, 210
Lindinger, W., 87, 88, 109, 114, 115, 129, 130, 145, 146, 163, 209, 212, 218, 427, 429, 430, 433, 485, 565, 567, 569, 573, 585
Lipeles, M., 19, 166, 170, 177, 178, 184, 185, 193, 219, 220
Lippincott, E. R., 243, 338
Liskow, D. H., 203, 223
Littlewood, I. M., 476, 486
Litton, J. F., 142, 215
Liu, B., 527, 583
Liu, C., 155, 169, 170, 171, 172, 173, 174, 188, 217, 221

Ljinse, P. L., 347, 396
Lo, H. H., 119, 152, 213
Loesch, H. J., 361, 392, 393, 397, 398
Loftus, A., 242, 338
Longmuire, M. S., 166, 177, 219
Lopez, F. O., 182, 219
Lorents, D. C., 119, 121, 152, 166, 168, 213, 216, 220, 519, 526, 527, 535, 551, 568, 583, 584
Lorquet, A. J., 136, 215
Lorquet, J. C., 136, 215
Los, J., 137, 157, 215, 217
Löwdin, P. O., 237, 338
Lowe, R. P., 169, 170, 187, 220
Lowry, J. F., 29, 30, 76
Lukirskii, A. P., 47, 49, 79
Luyken, B. F. J., 166, 179, 180, 219
Luyken, H. J., 32, 77
Lyash, A. V., 22, 75
Lynch, M. J., 33, 40, 41, 43, 78

Macek, J., 367, 380, 381, 383, 385, 397
McCarthy, I. E., 15, 34, 35, 72, 75, 77, 80
McCoy, B. V., 24, 80
McDaniel, E. W., 83, 207
McDonald, J. D., 557, 584
McDowall, C. A., 25, 26, 30, 76
McDowell, M. R. C., 174, 221
Macek, J., 367, 380, 381, 383, 385, 397
McEwan, M. J., 146, 215
McFarland, M., 87, 88, 110, 114, 115, 129, 130, 145, 146, 162, 209, 212, 218
McGillis, D. A., 347, 396
McGowan, J. W., 85, 87, 88, 92, 94, 95, 101, 106, 136, 137, 138, 140, 141, 142, 152, 165, 168, 174, 175, 207, 208, 211, 212, 216, 218, 343, 395
McGuire, P., 391, 398
Mackay, G. I., 146, 215
McLafferty, F. W., 141, 215
Macleanner, D. A., 581
McLure, G. W., 83, 136, 207, 215
McMahon, T. B., 117, 215
McMillan, J. H., 25, 76
McNeil, J. R., 167, 219
McVicar, D. D., 29, 30, 76
Maecker, H., 519, 582
Magnuson, G. D., 86, 94, 110, 111, 112, 133, 164, 207, 210, 212, 218, 453, 469, 485, 486, 561, 563, 564, 584

Mahan, B. H., 86, 88, 89, 111, 123, 124, 125, 136, 137, 157, 158, 164, 197, 198, 201, 202, 203, 208, 209, 214, 217, 218, 222, 223

Maier, W. B., II, 138, 139, 140, 168, 169, 180, 184, 187, 208, 219, 220

Maisch, W. G., 265, 288, 338, 339

Mandl, F., 357, 397

Mannkopf, R., 343, 394

Manson, S. T., 42, 43, 78

Manus, C., 427, 437, 452, 453, 454, 485, 565, 566, 567, 568, 584

March, R. E., 146, 215

Marchi, R. P., 551, 568, 584

Marcotte, R. E., 88, 89, 111, 136, 138, 139, 141, 142, 209

Marcus, A. B., 428, 430, 485

Markin, M. I., 87, 208

Marmet, P., 87, 95, 208, 211

Marquardt, D. W., 522, 526, 527, 583

Marr, G. V., 33, 40, 41, 42, 43, 55, 67, 76, 78, 80

Martin, D. W., 520, 530, 545, 548, 551, 552, 555, 567, 568, 570, 571, 572, 576, 579, 583, 584

Martin, P. J., 392, 398

Martin, R. M., 573, 585

Marx, R., 119, 165, 168, 169, 172, 207, 213, 219, 220, 221

Mash, H., 55, 79

Mason, E. A., 243, 265, 288, 338, 339

Massey, H. S. W., 13, 14, 15, 75, 120, 177, 213, 301, 339, 351, 396, 497, 518, 525, 526, 531, 539, 582

Mathis, R. F., 85, 88, 89, 110, 111, 112, 209, 212

Mauclaire, G., 119, 168, 169, 172, 213, 220

Maylotte, D. H., 128, 214

Meador, W. E., 288, 339

Medved, M., 155, 217

Mehlhorn, W., 41, 80

Meisels, G. G., 19, 150, 151, 208, 216

Melius, C. F., 182, 183, 220

Melton, C. E., 87, 89, 126, 131, 208, 210

Melzer, J. E., 338

Mentall, J. E., 119, 121, 171, 172, 213, 393, 398

Menzinger, M., 195, 222

Metzger, P. H., 40, 41, 50, 52, 78, 79

Meyer, H. D., 495, 581

Meyer, W., 389, 398

Meyerott, R. E., 227, 337

Miasek, P. B., 150, 216

Miasek, P. G., 89, 126, 210

Micha, D. A., 494, 499, 581

Michels, H. H., 227, 228, 230, 238, 239, 240, 241, 265, 268, 302, 318, 337, 338

Mielczarek, S. R., 38, 77

Miescher, E., 241, 288, 318, 338, 339, 340

Milczarek, S. R., 40, 78

Miles, B. M., 106, 212

Miller, T. M., 491

Miller, W. F., 4, 46, 74

Miller, W. H., 265, 339, 403, 404, 406, 412, 417, 439, 440, 442, 484, 493, 494, 497, 501, 502, 506, 508, 554, 567, 568, 581, 582, 584

Mims, C. A., 347, 396

Mitchell, K. B., 146, 215

Mitchum, R. K., 151, 216

Mittmann, H. U., 437, 474, 485, 568, 585

Moak, C. D., 34, 77

Mohler, F., 489, 521, 580

Möhlmann, G. R., 107, 212

Molnar, J. P., 472, 486, 489, 521, 536, 581, 583

Molof, R. A., 491

Monahan, K. M., 87, 96, 104, 209

Monge, A. A., 152, 153, 154, 167, 179, 217

Moore, C. E., 207, 491

Moore, J. H., Jr., 119, 151, 154, 155, 158, 169, 170, 173, 185, 186, 188, 189, 213, 216, 217, 220, 222

Moran, T. F., 87, 88, 89, 92, 94, 95, 96, 101, 107, 111, 116, 122, 123, 133, 136, 141, 157, 158, 161, 163, 169, 170, 185, 188, 189, 208, 209, 211, 212, 213, 214, 217, 218, 220, 221

Morgenstern, R., 519, 535, 583

Morgner, H., 406, 412, 413, 415, 417, 422, 424, 437, 438, 441, 442, 443, 444, 449, 454, 456, 457, 474, 476, 484, 485, 486, 494, 509, 545, 546, 560, 581, 583

Mori, Y., 352, 396

Morrison, J. D., 87, 208

Moseley, J. T., 93, 210

Moser, C. M., 318, 340, 289, 339

Mosesman, M., 181, 190, 222
Moss, H. W., 491
Mott, N. F., 13, 14, 15, 75, 497, 582
Moursund, A. L., 495, 501, 581
Muckerman, J. T., 206, 224
Mueller, C. R., 519, 582
Mullen, J. M., 476, 486
Mulliken, R. S., 242, 302, 338, 339
Munson, M. S. B., 87, 88, 89, 126, 131,
    191, 208, 209, 210
Münzer, A., 471, 486
Murphy, K., 197, 222
Murray, L. E., 167, 219
Murray, P. T., 104, 210
Muschlitz, E. E., Jr., 428, 430, 469, 485,
    486, 490, 573, 574, 581, 585
Myers, B. F., 111, 112, 212
Myher, J. J., 89, 126, 210

Nakamura, H., 403, 406, 473, 484, 486,
    493, 500, 502, 581
Nakamura, M., 28, 55, 76, 79
Nathan, R. D., 166, 168, 180, 219, 220
Neff, S. H., 152, 153, 154, 155, 167, 169,
    170, 173, 179, 216, 217, 219
Neumann, D., 303, 339
Neynaber, R. H., 86, 94, 110, 111, 112,
    133, 164, 207, 210, 212, 218, 420, 453,
    469, 484, 485, 486, 490, 561, 563, 564,
    581, 584
Ngo-Trong, C., 24, 76
Niehaus, A., 30, 76, 403, 404, 408, 411,
    413, 420, 422, 425, 426, 427, 428, 429,
    430, 431, 432, 433, 434, 435, 437, 438,
    441, 442, 443, 445, 446, 447, 448, 449,
    450, 451, 456, 457, 458, 459, 460, 465,
    466, 469, 470, 471, 474, 476, 477, 479,
    480, 482, 483, 484, 485, 486, 492, 494,
    565, 566, 567, 568, 581, 584
Niehause, A., 403, 404, 484
Nielsen, A. K., 152, 167, 217
Nienhuis, G., 377, 379, 397
Nikitin, E. E., 345, 352, 354, 355, 389,
    395, 396, 397, 398, 582
Nip, W., 493, 581
Nishikawa, S., 69, 73, 80
Nishimura, H., 47, 79
Noller, H. G., 26, 76
Norbeck, J., 99, 211
Norrish, R. G. W., 347, 379, 396

North, A. M., 145, 215
North, G. R., 222
Novick, R., 19, 119, 152, 166, 177, 178,
    213, 216
Numrich, R. W., 356, 396

Oda, N., 31, 77
Odiorne, T. J., 128, 214
Odom, R., 119, 152, 167, 168, 171, 213,
    217, 220
Oertzen, W. V., 525, 583
Oesterlin, P., 520, 545, 552, 554, 555, 570,
    571, 573, 574, 575, 578, 583, 585
Ogawa, M., 39, 40, 55, 57, 66, 68, 69, 78,
    79, 80
Ogilvie, K. W., 170, 220
Ogurtsov, G. N., 476, 486
Ogurtsov, V. I., 152, 216
Öhrn, Y., 318, 340
Ohtani, S., 47, 79
Oksyuk, A. A., 170, 220
Olnso, K., 265, 339
Olsen, K. J., 152, 167, 217
Olson, R. E., 485, 510, 519, 535, 582, 583
O'Malley, R. M., 89, 126, 210
O'Malley, T. F., 162, 218, 585
O'Neill, S. V., 223
Ong, P. P., 122, 214
Opal, C. P., 31, 32, 76
Orchard, A. F., 4, 22, 75
Orr, B. J., 22, 23, 75
Ottinger, C., 86, 105, 119, 124, 155, 169,
    170, 171, 176, 185, 193, 195, 196, 203,
    207, 212, 214, 222
Oxley, C. L., 152, 166, 216
Ozenne, J. -B., 210

Pacak, V., 222
Pacala, T. J., 190, 207
Pack, J. L., 301, 339
Padial, N., 24, 80
Panev, G. S., 152, 217
Pang, K. D., 87, 96, 105, 209
Parilis, E. S., 476, 486
Park, J. T., 155, 217
Parrano, C., 345, 395
Parson, J. M., 163, 218
Paul, E., 347, 396
Paul, W., 347, 396
Paulson, J. F., 89, 210

Pauly, H., 345, 358, 392, 395, 397, 398, 554, 556, 567, 584
Peach, G., 526, 583
Pearl, A. S., 491
Pearson, P. K., 203, 223
Peatman, W. B., 167, 219
Peek, J. M., 83, 137, 142, 207, 215, 437, 485
Penning, F. M., 349, 401, 402, 484, 489, 580
Percival, I. C., 381, 397
Perel, V. I., 152, 216
Perie, S., 265, 339
Pernot, C., 210
Peshkin, M., 20, 22, 75
Pesnelle, A., 427, 437, 442, 452, 453, 455, 486, 566, 567, 568, 569, 585
Petersen, A. B., 343, 344, 347, 351, 359, 394, 395
Peterson, J. R., 93, 119, 121, 166, 168, 210, 213, 220, 519, 535, 583
Peterson, W. K., 31, 32, 76
Petty, F. C., 87, 88, 89, 92, 136, 141, 157, 158, 161, 208, 217
Peyerimhoff, S. D., 265, 339
Pham Dong, 87, 209
Phelps, A. V., 32, 77, 301, 339, 521, 536, 583
Phillips, L. F., 146, 215
Pieper, R. J., 168, 169, 170, 174, 175, 190, 191, 192, 193, 220
Pijnse, P. L., 345, 347, 352, 357, 395
Pinizzotto, R. F., 98, 211
Pipano, A., 203, 223
Piper, L. G., 119, 168, 169, 174, 175, 184, 190, 193, 213, 572, 574, 579, 585
Pipkin, F. M., 539, 540, 583
Platas, O. R., 227, 337
Platzmann, R. L., 4, 46, 74
Plummer, E. W., 73, 80
Plummer, W., 7, 23, 33, 55, 62, 67, 68, 75
Pobo, L. G., 127, 214
Poe, R. T., 46, 77
Polanyi, J. C., 128, 134, 214, 223, 224, 344, 347, 350, 359, 395
Polaschegg, H. D., 26, 76
Polyakova, G. N., 168, 169, 170, 184, 220, 221
Pomerance, W., 116, 213

Porter, R. N., 135, 203, 205, 214, 223, 224
Posthusta, R. D., 205, 223
Potts, A. W., 32, 61, 77, 79
Powell, C. J., 40, 78
Pressley, R. J., 490, 581
Preston, R. K., 83, 204, 205, 206, 207, 223, 224, 554, 584
Pretzer, D., 19, 166, 168, 220
Price, W. C., 32, 77
Pritchard, D. E., 345, 395
Pritchard, R. H., 265, 339
Puhl, F., 393, 398
Purcell, E. M., 25, 76
Pust, D., 393, 398

Raabe, C. P., 554, 584
Rabik, L. L., 22, 75
Rabinovitch, B. S., 150, 215
Raff, L. M., 198, 206, 222
Ranf, 62, 106, 133, 134, 142, 153, 184, 203
Rangarajan, R., 198, 206, 222
Ranjbar, F., 175, 187, 221, 222
Rapp, D., 120, 185, 213, 221
Rayermann, P., 88, 96, 107, 209, 211
Razumovskii, L. A., 152, 216
Read, F. H., 15, 26, 29, 30, 47, 75, 76, 79
Rebick, C., 137, 197, 215
Rees, A. L. G., 243, 338
Reese, R. M., 50, 79, 102, 210
Refaey, K., 94, 103, 115, 147, 210
Reid, R. D., 167, 219
Reiland, W., 367, 381, 383, 384, 385, 397
Rescigno, T. N., 24, 80
Reuben, B. G., 127, 158, 214
Rhodes, C. K., 490, 581
Richardson, W. C., 468, 486, 561, 584
Riola, J. P., 426, 427, 441, 443, 484, 485, 511, 565, 566, 567, 568, 582, 585
Risley, J. S., 564, 565, 584
Ritchie, B., 23, 75
Roach, A. C., 223
Robbins, M. F., 152, 153, 154, 167, 179, 217
Roberts, J. R., 88, 89, 209
Robertson, W. W., 172, 173, 174, 175, 191, 221
Robinson, P. J., 199, 222
Rodgers, W. E., 519, 582
Rojansky, V., 25, 76

Rolfe, J., 302, 339
Ron, A., 24, 76
Roothan, C. C. J., 230, 338
Rosen, B., 207, 338
Rosenberg, B. J., 203, 223
Rosenstock, H. M., 222
Rosenthal, H., 152, 153, 216
Rosner, S. D., 93, 210, 361, 397
Rosner, S. K., 539, 540, 583
Ross, J., 128, 214, 495, 501, 581, 582
Rost, K. A., 295, 345, 360, 361, 368, 373, 385, 387, 394, 397, 398
Rothe, E. W., 420, 484, 490, 564, 581
Roueff, E., 203, 223
Roussel, J., 512
Rowe, D. J., 24, 76
Rozett, R. W., 85, 86, 89, 111, 124, 207, 209
Rudolph, K., 160, 161, 218
Ruf, M. W., 420, 476, 477, 479, 480, 482, 483, 484, 485, 486
Ruhlar, D. G., 356, 396
Rumble, F. R., Jr., 345, 377, 387, 389, 395
Rundel, R. D., 207, 426, 427, 441, 443, 484, 485, 511, 514, 565, 566, 567, 568, 582, 585
Ruska, W. E. W., 124, 164, 201, 202, 214, 218
Russell, M. E., 87, 94, 102, 103, 115, 147, 208, 210
Rutherford, J. A., 84, 85, 86, 87, 88, 89, 90, 91, 92, 94, 106, 111, 112, 120, 121, 122, 128, 129, 133, 162, 207, 208, 209, 210
Ryan, K. R., 87, 88, 107, 116, 122, 132, 208, 212
Ryan, P. W., 97, 111, 135, 148, 163, 211, 212
Rydberg, R., 243, 338

Saban, G. H., 169, 170, 185, 188, 220
Sackett, F. B., 344, 347, 351, 359, 395
Sadowski, C. M., 351, 396
Safron, S. A., 223
Salop, A., 119, 121, 166, 168, 213, 219, 220
Samson, J. A. R., 9, 11, 23, 29, 30, 31, 33, 39, 40, 41, 42, 43, 47, 49, 50, 52, 54, 55, 61, 62, 63, 65, 67, 68, 73, 75, 76, 77, 78,

79, 80
Sanders, R. A., 573, 585
Sandner, N., 41, 46, 78, 80
Sando, K. M., 518, 526, 527, 531, 539, 582, 583
Saporoschenko, M., 87, 131, 208
Sar-El, H. Z., 25, 76
Sathyamurthy, N., 198, 206, 222
Sawyer, W., 164, 218
Saxon, R. P., 527, 583
Sbar, N., 109, 212
Schadlich, E., 392, 398
Schaefer, H. F., III, 227, 240, 265, 337, 339, 403, 404, 412, 417, 484, 494, 502, 581
Schaefer, J., 203, 223
Schäfer, T. P., 163, 218
Schearer, L. K., 539, 540, 583
Schepper, W., 393, 398
Schiff, H. I., 87, 108, 110, 209, 212, 351, 396
Schillalies, H., 26, 76
Schlier, C., 554, 584
Schlumbohm, H., 166, 167, 169, 170, 171, 172, 219, 220
Schmeltekopf, A. L., 87, 88, 108, 109, 110, 113, 114, 115, 129, 130, 145, 146, 162, 166, 168, 169, 209, 212, 218, 220, 318, 340, 403, 404, 422, 426, 427, 429, 430, 433, 434, 458, 468, 484, 485, 486, 565, 567, 569, 573, 585
Schmidt, H., 160, 161, 218
Schmidt, K., 420, 421, 422, 423, 424, 425, 426, 432, 433, 437, 453, 455, 484, 519, 520, 545, 546, 547, 548, 550, 551, 552, 554, 555, 564, 565, 566, 568, 569, 570, 573, 574, 575, 578, 582, 583, 585
Schmidt, V., 41, 46, 78, 80
Schmoranzer, H., 29, 38, 76, 77
Schneider, B., 527, 583
Schneider, B. S., 171, 221
Schneider, F., 198, 222
Schoen, R. L., 52, 63, 80
Schoonover, D. R., 155, 217
Schopman, J., 136, 137, 215
Schowengerdt, F. D., 155, 217
Schuebel, W. K., 166, 219
Schulte, H., 104, 210
Schultz, M., 164, 218

Schulz, G. J., 32, 76, 77, 87, 106, 131, 208, 301, 339, 356, 396
Schut, T. G., 402, 484
Schwartz, H. L., 491
Schweid, A. N., 573, 585
Schweig, A., 22, 69, 71, 75, 80
Schweinler, H. C., 100, 211
Searcy, J. Q., 519, 582
Seaton, M. J., 381, 397
Seibt, W., 136, 215
Sem, M. F., 486
Setser, D. W., 119, 168, 169, 174, 175, 184, 190, 193, 213, 468, 486, 490, 493, 561, 572, 574, 579, 581, 584, 585
Sevier, K. D., 25, 76
Shannon, J., 490, 581
Sharma, R. D., 205, 224
Shaw, G. B., 23, 76
Shaw, M. J., 122, 214
Shay, T., 166, 167, 181, 182, 183, 219
Shearer, L. D., 457, 485
Sheridan, J. R., 119, 168, 169, 213, 220
Shirley, D. A., 4
Shobatake, K., 494, 495, 545, 581
Sholette, W. P., 490, 581
Shortrige, R. G., 344, 347, 350, 359, 395
Shöttler, J., 157, 158, 217
Shpenik, O. B., 152, 217
Shugart, H. A., 491
Shuler, K. E., 23
Sichel, J. M., 22, 23, 75
Sieck, L. W., 98, 99, 115, 127, 211, 214
Siedler, D., 119, 171, 213
Siegbahn, K., 4, 33, 34, 77
Siegel, J., 23, 76
Siegenthaler, K. E., 171, 221
Silfvast, W. T., 166, 167, 219
Silverman, S. M., 47, 49, 79
Silvers, J. A., 353, 372, 398
Sim, W., 172, 173, 189, 221
Simms, D. L., 152, 153, 154, 167, 179, 217
Simonis, J., 86, 119, 124, 155, 169, 170, 171, 176, 185, 193, 195, 196, 207, 214, 222
Simpson, J. A., 26, 76
Sinfailam, A. L., 23, 76
Siska, P. E., 163, 218, 520, 522, 533, 545, 548, 551, 552, 555, 557, 560, 561, 562, 567, 568, 570, 571, 572, 573, 576, 579,

583, 584, 585
Skalko, O. A., 152, 217
Skardis, G. M., 351, 396
Skerbele, A., 4, 5, 8, 19, 26, 29, 38, 39, 74, 75, 77
Skodje, R. T., 203, 223
Slanger, T. G., 170, 171, 172, 221
Slater, J. C., 228, 338
Sloane, T. M., 128, 214
Slocomb, C. A., 403, 404, 484, 494, 502, 581
Small-Warren, N. E., 491
Smit, J. A., 402, 484
Smith, A. C. H., 565, 584
Smith, A. L., 171, 221
Smith, D., 104, 210
Smith, D. L., 87, 97, 106, 117, 122, 135, 144, 148, 149, 208, 211, 213, 214, 215
Smith, F. T., 120, 152, 177, 179, 213, 216, 219, 510, 551, 568, 582, 584
Smith, G. P. K., 164, 205, 218, 223, 356, 357, 373, 396
Smith, I. W. M., 344, 345, 347, 350, 359, 363, 365, 395
Smith, M., 106, 212
Smith, R. C. H., 565, 584
Smith, R. D., 87, 144, 148, 208, 211, 215
Smith, S. J., 545, 583
Smith, W. M., 347, 379, 396
Smith, W. W., 119, 152, 166, 213, 216
Snow, W. R., 110, 111, 212
Sourisseau, C., 119, 168, 169, 213
Specht, L. T., 469, 486
Speier, F., 146, 215
Spicer, L. D., 150, 215
Squires, L., 96, 98, 104, 122, 123, 142, 143, 144, 210
Stackhouse, F. A., 585
Stamatovic, A., 367, 381, 383, 384, 385, 397
Starace, A. F., 24, 41, 76, 78
Stebbings, R. F., 85, 86, 87, 111, 128, 152, 165, 166, 168, 174, 175, 207, 212, 216, 218, 426, 427, 441, 443, 484, 485, 511, 514, 565, 566, 567, 568, 582, 585
Steckelmacher, W., 25, 76
Steinberger, S., 519, 582
Steubing, W., 472, 486
Stewart, A. L., 47, 50, 79
Stewart, D. T., 40, 78

Stewart, W. B., 25, 26, 30, 76
Stine, J. R., 206, 224
Stock, H. M., 88, 132, 208
Stockbauer, R., 37, 71, 77, 80, 104, 210
Stockdale, J. A. D., 100, 127, 211, 214
Stockton, M., 190, 207, 221
Stoll, W., 7, 17, 18, 20, 21, 23, 29, 31, 32, 33, 36, 37, 54, 55, 60, 61, 62, 63, 64, 65, 66, 67, 68, 69, 70, 71, 72, 75, 79, 80, 344, 345, 365, 367, 380, 381, 383, 384, 395, 397
Streets, D. J., 32, 77
Stuckelberg, E. C. G., 503, 504, 582
Suchard, S. N., 338
Sukumar, C. V., 358, 397
Sullivan Wilson, P., 85, 111, 209
Sun, H., 40, 41, 78
Suter, R., 288, 339
Suzuki, H., 47, 79
Suzuki, I., 351, 396
Svec, H. J., 87, 131, 208
Swift, C. D., 55, 79
Swift, R. D., 166, 177, 219
Swingle, J. C., 490, 581
Szabo, I., 88, 89, 106, 126, 131, 210

Taguchi, R. T., 344, 347, 350, 359, 395
Tal'roze, V. L., 87, 208
Tam, W. C., 26, 76
Tan, K. H., 43, 73, 80
Tanaka, I., 96, 115, 210, 212
Tanaka, K., 96, 115, 210, 212
Tang, K. C., 197, 222
Tang, S. Y., 128, 214, 428, 430, 485
Taylor, A. J., 491
Taylor, H. S., 356, 396, 585
Taylor, K. T., 42, 78
Taylor, R. L., 343, 394
Tebra, W., 36, 37, 75, 77
Teller, E., 491
Tellinghuisen, J. B., 62, 106, 133, 134, 137, 142, 146, 153, 184, 203, 215
Terenin, A., 346, 396
Teubner, P. J. O., 15, 72, 75, 80
Theard, L. P., 95, 133, 147, 210, 211
Thiel, W., 22, 69, 71, 75, 80
Thomas, E. W., 152, 164, 216
Thompson, D. L., 224
Thrush, B. A., 561, 584
Thulstrup, E. W., 140, 215, 268,
318, 339, 340
Thulstrup, P. W., 289, 339
Tiernan, T. O., 83, 85, 86, 88, 89, 90, 92, 97, 106, 111, 112, 115, 119, 126, 127, 136, 138, 139, 141, 142, 144, 163, 166, 168, 171, 179, 180, 200, 203, 207, 209, 210, 211, 212, 213, 214, 215, 218, 219, 220, 221, 223, 885
Todd, J., 237, 238
Toennies, J. P., 157, 158, 160, 161, 217, 218, 358, 394, 495, 581
Tol, R. R., 7, 16, 19, 29, 30, 38, 46, 47, 49, 50, 51, 57, 69, 71, 72, 75, 77, 79
Tolk, N. H., 19, 119, 152, 153, 154, 166, 167, 177, 178, 179, 213, 216, 217
Tolstolutskaya, G. D., 170, 184, 220, 221
Tolstolutskii, A. G., 169, 170, 220
Tomboulian, D. H., 29, 30, 47, 50, 76, 79
Tomcho, L., 184, 185, 187, 221
Topouzkhanian, A., 132, 173, 214
Truhlar, D. G., 197, 222
Trujillo, S. M., 111, 112, 212, 420, 484, 490, 538, 564, 565, 581, 583, 584
Tsai, B. P., 104, 210
Tsai, S. -C., 141, 215
Tsao, C. W., 88, 89, 111, 136, 137, 157, 158, 164, 209, 217, 218
Tsuchija, S., 351, 396
Tuckwell, H. C., 23, 27, 75
Tully, F. P., 163, 218
Tully, J. C., 83, 152, 153, 154, 167, 179, 204, 205, 206, 207, 217, 223, 224
Turner, B. R., 84, 85, 86, 87, 88, 89, 90, 91, 92, 111, 112, 128, 207, 209
Turner, G. S., 157, 161, 217
Turner-Smith, A. R., 166, 167, 219
Twiddy, N. D., 122, 214, 427, 485

Uckotter, M., 141, 215
Udseth, H., 159, 160, 217
Ury, G. B., 554, 555, 584
Utterback, N. G., 88, 96, 164, 173, 174, 184, 207, 211, 218, 221

Vance, D. W., 92, 209
Van den Bergh, H. E., 160, 161, 218
van Den Bos, J., 152, 216
Van der Leeuw, P. E., 57, 73, 80
Vanderslice, J. T., 243, 265, 288, 338, 339

Van de runstraat, C. A., 61, 68, 79, 88, 96, 119, 136, 137, 164, 168, 169, 184, 213, 215
Van der Wiel, M. J., 5, 7, 16, 17, 18, 19, 20, 21, 25, 28, 29, 30, 31, 32, 33, 34, 36, 37, 38, 39, 40, 41, 43, 46, 47, 48, 49, 50, 51, 52, 53, 54, 55, 56, 57, 58, 59, 60, 61, 62, 63, 65, 66, 67, 68, 69, 70, 71, 72, 73, 75, 76, 77, 78, 79, 80
Van Dop, H., 157, 217
Van Dyck, R. S., Jr., 491
van Eck, J., 152, 166, 216
van Wingerden, B., 32, 77
Van Zandt, T. E., 162, 218
Van Zyl, B., 19, 166, 168, 220
Velazco, J. E., 119, 168, 169, 174, 175, 184, 190, 193, 213, 493, 581
Veldre, Y. V., 22, 75
Vestal, M. L., 97, 111, 135, 148, 163, 211, 212, 221
Victor, G. A., 491, 518, 582
Viebock, F., 361, 397
Vlasov, M. N., 343, 395
Vogler, M., 136, 215
Volpi, G. G., 87, 208
Vonderschen, M., 392, 398
Von Koch, H., 94, 95, 133, 208, 211
Vroom, D. A., 85, 86, 88, 94, 106, 111, 112, 120, 121, 122, 129, 133, 162, 207, 208, 209, 210

Wagner, A. F., 197, 222
Wahl, A. C., 136, 163, 168, 203, 214, 218, 220, 223, 227, 229, 230, 237, 302, 303, 337, 338, 339
Wahrhaftig, A. L., 222, 563, 584
Wakiya, K., 47, 79
Walker, I. C., 16, 75
Walker, J. A., 102, 210
Walker, T. E. H., 24, 76
Wallace, L., 243, 265, 288, 318, 338
Wallenstein, M. B., 222
Waller, R. A., 119, 174, 181, 190, 193, 213, 221
Walters, G. K., 539, 540, 545, 583
Wang, S. W., 512
Wang, Z. F., 494, 495, 545, 581
Warneck, P., 89, 210
Warner, J., 345, 395
Watanabe, T., 69, 73, 80

Watel, G., 427, 437, 442, 452, 453, 454, 485, 512, 526, 527, 565, 566, 567, 568, 570, 571, 573, 574, 578, 582, 583, 584, 585
Watkins, H. P., 86, 131, 132, 207
Watson, W. S., 40, 46, 78
Webb, C. E., 166, 219, 476, 486
Webb, H. W., 402, 484
Webb, T. G., 47, 50, 79
Weber, W., 552
Wegener, P. P., 495, 581
Weigold, E., 15, 34, 35, 72, 75, 77, 80
Weihofen, W. H., 137, 142, 215
Weiner, E. R., 111, 212
Weiner, J., 119, 152, 167, 168, 171, 213, 217, 219, 220
Weingartshofer, A., 94, 212
Weinreb, M. P., 166, 177, 219
Weinstein, N. D., 223
Weise, H. P., 437, 474, 485, 568, 585
Weiss, M., 573, 585
Weissler, G. L., 40, 41, 78
Wellenstein, H. F., 29, 76
Wendell, K. L., 86, 111, 163, 203, 207, 212, 218
Wendin, G., 43, 78
Werner, A. S., 88, 89, 98, 104, 111, 136, 137, 142, 143, 144, 157, 158, 164, 209, 210, 217, 218
Werner, F., 549
West, J. B., 33, 40, 41, 42, 43, 44, 46, 76, 77, 78
Weston, R. E., Jr., 343, 344, 347, 349, 357, 395
Wexler, S., 127, 214
Wharton, L., 554, 555, 584
Wheeler, J. A., 16, 41, 78
White, C. W., 152, 153, 154, 166, 167, 179, 216, 217
White, E. R., 73, 80
Whitton, W. N., 198, 205, 206, 222, 223
Wiebes, G., 7, 36, 46, 47, 48, 49, 78
Wiese, W. L., 106, 212
Wight, G. R., 7, 16, 19, 28, 29, 30, 38, 39, 40, 43, 46, 47, 49, 50, 51, 52, 53, 54, 55, 57, 58, 59, 60, 61, 62, 63, 66, 67, 69, 70, 71, 72, 73, 75, 76, 77, 78, 79, 80
Wilcox, J. B., 87, 92, 111, 141, 208
Wilkinson, P. G., 243, 338
Willett, C. S., 207

Williams, T. A., 61, 79
Williamson, A. D., 88, 89, 96, 98, 99, 115, 116, 124, 127, 200, 209, 210, 211
Willmann, K., 356, 396
Wilmenius, P., 89, 106, 126, 210
Wilson, A. D., 376, 397
Wilson, J. F., 502, 582
Wilson, W., 197, 222
Winicur, D. H., 398, 496, 545, 546, 556, 573, 581, 583, 585
Winkelman, K., 495, 581
Winkler, C. A., 146, 215, 242, 338
Winn, J. S., 164, 218
Winter, G. J., 152, 216
Wittig, C., 190, 207, 343, 344, 347, 351, 359, 394, 395
Wodarczyk, F. J., 344, 347, 351, 359, 395
Wolf, F. A., 200, 223
Wolfgang, R., 164, 218
Wolfhard, H. G., 348, 396
Wolterbeek Muller, L., 119, 152, 166, 179, 180, 213, 216, 219
Wong, W. H., 134, 214
Wong, Y. C., 163, 218
Wood, A. R., 357, 397
Wood, P. M., 146, 215
Woodruff, P. R., 33, 41, 42, 43, 44, 46, 55, 67, 77, 78, 80
Woodworth, R. J., 491

Wren, D. J., 195, 222
Wright, A. N., 242, 338
Wu, R. L. C., 136, 163, 214, 215, 218
Wuilleumier, F., 41, 43, 46, 56, 78, 79, 80

Yamazaki, M., 288, 339
Yee, D. S. C., 25, 26, 30, 76, 465, 466, 486
Yinon, J., 115, 212
Yoshimine, M., 265, 339
Young, R. A., 152, 166, 168, 177, 179, 216, 219, 551, 568, 584
Young, W. S., 519, 582

Zapesochnyi, I. P., 152, 217
Zare, R. N., 4, 20, 22, 74, 318, 340
Zars, A. V., 170, 184, 220
Zastrow, A., 466, 486
Zats, A. V., 168, 169, 170, 184, 220, 221
Zavilopulo, A. N., 152, 217
Zeegers, P. J. T., 347, 395
Zehnle, L., 392, 398
Zemke, W. T., 302, 303, 339
Zener, C., 503, 582
Zetik, D. F., 205, 223
Zimkina, T. M., 47, 49, 79
Zipf, E. C., 343, 394
Zocchi, F., 127, 214
Zuhrt, C., 198, 222
Zulicke, L., 198, 222

SUBJECT INDEX

Afterglow, flowing, technique, 113-115, 427
Alkali atoms excitation, 391-393
Approximation, Born-Oppenheimer, 229, 403-405, 496
Aston banding, 92
Atmospheric systems, excited states, 237-340
Atom beams, crossed, 342-394
Atoms, laser excited, in crossed beams, 342-394
Auger electron-emission yields from Mo, 92

Beam, ion-neutral, apparatus, 108, 111-113
Born-Oppenheimer approximation, 229, 403-405, 496
  separation, 231-232, 239

Charge stripping, 141-142
Charge transfer, 105
  atomic systems, 177-189
  heavy particle transfer, 193-196
  molecular systems, 189-193
  reactions producing luminescence, 165-196
Chemiluminescence reactions in ion-neutral collisions, 165-196
  He^+-Ar, 178, 180
  He^+-Xe, 179
  see also Luminescence
CI, see Configuration-interaction methods
Collision mechanisms, 196-206
  C^+-H_2, 203
  $C_2H_4^+$-C_2H_4, 200-201
  H_2^+-He, 196-199
  N^+-H_2, 201-202
  N_2O^+, 203
Collisions:
  molecular, dynamics, 239-240
  neutral-neutral, 145-161
Configuration-interaction methods:
  electronic structure calculations, 230-231, 236-239
  see also Hartree-Fock

Continuum oscillator strength, 5, 41
  transitions, 20, 23
Crossed beam quenching, with Na* atoms, 358-379

Deactivation, collisional, of excited ions, 145-151
Dissociation limits:
  of N_2, 265
  of NO, 299
  of NO^+, 335
  of O_2, 286
  of O_2^-, 316

Electron analyzers, 24, 27, 30
Electron detectors, 33-38
  coincidence methods, 34-38
Electronic structure:
  of atmospheric systems, calculations, 229-231, 240-337
  Born-Oppenheimer separation, 231-232
  configuration-interaction methods, 230-231, 236-239
  Hartree-Fock, restricted, method, 230, 234-236
  iterative natural spin orbitals method, 237-238
  for N_2, potential energy curves, 241-267
  for NO, potential energy curves, 288-301
  for NO^+ ion, potential energy curves, 318-336
  for O_2, potential energy curves, 265-287
  for O_2^- ion, potential energy curves, 301-317
  self-consistent field methods, 234-236
  variational methods, 232-240
Electron impact:
  fractional abundance determination, 85-92
  in ion beams, detecting, 92-101
  ionization by, 84-111
Electron-impact spectroscopy, 8, 17, 26, 29-31
  excitation, 5-7
Electron lenses, 26

Electron scattering processes, 385-391
Electron spectra:
  of CH_4, 70
  of N_2, 28
Electron spectroscopy, 3-74
  Penning, 403, 494
Electron storage rings, 3
Endoergic reactions, 128-144
Energy exchange, 152-161
  by electronic excitation, 152-157
Energy transfer:
  collisional, 343-394
  E-V-R, 343-346, 350-351
  Hg^* + CO, 344
  Na^* + CO_2, 378
  Na^* + N_2O, 378
  spectra, for $Na(3^2 P_{3/2})$, 368-373
Excitation:
  electron-impact, 5-7
  H^+ + H_2, 159
  rotational, 157-161
  vibrational:
    in collision dissociations, 144, 157-161
    from ion-molecule reactions, 184-196
  vibrational and rotational, 157-161
Excitation-transfer systems, 503-505, 509-510, 539-540
  cross-sections, 573-574
  elastic scattering, 519-523
  product angular distributions, 574-579
Excited ionic states, lifetimes, 86-89, 106-107
  of (N_2^{+*}), 106-107
Excited ions:
  charge transfer, 128-131
  collisional dissociation, 135-144
    $CH_2 Br_2^+$, 143
    H_2^+ + He, 143-144
    N_2^+ + N_2, 140-141
    $(NO^+)^*$ + Ar, 141-142
    reactant ion vibrational excitation, 142-143
    vibrational excitation, 144
  endoergic reactions, 128-144
  excitation, electronic, 152-161
    N_2^+ + Ar, 155
  exoergic reactions, 120-127
    association reactions, 127
      $C_6 H_6$ dimer ion formation, 127
    charge transfer, 120-123

N^+ – Kr, 121
heavy particle transfer, 123-127
  C^+ – O_2, 124
  $C_2 H_4^+$ – $C_2 H_2$, 124
  $C_2 H_4^+$ – $C_2 H_4$, 124
  F^+ – H_2, 123
  N^+ – H_2, 123
  NH_3^+ – H_2O, 126
non-reactive scattering, energy transfer, 145-161
  collisional deactivation, 147-151
  ion-impact excitation, 152-157
  N_2^+ – Ar, 155
  oscillatory structure, 153-154
  reaction scattering, 120-144
  rotational excitation, 157-161
  vibrational excitation, 157-161
Excited neutrals, reactions with ions, 108, 161-163. See also Ions, reactions with excited neutrals
Excited products, see Ion-neutral collisions
Excited states:
  of atmospheric systems, 227-340
  in ion-neutral reactions, 83-124
Exoergic reactions, 120-127

Filter, Wien, 25
Flowing-afterglow technique, 113-115, 427

Hartree-Fock, restricted, calculations, 229, 234-236
Helium, metastables, 455-457. See also Noble gases

Ionic states, see Excited ionic states; Excited ions
Ionization:
  approximations, semiclassical, 417-420
  by electron impact, 84-111
  by electron transport, 475-483
    He-Ba system, 476-478
    He-Ca system, 478-480
    He-Hg system, 480-483
  Penning, 401-472
    elastic scattering, He-Ar systems, 420-426
    electron angular distribution, 445-452
    electron energy distribution, 434-445
    ionization cross-sections, total, 426-445

flowing-afterglow measurement, 427
quantum-mechanical relations, 412-417
spontaneous, 401-483
transfer, 401
true associative, 402, 472-475
Ion-molecule reactions, 105-106, 184-196
diatomic ions, 189-196
flowing-afterglow technique, 113-115
luminescence:
apparatus for study, 118-119
from diatomic reactions, 189-196
$He_2^+ + N_2$, 190-193
$He_2^+ + Ne_2^+$, 193
mass spectrometers for study, 115-117
neutral reactants effect on, 108-110, 112-113
vibrational excitation from, 184-196
Ion-neutral collisions:
charge transfer in, 180-183
in molecular systems, 184-196
$He^+ + N_2$, 190-192
$He_2^+ + N_2$, 190-192
chemiluminescent reactions, 165-196
collision mechanisms, 196-206
calculations, 201-206
theories, 199-201
heavy particle transfer, 193-196
$C^+ - H_2$, 195
$C^+ - O_2$, 195
$N^+ - NO$, 193
$O^+ - H_2$, 194
luminescence measurements, 164-183
Penning ionization excitation, 182
Ions:
excited, *see* Excited ions
reactions with excited neutrals, 108-110, 161-163
$Ar^+ + H_2$, 109
$He^+ + N_2$, 109
$Ne^+ + N_2$, 110
$NH_3^+ + H_2$, 109
$O^+ + N_2$, 109, 110
$O^- + O_2$, 110
$O_2^- + O_2$, 110
vibrationally excited, reactions, 94-100

Jesse effect, 489

Laser:
induced fluorescence, 93

photodissociation spectroscopy, 93
photofragment spectroscopy, 93
Laser-excited Na atoms, crossed beam quenching, 358-379
Laser optical pumping, Na atom beam, 365
Lenses, electron, 26
Luminescence:
apparatus, 118-119
measurements of products of ion-neutral collisions, 164-183
spectra from diatomic ion-molecule reactions, 189-196
see also Ion-neutral collisions

Mass spectrometer, photoionization, 115-117
Molecular-orbital electron configurations:
for N_2, 266
for NO, 299
for NO^+, 335
for O_2, 286
for O_2^-, 316
Molecular states, low-lying:
of N_2, 265
of NO, 299
of NO^+, 335
of O_2, 286
of O_2^-, 316

Neutrals:
collisions with ions, excited products, 163-196
interaction with excited ions, 120-161
see also Excited neutrals
Noble gases:
angular distributions, $He^* + Ne$, 574-579
asymmetric systems, 545-579
scattering:
in excitation transfer systems, 569-579
in Penning systems, 545-569
atoms, metastable, in molecular beams, 488-580
cross-sections, total, $He^* + He$, 438-539
excitation transfer, 540-541, 573-574
excimer states, 526-527
excitation transfer:
cross sections, 539-540
interference structure, 540-544
interactions, potentials, He^*, 567-569

photoelectron spectroscopy, 41-50
properties, 491-492
scattering:
  Ar* + Ar, 536
  Dr* + Kr, 536
  He* + He, 527-536
  Ne* + Ne, 536
  elastic, 569-573
    He* + Ne, 570-571

Optical pumping, *see* Laser optical pumping
Orbitals, iterative natural spin, 237-238
Oscillations in excitation functions for
    ion-neutral collisions, 179
Oscillator strength:
  calculation:
    by moment theory method, 24
    for CO and N_2, 24
    by MS-X method, 23, 24
  continuum, 41
  ionization, 40, 41
  optical, measurement, 3-74
  spectra:
    Ar, 48
    CH_4, 69-72
    CO, 54-60, 62, 66-68
    H_2, 50-54
    H_2O, 73
    HF, 73
    He, 47
    N_2, 54-55, 57-60
    Ne, 46
    NH_3, 40-41, 73
    SF_6, 73
  see also Spectroscopy, electron

Penning ionization, 105-106, 108, 401-472,
    489, 492, 496-497, 506-509
  associative, 452-454, 472-475
  electron angular distribution, 447-452
    $He(2^1S)/Hg$, 433
    $He(2^3S)$-Ar, 433, 447-449, 564-567
    $He(2^3S)$-CO, 448
    $He(2^3S)/Hg$, 433
    Ne*-Kr, 430
  electron energy distribution, 434-445
    $He(2^3D)$-Ar, 438
    $He(2^3S)$-Ar, 437
    $He(2^3S)$-H, 437, 439-441
  electron spectroscopy, 403, 494

model, optical, 502-503
potentials, $He(2^3S)/H$, 464
quantum mechanical relations, 412-417
scattering, elastic, 420-426
targets:
  atomic, 460-463
  molecular, 463-472
Penning systems, 545-573
  He* + Ar, 545-553
  He* + D_2, 554
  He* + Li, 555
  ion angular distribution, 557-564
    He* + Ar, 557-559
    He* + CH_4, 562
    He* + H_2, 560
    He* + O_2, 560
    Ne* + Ar, 563
  ionization cross sections, 426-445, 564-
    567
  scattering, elastic, 545-557
Photoelectrons, *see* Spectroscopy
Photoionization, 7, 41-73, 101-105, 115-
    117
  cross-section, NH_3 and NH_4^+, 126
  H_2, 103
  mass spectrometer, 115-117
  see also Spectroscopy, photoelectron
Photon simulation, 5, 7
Potential energy curves, electronic structure
    of atmospheric systems, 241-337.
    See also Electronic structure
Potential energy surfaces, 201-205
  H_3^+, 203
  $He_2^+ - N_2$, 205
  $C^+ - H_2$, 203
Pseudo-photons, 5, 6

Quenching, 342-394
  alkali atoms, 346-353
  by CO, 369-373
  crossed beam experiments, $Na(3^2P)$, 358-
    368
  by D_2, 369-373
  by H_2, 369-373
  inverse, 391
  of K*, 392-393
  models, theoretical, 251-258
  Na* + CO, 350
  Na* + CO_2, 378
  Na* + N_2, 352-358

Na* + NO$_2$, 378
Na* + O$_2$, 377
Na*(3p) + C$_2$H$_4$, 379
Quench lamp, 514-515

RHF, *see* Hartree-Fock, restricted

Scattering:
  apparatus, Freiburg, 515-516
  elastic, 521-523
    He* + Ne, 570-571
  excited state, 510-521
  He*–He system, 527-536
  potential, theory, 496-510
  signal, 364-365
  velocity distribution, He* + He, 516-521
Self-consistent field method, *see* Hartree-
    Fock
Separation, Born-Oppenheimer, 231-232
Sodium atoms:
  laser excited, in crossed beam quenching
      experiments, 358-393
  laser optical pumping, 365-368
  polarized light in, 380-393
  quenching:
    by diatomic molecules, 368-377
    by polyatomic molecules, 379-380
    by triatomic molecules, 377-379
  see also Alkali atoms excitation
Spectra, energy transfer, for Na($3^2 P_{3/2}$)
    quenching, 368-381
  by diatomic molecules, 369-377
  by polyatomic molecules, 379-380
  by triatomic molecules, 377-379
Spectrometers, ion-cyclotron resonance,
    117-118. *See also* Mass spectrometer

Spectroscopic constants for bound states:
  N$_2$, 267
  NO, 300
  NO$^+$, 336
  O$_2$, 287
  O$_2^-$, 317
Spectroscopy:
  CH$_4$, 69-72
  CO, 54-69
  CO$_2$, 72-73
  collisional energy transfer, 343-
    394
  electron, 3-74
  electron-impact, 5-7, 8, 17, 26, 29-31
  H, 50-54
  HF, 72-73
  H$_2$O, 72-73
  N$_2$, 54-69
  N$_2$O, 72-73
  NO$^+$, 104
  NH$_3$, 72-73
  noble gases, 41-50
  O$_2$, 72-73
  Penning, 403, 494
  photoelectron, 7, 29-33, 41-44,
    104
Structure, electronic, atmospheric systems,
    229-231, 240-337
  calculations, 231-232
  see also Electronic structure

Valence configuration interaction, 238
Vibrational-intensity distribution, NO$^+$, 104.
    See also Ions, vibrationally excited

Wien filter, 25